PHYSICS IS LOGIC PAINTED ON THE VOID:

Origin of Bare Masses and The Standard Model in Logic, U(4) Origin of the Generations, Normal and Dark Baryonic Forces, Dark Matter, Dark Energy, The Big Bang, Complex General Relativity, A Megaverse of *Universe Particles*

Stephen Blaha

Blaha Research

ISBN: 978-0-9893826-6-3

Cover Credits

The front cover contains a self-portrait by Rembrandt at age 63. The cover reflects the view that all universes, and their contents, are but a representation of reality colored by Physics just as Rembrandt's paintings are representations of reality colored by his artistic sense.

Rev. 00/00/01 August 20, 2015

To My Grandchildren:

Alexandre, Maxim, Milan, and Nicholas

Some Other Books by Stephen Blaha

All the Megaverse! Starships Exploring the Endless Universes of the Cosmos using the Baryonic Force (Blaha Research, Auburn, NH, 2014)

All the Universe! Faster Than Light Tachyon Quark Starships & Particle Accelerators with the LHC as a Prototype Starship Drive Scientific Edition (Pingree-Hill Publishing, Auburn, NH, 2011).

From Asynchronous Logic to The Standard Model to Superflight to the Stars (Blaha Research, Auburn, NH, 2011)

From Asynchronous Logic to The Standard Model to Superflight to the Stars; Volume 2: Superluminal CP and CPT, U(4) Complex General Relativity and The Standard Model, Complex Vierbein General Relativity, Kinetic Theory, Thermodynamics (Blaha Research, Auburn, NH, 2012)

The Algebra of Thought & Reality: The Mathematical Basis for Plato's Theory of Ideas, and Reality Extended to Include A Priori Observers and Space-Time; Second Edition (Pingree-Hill Publishing, Auburn, NH, 2009)

Quantum Big Bang Cosmology: Complex Space-time General Relativity, Quantum Coordinates, Dodecahedral Universe, Inflation, and New Spin 0, ½, 1 & 2 Tachyons & Imagyons™ (Pingree-Hill Publishing, Auburn, NH, 2004)

SuperCivilizations: Civilizations as Superorganisms (McMann-Fisher Publishing, Auburn, NH, 2010)

Standard Model Symmetries, And Four and Sixteen Dimension Complex Relativity; The Origin Of Higgs Mass Terms (Blaha Research, Auburn, NH, 2012)

The Bridge to Dark Matter; A New Sister Universe; Dark Energy; Inflatons; Quantum Big Bang; Superluminal Physics; An Extended Standard Model Based on Geometry (Blaha Research, Auburn, NH, 2013)

Universes and Megaverses: From a New Standard Model to a Physical Megaverse; The Big Bang; Our Sister Universe's Wormhole; Origin of the Cosmological Constant, Spatial Asymmetry of the Universe, and its Web of Galaxies; A Baryonic Field between Universes and Particles; Flatverse Extended Wheeler-DeWitt Equation (Blaha Research, Auburn, NH, 2014)

Available on bn.com, Amazon.com, Amazon.co.uk and other international web sites as well as at better bookstores (through Ingram Distributors).

Preface

Scientists and philosophers have wondered for hundreds of years about the ability of Science, and Physics in particular, to accurately be described by Mathematics. This book answers that question. The fundamental Physics theory of the universe – Elementary Particle Physics and Quantum Gravity, from which all Science is ultimately derived, is based on Logic – Asynchronous Logic in particular. Since Mathematics ultimately is based on Logic as well, we see a common source for Science and Mathematics. Their symbiotic relation is due to their common progenitor and they cannot disagree.

Our derivation begins by defining 4-valued logic pre-particles called iotas that we clothe in complex space-time coordinates making "bare" fermions. We then show how the four fermion species result.

From that point the book presents a detailed derivation/construction of The Extended Standard Model of the author which contains within it the known The Standard Model. It is a derivation that has a similarity of sorts with the geometry of Euclid: both introduce fundamental axioms and then proceed to develop constructs from which a complete theory of the four known interactions: Electromagnetic, Weak, Strong and gravitational emerge in a natural way from the geometry of complex space-time when it is mapped to real space-time coordinates by a group we call the Reality group. The geometry of space-time leads directly to the classification of fermions and Dark fermions each into four species: charged leptons, neutral leptons, up-type quarks and down-type quarks. It also provides a detailed description of Dark Matter: from its fundamental nature to its chemistry; and Dark Energy. It shows how to make perturbation theory calculations automatically finite (no divergences); and the beginning state of the Big Bang finite (no infinities); and how the universe's expansion is fueled.

We also point out that data suggests the existence of a baryonic force tied to baryon number conservation for which we calculate a coupling constant based on experimental determinations of the gravitational constant G. This new force leads us to consider Lepton number, and their Dark equivalents which we postulate are all conserved and which all have associated forces and gauge fields.

An analysis of the four particle number force laws lead to a new U(4) symmetry (broken), called the Generation group, that is the source of the four fermion generations. As part of the derivation we describe Higgs Mechanisms for the ElectroWeak sector, for tachyons, and for Generation group breaking contributions to fermion generations masses.

After describing complex Gravity, and discussing Quantum Gravity, the Wheeler-Dewitt equation in particular, we show that the restriction of the complexified Wheeler-DeWitt equation to real space-time automatically generates a cosmological constant term.

We next consider important Quantum Gravity reasons for having a Megaverse, an infinite region of high dimension, in which we can embed our universe and other universes as surfaces. Based on the extension of features of our universe to other universes, and to the Megaverse, we find the Megaverse should have 16 complex dimensions so that the Reality group of 4-dimensional universes is also the Reality group of the Megaverse.

Quantum Gravity considerations lead us to "make" universes into extended particles in a quantum field theory framework. We consider their interactions, consisting of 16-dimensional particle number force law gauge fields, in creating universes via vacuum fluctuations, in collisions of universes, and in amalgamations of universes.

Lastly we consider starships for travel within our universe and into the Megaverse using quark-gluon faster than light ion thrust. We show exit from our universe is possible via the baryonic force using an umbrella-shaped starship configuration.

In the past fourteen years we have developed the derivation/construction presented and extended here. This book provides a much deeper development, changes some of our earlier work, and corrects typos.

CONTENTS

DERIVATION OF THE EXTENDED STANDARD MODEL

1. CONSTRUCTION PRINCIPLES FOR A PHYSICAL THEORY OF UNIVERSES 3

1.1 THE PRINCIPLE OF LOGIC AS THE ORIGIN AND THE METHOD OF CONSTRUCTION 3

1.1.1 Requirement of Logic in Physics 3

1.1.2 The Logic Building Block of Particles – The Iota 4

1.1.3 Mass of an Iota 4

1.2 OCKHAM'S RAZOR 5

1.3 LEIBNIZ'S MINIMAX PRINCIPLE FOR PHYSICS THEORIES 5

1.4 A PHYSICAL SPACE EXISTS 6

1.5 NON-LOCALIZED PHYSICAL PROCESSES EXIST 6

1.6 CONSTRUCTION PRINCIPLES OF EUCLID'S GEOMETRY 6

2. ASYNCHRONOUS LOGIC AND PHYSICAL PROCESSES 7

2.1 FOUR-VALUED ASYNCHRONOUS LOGIC 7

2.2 PRINCIPLE OF ASYNCHRONICITY 8

3. IOTAS AS THE CORE OF FERMIONS 10

3.1 MATRIX REPRESENTATION OF ASYNCHRONOUS LOGIC 10

3.2 MATTER IS INSUBSTANTIAL NEGLECTING PARTICLE INTERACTIONS 11

APPENDIX 3-A. A MAP BETWEEN INTERACTING PARTICLES AND COMPUTER LANGUAGES 12

3-A.1 THE SIMILARITY OF DATA TO PARTICLES 12

3-A.2 PARTICLE PHYSICS LAGRANGIANS DEFINES A LANGUAGE 12

3-A.3 A LINGUISTIC REPRESENTATION OF THE STANDARD MODEL 13

3-A.3.1 Linguistic View of an Interaction 14

3-A.3.2 Computer Grammars 15

3-A.3.3 Generalized Input Chomsky Languages 16

3-A.4 PROBABILISTIC COMPUTER GRAMMARS 17

3-A.4.1 Probabilistic Computer Grammars™ 17

3-A.4.2 Quantum Probabilistic Grammar 19

3-A.4.3 Probability Amplitudes of Quantum Grammars 24

3-A.5 STANDARD MODEL QUANTUM GRAMMAR 25

3-A.5.1 Grammar Production Rules of Quantum Electrodynamics 25

3-A.5.2 Production Rules for the Weak and Strong Interactions 28

3-A.5.3 The Standard Model Quantum Grammar 30

3-A.5.4 The Standard Model Language is Surprisingly Simple *31*

4. SPACE-TIME COORDINATES **32**

4.1 Real or Complex Coordinates? 32
4.2 Lorentz Transformations 33
4.3 The Reality Group 34
4.4 The Reality Group as the Source of the Extended Standard Model and the Dark Gauge Fields in Particular 35
4.5. The Reality Group – U(4) 35
4.6 Consequences of the Unitary Nature of the Reality Group 36
4.7 A Constraint on Complex Coordinate Transformations 36
4.8 Why SU(3)⊗SU(2)⊗U(1)⊗SU(2)⊗U(1)? 37

APPENDIX 4-A COMPLEX LORENTZ GROUP DETAILS **39**

4-A.1 Transformations Between Coordinate Systems 39
4-A.2 The Lorentz Group 40
4-A.3 The Nature of Λ(Ω, u) for Complex Ω 42
4-A.4 Complex Lorentz Group 43
4-A.5 Faster-than-Light Transformations 43
4-A.6 Left-Handed Superluminal Transformations 44
4-A.6.1 Cosh-Sinh Representation of Left-Handed Superluminal Boosts *45*
4-A.6.2 General Velocity Transformation Law – Left-Handed Superluminal Boosts *46*
4-A.6.3 Left-Handed Transformations Multiplication Rules *47*
4-A.6.4 Inverse of Left-Handed Transformations *48*
4-A.7 Right-Handed Superluminal Transformations 48
4-A.8 Inhomogeneous Left-Handed Lorentz Group Transformations 50
4-A.9 Inhomogeneous Right-Handed Extended Lorentz Group 52
4-A.10 General Forms of Superluminal Boosts 52
4-A.11.1 "Lepton-like" Left-Handed Boosts *53*
4-A.11.2 "Lepton-like" Right-Handed Boosts *53*
4-A.11.3 "Quark-like" Left-Handed Boosts *54*
4-A.11.4 "Quark-like" Right-Handed Boosts *54*
4-A.11.5 "Quark-like" Boosts *54*
4-A.11.6 Conventional "Dirac" Boosts *55*

5. IOTA DIRAC-LIKE EQUATIONS FOR PARTICLES AND ANTIPARTICLES....................... 56

5.1 MATRIX REPRESENTATION OF COMPLEX LORENTZ GROUP L_C BOOSTS 57

5.2 LEFT-HANDED AND RIGHT-HANDED PARTS OF L_C 57

Left-handed Part of L_C 57

Right-handed part of L_C 58

5.3 DIFFERENCE BETWEEN THE PARTS OF L_C REDUCED TO PARALLELISM OF $\hat{U}_R$ AND $\hat{U}_I$ 58

5.4 FREE SPIN ½ PARTICLES – LEPTONS & QUARKS 58

5.4.1 Introduction *58*

5.4.2 First Step - Deriving the Conventional Dirac Equation *59*

5.4.3 Derivation of the Tachyon Dirac Equation *60*

Probability Conservation Law 62

Energy-Momentum Tensor 62

5.4.4 Tachyon Canonical Quantization *63*

Separation into Left-Handed and Right-Handed Fields 63

Further Separation into + and – Light-Front Fields 64

Left-Handed Tachyons 67

Right-Handed Tachyons 68

5.4.5 Interpretation of Tachyon Creation and Annihilation Operators *69*

Left-Handed Tachyon Creation and Annihilation Operators 70

Right-Handed Tachyon Creation and Annihilation Operators 70

5.4.6 Tachyon Feynman Propagator *71*

Dirac Field Light-Front Propagators 71

Tachyon Field Light Front Propagators 72

5.5 COMPLEX SPACE AND 3-MOMENTUM & REAL-VALUED ENERGY FERMIONS (QUARKS) 72

5.5.1 L_C Spinor "Normal" Lorentz Boosts & More Spin ½ Particle Types *73*

Case 1: Parallel Real and Imaginary Relative Vectors 73

Case 2: Anti-Parallel Real and Imaginary Relative Vectors 74

Case 3: Complexons: A New Type of Particle with Perpendicular Real and Imaginary 3-Momenta 74

A Global SU(3) Symmetry Revealed 78

Global SU(3) Spin ½ Complexon Fields 80

Lagrangian Formulation and Second Quantization of Complexons 80

Complexon Feynman Propagator 83

Case 4: Left-handed Tachyon Complexons 84

Global SU(3) Symmetry 86

Light-Front Quantization of Tachyonic Complexons 86

Separation into Left-Handed and Right-Handed Fields 87

Further Separation into + and – Light-Front Complexon Fields 88

Left-Handed Tachyonic Complexons 90

Left-handed Case 4: Tachyonic Complexon Feynman Propagator 92

Case 5: Right-Handed Tachyonic Complexons 93
Right-handed Case 5: Tachyonic Complexon Feynman Propagator 94
Other Cases? No 95
5.6 Spinor Boosts Generate 4 Species of Particles: Leptons and Quarks 95
Charged lepton fermions 95
Neutrinos 95
Up-type Color Quarks 96
Down-type Color Quarks 96
5.7 First Step towards The Standard Model 97
5.8 Dirac-like Equations of Matter from 4-Valued Logic 98
5.9 Why Second Quantization of Fields? 99
APPENDIX 5-A. LEPTONIC TACHYON SPINORS 100
APPENDIX 5-B. EXPERIMENTAL EVIDENCE FOR FASTER-THAN-LIGHT PARTICLES & PHYSICS 102
5-B.1 Instantaneous Quantum Mechanical Effects 103
5-B.2 Tritium Decay Experiments Yielding Neutrinos 103
5-B.3 LHC/Gran Sasso Direct Measurements of Neutrino Speeds 104
5-B.4 Tachyonic Behavior Within Black Holes 104
5-B.5 Higgs Fields are Tachyons 104
5-B.6 Conclusion: Faster-Than-Light Particles – Tachyons Exist in Nature 105
APPENDIX 5-C. PHENOMENA BEYOND THE LIGHT BARRIER 106
5-C.1 Superluminal (Faster-than-Light) Transformations 106
5-C.2 Length Dilations and Time Contractions 107
Superluminal Length Dilation/Contraction 107
Superluminal Time Contraction/Dilation 108
5-C.3 Tachyon Fission to More Massive Particles – Reverse Fission 108
5-C.4 Light Chasing Faster-than-Light Particles? 110
5-C.5 Electromagnetic Field of a Charged Tachyon – A Pancake Effect? 111
Sublight Charged Particle 111
Charged Tachyon 112
Are Tachyonic Cones in the Au-Au Scattering Quark-Gluon Plasma? 112
5-C.6 Superluminal (Tachyon) Physics is Different 112

APPENDIX 5-D. SUPERLUMINAL (FASTER THAN LIGHT) KINETIC THEORY AND THERMODYNAMICS 113

5-D.1 SUPERLUMINAL KINETIC THEORY 113

5-D.1.1 Relativistic Form of the Maxwell-Boltzmann Distribution 113

5-D.1.2 Superluminal Form of the Maxwell-Boltzmann Distribution 114

5-D.2 SUPERLUMINAL THERMODYNAMICS 116

5-D.3 APPROXIMATE CALCULATION OF KINETIC AND THERMODYNAMIC QUANTITIES 118

5-D.4 SUPERLUMINAL KINETICS AND THEMODYNAMICS ARE SIMILAR TO THE NON-RELATIVISTIC CASE 121

6. ELECTROWEAK DOUBLETS FROM ROTATIONS BETWEEN TACHYON AND NORMAL PARTICLES 122

6.0 INTRODUCTION 122

6.1 TRANSFORMATIONS OF DIRAC AND TACHYON EQUATIONS 122

6.2 DOUBLET EXTENDED DIRAC EQUATIONS 123

6.3 NON-INVARIANCE OF THE EXTENDED FREE ACTION UNDER A LEFT-HANDED EXTENDED LORENTZ TRANSFORMATION 125

6.4 THE DIRACIAN DILEMMA – TO WHAT DO LEFT-HANDED EXTENDED LORENTZ BOOST PARTICLES CORRESPOND? ANSWER: LEPTONS 126

6.5 TO WHAT DO COMPLEXONS CORRESPOND? QUARKS 127

6.6 QUARK DOUBLETS 127

Summary of L_C Boosts to Generate Spin ½ Equations 128

L_C Boosts between Complexons and Tachyonic Complexons 129

Doublet Dynamical Equation for Complexons 129

Non-Invariance of the Generalized Free Complexon Action under an L_C Boost 131

7. ELECTROWEAK SU(2)⊗U(1) FROM REAL SUPERLUMINAL VELOCITIES 132

8. DARK ELECTROWEAK SU(2)⊗U(1) FROM COMPLEX SUPERLUMINAL VELOCITIES 138

9. COLOR SU(3) 141

9.1 TWO POSSIBLE APPROACHES TO COLOR SU(3) 141

9.2 A GLOBAL SU(3) SYMMETRY OF COMPLEXON QUARKS 141

9.3 LOCAL COLOR SU(3) AND THE STRONG INTERACTIONS 144

9.4 INTERACTIONS RESULTING FROM COMPLEX SPACE-TIME PROJECTED TO REAL PHYSICAL SPACE-TIME 146

10. DERIVATION OF ONE GENERATION LEPTONIC STANDARD MODEL FEATURES 147

10.1 THE STANDARD MODEL LEPTONIC SECTOR AS A CONSEQUENCE OF THE COMPLEX LORENTZ GROUP COORDINATES PROJECTED TO REAL VALUES 147

10.2 Assumptions for the Leptonic Sector of the Standard Model 147
10.3 Derivation of the Leptonic Sector of the Standard Model 148

11. CAN NEUTRINOS BE TACHYONS? 151

12.DERIVATION OF ONE GENERATION QUARK SECTOR STANDARD MODEL FEATURES 153

12.1 Quark Sector of the Standard Model 153
12.2 Quark Sector Assumptions 153
12.3 Derivation of the Form of the Standard Model Quark Sector 153
12.4 Color SU(3) 157
12.5 Pure Complexon Gauge Groups 158
12.6 Pure Gauge Complexon Path Integral Formulation and Faddeev-Popov Method 160
Faddeev-Popov Application to Complexon Condition 161
12.7 Complexon Quark-Gluon Perturbation Theory 162
12.8 Complexon Quark ElectroWeak Interactions 163
12.9 Complexon Quarks with Color SU(3) 164

13. SU(3)⊗SU(2)⊗U(1)⊗SU(2)⊗U(1) EXTENDED STANDARD MODEL 165

13.1 Two-Tier Formulation of One Generation Standard Model 165
Simple Two-Tier Formalism 166
Two-Tier L_C-based Standard Model Theory 167
Feynman Propagators 168
Complexon Feynman Propagator 169
13.2 The New Extended Standard Model Lagrangian 169
13.2.1 Two-Tier Lepton Sector with Dark Leptons 170
13.2.2 Quark Sector with Dark Quarks 170
13.3 Renormalization and Divergence Issues 173

APPENDIX 13-A MULTI-QUARK PARTICLES AND QUARK CHEMISTRY 174

13-A.1 New Penta-Quark Particles 174
13-A.2 Multi-Quark Molecules – Quark Chemistry 174
13-A.3 Large Multi-Quark "Molecules" and Eventually Quark Stars 174

14. DARK MATTER FEATURES: PARTICLE SPECTRUM AND BASIC CHEMISTRY OF DARK MATTER, DARK MATTER BODIES 175

14.1 Fundamental Dark Particles 175
14.2 Dark Particle Chemistry 176
14.3 Gravitation and Dark Matter Bodies: Galactic Clusters, Galaxies 178
14.4 Dark Stars, Dark Planets, Dark Globular Clusters—Or Just Darkened? 179
14.4.1 Open Questions on Dark Matter in Our Solar System 179
14.4.2 Open Questions on Dark(ened) Stars and Dark(ened) Globular Clusters of Stars 179
14.4.3 Comments on these Questions 179

15. BARYON, LEPTON, DARK BARYON, AND DARK LEPTON CONSERVATION LAWS AND NEW U(4) GAUGE SYMMETRY 181

15.1 Baryon Number Conservation and a Possible Baryonic Force 181
15.1.1 Estimate of the Baryonic Coupling Constant 182
15.1.2 A Baryonic Gauge Field 183
15.1.3 Planckton Second Quantization 184
15.1.4 Bary-Electric Fields and Bary-Magnetic Fields 186
15.1.5 The Baryonic "Coulombic" Gauge Field 188
15.1.6 The Baryonic Force on Baryonic Objects 188
15.2 Lepton Number, Dark Baryon Number, and Dark Lepton Number Conservation Laws 189
15.2.1 Other Conserved Particle Numbers 189
15.3 U(4) Number Symmetry 190

16. FERMION GENERATIONS AND BROKEN U(4) 191

16.1 Four Generation Extended Standard Model 191
16.1.1 Two-Tier Lepton Sector 191
16.1.2 Quark Sector 192
16.2 U(4) Gauge Symmetry Breaking and Long Range Forces 195
16.3 Higgs Mass Mechanism for U(4) Gauge Fields 196
16.4 Impact of this U(4) Higgs Mechanism on Fermion Generation Masses 199
16.5 Generation Group Higgs Mechanism for Fermion Masses 199

17. HIGGS MECHANISM IN THE ELECTROWEAK SECTOR 203

17.1 ElectroWeak Higgs Mechanism Generation of ElectroWeak Gauge Field Masses ... 203
17.2 ElectroWeak Higgs Mechanism Generation of Fermion Masses 204
17.3 Combined Effect of Generation Group and ElectroWeak Higgs Mechanisms on Fermion Masses 207
17.4 Generalization of the ElectroWeak Higgs Mechanism for Gauge Field Masses to Include Dark Gauge Fields 207
17.5 The Reality Group Compared to the Generations Group U(4) 209

17.6 COMPLETION OF CONSTRUCTION/DERIVATION OF THE FORM OF THE EXTENDED STANDARD MODEL AND QUANTUM GRAVITY - UNIFICATION 209

APPENDIX 17-A. HIGGS MECHANISM FOR TACHYONS 211

18. TWO-TIER QUANTUM FIELD THEORY AND RENORMALIZATION 212

18.1 PERTURBATION THEORY DIVERGENCES 212
18.2 QUANTIZATION OF COORDINATE SYSTEMS 213
18.2.1 Non-commuting Coordinates 213
18.2.2 New Approach to Non-Commuting Coordinates 213
18.2.3 Quantization Using a C-Number X^μ 214
18.2.4 Coordinate Quantization 216
18.2.5 Gauge Invariance 217
18.2.6 Bare ϕ Particle States 219
18.2.7 Y Fock Space Imaginary Coordinate States 220
18.2.8 Y Coherent Imaginary Coordinate States 221
18.2.9 The Dynamical Generation of New Dimensions 221
18.2.10 Generation of Quantum Dimensions™ by the $\phi(X)$ field 222
18.2.11 Hamiltonian for Particle and Coordinate States 222
18.2.12 Second Quantized Coordinates 223
18.3 SCALAR TWO-TIER QUANTUM FIELD THEORY 224
18.3.1 Introduction 224
18.3.2 "Two-Tier" Space 224
18.3.3 Vacuum Fluctuations 225
18.3.4 The Feynman Propagator 226
18.3.5 Large Distance Behavior of Two-Tier Theories 227
18.3.6 Short Distance Behavior of Two-Tier Theories 227
18.3.7 String-like Substructure of the Theory 228
18.3.8 Parity 230
18.3.9 X Parity 230
18.3.10 y Parity 231
18.3.11 Forms of the Parity Transformations 231
18.3.12 Charge Conjugation 232
18.3.13 Time Reversal 232
18.4 INTERACTING QUANTUM FIELD THEORY – PERTURBATION THEORY 233

18.4.1 Introduction 233
18.4.2 An Auxiliary Asymptotic Field 233
18.4.3 Transformation Between $\Phi(y)$ and $\phi(X(y))$ 234
18.4.4 Model Lagrangian with ϕ^4 Interaction 236
18.4.5 In-states and Out-States 237
18.4.6 ϕ In-Field 237
18.4.7 ϕ Out-Field 238
18.4.8 The Y Field 239
18.4.9 S Matrix 241
18.4.10 LSZ Reduction for Scalar Fields 241
18.5 THE U MATRIX IN PERTURBATION THEORY 243
18.5.1 Reduction of Time Ordered φ Products 244
18.5.2 The $\int d^3X$ Integration 247
18.5.3 Y In and out states 248
18.5.4 Unitarity 248
18.5.5 Finite Renormalization of External Legs 250
18.5.6 Perturbation Expansion 250
18.5.7 Higher Order Diagram With a Loop 252
18.5.8 Finite Renormalization of External Particle Legs & Unitarity Example 254
18.5.9 General Form of Propagators 256
18.5.10 Scalar Particle Propagator 257
18.5.11 Spin ½ Particle Propagator 258
18.5.12 Massless Spin 1 Particle Propagator 259
18.5.13 Spin 2 Particle Propagator 260
18.6 FINITE QUANTUM GRAVITY 261
18.6.1 Introduction - Two-Tier Quantum Gravity: Finite! 261
18.6.2 Two-Tier General Relativity 262
18.6.3 Lagrangian Formulation 263
18.6.4 Unified Standard Model and Quantum Gravity Lagrangian 263
18.6.5 Why Are the Y Field Dynamics Independent of the Gravitational Field? 264
18.6.6 No "Space-time Foam" 264
18.6.7 Quantum Gravity – Scalar Particle Model Lagrangian 265
18.6.8 A Justifiable Weak Field Approximation for Quantum Gravity 265
18.6.9 Quantization of Quantum Gravity – Scalar Particle Model 266
18.6.10 Quantum Gravity–Scalar Particle Model Path Integral 269
18.6.11 Finiteness of Quantum Gravity–Scalar Particle Model 272
18.6.12 Unitarity of Quantum Gravity–Scalar Particle Model 272
18.6.13 The Mass Scale M_c 273
18.6.14 Planck Scale Physics 273

18.6.15 Quantum Foam *274*
18.6.16 Measurement of the Quantum Gravity Field *274*
18.6.17 Measurement of Time Intervals *274*
18.6.18 Vacuum Fluctuations in the Gravitation Fields *275*
18.6.19 The Two-Tier Gravitational Potential vs. Newton's Gravitational Potential *275*
18.6.20 Black Holes *278*
18.7 CURVED SPACE-TIME GENERALIZATION OF TWO-TIER QUANTUM GRAVITY 279
18.8 WHY ARE THE Y FIELD DYNAMICS INDEPENDENT OF THE GRAVITATIONAL FIELD? 282
18.9 RENORMALIZATION OF THE STANDARD MODEL 282
18.10 ADLER-BELL-JACKIW ANOMALIES 283
18.11 GRAVITY AND Y^{μ} (Y) 285

APPENDIX 18-A. NEW PARADIGM: COMPOSITION OF EXTREMA IN THE CALCULUS OF VARIATIONS 286

18-A.1. A NEW PARADIGM IN THE CALCULUS OF VARIATIONS 286
18-A.1.1. A Classification of Variational Problems *286*
18-A.2. SIMPLE PHYSICAL EXAMPLE – STRINGS ON SPRINGS 287
18-A.2.1 A Strings on Springs Mechanics Problem *287*
18-A.3. THE COMPOSITION OF EXTREMA – A LAGRANGIAN FORMULATION 289
18-A.3.1. Example: a hyperplane *292*
18-A.3.2 Coordinate Transformation Determined as an Extremum Solution *292*
18-A.3.3. Separable Lagrangian Case *292*
18-A.3.4 Klein-Gordon Example *293*
18-A.4. THE COMPOSITION OF EXTREMA – HAMILTONIAN FORMULATION 294
18-A.5. TRANSLATIONAL INVARIANCE 296
18-A.6. LORENTZ INVARIANCE AND ANGULAR MOMENTUM CONSERVATION 297
18-A.7. INTERNAL SYMMETRIES 299
18-A.8. SEPARABLE LAGRANGIANS 299
18-A.9. SEPARABLE LAGRANGIANS AND TRANSLATIONAL INVARIANCE 302
18-A.10. SEPARABLE LAGRANGIANS AND ANGULAR MOMENTUM CONSERVATION 304
18-A.10.1. Angular Momentum and $\mathscr{L}_F$ *305*
18-A.11. SEPARABLE LAGRANGIANS AND INTERNAL SYMMETRIES 307

COSMOLOGY

19. DARK ENERGY AND THE BIG BANG 311

19.1 DARK ENERGY 311
19.2 THE BIG BANG EXPERIMENTALLY 311
19.3 THE STATE OF THE UNIVERSE AT T = 0 312
19.4 TWO-TIER QUANTUM MODEL FOR THE BEGINNING OF THE UNIVERSE 315
19.4.1 Einstein Equations Near t = 0 .... *315*
19.4.2 Y Black-Body Coherent States *317*
19.4.3 Expectation Value of Y in Coherent States *318*
19.4.4 Expectation Values of the Scale Factor A(t, X) and the Invariant Interval $d\tau^2$ *319*
19.4.5 Representation and Approximations for $Y_{BB}(t, z)$ *320*
19.4.5.1 Some Representations of Y_{BB} 321
19.4.5.2 Approximate Solution for Y_{BB} 321
19.5 THE SCALE FACTOR A(T) NEAR T = 0 322
19.6 A COMPLEX BLACKBODY TEMPERATURE NEAR T = 0 323
19.7 THE NATURE OF THE UNIVERSE NEAR T = 0 323
19.7.1 Behavior of the Complete Scale Factor B(t, y) Near t = 0 .... *324*
19.7.2 The Expectation Value of the Scale Factor $A_{BB}(0, y)$ near t = 0 .... *325*

APPENDIX 19-A. DERIVATION OF THE EXTENDED ROBERTSON-WALKER MODEL.... 326

19-A.1 THE ROBERTSON-WALKER METRIC 326
19-A.2 A GENERALIZATION OF THE ROBERSON-WALKER METRIC 327
19-A.3 THE EINSTEIN EQUATIONS FOR THE GENERALIZED ROBERTSON-WALKER METRIC 327
19-A.4 THE DIFFERENTIAL EQUATIONS FOR THE GENERALIZED SCALE FACTOR A(T, R) 328
19-A.5 THE SOLUTION FOR THE GENERALIZED SCALE FACTOR A(T, R) 330
19-A.6 THE SOLUTION EXPRESSED IN THE ORIGINAL RADIAL COORDINATE 331
19-A.7 EQUIVALENCE OF THE GENERAL SOLUTION WITH THE ORIGINAL ROBERTSON-WALKER SOLUTION 332
19-A.8 HUBBLE'S LAW IN THE GENERALIZED ROBERTSON-WALKER MODEL 333

20. BIG BANG SCALE FACTOR FROM T = 0 TO THE PRESENT 334

20.1 INTRODUCTION 334
20.2 THE BEHAVIOR OF THE SCALE FACTOR A(T) AFTER THE BIG BANG PERIOD 335
20.2.1 General Form of a(t) Scale Factor Einstein Equation *335*
20.2.2 The Explosive Growth Epoch *337*
20.2.3 The Expanding Epoch *338*
20.3 TWO-TIER QUANTUM BIG BANG MODEL AT THE BEGINNING OF TIME 342
20.3.1 The CMB Temperature *343*
20.3.2 The Generalized Robertson-Walker Scale Factor *343*

20.3.3 The Temperature of the Early Universe in the Generalized Robertson-Walker Metric *346*
20.3.4 Consistency of Y_{BB} Approximation with the Resulting Temperature near t = 0 .... *347*
20.3.5 Plots of the Scale Factor from t = 0 to the Present *348*
20.4 THE INTERPRETATION OF THE COMPLEX SCALE FACTOR 352
20.5 TIME EVOLUTION OF THE HUBBLE RATE 353

21. THE BIG BANG EPOCH 354

21.1 THE T = 0 BIG BANG SCENARIO... 354
21.2 THE RADIUS OF THE UNIVERSE IN THE BIG BANG EPOCH 358
21.2.1 Localization of Particles at t = 0 .... *360*
21.3 A QUANTUM BIG BANG AND EVOLUTIONARY THEORY 360

22. QUANTUM GRAVITY AND THE WHEELER-DEWITT EQUATION EXTENDED TO COMPLEX COORDINATES 362

22.1 INTRODUCTION 362
22.2 ANALYTICALLY CONTINUED WHEELER-DEWITT EQUATION TO COMPLEX METRICS UNDER A FADDEEV-POPOV METHOD RESTRICTION 362
22.3 POSSIBLE SOURCE OF THE COSMOLOGICAL CONSTANT IN THE COMPLEX SPACE-TIME – FADDEEV-POPOV CONSTRAINT TERM 365
22.4 IMPACT OF THE FADDEEV-POPOV COMPLEXITY TERM ON THE WHEELER-DEWITT EQUATION.... 366

23. THE MEGAVERSE: A 16-DIMENSIONAL SPACE 368

23.1 INTRODUCTION TO THE MEGAVERSE 368
23.2 REASONS FOR A MEGAVERSE 368
23.2.1 Universe Clocks *369*
23.2.2 Quantum Observer *370*
23.2.3 Asynchronous Logic is a Requirement of Universes *371*
23.3 GENERAL NATURE OF THE MEGAVERSE 371
23.3.1 The Universes Within the Megaverse *371*
23.3.2 Megaverse characteristics *372*
23.4 SUMMARY OF FEATURES OF THE MEGAVERSE 373
23.5 EXTERNAL PROPERTIES OF UNIVERSES WITHIN THE MEGAVERSE 374
23.6 THE SURFACE OF A UNIVERSE – A TYPE OF HORIZON 375
23.7 FULLY REDUCIBLE REPRESENTATION OF 16-DIMENSIONAL REALITY GROUP 376
23.8 MEGAVERSE REALITY GROUP VS. OUR UNIVERSE'S REALITY GROUP 377

23.8.1 Restriction of Complexity of the 16-dimensional Megaverse ... *378*
23.9 Covariant Derivatives and Connections in the Megaverse ... 379
23.10 Covariant Derivatives in 4-Dimensional Lorentzian Universes such as Our Universe ... 379
23.11 Corresponding Lorentzian Transformations in a Universe and the Megaverse ... 380
23.11.1 Special Case of Flat Universes ... *380*
23.11.2 Corresponding General Coordinate Transformations in a Universe and the Megaverse .. *381*
23.12 Quantum Gravity in the Megaverse and its Universes ... 382
23.13 Megaverse Wheeler-Dewitt Equation ... 383
23.15 Megaverse Metric Functional Integral in a Universe ... 384
23.16 Megaverse Wheeler-DeWitt Solutions ... 385
23.16.1 Tachyonic Solutions of Wheeler-DeWitt Equation ... *385*
23.16.2 Problems in the Solutions of the Wheeler-DeWitt Equation ... *385*
23.17.2.1 Negative Frequencies and Probabilities – Anti-Universes ... 386
23.17 General Types of Universes within the Megaverse ... 386
23.18 Quantum Aspects of the Megaverse ... 388

24. THE PARTICLE INTERPRETATION OF UNIVERSES ... 389

24.1 The Hierarchy of the Cosmos ... 389
24.2 The Particle Interpretation of Extended Wheeler-DeWitt Equation Solutions ... 390
24.3 "Free Field" Dynamics of Fermionic Universe Particles ... 391
24.3.1 Four Types of Fermionic Universe Particles ... *392*
24.3.1.1 Dirac-like Equation – Type I universe Particle ... 393
24.3.1.2 Complex Boosts ... 394
24.3.1.3 Tachyon Universe particle Dirac Equation ... 395
24.3.1.4 Type IIa Case: Left-Handed Tachyonic Universe Particles ... 395
24.3.1.5 Type IIb Case: Right-Handed Tachyonic Universe Particles ... 396
24.3.1.6 Type III Case: "Up-Quark-like" Universe Particles ... 396
24.3.1.7 Type IVa Case: Left-Handed "Down-Quark-like" Tachyonic Universe Particles ... 397
24.3.1.8 Type IVb Case: Right-Handed Down-Quark-like Tachyonic Universe Particles ... 398
24.3.2 Lagrangians ... *399*
24.3.2.1 Type I Universe Particle Lagrangian ... 399
24.3.2.2 Type II Tachyon Universe Particle Lagrangian ... 399
24.3.2.3 Type III "Up-Quark-like" Universe Particle Lagrangian ... 399
24.3.2.4 Type IV "Down-Quark-like" Tachyon Universe Particle Lagrangian ... 400
24.3.3 Form of The Megaverse Quantum Coordinates Gauge Field ... *400*
24.3.3.1 Y_u Fock Space Imaginary Coordinate States ... 402
24.3.3.2 Y_u Coherent Imaginary Coordinate States ... 403
24.3.3.3 Quantization of the Type I Free Universe Particle Dirac Field ... 404
24.3.3.4 Feynman Propagators for the Type I Free Universe Particle Dirac Field ... 405

24.3.3.5 Feynman Propagators for the Types II, III, and IV Free Universe Particle Dirac Fields................405
24.3.4 Expanding and Contracting Universes: Impact of Time Dependent Universe Particle Masses 405
24.3.5 Left-Handed and Right-handed Universe Particles 408
24.3.6 Internal Structure of Universe Particles 408
24.4 WHEN UNIVERSES COLLIDE: INTERACTIONS AND COLLISIONS OF UNIVERSE PARTICLES 410
24.4.1 Gravitation and the Fifth Force 410
24.4.2 Universes in Collision 410
24.5 BOSONIC UNIVERSE PARTICLES 411
24.6 PHYSICAL MEANING OF UNIVERSE PARTICLE SPIN 411
24.7 ELEMENTARY PARTICLES WITH TIME DEPENDENT MASSES 412
24.8 IMPACT OF UNIVERSE PARTICLE ACCELERATION – LOPSIDED INTERNAL STRUCTURE OF UNIVERSE 412
24.9 MEGAVERSE BARYONIC GAUGE FIELD - PLANCKTONS 412
24.10 BEYOND THE PLANCKTON 413
24.11 PLANCKTON INTERACTIONS WITH UNIVERSE PARTICLES AND INDIVIDUAL BARYONS 413
24.12 CREATION OF UNIVERSES THROUGH BARYONIC GAUGE FIELD FLUCTUATIONS 414
24.13 WHEN UNIVERSES COLLIDE: COALESCENCE OF UNIVERSES 415
24.14 FISSION OF UNIVERSES 415
24.14.1 Fission of Normal universes 415
24.14.2 Tachyon Universe Particle Fission to More Massive Universe Particles 416
24.15 UNIVERSE PARTICLE – PLANCKTON INTERACTIONS 418
24.16 INTERNAL STRUCTURE OF UNIVERSE PARTICLES 418
24.16.1 Planckton Probes 419
24.16.2 Internal Structure of a Universe Particle 419
24.17 CENTRAL ROLE OF THE BARYONIC FORCE FOR TRAVEL INTO THE MEGAVERSE 420

25. STARSHIPS IN THE UNIVERSE AND MEGAVERSE 421

25.1 PHASE ONE: STARSHIP ENGINE DEVELOPMENT 422
25.1.1 High Speed 424
25.1.2 From Light Speed to Enormous Speeds 425
25.1.3 The Acceleration Experienced on the Starship 426
25.1.4 Travel Time on the Starship – Suspended Animation 427
25.1.5 Constant Superluminal Starship Travel 427
25.2 PHASE TWO: STARSHIP SLINGSHOT INTO THE MEGAVERSE 430

25.2.1 The Characteristics of a Rapidly Spinning Neutron Star *430*
25.2.2 The Baryonic Fields of a Rapidly Spinning Neutron Star *430*
25.2.3 Baryonic Current of a Neutron Star *431*
25.2.4 The Sixteen Component Baryonic Vector Potential *431*
25.2.5 The Baryonic Electric and Magnetic Field Strengths *433*
25.2.6 The Baryonic Coulomb Force Slingshot into the Megaverse *433*
25.2.7 Point-like Uniship Slingshots *435*
25.2.8 The Uniship Slingshot Trajectory *435*
25.2.9 Uniship Neutron Star Slingshot Dynamics *436*
25.2.10 Umbrella-Shaped Uniship Slingshots *437*
25.2.11 Scenario for the Opening of a Uniship Umbrella *438*
25.2.12 Equations for the Motion of the Thrust Tubes Entering the Megaverse *438*

26. SOME EXPERIMENTAL TESTS OF OUR EXTENDED STANDARD MODELTHEORY . 440

26.1 ARE NEUTRINOS TACHYONS? 440
26.2 ARE QUARKS COMPLEXONS? 440
26.3 DARK MATTER STRUCTURE 440
26.4 DARK ENERGY – WHAT IS IT? TWO-TIER Y PARTICLES? 440
26.5 A BARYONIC FORCE? 441
26.6 A 4TH GENERATION? 441

27. SUMMARY OF THE AUTHOR'S PARTICLE PHYSICS BOOKS (2001- 2014) 442

28. SUMMARY OF THE AUTHOR'S BOOKS ON SPACE TRAVEL, AND STARSHIP TRAVEL 444

APPENDIX A. C, P AND T FOR TACHYONS AND DIRAC FERMION FIELDS 445

A.1 PARITY FOR DIRAC AND TACHYONIC FERMIONS 445
A.1.1 Parity for Dirac Fermions *445*
A.1.2 Parity for Tachyonic Fermions *445*
A.2 CHARGE CONJUGATION FOR DIRAC AND TACHYONIC FERMIONS 447
A.2.1 Charge Conjugation for Dirac Fermions *447*
A.2.2 Charge Conjugation for Tachyonic Fermions *447*
A.3 TIME REVERSAL FOR DIRAC AND TACHYONIC FERMIONS 448
A.3.1 Time Reversal for Dirac Fermions *448*
A.3.2 Time Reversal for Tachyonic Fermions *449*

APPENDIX B. CP AND CPT FOR NEUTRINO AND QUARK TACHYONS 451

B.1 CP AND CPT IN TACHYON QUANTUM FIELD THEORY 451
B.2 CP SYMMETRY, TACHYON PARTICLE STATES, AND THE LORENTZ INVARIANCE OF TACHYON PARTICLE NUMBER 451

B.2.1 Tachyon Quantum Fields *451*
B.2.2 Tachyon CP Eigenstates *452*
B.2.3 Neutrino CP Eigenstates, Neutrino-Anti-Neutrino Oscillations *453*
B.2.4 CP and Tachyonic Quarks *454*
B.3 CPT SYMMETRY, AND TACHYONS AND MICROCAUSALITY 456
B.3.1 Tachyon CPT Non-Invariance *456*
B.3.2 Microcausality and Tachyons *456*
Consequently, free left-handed (or free right-handed) tachyons with light-front quantization separately satisfy the microcausality condition. 458
REFERENCES **459**
INDEX **463**
ABOUT THE AUTHOR **470**

FIGURES and TABLES

Figure 3-A.1. The first few terms in the approximate perturbation theory calculation of the scattering of two electrons. The dotted lines represent photons that carry the electromagnetic force between the electrons. 13

Figure 3-A.2 Generating an output string of terminal characters from an input string of terminal characters using the production rules of a grammar. Inside the black box, transformations of the input string take place and non-terminal symbols may appear and disappear. Non-terminal symbols, by definition, cannot appear in the input or output strings of characters. 16

Figure 3-A.3. Two slit photon experimental setup. 19

Figure 3-A.4. Diagram for $\phi\phi \rightarrow \phi$. The input states are always on the left and the output states are always on the right in Feynman and Feynman-like diagrams. 21

Figure 3-A.5 The true Feynman diagram is the "sum" of the two time-ordered Feynman-like diagrams. 21

Figure 3-A.6 Slicing Feynman-like diagrams. A slice is made after each emission or absorption. As you read down a slice the particles are listed in the same order as the corresponding string. Each slice is numbered starting from the left. 21

Figure 3-A.7. Diagrams for an elastic collision with two incoming particles and two outgoing particles. 22

Figure 3-A.8. The power of g for some simple diagrams. 23

Figure 3-A.9. A diagram showing how two electrons interact by exchanging a photon. As time increases the electrons move from left to right. The upper electron emits the photon. This corresponds to the left e transitioning to eA using the grammar rule e → eA. (There is a similar diagram Fig. 3-A.10 in which the lower electron emits the photon.) 26

Figure 3-A.10. Another diagram with two electrons interacting by exchanging a photon. In this case the lower electron emits the photon. This corresponds to the right e in the initial ee string transitioning to Ae using the grammar rule e → Ae. 26

Figure 3-A.11. A Feynman diagram represents several of our Feynman-like diagrams with different time orderings of particle emission and absorption. 27

Figure 3-A.12. A diagram for the collision of two electrons that produce a new electron-positron pair: ee → eepe. 27

Table 5-B.1 Electron neutrino mass squared values found in various tritium decay experiments. (Masses are in units of eV.) The average mass squared is negative suggesting electron neutrinos are tachyons. 103

Figure 5-C.1. Two coordinate systems having a relative speed v in the x direction. 106

Figure 5-C.2. Two particle decay of a tachyon. 108

Figure 5-C.3. Two coordinate systems having a relative speed v in the x direction. A pulse of light is displayed as a thick arrow. 111

Figure 7.1. Depiction of two coordinate systems. The "primed" coordinate system is moving with velocity v in the positive x direction with respect to the "unprimed" coordinate system. We choose parallel axes for convenience. 132

Figure 8.1. Depiction of two coordinate systems. The "primed" coordinate system is moving with velocity $u = u_x i + i u_y j$ with respect to the "unprimed" coordinate system. We choose parallel axes for convenience. 138

Figure 12.1. Quark self-energy diagram in which a quark emits and subsequently absorbs a gluon. The quark momentum is $p = (p^0, p_r + ip_i)$ and the gluon momentum is $q = (q^0, q_r + iq_i)$ with conservation of energy, and separate conservation of real spatial momentum and imaginary spatial momentum at each vertex. 163

Figure 12.2. Quark self-energy diagram in which a quark emits and subsequently absorbs a photon. The quark momentum is $p = (p^0, p_r + ip_i)$ and the photon momentum is $q = (q^0, q_r)$ with conservation of energy, and separate conservation of real spatial momentum and imaginary spatial momentum at each vertex. 163

Figure 12.3. Photon self-energy diagram with a quark loop. The photon momentum is $q = (q^0, q_r)$ and quark loop momentum is $p = (p^0, p_r + ip_i)$. 164

Figure 14.1 "Periodic Table" of four generations of normal and Dark fundamental fermions. 176

Figure 14.2. Example of two Dark atoms binding in a manner similar to the binding of hydrogen molecules in H_2. Considering the possible combinations of quarks and leptons there are 64 bound varieties of this type. Chains of these molecules are also possible in principle. 177

Figure 14.3. Example of a simple Dark Matter chain. 178

Figure 14.4. A segment of two strands interacting with each other. Other combinations are possible using other Dark charged fermions. This diagram is suggestive of a Dark form of DNA. Dark Life? 178

Figure 18.2.1. The change from purely real space to a slightly complex space with imaginary quantum fluctuations for each spatial axis in the Coulomb gauge of the Y field. 214

Figure 18.3.1. Feynman diagram for conventional and cloaked Two-Tier propagators. 229

Figure 18.5.1. Lowest order quartic interaction diagram. 250

Figure 18.5.2. Lowest order loop scattering diagrams. 252

Figure 18.6.1. Plot of Two-Tier gravitational potential for $M_c = 1$ TeV/c^2 and Newton's gravitational potential. The potentials are measured in units of 10^{-36} GeV^{-1}. The radial distance is measured in units of 10^{-5} GeV^{-1}. The plot of the Two-Tier potential shows the force of gravity is repulsive for small $r < 5.7 \times 10^{-4}$ GeV^{-1} 277

Figure 18.6.2. Plot of Two-Tier gravitational potential for $M_c = 1.22 \times 10^{19}$ GeV/c^2 (the Planck mass) and Newton's gravitational potential. Potentials are measured in units of 10^{-18} TeV^{-1}. The radial distance is measured in units of 10^{-18} TeV^{-1}. 278
Figure 18.10.1. The V-V-A triangle diagrams. 284
Figure 18-A.1. An oscillating string attached to a spring........ 287
Figure 20.2.1.1. A plot of a(t) generated from eq. 20.2.1.4 through numerical integration. 337
Table 20.2.1.1. Epochs and phases of the universe after the Big Bang Epoch........ 337
Figure 20.2.3.1. A plot of a(t) (horizontal axis) vs. time in seconds. The thick line is the plot of a(t) obtained by direct numerical integration of eq. 20.2.1.4 including the three density terms and the curvature constant term. The thin line is a plot of a(t) calculated directly from the approximation eq. 20.2.3.6. The approximation becomes increasingly better for small times as $t \to 0$. (Reader: please rotate page 90 degrees clockwise.)........ 340
Figure 20.2.3.2. A log-log (base 10) plot of a(t) vs. t (in seconds) for small times calculated from eq. 20.2.3.6........ 341
Table 20.3.2.1 Epochs and phases of the Universe since t = 0........ 345
Figure 20.3.5.1. A log-log (base 10) plot of the real and imaginary parts of a_{BBRW} and the Robertson-Walker scale factor a(t) versus the log (base 10) of time in seconds. Note the imaginary part of a_{BBRW} is very slowly growing. The real part of a_{BBRW} is growing slowly until $t_c \approx$ 10–167 s and thereafter equals a(t) to good approximation........ 349
Figure 20.3.5.2. A plot of the real and imaginary parts of a_{BBRW}, and $10^{39} \times$ a(t), versus time from t = 0 to $t = 1.2 \times 10^{-246}$ s. Note they are slowly varying and well behaved in the neighborhood of t = 0 with only a(t) having the value of zero. "E" indicates a power of ten (for example: 2.0E-248 = 2.0×10^{-248} s)........ 350
Figure 20.3.5.3. A plot of the real part of a_{BBRW} and a(t) vs. time in seconds around the time 10^{-167} s. Note Re a_{BBRW} quickly changes from slowly growing to growing like a(t)........ 351
Figure 21.1.1. A plot of the Two-Tier gravitational potential (solid line) and the Newtonian gravitational potential (dashed line). Note anti-gravity at short distances. See Fig. 7.3.9.3 of Blaha (2004) for more details........ 357
Figure 23.1. The fully reduced block structure of R matrix representation. 377
Figure 23.2. 16-dimensional representation of the SU(3) generators........ 377
Figure 23.3. A diagram illustrating the various functions relating the universes, and representing the coordinate transformations........ 382
Figure 23.4. A symbolic view of part of the Megaverse with universes depicted as circles. Each universe has its own curvilinear coordinate 4-vector denoted x. The Megaverse coordinates are a 16-vector y. A Wheeler-DeWitt equation applies within each universe and a comprehensive equation describes the entire Megaverse transitioning to each universe's wave function within its domain. 384
Table 4.1. Space-time dimensionality and number of spinor components corresponding to various n-valued logics. 387

Figure 24.1. A symbolic view of a high resolution (high energy) probe from a universe to a specific baryonic part of another universe. 409
Figure 24.2. A symbolic view of a low resolution (lower energy) probe from a universe to an entire universe. 409
Figure 24.3. Generation of a universe – anti-universe pair as a vacuum fluctuation. 415
Figure 24.4. Fission of a universe particle into two universe particles. 416
Figure 24.5. Two universe particle decay of a tachyon universe particle. 416
Figure 24.6. A Feynman diagram illustrating the continuity of a universe particle mass through a Planckton interaction. 418
Figure 25.1. A plot of the real part of the velocity of a starship on its 29th and 30th earth day of travel up to 5,000c. The dynamics of this case are described in the text where the real speed reaches 14572c and beyond. Time is measured in earth days. Note: as the speed of the starship increases rapidly near the singularity point, time on the starship also passes more quickly so that the starship occupants do not experience very high acceleration. Starship time $t' \approx \beta t$ when $\beta \gg 1$ where t is earth time. 426
Figure 25.2. "Coasting" part of travel time to various destinations at a real velocity of 30,000c. 428
Figure 25.3. A visualization of a starship. The outer disk contains the colliding hadron ring(s) which generate quark-gluon fireballs in a "combustion chamber." The fireball expands through a "rocket nozzle" generating a complex-valued thrust that enables the starship to exceed the speed of light. The four nuclear engines depicted on the underside of the ship provide "intra-solar system" speeds. 429
Figure 25.4. The trajectory of a uniship in a slingshot maneuver with a neutron star (the dark circle). The repulsive baryonic force causes the "turn" away from the star into the Megaverse. The solid line corresponds to the time in which the uniship is wholly within the universe and dominated by gravity. The dotted line reflects the transition of the uniship into the Megaverse. 434
Figure 25.5. Depiction of the uniship exposure interval of closest approach to a neutron star. The time interval spent in this region might be about three second or so. The uniship speed as it approaches might be about 0.7c or so to avoid capture by the neutron star. When the uniship totally exits our universe it disappears from view since electromagnetic radiation (light) from the uniship (or any object) "cannot penetrate" the boundary of our universe. The dashed line indicates the partial exit of the uniship from our universe into the Megaverse with Megaverse velocity and momentum components. 435
Figure 25.6. Tentative umbrella-like uniship design with the spokes of the umbrella forming a fan. The thrust tubes (umbrella spokes) extend kilometers from the thrust power generator(s) core to enable the baryonic force to maneuver them in the 15 different Megaverse directions. The thrust tubes are

able to swivel into all 15 spatial directions in response to the baryonic force as the uniship enters the Megaverse. Two fuel spheres are depicted under the assumption that one holds hydrogen and the other holds anti-hydrogen since they are presently the most powerful known possible energy source. The black forward part is for crew, supplies, and cargo. The rear gray part holds the engine apparatus and other related engine components. 438
Figure 25.7. Uniship slingshot past neutron star. Note the umbrella spokes which are attached to the uniship but moveable will respond differently to the baryonic force since they are at different radial distances from the neutron star and thus feel differing amounts of force. The baryonic force will twist them in different directions. They then can be re-oriented by the uniship computer to provide mobility in all Megaverse directions. 439

Derivation of The Extended Standard Model

1. Construction Principles for a Physical Theory of Universes

Euclid was fortunate in the creation of Euclidean geometry because he knew the method with which to construct geometry, and the primitives, straight lines, curves, and angles in space with which to define his axioms. The development of a fundamental physical theory is not so fortunate. The fundamental principles and axioms are open to debate. Many proposals have been put forward. They can be classified into two general categories: fundamental theories which simply assume a set of fundamental particles and symmetries; or fundamental theories which assume a set of fundamental artifacts such as strings and symmetries that ultimately form the known elementary particles and symmetries of the Standard Model. Both approaches are open to the same questions: What is their justification? What is their origin? How do we know that there is not yet a more fundamental level?

We think that we have derived a fundamental theory of the most primitive possible type that directly leads to the Standard Model with a natural extension for Dark Matter, and Quantum Gravity.

In this chapter we will consider certain fundamental general concepts which will set directions at points in the construction where there is a choice.

1.1 The Principle of Logic as the Origin and the Method of Construction

1.1.1 Requirement of Logic in Physics

Every physical theory is developed by logical deduction. Questions have been raised about the validity of logic. In particular, Gödel's Undecidability Theorem[1] is often cited as proof of a fundamental problem in logic. As we pointed out in Blaha (2011c), "We show these "paradoxes" and the Undecidability Theorem are resolved if we understand logic statements to be a form of function whose function arguments have a domain of validity. The verbal and symbolic statements that are at the heart of paradoxes use 'function' arguments that are outside the domain of the functions embodied in their statements.[2] Invalid arguments are the source of these seeming paradoxes and Gödel's Undecidability Theorem." Thus Logic is valid and not internally inconsistent.

---

[1] Gödel's Undecidability Theorem is not incorrect. However its interpretation as a flaw in the nature of Logic is incorrect. It really is a proof of the existence of statements that are both true and false because of the choice of subjects outside the domain of predicates. We view statements as a 'verbal' form of function.

[2] Most scientists, Logicians and Mathematicians are familiar with mathematical functions that have numerical arguments and calculate a numerical value or set of numerical values. However, those familiar with Computer Science know that a function, in general, has two types of output: its numeric value(s), and a status value indicating whether the computation that the function performed was successfully done or failed. Typically the status value is a zero or one, but it is understood as true or false also. (True – computation successful; false – computation failed for some reason) Thus we categorize functions as being in one of three broad categories:

Types of Functions

A Mathematical	A Function	A Function

We therefore adopt the principle that a fundamental theory must be logical.

1.1.2 The Logic Building Block of Particles – The Iota

Furthermore if we consider all possible 'things' that might constitute a fundamental building block for a fundamental theory they are all, at best, *ad hoc* and raise questions of their necessity and whether they are composed of yet a more fundamental substructure.

There is only one choice of building block that avoids these issues – a logic unit or bit. A unit of logic is a fundamental entity that is known to have an energy or equivalently a mass, and has no constituents of a more primitive form.[3] We will call a unit of logic that will form the core of a particle an *iota*.It has a conceptual value. But, in itself, it has no physical form or material existence in the sense of lumps of matter unless features, such as coordinates, supporting interactions are introduced by construction. Later we will introduce physical features that will cloak iotas with properties and interactions.

1.1.3 Mass of an Iota

Recent experiments have shown that a logical value (a bit or iota).as an energy.associated with it. One bit of information has about 3 x 10^{-21} joules of energy[4] or a rest mass, m_0, or about 0.02 eV using $m_0 = E/c^2$. This result was confirmed by Eric Lutz et al.[5] who showed that there is a minimum amount of heat produced per bit of erased data. This minimal heat is called the *Landauer[6] limit*. The equivalent mass we will call the *Landauer mass* and denote it as m_0. We will assume that a fundamental Landauer mass exists in our discussions although the precise value of the mass will not be used since we may expect all physical particle masses to be renormalized to different values when interactions are taken into account.

However, it is intriguing that the mass of the electron neutrino has been measured in a variety of experiments and found to be within an order of magnitude or so larger than our estimate of the Landauer mass as we would expect since particles acquire a 'cloud of virtual particles' due to interactions that can be expected to increase their mass above the Landauer mass. Since neutrinos only have the weak interaction it is not surprising that the increase due to interactions should not be large. The Mainz Neutrino Mass Experiment, for example, estimates the electron neutrino mass to be less than 2 eV.

A number of astronomical studies have also generated estimates of neutrino masses. In July 2010 the 3-D MegaZ DR7 galaxy survey found a limit for the combined mass of the three neutrino

---

Function Producing A Value(s) Only; Example: sin function	Producing a Value(s) and a Status Value; Example: a Programming Function	Producing a Status Value Only; Example: a logical statement

[3] Ab iota is a physical manifestation of a logical value. The relation of an iota to a logical value is analogous to the relation of a penciled point placed on paper to the concept of a point as a primitive in geometry.

[4] E. Muneyuki et al, *Nature Physics*, DOI: 10.1038/NPHYS1821.

[5] E. Lutz et al, *Nature* **483** (7388): 187–190,10.1038/nature10872, (2012).

[6] R. Landauer, "Irreversibility and heat generation in the computing process", *IBM Journal of Research and Developm
ent* **5** (3): 183–191, (1961).

varieties to be less than 0.28 eV.[7] A smaller upper bound for the sum of neutrino masses, 0.23 eV, was found in March 2013 by the Planck collaboration,[8] In February 2014 a new estimate of the sum was found to be 0.320 ± 0.081 eV due to discrepancies between the Planck's measurements of the Cosmic Microwave Background, and other predictions, combined with the assumption that neutrinos are the cause of weaker gravitational lensing than implied by massless neutrinos.[9]

Thus the experimentally measured values of neutrino masses are consistent with the iota Landauer mass estimate of 0.02 eV given above. We can thus assume that *a particle consists of an iota with a certain mass*[10] *that is renormalized, and having other features, that will emerge in the construction of the complete theory.*[11] We view reality as ultimately a representation (or painting) of logic values evolving through interactions in time and space.[12]

1.2 Ockham's Razor

In the construction of a theory there are occasions when a choice must be made between several alternatives. William of Ockham proposed a Law of Parsimony that is called *Ockham's Razor* which states that the simplest choice is to be preferred in a multiple choice situation. This principle is often stated as 'the simplest solution is usually the correct solution.'[13]

The best rationale for this principle is that it generally reduces the complexity that follows such a choice. Since many physics calculations are extremely difficult, picking the simplest choice.would generally tend to make subsequent theorems/calculations less difficult. This point of view might be thought to be ad hoc or anthropomorphic. But it reflects the reality of scientific calculation and of theory construction.

Thus we will assume Ockham's Law of Parsimony in the construction of our theory with the proviso that Leibniz's Minimax Principle (section 1.3) takes precedence if there is a conflict in the implications of the choices.

1.3 Leibniz's Minimax Principle for Physics Theories

Leibniz[14] developed a Minimax Principle that can be phrased for our purposes as, "The universe is based on the smallest set of properties or features that lead to the greatest variety of phenomena." This principle reflects the spirit of the minimum/maximum criteria of the Calculus of Variations[15] that play a

---

[7] S. Thomas et al, "Upper Bound of 0.28 eV on Neutrino Masses from the Largest Photometric Redshift Survey", Physical Review Letters **105**: 031301 (2010).

[8] Planck Collaboration, arXiv:1303.5076 (2013).

[9] R. A. Battye et al, "Evidence for Massive Neutrinos from Cosmic Microwave Background and Lensing Observations", Phys. Rev. Lett. **112,** 051303 (2014).

[10] Leibniz first proposed the idea of logic 'particles' which he called monads. Our definition of a logic 'particle' does not include (or exclude) the presence of a spiritual part which was part of the definition of Leibniz's monads.

[11] A recent experiment claims to separate the spin part (which we identify as a logical value later) of a molecule from the rest of the molecule.

[12] Those who might suggest matter is substantial, and logic values are not, should remember that matter would be completely insubstantial if there were no forces in nature. Neutrinos which are close to insubstantial would be completely insubstantial if there were no weak interactions.

[13] William of Ockham – Law of Parsimony – "Pluralitas non est ponenda sine necessitate" or "Plurality should not be posited without necessity." Ockham's Law was first stated by Durand De Saint-Pourçain (1270-1334 A.D.). In simple terms the principle states the simplest solution to a problem is most likely to be the correct solution.

[14] See Rescher (1967).

[15] Leibniz was one of the founders of the Calculus of Variations.

central role in many physics theories. This principle somewhat overlaps Ockham's Law of Parsimony. Given a choice of possible theoretical lines of construction there is a possibility that the Law of Parsimony and the Minimax Principle would suggest different choices. Fortunately, we will see that the construction of the Standard Model does not seem to present this potential dilemma.

An important, unremarked, aspect of Leibniz's Principle is the decision between a set of choices depends on the future part of the construction or theory. Thus future constructs determine past constructs in minimax decisions.[16]

1.4 A Physical Space Exists

We will postulate that a space exists of as yet undetermined dimensions and parametrized by coordinates. We will assume that an invariant distance measure can be defined on that space. Physically measured coordinates are necessarily real-valued because time and ruler measurements always yield a real value.

1.5 Non-Localized Physical Processes Exist

Physical processes can exist with spatially separated parts that may be intimately coordinated. Processes with spatially separated parts can have instant communication between the parts or can have parts communicate with time delays.

1.6 Construction Principles of Euclid's Geometry

In the above sections we have discussed construction principles for a fundamental physics theory. One might ask why does not Euclid's geometry have analogous construction principles? Normally we think of Euclid's geometry as based on primitive terms and five postulates.[17]

With small thought we see that Euclid's geometry has construction principles as well *at an intuitive level.* These principles include the definition of angles, the construction of simple geometric figures such as rectangles, triangles, trapezoids, and so on. The more advanced developments in geometry such as the geometry of Ptolomaic astronomy with cycles and epicycles also have implicit principles for construction.

Thus our introduction of additional principles 'along the way' to guide the construction of fundamental physics is analogous to the implicit construction/derivation principles of Euclid's Geometry.

---

[16] The knowledgable reader will remember Feynman's speculation that the physical universe may be evolving from the future into the past. Quantum field theories support such an interpretation of their mathematics. The similarity of this fundamental minimax principle's feature with the corresponding feature of physics theories encourages support for the minimax principle as a 'design' law of fundamental physical theory.

[17] Construction principles such as building triangles and other geometric figures as well as dissecting their features are implicit. Also there are some hidden assumptions, beyond the original five, that appear in the diagrams used in the proof of theorems.

2. Asynchronous Logic and Physical Processes

In sections 1.1 and 1.5 of chapter 1 we discussed the central role of Logic and the need for synchronization of non-local physical processes. The need for synchronized non-local physical processes requires the introduction of a new principle: the Principle of Asynchronicity.[18] When processes take place in parallel whether it is Quantum Mechanical entangled processes at small/large distances, or in high order Feynman diagrams (or their old fashioned time ordered perturbation theory predecessor) the synchronicity of a process is a physical requirement. It is implicitly resolved by physical laws which prevent asynchronicities (situations when parallel processes get "out of sync" resulting in the failure of an entire physical process to complete properly.) The Principle of Asynchronicity is described in the following pages. Asynchronicity can be briefly described as:

> In computation asynchronicity issues can arise. For example parallel computations or computer processes on a chip or set of chips have to be carefully managed for a parallel computer process to complete properly. In the case of computer chip design (VLSI chips and so on) techniques have been developed for the design of chips based on multi-valued logic. One conceptual approach uses 4-valued logic to define clockless computer logic circuits. The 4-valued logic developed by Fant (2005) has the four logic values TRUE, FALSE, NULL, and INTERMEDIATE. It is an extension of Boolean Logic that can accommodate time asynchronicities in asynchronous computer circuits. It enables circuits to avoid the use of system clocks to implement synchronization.[19] Thus the synchronization is explicit in 4-valued logic and non-logical constructs are not needed.[20] Concurrent transitions are coordinated solely by logical relationships with no need for any time constraints or relationships.

Now, realizing that The Standard Model, and physical theories that are ultimately derived from it such as Quantum Mechanics, potentially contain asynchronicities, we suggest that a Principle of Asynchronicity is embodied in the fundamental theory of Physics that leads to Dirac-like equations for the fundamental fermions – the leptons and quarks of The Standard Model.

2.1 Four-Valued Asynchronous Logic

The basic defining features of asynchronous circuits and Asynchronous Logic are:

---

[18] Much of this chapter is abstracted from Blaha (2011c) and printed in smaller type. Some might argue that it should be called the principle of synchronicity since the goal is synchronization of the parts of an evolving process. We chose to follow the terminology in the field of Asynchronous Logic as exemplified by Fant (2005) – a classic in that field.

[19] Remarkably Bjorken (1965) pp 220-226 presents an analogy of Feynman diagrams with electrical circuits where momenta map to currents, coordinates to voltages, Feynman parameters to resistance, and free particle equations of motion to Ohm's Law plus the equivalent of Kirchhoff's Laws. Thus Feynman diagrams and computer circuits are analogous.

[20] A two-valued asynchronous logic is also possible – just as the Dirac equation can be expressed as two 2-dimensional equarions. See Fant (2005) and Bjorken (1965).

1. An *asynchronous circuit* is a circuit in which the component parts are autonomous and can act in parallel at various rates of time evolution. They are not controled by a clock mechanism but proceed or wait for signals indicating that they can proceed.
2. *Asynchronous logic* is the logic used in the design of asynchronous circuits. The logic embodies the asynchronicity, and so the circuits built using the logic do not use a clock to control the execution speed of the various parts of an asynchronous circuit. Consequently logic elements do not necessarily have a distinct true or false state at any given point in time. 2-valued Boolean Logic is not sufficient and so asynchronous logic is multi-valued. The logic embodies states that allow for "stop and go" states within an executing asynchronous circuit.

In Fant's asynchronous 4-valued logic the four possible truth values of a state are:

True – status is true and all data is current
False – status is false and all data is current
Intermediate – status is indefinite with some data current
NULL – status is indefinite with no data present – results in a suspension of processing of the circuit part in a NULL state until current data becomes present

"Data" is the information flowing through all or part of a circuit. Using these truth values the evolution in time of the parts of an asynchronous circuit are effectively synchronized by the logic without the use of a clock mechanism. (A clock mechanism effectively is a subsidiary time constraint or set of time constraints.) See Fant (2005) for further details.

An implicit aspect of asynchronous logic is the coordination of spatially separated parts of a circuit. Since spatial separations in a circuit can be mapped to time delays using the speed of data propagation between parts, spatial asynchronicites are subsumed under time asynchronicities. This is particularly true for computer chips which are kept small to minimize delays.

2.2 Principle of Asynchronicity

An obvious feature of elementary particle phenomena is the coordination of the parts of a physical process in time and space. Complex Feynman diagrams embody the coordination of the parts of interacting particles. Quantum entanglement phenomena embody the coordination of the parts of a physical phenomena separated by large distances and perhaps times. Examples of these types, which could be multiplied indefinitely, lead to a Principle of Asynchronicity.

Principle: Nature requires asynchronicity. This asynchronicity is coordinated by 4-valued physico-logical structures for matter.

Elaboration: Elementary particle physical phenomena must support extended coordinated physical phenomena in space and time. The fundamental laws of particle physics must be such as to permit coordinated physical phenomena with coordination between the parts of a physical phenomenon at small/large distances and small/large time intervals. The coordination must be embodied within physical laws.

This principle will be applied below (and later in this book in more detail) to justify Dirac-like equations for particle dynamics.

Coordination is an obvious feature of physical phenomena. This principle goes beyond that by asserting that extended coordinated physical phenomena must exist. If particles exist, then their antiparticles must also exist to provide asynchronous behavior in interaction regions. If only particles existed then all interactions would proceed forward in time and the state of the interaction at any point in time would be known. With the addition of antiparticles, asynchronicity issues are introduced and at various 'time slices' (if one thinks in terms of old-fashioned time ordered perturbation theory) of the progress of an interaction, the state can be ambiguous since antiparticles are negative energy particles moving backward in time.

Asynchronicities are common in the many subcircuits of a computer chip. Asynchronicities are also common in the many interaction subregions of a set of particles in interaction. Page 7 of Fant (2005) has a diagram of a circuit with a set of subcircuits with five time slices of the interacting subcircuits showing five states of the "'data' wavefront" at five points in time. This diagram is similar to the time sliced diagram of an interacting system of particles in "old fashioned" time-ordered perturbation theory. Page 29 of Blaha (2005b) displays a similar diagram (Fig. 5.1.4) in a description of a Standard Model Quantum Langauge Grammar – a language representation of particle physics. Blaha's diagram is remarkably similar to Fant's diagram in overall features as one might expect since both address time asynchronicity.

The asynchronicity that appears in perturbation theory diagrams is intimately related to the appearance of antiparticles in diagrams. As noted earlier antiparticles are interpretable as negative energy particles traveling backwards in time. The time orderings, which are implicit in the Feynman diagram approach, and explicit in old fashioned perturbation theory, show time asynchronicity and the effect of the dynamics to coordinate the asynchronicities so that correct results follow from perturbative calculations.

3. Iotas as the Core of Fermions

3.1 Matrix Representation of Asynchronous Logic

The four possible logic states of Asynchronous Logic can be mapped to a matrix representation with four component columns and 4×4 matrices that transform between logic values.[21] The basic four pure logic states can be labeled using a notation that connects with physics:

$$u(+\tfrac{1}{2}) = \begin{bmatrix} 1 \\ 0 \\ 0 \\ 0 \end{bmatrix} \qquad u(-\tfrac{1}{2}) = \begin{bmatrix} 0 \\ 1 \\ 0 \\ 0 \end{bmatrix} \tag{3.1}$$

and

$$v(-\tfrac{1}{2}) = \begin{bmatrix} 0 \\ 0 \\ 1 \\ 0 \end{bmatrix} \qquad v(+\tfrac{1}{2}) = \begin{bmatrix} 0 \\ 0 \\ 0 \\ 1 \end{bmatrix} \tag{3.2}$$

The arguments of u and v will become physically values of particle spin: +½ represents an "up" spin state and −½ represents a "down" spin state. Linear combinations of these four states obtained using linear combinations of the sixteen 4×4 matrices with complex coefficients form *qubits*.[22] Any bit or qubit transformation can be constructed from a sum of the sixteen Dirac matrices[23] multiplied by complex coefficients. The iota states listed above are constants. They will become variable through the use of Lorentz transformations considered later.

A set of sixteen independent 4×4 matrices can be constructed from the four basic Dirac matrices, usually denoted γ^μ, by multiplications and summations. They can be found in most books on quantum field theory.

Applying these basic facts about the 4×4 matrix representation of Asynchronous Logic to iotas we can develop the Dirac equation formulation of spin ½ particles after introducing space-time coordinates in the next chapter.

---

[21] See Blaha (2010a) for more details.

[22] S. Weisner, "Conjugate coding". Association for Computing Machinery, Special Interest Group in Algorithms and Computation Theory **15**, 78–88 (1983). We will not discuss qubits in this construction.

[23] See Bjorken (1964).

We can view fermion particles as iotas parametrized by coordinates and interacting via interactions (forces) that primarily result from the need to impose symmetry requirements on coordinate transformations.[24]

3.2 Matter is Insubstantial Neglecting Particle Interactions

Philosophers and physicists have debated the nature of matter for milleniums. Differing opinions have been the norm. Perhaps one of the most interesting expressions of opinion was that of Dr. Johnson, the 19th century author of the Dictionary. Upon hearing of Bishop Berkeley's philosophic view that matter was insubstantial and "not real", Dr. Johnson proceeded to kick a rock while exclaiming "I refute it thus" according to his biographer Boswell. While Dr. Johnson's riposte cannot be denied in its succinctness, our theoretical development is based on the concept of iotas as the core of fermion particles, and the further development of our construction later leads to a refinement of the Berkeley-Johnson conflict as well as that of many philosophers and physicists.

For we propose that matter is truly insubstantial with Bishop Berkeley, and yet gains substantiality through interactions (forces) without which all matter would be interpenetrable and could reside at a single point.[25] Thus Dr. Johnson does not truly refute Dishop Berkeley but does show that forces exist amongst matter that gives it "substantiality."

Since all matter might have been concentrated[26] at an 'essentially' mathematical point in the absence of interactions we can call that point the *void* and view it as the 'location' of the Big Bang that presumably existed at the Beginning. Coordinates and interactions may well have originated at that point as the result of quantum fluctuations.

---

[24] We note that Asynchronous Logic can also be formulated with two dimensional vectors and 2×2 matrices just as the Dirac matrix formalism for spin ½ particles can be expressed in a 2×2 matrix formalism.

[25] To some extent we see an approximation of this proposal in the approximate interpenetrability of normal matter and Dark Matter which would be exact if there were not a very weak force between matter and Dark Matter as well as the gravitational interaction.

[26] This comment is based on our version of quantum field thory in which all interactions (including gravity) go to zero at zero distance. See Blaha (2005a) for a detailed discussion of our divergence-free form of quantum field theories. With zero forces, all particles can concentrate at a single point.

Appendix 3-A. A Map between Interacting Particles and Computer Languages

This appendix appears as Appendix B in Blaha (2013b). It is also presented in more detail in Blaha(2005b) *The Metatheory of Physics Theories, and the Theory of Everything as a Quantum Computer Language* and in Blaha (2005c*) The Equivalence of Elementary Particle Theories and Computer Languages: Quantum Computers, Turing Machines, Standard Model, Superstring Theory, and a Proof that Gödel's Theorem Implies Nature Must Be Quantum.*

This appendix can be skipped by the reader not interested in a discussion of a fairly detailed comparison of elementary particle theory "in the raw" and computation.

In this appendix we will provide detailed support for the view that The Standard Model is a particular implementation of a general specification for a computation. In this view particles are skeletonized to data and interactions to the execution of a computation. Thus we provide a complementary view to the current effort to build quantum computers from atoms and molecules using Quantum Theory. The discussion in this chapter is largely condensed extracts of parts of Blaha (2005b). Our earlier works, beginning with Blaha (1998), contain the basic ideas and their elaboration.

3-A.1 The Similarity of Data to Particles

Some years ago we pointed out that true creation and annihilation in our universe appears in only two arenas:[27] data transformations within computers and particle transformations in Nature. Both subatomic particles and data can be created or destroyed or combined to produce new particles or new data. The similarity is compelling! In the following sections we will amplify these ideas based on the theory of computer languages and computer grammars.

We will also consider particle interactions at a more abstract level and show how to create "subatomic particle quantum Turing machines" for both the Standard Model and Superstring theories. (In this appendix we take a neutral stance on whether the Theory of Everything is a Superstring theory or our Extended Standard Model. The possibility of a deep connection of these two theoretic approaches cannot be ruled out at present.)

3-A.2 Particle Physics Lagrangians Defines a Language

Physicists use perturbation theory to perform computations in quantum field theories such as The Standard Model. Perturbation theory takes the interaction terms of the Lagrangian and performs approximate calculations of the probabilities of particle interactions. Perturbation theory calculations are normally visualized using Feynman diagrams. For example the collision of two electrons to produce two electrons with different energies and momenta (called electron-electron elastic scattering) can be visualized as a sum of terms corresponding to the various ways the electrons can interact. The number of

---

[27] Blaha (1998). Please note that we will discuss only Nature in this book since theological constructs are beyond the scope of our inquiry.

terms in this example is infinite in principle. Since this infinite sum cannot be calculated the sum is approximated by a finite number of terms. The simplest and, as it turns out, the dominant terms in electron-electron scattering are:

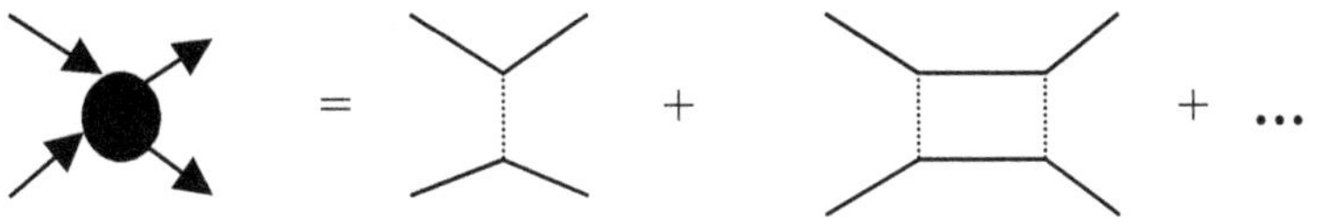

Figure 3-A.1. The first few terms in the approximate perturbation theory calculation of the scattering of two electrons. The dotted lines represent photons that carry the electromagnetic force between the electrons.

We will see that the individual terms (diagrams) in a perturbation theory calculation can be viewed as words. The words are part of a language with an alphabet and grammar.

The fundamental particles of the Standard Model constitute the set of symbols or the alphabet of a language with 52 letters in The Complexon Standard Model. (It is an interesting, but meaningless, coincidence that most alphabet-based human languages have 20 to 40 letters in their alphabets. English has 26 and so on.) As we shall see, the particle language grammar is a quantum extension of a type of computer language (developed by Chomsky and others) that uses *production rules*. The production rules for the grammar of the Standard Model are easily derived from the interaction terms of the Standard Model.

The concept of the language of the Standard Model is very simple. The technical details of the language description require a discussion of computer languages and Quantum Turing Machines.

The following sections describe the basic idea of this linguistic representation of the Standard Model. They show how particle interactions can be viewed as transformations (processes) within a Quantum Computer that accepts the computer language generated by the Standard Model Lagrangian interaction terms.

3-A.3 A Linguistic Representation of the Standard Model

In this book and earlier works we explored the features of the Standard Model. We saw that it was consistent with almost all the known properties of elementary particles. This section introduces a new view or representation of the Standard model that focuses on its interactions and *shows the Standard Model defines a language similar to a computer language.*

The *alphabet* of the language is the set of elementary particles of the Standard Model. The *words* of the language are quantum states consisting of elementary particles. A bound state of several particles such as a proton (three quarks bound together) is a word. A quantum state consisting of several free particles – particles that are not bound together and that are some distance from each other also constitute a word. In fact the entire universe constitutes one mighty word.[28]

The collision or scattering of particles can be viewed as beginning with a combination of letters corresponding to the set of initial particles – the input string or word. This input string undergoes transformations specified by grammar rules to produce an output string (word) of letters corresponding to the outgoing particles after the collision.

[28] Blaha (1998).

In describing this new view of the Standard Model we will focus on the essentials of the processes of creation, transformation and annihilation of matter ignoring (for the moment) particle spin, momentum, angular momentum and other details that are important in the complete theory of the Standard Model. This approximation may have been valid prior to the Big Bang when the universe might have been a mathematical point. Incorporating particle spin, momentum, angular momentum and so on into a "particle" language is not difficult.

The idea of associating physics with computers is not as unconventional as it might appear at first. Feynman[29] viewed computers as relevant for Physics: "If we suppose we know all the physical laws perfectly, of course we don't have to pay any attention to computers. It's interesting anyway to entertain oneself with the idea that we've got something to learn about physical laws; and if I take a relaxed view here ... I'll admit that we don't understand everything." Feynman wanted to simulate physics computations on a quantum computer in the hope that it would be faster than a conventional computer. We will show the Standard Model itself actually defines a specific (theoretical) quantum computer – a far more exciting possibility – because it gives a new view of Reality. Nature itself is a form of computer.

SuperString theory can also be formulated within a Quantum Computer framework.

A computer language representation of particle physics is of great interest in itself. It may generate new insights into the process of matter creation and transformation. It may lead to a new understanding of the fundamental nature of the universe. And it appears to suggest a rationale for approaches such as the currently popular SuperString theories of elementary particles.

3-A.3.1 Linguistic View of an Interaction

We will begin by looking at the simple interaction term:

$$\bar{e}Ae$$

From a computer language perspective this Lagrangian interaction term can be viewed as specifying a set of grammar rules called *production rules*.

In fact each interaction term in the Standard Model Lagrangian can be viewed as specifying a set of grammar rules. The combined set of grammar rules for all Standard Model interaction terms defines a grammar with particles constituting the alphabet (letters or symbols) of the grammar.

To appreciate the mapping (or analogy) between particles and alphabetic letters, and of interaction terms and computer grammar, we have to understand the process of data characters (or letters) flowing through a computer. It is an interesting and little noted fact (because it is viewed as trivial) that a computer can generate (or absorb) data as part of the computation process. For example we might write a computer program that takes a set of letters input into a computer and outputs each input letter twice:

In a sense the computer has created data characters just like particle interactions can create particles. Computers can also absorb (or annihilate) data (usually to our dismay). So we can see an analogy

[29] R. P. Feynman, International Journal of Theoretical Physics, **21**, 467 (1982).

between the transformations of data characters in a computer, and particle annihilation and creation. *Nothing else in Nature is so directly analogous to particle creation and annihilation.*

This observation leads us to take the view that particles are data packets that we denote with letters (symbols). They contain quantum numbers and other properties (mass, spin, momentum, and so on) that certainly are data. And they have a grammar that we summarize with a Lagrangian.

3-A.3.2 Computer Grammars

The Standard Model Lagrangian in our view specifies a *grammar* in the sense of Naom Chomsky. Chomsky's concept of a language, and of a grammar, has important applications in the theory of computation and computers.

There are four basic types of languages in the Chomsky approach: called type 0, type 1, type 2 and type 3. They differ in the allowed forms of their grammar rules (also called *production rules*). We will be interested in type 0 languages. A type 0 language (also called an *unrestricted rewriting system*) is the most general type of language. It allows any grammar production rule of the form

$$x \longrightarrow y$$

where x and y are strings of characters.

Production rules specify how one string of characters transforms into another string of characters. Calculations in computers using computer languages are reducible to sets of grammar rules for string manipulation that are similar to the one shown above.

Each term in the interaction part of the Standard Model Lagrangian is equivalent to one or more production rules where the characters are particles. *The Standard Model can be viewed as generating a type* 0 *language.* This language goes beyond current types of grammars because it is inherently quantum probabilistic in nature. Quantum aspects of these rules will be described later.

Before looking at the production rules generated by an interaction term in a Lagrangian we will discuss a formal grammar. A grammar is a quadruple of items that is usually symbolized by the expression

<N, T, S, P>

where N is a set of variables called *nonterminal symbols*, T is a set of *terminal symbols*, S is a special nonterminal symbol called the *head* or *start symbol*, and P is a finite set of production rules. The angle braces < and > are merely a mathematician's way of saying these items are grouped together to constitute (or make) a grammar.

The *terminal* symbols are the set of characters that are allowed in input strings or output strings. The *nonterminal* symbols are the set of characters that appear in intermediate steps that lead from the input string to the output string. They are like internal variables or symbols.[30] The combined set of terminal and nonterminal symbols make up the *vocabulary* (alphabet) of a language.

[3030] Their particle analogues are virtual particles that only appear in an evolving interaction but not in the input or output. One could view quark-partons as equivalent to nonterminal particles.

Chomsky's definition of a language is the set of all strings of terminal symbols that can be generated by applying the production rules to the *head* symbol (or *start* symbol) S.[31] The head symbol is the symbol that begins all strings of symbols that can be generated in a language.

A simple example of a language in this approach is a vocabulary or alphabet consisting of the ABC's with words created from these letters according to some set of production rules.

3-A.3.3 Generalized Input Chomsky Languages

We will generalize Chomsky's idea of language to be the set of all strings that can be generated from all finite input strings of terminal symbols as well as the *head symbol.* We can also view all particles as generated directly or indirectly at the beginning on the universe. The "Big Bang" (the beginning of the universe) then becomes the primeval head symbol.

We can visualize the application of production rules to transform an input string of terminal characters into an output string of terminal characters as:

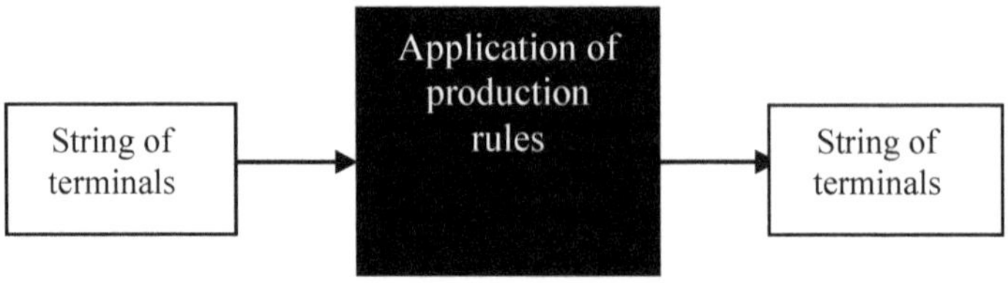

Figure 3-A.2 Generating an output string of terminal characters from an input string of terminal characters using the production rules of a grammar. Inside the black box, transformations of the input string take place and non-terminal symbols may appear and disappear. Non-terminal symbols, by definition, cannot appear in the input or output strings of characters.

In order to make these grammar concepts more concrete we will look at a simple artificial grammar before looking at the grammar generated by interaction terms in the Standard Model. The nonterminal symbols will be the letters S (the head symbol), A and B. The terminal symbols will be the letters x and y. The production rules will be

$S \rightarrow AB$ Rule I
$A \rightarrow y$ Rule II
$A \rightarrow Ay$ Rule III
$B \rightarrow x$ Rule IV
$B \rightarrow Bx$ Rule V

The Chomsky computer language that this grammar generates consists of all strings containing any number of y's followed by any number of x's since any of these strings can be generated from the head symbol S using the production rules. Because of rule I the y's are placed to the left and x's are

[31] The head symbol is analogous to the vacuum state in quantum field theory. Chomsky's definition specifies the possible vacuum fluctuations that can occur. Each string is a specific vacuum fluctuation. An example is an electron-positron pair momentarily popping out of the vacuum—a vacuum fluctuation.

placed to the right. The order of the symbols matters just as it does in human language – consider the words ides and dies which differ only in the order of i and d!

An example of generating a string yyxxx from the head symbol is:

S	→ AB	Rule I
AB	→ AyB	Rule III
AyB	→ yyB	Rule II
yyB	→ yyBx	Rule V
yyBx	→ yyBxx	Rule V
yyBxx	→ yyxxx	Rule IV

The production rule used to make each transition is listed above on the right.

Our generalization of the Chomsky definition of language would allow any string to be the starting point – not just the head symbol S. Using the sample grammar described on the previous page the generalized language becomes any string of x's and y's.

A more interesting language can be created by adding two new rules to the rules on the preceding page:

y → A	Rule VI
x → B	Rule VII

The resulting language – the set of strings of terminal symbols – remains the same despite the addition of these new grammar rules. However the number and variety of transitions becomes much larger. For example the following chain of transitions is allowed,

yx → AB → AyBx → AyyBx→ AyyyBxx → yyyyxx

In the next section we will extend the concept of computer grammars by allowing probabilistic grammar rules – production rules which have an associated probability of executing.

3-A.4 Probabilistic Computer Grammars

Grammar, which knows how to control even kings.
Molière - Les Femmes Savantes (1672) Act II, Scene 6

3-A.4.1 Probabilistic Computer Grammars™

The preceding section described the production rules for a *deterministic grammar*. The left side of each production rule has one, and only one, possible transition.

Non-deterministic grammars allow two or more grammar rules to have the same left side, and different right sides. For example,

A → y
A → x

could both appear in a non-deterministic grammar.

Non-deterministic grammars can be easily (almost "naturally") associated with probabilities. The probabilities can be classical probabilities or quantum probabilities. An example of a simple non-deterministic grammar is specified by the production rules:

S → xy Rule I
x → xx Rule II Relative Probability = .75
x → xy Rule III Relative Probability = .25
y → yy Rule IV

where the head symbol is the letter S, and the terminal symbols are the letters x and y. The relative probability of generating the string xxy vs. the relative probability of generating the string xyy from the string xy is

xy → xxy relative probability = .75
xy → xyy relative probability = .25

The string xxy is three times more likely to be produced than the string xyy.

For each starting string one can obtain the relative probabilities that various possible output strings will be produced.

A more practical example of a Probabilistic Grammar™ can be abstracted from flipping coins – heads or tails occur with equal probability – 50-50. From this observation we can create a little Probabilistic Grammar™ for the case of flipping two coins. Let h represent heads and t represent tails. Then consider the grammar:

S → hh
S → tt
S → ht
S → th
h → t probability = .5 (50%)
h → h probability = .5 (50%)
t → h probability = .5 (50%)
t → t probability = .5 (50%)

The last four rules above embody the statement that flipping a coin yields heads or tails with equal probability (50% or .5).

Now let us consider starting with two heads hh. The possible outcomes and their probabilities are:

hh → hh probability = .5 * .5 = .25
hh → th probability = .5 * .5 = .25
hh → ht probability = .5 * .5 = .25
hh → tt probability = .5 * .5 = .25

If we don't care about the order of the output heads and tails, then the probability of flipping two heads and getting a head and tail (hh → ht or hh → th) is .25 + .25 = .5.

This simple example shows the basic thought process of a non-deterministic grammar with associated probabilities.

The combination of a non-deterministic grammar and an associated set of probabilities for transitions can be called a *Probabilistic Grammar*. We will see that the grammar production rules for the Standard Model must be viewed as constituting a Probabilistic Grammar with one difference. The "square roots" of probabilities – probability amplitudes – are specified for the transitions in the grammar. The Standard Model requires probability amplitudes since it is a quantum theory. Therefore we will describe probabilistic grammars with associated probability amplitudes (such as that of the Standard Model) as *Quantum Probabilistic Grammars*.

3-A.4.2 Quantum Probabilistic Grammar

An example of a Quantum Probabilistic Grammar can be constructed based on an analogy with a two slit photon experiment. Imagine a wall with two slits. A source shoots photons at the wall. A photon can go through either slit with equal quantum probability. An illustration of this experimental arrangement is:

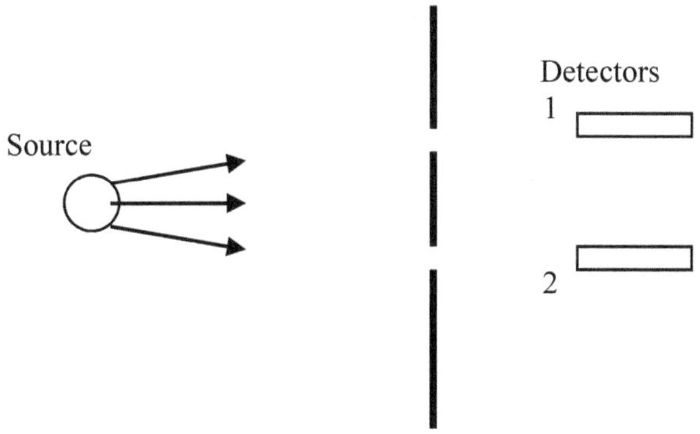

Figure 3-A.3. Two slit photon experimental setup.

A simple Quantum Probabilistic Grammar can be constructed corresponding to this experimental setup:

S → 1 probability amplitude = $1/\sqrt{2}$
S → 2 probability amplitude = $1/\sqrt{2}$

The head symbol S represents the source. The digit 1 represents a photon going through slit 1. The digit 2 represents a photon going through slit 2.

The values of the probability amplitudes $1/\sqrt{2}$ can be calculated using Quantum Mechanics. The probability for a photon to go through slit 1 is the absolute value squared of the probability amplitude:

$$\text{Probability to go through slit 1} = (1/\sqrt{2})^2 = .5$$

and the probability for a photon to go through slit 2 is

$$\text{Probability to go through slit 2} = (1/\sqrt{2})^2 = .5$$

This simple example illustrates the basics of a Quantum Probabilistic Grammar.

Before applying these concepts to the Standard Model we will look at a simpler Quantum Field Theory called a ϕ^3 ("phi cubed") theory (ϕ is the Greek letter phi). This theory describes a self-interacting spin 0 particle with no internal symmetries. This theory is a stepping stone to the far more complex Standard Model Quantum Field Theory. We are only interested in it as a simple example of quantum probabilistic grammar rules.

The ϕ^3 theory is so named because it has a cubic Lagrangian interaction term. (Note the exponent 3.) The grammar rules for the ϕ^3 theory are:

$$\phi \rightarrow \phi\phi \qquad \text{Rule I}$$
$$\phi\phi \rightarrow \phi \text{ Rule II}$$

Rule I corresponds to the emission of a ϕ particle and rule II corresponds to the absorption of a ϕ particle. We will not introduce a start symbol. Instead we will consider the transitions from an input state of a number of ϕ particles to an output state of (possibly) a different number of ϕ particles. *We will ignore the momenta of the particles.* (This assumption is equivalent to assuming the ϕ particles have infinite mass.)

We will assume either transition above takes place with a "relative probability amplitude" g. We will call this simplified theory the *modified* ϕ^3 *theory*. We will view g as a measure of the probability amplitude for an absorption or emission of a ϕ particle. (g is similar to a coupling constant in Quantum Field Theory.) The probabilities have to be normalized or rescaled so that the sum of all probabilities equals one.

To get a feel for the Quantum Probabiistic Grammar approach we will look at the case of an input state consisting of two ϕ particles. The output states can have one ϕ particle, two ϕ's, three ϕ's, and so on. Each possible output state has a certain probability of occurring. The sum of the probabilities for producing all possible output states must equal one. (Remember that the sum of all possible outcomes of flipping a coin is one. Having it come up heads has probability ½ and having it come up as a tails has probability ½ also.)

The simplest string transition from a two ϕ "input" state to a one ϕ "output" state is:

$$\phi\phi \rightarrow \phi$$

using Rule II. The probability amplitude of this transition is g by assumption.

The transitions between strings can be visualized with diagrams that are like the Feynman diagrams that used in Quantum Field Theory perturbation theory calculations. These diagrams are not the same as Feynman diagrams because they embody time orderings of emissions and absorptions of ϕ particles. (They actually harken back to the time-ordered diagrams that were used by physicists prior to 1950.) **This feature supports the need for the Principle of Asynchronicity described earlier.**

In some simple cases the time ordering is irrelevant. For example, the Feynman-like diagram for the simplest case of a two ϕ input state transitioning *directly* to a one ϕ output state is the same as the Feynman diagram:

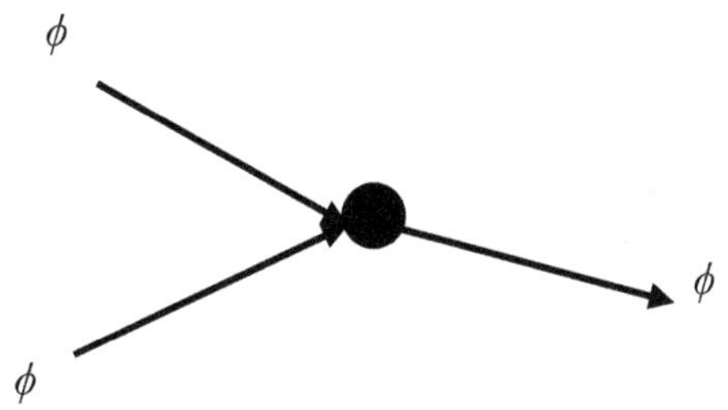

Figure 3-A.4. Diagram for $\phi\phi \rightarrow \phi$. The input states are always on the left and the output states are always on the right in Feynman and Feynman-like diagrams.

The time order of emission and absorption of ϕ particles can be symbolized using parentheses. For example,

$$(\phi)\phi \rightarrow (\phi\phi)\phi = \phi(\phi\phi) \rightarrow \phi(\phi) = \phi\phi \qquad \text{Diagram A (Fig. 3-A.5)}$$

and

$$\phi(\phi) \rightarrow \phi(\phi\phi) = (\phi\phi)\phi \rightarrow (\phi)\phi = \phi\phi \qquad \text{Diagram B (Fig. 3-A.5)}$$

These string transitions correspond to different time-ordered Feynman-like diagrams:

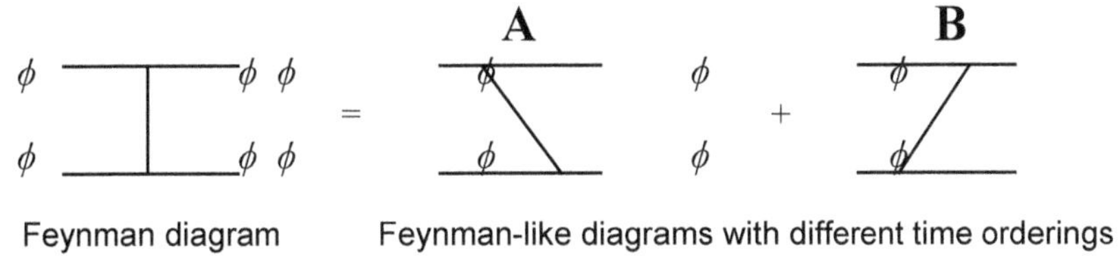

Figure 3-A.5 The true Feynman diagram is the "sum" of the two time-ordered Feynman-like diagrams.

The correspondence between Feynman-like diagrams and the transitions between strings based on the Quantum Probabilistic Grammar can be seen by taking vertical slices on diagrams A or B above after each emission or absorption. For example,

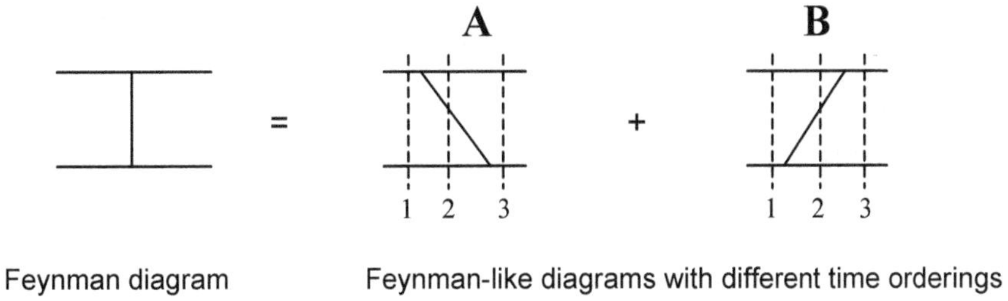

Figure 3-A.6 Slicing Feynman-like diagrams. A slice is made after each emission or absorption. As you read down a slice the particles are listed in the same order as the corresponding string. Each slice is numbered starting from the left.

The string corresponding to each numbered slice in the above figures is similarly numbered in the following transitions:

Slice:	1		2		3	
	$(\phi)\phi$	$\rightarrow$	$(\phi\phi)\phi = \phi(\phi\phi)$	$\rightarrow$	$\phi(\phi) = \phi\phi$	**A**
Slice:	1		2		3	
	$\phi(\phi)$	$\rightarrow$	$\phi(\phi\phi) = (\phi\phi)\phi$	$\rightarrow$	$(\phi)\phi = \phi\phi$	**B**

Parentheses on the left side of an arrow indicate the particle(s) that emits a new particle(s) appearing within the corresponding parentheses on the right side of the arrow.

A transition from an input state containing ϕ particles to an output state containing ϕ particles always has an infinite number of ways of taking place and thus an infinite number of Feynman-like diagrams. Readers familiar with the perturbation theory of Quantum Field Theory will remember that these diagrams are the same as the Feynman diagrams generated by perturbation theory with the additional feature of having time orderings.

We will now look at the transition of two ϕ particles to two ϕ particles: $\phi\phi \rightarrow \phi\phi$. There are an infinite number of Feynman-like diagrams for this transition. Some of the simpler Feynman diagrams are:

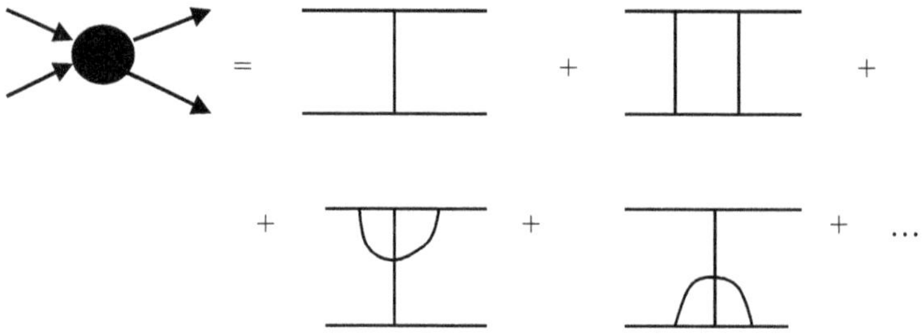

Figure 3-A.7. Diagrams for an elastic collision with two incoming particles and two outgoing particles.

Each Feynman diagrams corresponds to several time-ordered Feynman-like diagrams.

In evaluating these diagrams to calculate probabilities we must remember that we are ignoring space-time aspects such as particle propagators and momenta. So the calculation of the probability amplitude for this process becomes a counting problem of the number of diagrams that exist for each power of g^2. The probability amplitude for each diagram is a power of g^2.

Counting diagrams is a combinatorial mathematics problem that we will not explore in detail because it is peripheral to our interests. Consequently we will simply express the probability amplitude as:

$$A_2(g) = \sum_{n=1}^{\infty} a_n g^{2n}$$

where the mathematical expression on the right represents a sum from n = 1 to infinity and where the numbers a_n are integer numbers equal to the number of different diagrams having a power of g^{2n} as its probability amplitude. Each intersection of lines (called a vertex) contributes a factor of g to the amplitude for that diagram. The powers of g for the simpler diagrams that appear on the previous page are:

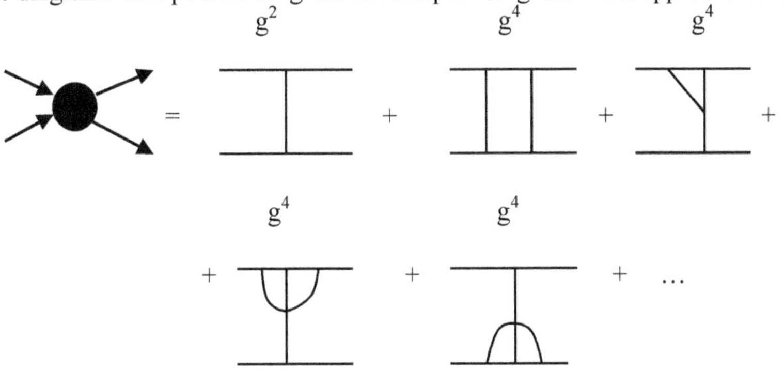

Figure 3-A.8. The power of g for some simple diagrams.

The value of the first constant a_1 is 1 since there is only one Feynman diagram with amplitude value g^2 for this process – the first diagram to the right of the = sign in Fig. 3-A.8. We treat all time-ordered variations of a Feynman diagram as contributing one to the value of a_n. The value of a_n grows rapidly as n increases. For large n the value of a_n is of the order of $(n!)^2$. Consequently the sum is an asymptotic power series. Here again the details are not important for us. We are not looking for a numerical result.

The (unnormalized) relative probability for the transition $\phi\phi \rightarrow \phi\phi$ is:

$$P_2 = |A_2(g)|^2$$

where |...| represents the absolute value of the complex probability amplitude. In quantum theories a probability is always the square of the absolute value of its probability amplitude. The probability P_2 is a relative probability that must be normalized – multiplied by a factor that makes the sum of the probabilities of all possible outcome states equal to one. To calculate this probability we must calculate the sum of all the relative probabilities P_n to produce any number of ϕ particles from a two ϕ input state.

$$\phi\phi \rightarrow \phi \ldots$$

The total of the relative probabilities is:

$$P = \sum_{n=1}^{\infty} P_n$$

where P_n is the relative probability to produce an output state with n ϕ particles.

The calculation of the relative probabilities P_n for n ϕ particles output states is similar to the calculation P_2. For example, for three particles

$$A_3(g) = \sum_{n=1}^{\infty} b_n g^{2n+1}$$

where the numbers b_n count the number of distinct diagrams with the power g^{2n+1} and

$$P_3 = |A_3(g)|^2$$

The absolute (normalized) probability to produce an n ϕ particle output state is

$$Q_n = P_n/P$$

The sum of all possible output state probabilities equals one:

$$1 = \sum_{n=1}^{\infty} Q_n$$

The modified ϕ^3 Quantum Field Theory provides a simple example of a Quantum Probabilistic Grammar. We will now turn to the Standard Model and examine its Quantum Probabilistic Grammar. Because it encompasses a much larger number of different particles (letters) and interactions (grammar rules) it will be significantly more complex.

3-A.4.3 Probability Amplitudes of Quantum Grammars

Quantum Grammar rules associate a probability amplitude with each production rule. To find the probability for a transition from an initial state to a final state we must calculate the probability amplitude for the transition through the repeated application of the production rules for each possible path from the initial state to the final state. We assume each initial state of the Quantum Turing machine begins with probability amplitude one. (This is a normalization condition for the initial state in reality.)

When a production rule is applied to a state to produce a transition to a new state the current probability amplitude is multiplied by the probability amplitude of the production rule. Thus the total relative probability for the passage from an initial state i to a specific final state f is

$$P_{fi} = \left|\sum_{\text{paths}} a_1\, a_2\, a_3\, \ldots\, a_{n(\text{path})}\right|^2$$

where the sum is over all finite paths that lead from the initial state to the final state through the application of all relevant production rules, and where $a_1 a_2\, a_3\, \ldots\, a_{n(\text{path})}$ is the product of the probability amplitudes of the production rules for each individual path. P_{fi} is similar in form to a Feynman path integral expression (See Feynman (1965)).

The value of n is path dependent and thus denoted n(path). The relative probability is the absolute value squared of the sum of products of the amplitudes. The relative probability must be normalized to produce an absolute probability such that the sum of the absolute probabilities of all possible final states is one:

$$P_{\text{absolute-}f\,i} = NP_{fi}$$

$$N = \sum_f P_{fi}$$

$$1 = \sum_f P_{\text{absolute-}f\,i}$$

where the sums are over all possible final states f.

Thus we have a well-defined method for calculating the probability of a transition from an initial state to a final state that is illustrated by the preceding examples. The fact that it bears some resemblance to the path integral methods for quantum mechanics pioneered by Feynman suggests that Quantum Grammars are of interest to physics. This author was first struck by the similarity of string transitions via production rules to path integrals in 1981. After all, a jagged (discrete) Feynman path is really a string of coordinates marking the end points of each line segment of which the path is composed. Thus each path in a Feynman sum over paths can be represented by a string. The evolution of a path from line segment to line segment can be viewed as the repeated application of a probabilistic production rule. The path sum equivalent of the probability amplitude of a production rule is an exponential Hamiltonian factor that is a function of the change in string coordinates "due to the production rule."

3-A.5 Standard Model Quantum Grammar

3-A.5.1 Grammar Production Rules of Quantum Electrodynamics

We now consider the grammar production rules of the Quantum Electrodynamics (electromagnetism) sector of the Standard Model. The production rules corresponding to the electromagnetic interaction term for electrons and positrons in the Standard Model

$$\bar{e}Ae$$

are:

ELECTRON-POSITRON QED PRODUCTION RULES

$$e \to eA$$
$$e \to Ae$$
$$eA \to e$$
$$Ae \to e$$
$$p \to pA$$
$$p \to Ap$$
$$Ap \to p$$
$$pA \to p$$
$$ep \to A$$
$$pe \to A$$
$$A \to ep$$
$$A \to pe$$

where e represents an electron, p represents a positron, and A represents a photon. The production rules describe the emission and absorption of photons by electrons and positrons as well as the annihilation of an electron and positron to produce a photon, and the decay of a photon into an electron-positron pair.

An example of an interaction between two electrons in the linguistic approach is:

$$\overset{1}{ee} \to \overset{2}{eAe} \to \overset{3}{ee}$$

where the electrons interact by exchanging one photon. One Feynman-like diagram for these transitions is:

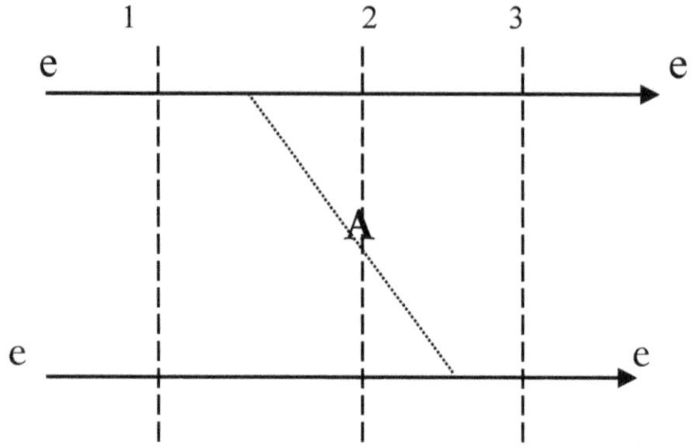

Figure 3-A.9. A diagram showing how two electrons interact by exchanging a photon. As time increases the electrons move from left to right. The upper electron emits the photon. This corresponds to the left e transitioning to eA using the grammar rule e → eA. (There is a similar diagram Fig. 3-A.10 in which the lower electron emits the photon.)

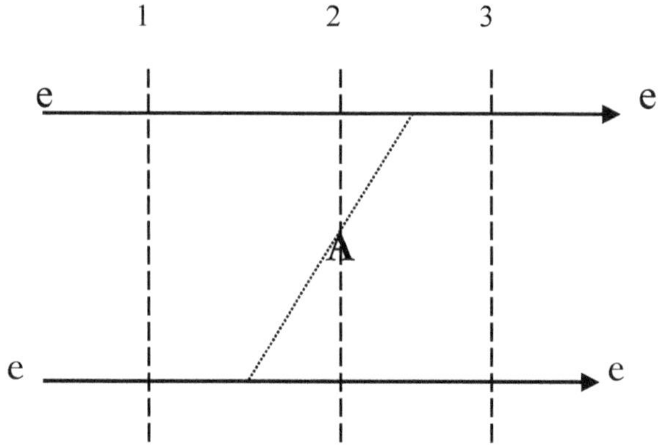

Figure 3-A.10. Another diagram with two electrons interacting by exchanging a photon. In this case the lower electron emits the photon. This corresponds to the right e in the initial ee string transitioning to Ae using the grammar rule e → Ae.

The vertical slices in the Feynman diagram which are numbered 1, 2 and 3 correspond to the three numbered strings in the transitions generated from the production rules. Each string has an ordering that corresponds to the order of particles as you descend a slice. For example slice 2 in Fig. 3-A.10 has an electron, photon, and another electron in that order as you descend corresponding to string 2 above.

Another Feynman-like diagram that contributes to this process has the lower electron emitting a photon that is then absorbed by the upper electron. The Feynman diagram for this process represents the sum of both of the previous diagrams:

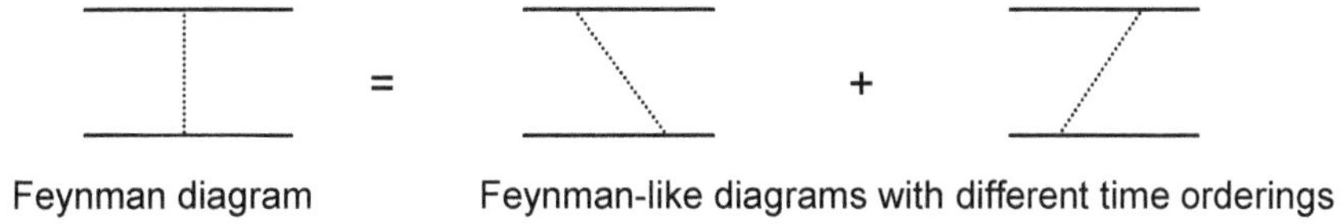

Figure 3-A.11. A Feynman diagram represents several of our Feynman-like diagrams with different time orderings of particle emission and absorption.

A more complex example of a Feynman-like diagram appears in Fig. 3-A.12. Six slices appear corresponding to the various intermediate states in this complex electron-electron interaction. The production rules can be used to generate a sequence of strings that correspond to the slices. As you descend each slice the particles are ordered in the same way as the corresponding string. For example, as you descend slice 5 the order of the particles is electron, electron, positron, photon and electron, and the string is eepAe.

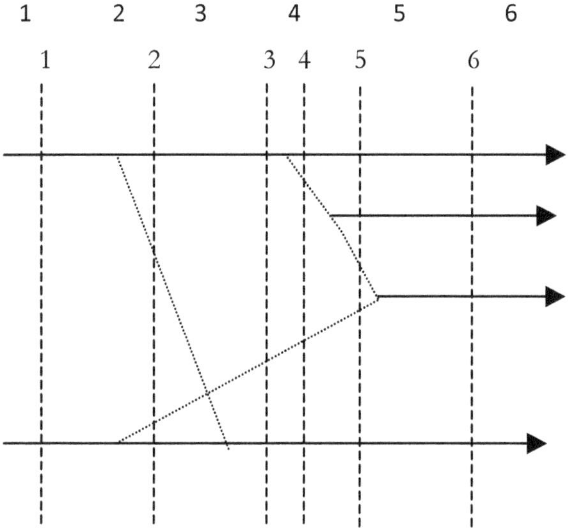

$$ee \rightarrow eAAe \rightarrow eAe \rightarrow eAAe \rightarrow eepAe \rightarrow eepe$$

Figure 3-A.12. A diagram for the collision of two electrons that produce a new electron-positron pair: $ee \rightarrow eepe$.

The transitions between character strings have an ambiguity. For example in the above transition

$$\underset{2}{eAAe} \rightarrow \underset{3}{eAe}$$

could have taken place through (eA)Ae → (e)Ae with the left (upper) e absorbing an A or through eA(Ae) → eA(e) with the right (lower) e absorbing an A. (Parentheses are used for grouping to show which electron absorbed the photon.) This ambiguity reflects the fact that there are several possible time orderings. The preceding diagram actually corresponds to eA(Ae) → eA(e). The right (lower) electron absorbs the photon.

To calculate the probability of an actual physical transition occurring such as ee → eepe we must take account of all diagrams with all possible time orderings for the specified input and output states. This is a monumental chore since an infinite number of diagrams are involved. Normally only the simplest diagrams are evaluated since they dominate the electromagnetic interactions of electrons and positrons.

The above examples show how the electromagnetic interaction part of the Standard Model lagrangian can be viewed as defining a grammar. The grammar has corresponding Feynman-like diagrams. As we pointed out earlier, these types of diagrams have time orderings that are similar to the time orderings that appeared in perturbation theory calculations before 1950.

The Weak Interaction and Strong Interaction parts of the Standard Models also define grammars. Consequently we can view the complete Standard Model Lagrangian as defining a grammar where the "letters" (alphabet or vocabulary) are the elementary particles of the model and the Feynman-like diagrams corresponding to the Standard Model are a sequence of strings generated by applying the production rules specified by the Standard Model Lagrangian.

3-A.5.2 Production Rules for the Weak and Strong Interactions

The Weak and Strong interaction terms in the Standard Model Lagrangian are also easily translated into grammar production rules (although the process is laborious since there are so many of them). We will illustrate these cases using the Weak interaction terms:

$$\bar{\nu}_e W^- e$$

and

$$\bar{\nu}_\mu W^- \mu$$

where ν_e represents an electron neutrino, ν_μ represents a muon neutrino, μ represents a muon, W^- is a gauge field of the Weak interaction and e is an electron; and the Strong interaction term

$$\bar{u} G u$$

where u is a u quark and G represents gauge fields of the Strong interaction.

Notice that there are several types of neutrinos: electron neutrinos, muon neutrinos and tau neutrinos. The three kinds of neutrinos have different internal quantum numbers that distinguish them. Neutrinos do not have electromagnetic charge. They are neutral as their name suggests. Each kind of neutrino has a corresponding charged partner. We are familiar with the electron. The other charged partners are the muon and tau particle. These charged particles are like heavy electrons for the most part. The three charged leptons also have distinguishing internal quantum numbers.

The preceding interaction terms imply production rules such as:

$$e \to W^- \nu_e$$

$$e \to \nu_e W^-$$

$$W^- \to e\,\nu_e$$

$$W^- \to \nu_e\, e$$

$$\mu \to \nu_\mu W^-$$

$$\mu \to W^- \nu_\mu$$

$$u \to Gu'$$

$$u \to u'G$$

and so on where e is an electron, p is a positron, W^- is a negative W gauge boson, ν is a neutrino, G is a Strong interaction gauge boson and u and u' are u quarks which may have different color quantum numbers.

These production rules generate string equivalents of Weak interaction transitions such as muon decay:

$$\overset{1}{\mu} \;\to\; \overset{2}{W^-\nu_\mu} \;\to\; \overset{3}{e\nu_e\nu_\mu}$$

which has the corresponding Feynman-like diagram:

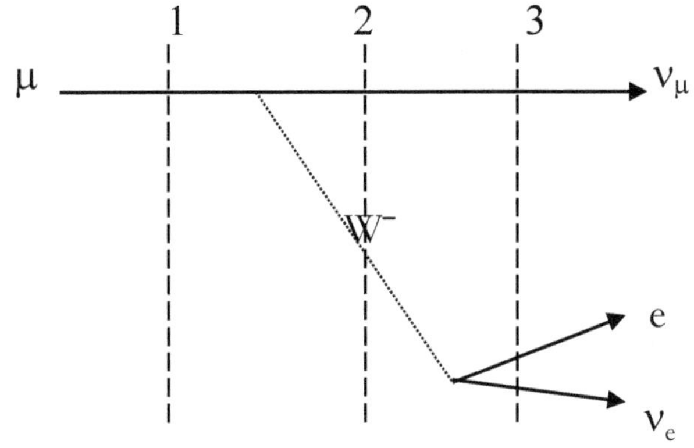

They also generate string equivalents of Strong interaction transitions between quarks and color gauge fields such as:

$$\overset{1}{uu} \to \overset{2}{uGu} \to \overset{3}{u'u'}$$

with the corresponding Feynman-like diagram:

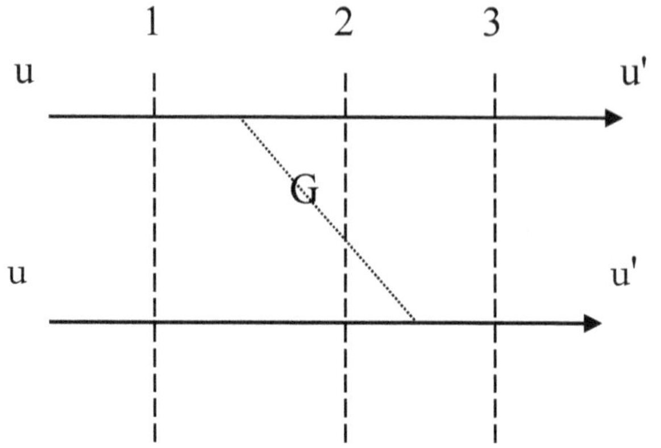

The preceding examples are among the simplest cases of the infinite variety of Feynman-like diagrams that can be generated from the Standard Model production rules.

3-A.5.3 The Standard Model Quantum Grammar

At this point we have created a view of the Standard Model of elementary particles in which the particles are an alphabet of perhaps 52 letters. The particle alphabet can be combined into strings that represent input or output states of scattering particles as well as bound states. Transitions between strings take place through quantum grammar production rules and correspond to time-ordered Feynman-like diagrams.

So we now have a map between the Standard Model and a language, with letters, words and a grammar; and an interpretation of the language in terms of rules of calculation and experimental setups.

The linguistic representation of the Standard Model that we have developed omits many important calculation details as well as important particle properties such as spin and particle momenta. These features could be added in a direct way. The focus of our investigation is on the essentials of the acts of creation and annihilation of particles in particle interactions.

The character string transition approach based on production rules is equivalent to the Feynman diagram approach. It does however provide a different, and simpler, view. Interestingly the Feynman diagram approach to calculations is very difficult and tedious – but it often leads to a simple result due to the massive cancellation of many complex terms with each other. (An attempt to simplify perturbation theory diagram calculations was made by Cheng and Wu in the early 1970's and by this author privately.) The simple linguistic approach might be a hint of a more efficient way of calculating in quantum field theories like the Standard Model where complications are absent from the very beginning.

Whether or not the linguistic approach leads to a less complicated theory or method of calculation remains to be seen. However we have now obtained a rather amazing result. After 2500 years of speculation on the nature of matter we have developed a surprisingly simple theory (for everything except gravitation) called the Standard Model that can be viewed, in part, as a quantum type 0 computer language. It has an alphabet (vocabulary) of roughly 52 particles, and a set of production rules specified by the interaction part of the Standard Model Lagrangian.

This situation represents something of a miracle. There is no reason that Nature should have so few particles that interact with each other through a simple set of rules. A computer language theorist would call the language of the Standard Model a language with a finite representation. Simply put, this means the words of the language can be generated from a finite vocabulary (alphabet or set of particles) and a finite set of production rules.

3-A.5.4 The Standard Model Language is Surprisingly Simple

Many physicists have felt that there are too many elementary particles in the Standard Model. From the point of view of a language theorist *the finite language representation of the Standard Model is a very special situation.* As Hopcroft and Ullman[32] point out, "there are many more languages than finite representations." Languages can have infinite alphabets or infinite sets of production rules or other complications.

The physical equivalent of an infinite alphabet would be a universe with an infinite number of different types of matter. Every particle of matter could have a different mass and differ in other properties. From this point of view the Standard Model is truly a marvel of simplicity.

The Standard Model, in fact, is a very compact, finite description of most of the known features of our universe. The linguistic view of the Standard Model suggests we should view elementary particles as symbols or clumps of data – a vocabulary. The interactions of the elementary particles involve the creation or annihilation of particles – creation in the deepest form seen by man.

Remarkably, the data flowing through a computer can be viewed as being transformed from one form to another and being output to different destinations. Data flowing through a computer can be divided into streams that can be sent to different output channels such as a printer or the computer screen. For example, the character, 'a', can be a data item in a stream of data that is sent to both a printer and the screen:

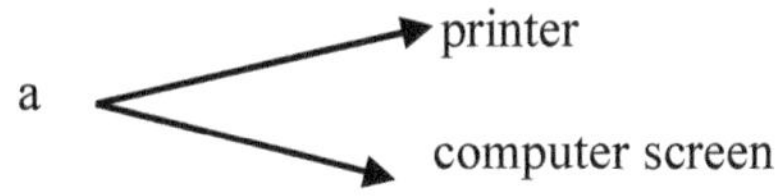

The streaming of data characters to different output channels is quite analogous to the output of particles from particle scattering as we have seen.

The simple production rules that describe particle scattering in the Standard Model suggest a fundamental simplicity at the core of Reality that goes far beyond the speculations of philosophers and scientists of earlier ages.

[32] J. E. Hopcroft and J. D. Ullman, *Formal Languages and Their Relation to Automata*, (Addison-Wesley, Reading, MA, 1975) page 2.

4. Space-Time Coordinates

A physics that is solely based on logic transformations is possible and can be implemented by a Turing machine as the author showed in Blaha (2005b) and (2005c) as well as in other books. However it is clear that the physics and universe with which we are familiar is more than logic transformations – it requires coordinate systems.

Thus our construction must define fermion particles[33] as iotas combined with coordinates and possibly other internal quantum nunbers. The first question that appears is the dimensionality of the coordinate system. Since iotas are four component vectors in a 4×4 matrix formulation of asynchronous logic they can be taken to be Dirac spinors. The matrices of the representation are linear combinations with complex coefficients of the set of sixteen independent 4×4 matrices that can be defined in terms of sums and products of the four Dirac γ matrices.

This representation is a spin ½ representation and thus the space-time dimensionality of associated coordinate system must be four as pointed out in Weinberg (1995) and other works.

4.1 Real or Complex Coordinates?

The next question is the nature of the coordinates: complex or real? To the extent that we know the universe it would appear that coordinates are real-valued. However a cautionary hint appears: the reality of coordinates is a result of our measuring instruments: clocks and yardsticks can only yield real number values. We are initially led to real coordinates by our measuring instruments. The possibility of complex coordinates is not ruled out in principle. In fact, a rigorous approach to quantum field theory, axiomatic quantum field theory, requires complex coordinates in many of its proofs.[34]

In addition following Leibniz's Principle: the choice of complex fundamental coordinates which are made real (as we see coordinates in nature) using a group, which we call the Reality group, that transforms the coordinates at any point to real-valued coordinates gives major results that would not be *directly* found using any other known approach to understanding the form of The Standard Model. We obtain tachyons,[35] ElectroWeak doublets, and most of the form of the part of The Standard Model currently established by experimentalists and theorists. The Higgs Mechanism is not implied but not ruled out. An addition to The Extended Standard Model also appears with a sector that is natural to associate with Dark Matter.

The choice of complex coordinates is one of the two most significant uses of Leibniz's Minimax Principle in the construction of the Extended Standard Model. (The other significant use is the use of Two-Tier Quantum Field Theory to eliminate divergences that appear in ordinary quantum field theory and require renormalization to obtain finite results in calculations.[36])

---

[33] Bosons do not contain iotas; they are effectively the "wires" between fermions.

[34] See Streater, R. F. and Wightman, A. S., (2000).

[35] Those troubled by the tachyon concept will hopefully be reassured later when we point to clear experimental evidence for tachyons that we showed in earlier nooks.

[36] Described later in this book and presented in Blaha (2005a) and our earlier books.

4.2 Lorentz Transformations

Upon the introduction of coordinates for iotas, a distance measure becomes necessary to determine the distances within and between processes. We define a distance measure to have a conventional quadratic form – but using complex coordinate distances. In complex rectangular coordinates, we will define the infinitesimal distance measure ds by

$$ds^2 = g_{\mu\nu}dz^\mu dz^\nu \tag{4.1}$$

with metric $g_{\mu\nu}$.and complex infinitesimal intervals dz^μ for $\mu,\nu = 0, 1, 2, 3$. The physical invariant distance is $ds = |ds^2|^{1/2}$ as will be seen later.

The reader will note that each coordinate appears as a square rather than as its absolute value squared. An immediate consequence of this choice is that letting $dz^0 = idx^0$ where x^0 is real and $dz^i = dx^i$ for i = 1, 2, 3 we recover the invariant distance for conventional special and general relativity, and thence their dynamic equations. *We can obtain the equations of complex general relativity by analytically continuing all coordinates to complex values as is well known in complex variable theory.*[37] Lastly we note that the relative signs of the terms in eq. 4.1 are arbitrary since we can rotate the signs of terms arbitrarily. (The complex terms appear as squares – not absolute values.) When transformed to purely real coordinates this degree of freedom is lost.

An alternative infinitesimal distance measure for complex coordinates is

$$ds^2 = g_{\mu\nu}dx^\mu dx^{\nu *} \tag{4.2}$$

with metric $g_{\mu\nu}^* = g_{\nu\mu}$. This possibility was briefly explored in Blaha (2013b). The impact on the results of this book is little changed from those that would have been found using eq. 4.1 above as the invariant distance measure. The major differences are in the detailed study of each complex General Relativity since *eq. 4.2 does not support analytic continuation from real-valued coordinates.* Based on Ockham's Razor and Leibniz's Minimax Principle the simplest choice for invariant distance is eq. 4.1 above as was done in Blaha (2012a). Analytic continuation dramatically simplifies the formulation of complex Geneneral Relativity compared to the formulation based on eq. 4.2.

Assumption: Flat space-time has the metric $\eta_{\mu\nu}$ = diag(1, -1, -1, -1) in rectangular coordinates.

This assumption plays the role of a primitive in our construction just as the specification of a 'straight line' is a primitive in Euclidean geometry.

An inertial reference frame can be defined at any (non-singular) point of an arbitrary metric. We consider a complex coordinates point ζ^α in an arbitrary coordinate system. The metric $g_{\mu\nu}$ at this point for an inertial reference frame parametrized by the coordinates z^ρ is

$$g_{\mu\nu} = \partial\zeta^\alpha/\partial z^\mu\, \partial\zeta^\beta/\partial z^\nu\, \eta_{\alpha\beta} \tag{4.3}$$

The invariant distance for complex coordinates is invariant under complex Lorentz group transformations. The general form of a homogeneous complex Lorentz transformation for complex or real-valued coordinates is

$$\Lambda_C = \exp[i(\omega_r\hat{\mathbf{u}}_r + i\omega_i\hat{\mathbf{u}}_i)\cdot\mathbf{K} + i\theta_c\cdot\mathbf{J}] \tag{4.4}$$

---

[37] See Blaha (2004).

where the vector $\boldsymbol{\theta}_c$ is a complex 3-vector, $\omega_r \geq 0$ and $\omega_i \geq 0$ are real numbers, and $\hat{\mathbf{u}}_r$ and $\hat{\mathbf{u}}_i$ are real normalized 3-vectors such that $\hat{\mathbf{u}}_r \cdot \hat{\mathbf{u}}_r = 1 = \hat{\mathbf{u}}_i \cdot \hat{\mathbf{u}}_i$. The generators of the homogeneous group are the boost generators denoted **K**, and and the rotation generators **J.**

If $\omega_i = 0$ and the angle constants are real, then Λ_C is a real Lorentz group transformation. If either of these conditions are not met then Λ_C is a complex Lorentz group generator.

Lorentz transformations satisfy the condition

$$\Lambda_C{}^T G \Lambda_C = G \tag{4.5}$$

if expressed in matrix form where G is the metric $g_{\mu\nu}$ expressed as a 4×4 matrix and where the superscript ^{T} specifies the transpose of the matrix. This condition is necessary if physical processes are to have the same character in all inertial reference frames. Thus we require the assumption:

Assumption: All physical processes take place in the same manner in all inertial reference frames.

Again we view this assumption as a 'primitive' in our construction of the Extended Standard Model.

4.3 The Reality Group

In a following chapter we will examine the four special cases of complex Lorentz transformation boosts

$$\Lambda_{CB} = \exp[i(\omega_r \hat{\mathbf{u}}_r + i\omega_i \hat{\mathbf{u}}_i) \cdot \mathbf{K}] \tag{4.6}$$

that lead to the four different species of fermions: neutrinos, charged leptons, up-type quarks and down-type quarks – the major form of the fermion spectrum that has never been explained until our researches. This is the first step in the detailed construction of the form of The Standard Model.

Before that we will consider the Reality group – the group that maps complex valued coordinates to real values – the physical values that we measure in nature. In earlier books, Blaha (2011c) through (2012c),[38] we showed that the Reality group was R = SU(3)⊗SU(2)⊗U(1)⊗SU(2)⊗U(1). Elements of this group can map complex 4-dimensional coordinates to real-valued coordinates. We noted that the known part of the symmetry group of The Standard Model is SU(3)⊗SU(2)⊗U(1). The Reality group contains an additional pair of factors SU(2)⊗U(1) which we identified as the source of the Dark Matter sector of the full fermion spectrum. Blaha (2013b) has a detailed discussion of the Dark Matter fermion spectrum and aspects of the basic chemistry of Dark Matter.

A weak interaction between the known and Dark fermion sectors is described in chapter 1 of Blaha (2013b) particularly in eqs. 1.12 – 1.16. This interaction is presumably the source of the Dark Matter particles apparently generated by collisions of 'normal' matter at the CERN LHC.

The Reality group is the origin of the gauge field interactions of The Standard Mode extended to include Dark Matter.

---

[38] In particular see Blaha section 15.12 and Appendix 18A of (2011c); and chapter 3 of Blaha (2012a).

4.4 The Reality Group as the Source of the Extended Standard Model and the Dark Gauge Fields in Particular

The Reality group is the source of the known and Dark gauge bosons of our Extended Standard Model.

If we transform coordinates from complex to real-valued we must require any dynamical equations (shown later) to transform covariantly to preserve the physical content of the equations. Introducing connections (gauge fields) leads to covariant dynamical equations. If the equations can be derived from a lagrangian then the lagrangian must be invariant under Reality group transformations. We regard a lagrangian formulation as required by Leibniz Minimax Principle for maximal effects most simply achieved.

Gauge fields have spin 1 and thus they have the logic values true, false and intermediate (during transit between other particles). They act as a spin exchange (logic value exchange) mechanism.

While we have not as yet seen the generation of fundamental fermions using Lorentz group boosts (and other relevant features) it appears reasonable to show the Extended Standard Model lagrangian in this chapter to exhibit the role of its gauge fields which we will do by reproducing chapter 6 of Blaha (2013b),[39] Gauge fields are required for Reality Group covariance (Blaha (2012b)).

4.5. The Reality Group – SU(3)⊗SU(2)⊗U(1)⊗SU(2)⊗U(1)

In earlier work we showed that R = SU(3)⊗SU(2)⊗U(1)⊗SU(2)⊗U(1), which has 16 generators, tcan make any complex 4×4 matrix into a purely real-valued matrix. Consider the matrix product

$$D(x'', x)^{\rho}{}_{\nu} = S(x'', x')^{\rho}{}_{\mu}(\partial x^{\mu}/\partial x'^{\nu}) \qquad (4.7)$$

where $\partial x^{\mu}/\partial x'^{\nu}$ has 16 complex entries (32 real entries) and $S(x'', x')^{\rho}{}_{\mu}$ is a local 4×4 R group matrix with sixteen local parameters. Eq. 4.7 can be viewed as 32 real equations. Sixteen of these equations fix the imaginary part of each D matrix element to zero:

$$\text{Im } D(x'', x)^{\rho}{}_{\nu} = 0 \qquad (4.8)$$

making the elements of D(x", x) real-valued. The 16 parameters of S(x", x') are fixed by eq. 4.8. Therefore there is a unique S(x", x') for each complex general coordinate transformation $\partial x^{\mu}/\partial x'^{\nu}$ and thus a unique D(x", x) for each transformation $\partial x^{\mu}/\partial x'^{\nu}$.

U(4) has the subgroups SU(3), SU(2), U(1), SU(2) and U(1).[40] These subgoups do not commute so they are <u>not</u> the direct product SU(3)⊗SU(2)⊗U(1)⊗SU(2)⊗U(1).

However the direct product R is of interest. The direct product SU(3)⊗SU(2)⊗U(1) is the symmetry group of the Standard Model. An additional SU(2)⊗U(1) factor could enlarge the Standard Model to include Dark Matter.

In the past we derived The Standard Model with SU(3)⊗SU(2)⊗U(1) symmetry by assuming neutrinos and down-type quarks were tachyons. Their tachyon nature requires complex Lorentz group

---

[39] We use the word 'Complexon' instead of 'Extended' in this reproduced chapter. **We prefix page, section and equation numbers with a ' in this insertion.**

[40] Blaha (2011c).

transformations, which in turn requires a supplemental group, the Reality group, to transform complex Lorentz group transformations to real-valued transformations. This process exactly parallels the role of the Reality group in Complex General Relativity.

4.6 Consequences of the Unitary Nature of the Reality Group

Since S(x", x') is a Reality group unitary transformation, eq. 4.7 is invertible:

$$\partial x^{\mu}/\partial x^{\prime\nu} = S^{-1}(x'', x')^{\mu}{}_{\rho}\, D(x'', x)^{\rho}{}_{\nu} \tag{4.9}$$

so one can determine the complex coordinate transformation – given D(x", x) and S(x", x') or S^{-1}(x", x').

The inverse of S(x", x') expressed in matrix notation is

$$S^{-1}(x'', x') = S^{\dagger}(x'', x') \tag{4.10}$$

where † signifies hermitean conjugation.

4.7 A Constraint on Complex Coordinate Transformations

In eq. 4.7 a complex coordinate transformation $\partial x^{\mu}/\partial x^{\prime\nu}$ was transformed into a real coordinate transformation $D(x'', x)^{\rho}{}_{\nu}$ using a Reality group transformation. A limitation[41] on the allowed *complex* General Relativistic transformations[42] appears if we consider the complex conjugate of eq. 4.7:

$$\begin{aligned} D^*(x'', x)^{\rho}{}_{\nu} &= S^*(x'', x')^{\rho}{}_{\mu}(\partial x^{\mu}/\partial x^{\prime\nu})^* \\ &= D(x'', x)^{\rho}{}_{\nu} \end{aligned} \tag{4.11}$$

Combining eq. 4.7 and 4.11 in matrix form we find

$$SE = S^*E^* \tag{4.12}$$

where E is a matrix whose elements are $\partial x^{\mu}/\partial x^{\prime\nu}$,

$$E = [\partial x^{\mu}/\partial x^{\prime\nu}] \qquad 4.13)$$

with μ as the column index and ν as the row index. Since S is unitary we see

$$E^*E^{-1} = S^T S \tag{4.14}$$

where ^{T} represents the transpose. Now

$$S^T S(S^T S)^{\dagger} = S^T S S^{\dagger} S^* = S^T S^* = I \tag{4.15}$$

and so $S^T S$ is unitary, and is also in the Reality group R. In addition

---

41 This section is similar to the corresponding discussion of Blaha (2012).

42 Real general relativistic transformations are not restricted by this constraint.

$$E^*E^{-1}\ \varepsilon\ R \tag{4.16}$$

Consequently complex (or real) coordinate transformations of the form E^*E^{-1} are R transformations. Complex transformations $E = [\partial x^{\mu}/\partial x'^{\nu}]$ may, or may not, be R transformations. However, a Reality transformation S always exists such that D = SE, of the form of eq. 4.7, is a real-valued transformation and thus a physically acceptable general coordinate transformation.

Further every complex (or real) general coordinate transformation E has a corresponding R general coordinate transformation E^*E^{-1}. Real-valued coordinate transformations (where $E = E^*$, and thus $E^*E^{-1} = I$) trivially satisfy eq. 4.16. Thus all real-valued general relativistic coordinate transformations are also allowed.

We thus have a classification of general coordinate transformations in our universe into real-valued coordinate transformations, or into R complex-valued coordinate transformations, or into non-R complex coordinate transformations. Treating coordinate transformations as matrices we find:

1. Real-valued general coordinate transformations.
   Matrix elements $[\partial x^{\mu}/\partial x'^{\nu}]$ are all real-valued.

2. R complex-valued general coordinate transformations.
   Matrices satisfy:

   row column row column

$$[\partial x^{\mu}/\partial x'^{\nu}]^{\dagger} = [\partial x^{\nu}/\partial x'^{\mu}]^{*} = [\partial x^{\mu}/\partial x'^{\nu}]^{-1}$$

3. Non-R complex-valued general coordinate transformations.
   Matrices satisfy:

   row column row column

$$[\partial x^{\mu}/\partial x'^{\nu}]^{\dagger} = [\partial x^{\nu}/\partial x'^{\mu}]^{*} \neq [\partial x^{\mu}/\partial x'^{\nu}]^{-1}$$

4.8 Why SU(3)⊗SU(2)⊗U(1)⊗SU(2)⊗U(1)?

Having developed an SU(3)⊗SU(2)⊗U(1)⊗SU(2)⊗U(1) Reality group, the question arises, "Why this specific group?" Clearly it is of interest since it is the symmetry group of our enhanced version of The Standard Model.

We believe the answer is the following:

1. We require a group with at least 16 generators to enable the R Reality group of our universe to have a counterpart in the flatverse.

2. More than 16 generators would make the relation between real and complex coordinates ambiguous. Thus the Megaverse Reality group must have exactly 16 generators for a unique solution. (See section 23.8.)

3. While one can embed our 4-dimensional space-time in 10 flat dimensions, the simplest block structure representation that leads to commuting factors, and 16 generators, in the direct product of subgroups is a 16-dimensional flat complex space.

4. The possible subgroup direct products that have a 16×16 block structure, and 16 generators, are SU(3)⊗SU(2)⊗U(1)⊗SU(2)⊗U(1), SU(3)⊗SU(3), ... , and $[U(1)]^{16}$. SU(3)⊗SU(3) is ruled out because it's does not have a simple 10×10 submatrix format to handle a minimal embedding of our universe in the flatverse. SU(3)⊗SU(2)⊗U(1)⊗SU(2)⊗U(1) does have a simple 10×10 submatrix format consisting of the blocks of SU(3)⊗U(1)⊗U(1). As for other possibilities we take a minimalist position (Ockham's Razor) and choose the direct product with the minimal number of factors. This selects SU(3)⊗SU(2)⊗U(1)⊗SU(2)⊗U(1) uniquely.

Thus we find SU(3)⊗SU(2)⊗U(1)⊗SU(2)⊗U(1) to be the Reality group of the universe conforming to Leibniz's Minimax Principle. This group is also the Reality group of our version of the Standard Model described in Blaha (2011c) and (2012). And the factors in this direct product appear as the (non-commuting) subgroups of U(4) the Reality group of our universe.

Appendix 4-A Complex Lorentz Group Details

4-A.1 Transformations Between Coordinate Systems

The measurement of time and space is simple in practice but raises weighty questions when their underlying basis is examined. We shall begin by measuring spatial distances with a ruler, and by measuring time with a clock. Earlier we determined that four dimensions: one space dimension and three space dimensions were required. We now define rectangular coordinate systems with x, y, and z axes as pictured in Fig. 4-A.1 below. We then postulate:

Postulate 4-A.1. Any observer can define a set of time and space coordinates called a coordinate system in which the observer is at rest. One can define a transformation that relates the coordinate systems of two observers traveling at a any constant velocity with respect to each other.

One can always relate the coordinates of two coordinate systems by having an observer in each coordinate system specify the coordinates of objects located at each spatial point, and then creating a map between the coordinates of corresponding spatial locations.

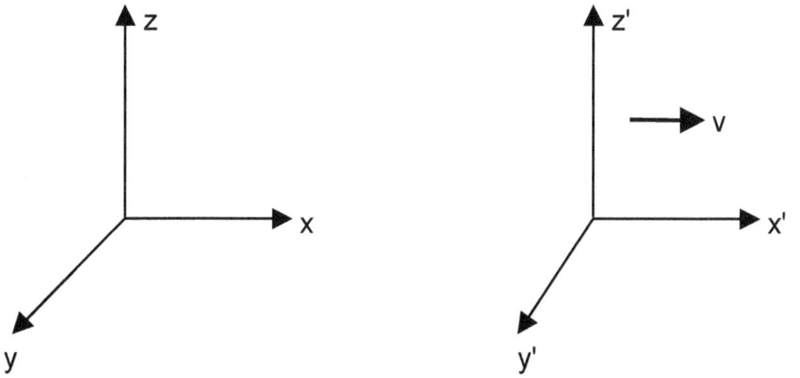

Figure 4-A.1. Depiction of two coordinate systems. The "primed" coordinate system is moving with velocity **v** in the positive x direction with respect to the "unprimed" coordinate system. We choose parallel axes for convenience.

If space is flat the relation between the respective coordinates is linear. (One could reverse the logic of that statement by defining a flat space to be one in which the coordinates of a point in any coordinate system are linearly related to the coordinates of any other coordinate system moving at a constant velocity with respect to it.) Thus we can express the relation between the coordinates in the "unprimed" system to the coordinates in the "primed" system as a transformation between coordinate systems:

$$\mathbf{a}' = \mathbf{A}\mathbf{a} + \mathbf{B}t + \mathbf{C} \qquad (4\text{-A}.1)$$

$$t' = Dt + \mathbf{E}\cdot\mathbf{a}$$

where A is a 3×3 matrix, **B**, **C** and **E** are 3-vectors, and D is a number (scalar value).

Having restricted the set of transformations between coordinate systems to the form of eq. 4-A.1 we now assert postulates that restrict the form of the transformation to Lorentz transformations and transformations similar to Lorentz transformations.

Postulate 4-A.2. The speed of light, c, is the same in all coordinate systems.

Postulate 4-A.3. The invariant interval or distance $d\tau$ is defined by

$$d\tau^2 = g_{\mu\nu}dx^\mu dx^\nu \qquad (4\text{-A.2})$$

It is invariant under a change of coordinate systems. The 16 quantities $g_{\mu\nu}$ are known as the metric tensor.[43] *The four quantities dx^μ are infinitesimal displacements in space and time.*

If we expand eq. 4-A.2 in rectangular coordinates it is equivalent to

$$d\tau^2 = g_{00}dx^0dx^0 + g_{11}dx^1dx^1 + g_{22}dx^2dx^2 + g_{33}dx^3dx^3 \qquad (4\text{-A.3})$$

which equals

$$d\tau^2 = c^2dt^2 - dx^2 - dy^2 - dz^2 \qquad (4\text{-A.4})$$

using the familiar form of the time and rectangular space coordinates.

4-A.2 The Lorentz Group

The metric tensor $g_{\mu\nu}$ for rectangular coordinates has the matrix form G = diag(1, –1, –1, –1):

$$G = \begin{bmatrix} 1 & 0 & 0 & 0 \\ 0 & -1 & 0 & 0 \\ 0 & 0 & -1 & 0 \\ 0 & 0 & 0 & -1 \end{bmatrix} \qquad (4\text{-A.5})$$

The invariant interval under a transformation between rectangular coordinate systems (with "primed" and "unprimed" coordinates) has the form of eq. 4-A.4 for the unprimed coordinates and the same form for the primed coordinates:

$$d\tau^2 = c^2dt'^2 - dx'^2 - dy'^2 - dz'^2 \qquad (4\text{-A.6})$$

---

[43] The repeated indices indicate a summation. In this case from 0 to 3 as shown in eq. 4-A.3

In matrix form we can define an "unprimed" coordinate column vector with

$$a = \begin{bmatrix} t \\ x \\ y \\ z \end{bmatrix} \tag{4-A.7a}$$

and its corresponding "primed" coordinate with

$$a' = \begin{bmatrix} t' \\ x' \\ y' \\ z' \end{bmatrix} \tag{4-A.7b}$$

If a and a' are the coordinates of the same point in the respective coordinate systems then, by postulates 4-A.2 and 4-A.3, they are related by a boost Lorentz transformation $\Lambda(\mathbf{v})$ with the form

$$a' = \Lambda(\mathbf{v})a \tag{4-A.8}$$

(and possibly a spatial rotation matrix factor), where **v** is the relative velocity of the coordinate systems. The form of the transformation eq. 4-A.8, which is called a Lorentz *boost*, is constrained by postulates 4-A.2 and 4-A.3 to be[44]

$$\Lambda(\mathbf{v}) = \begin{bmatrix} \gamma & -\gamma v_x & -\gamma v_y & -\gamma v_z \\ -\gamma v_x & 1 + (\gamma - 1)v_x^2/v^2 & (\gamma - 1)v_x v_y/v^2 & (\gamma - 1)v_x v_z/v^2 \\ -\gamma v_y & (\gamma - 1)v_x v_y/v^2 & 1 + (\gamma - 1)v_y^2/v^2 & (\gamma - 1)v_y v_z/v^2 \\ -\gamma v_z & (\gamma - 1)v_x v_z/v^2 & (\gamma - 1)v_y v_z/v^2 & 1 + (\gamma - 1)v_z^2/v^2 \end{bmatrix} \tag{4-A.9}$$

where $\gamma = (1 - v^2)^{-1/2}$, $\mathbf{v} = (v_x, v_y, v_z)$, $v = |\mathbf{v}|$ and we set c = 1 for convenience.[45] The set of all matrices of the form of $\Lambda(\mathbf{v})$, or $\Lambda(\mathbf{v})\mathcal{R}(\boldsymbol{\theta})$ or $\mathcal{R}(\boldsymbol{\theta})\Lambda(\mathbf{v})$ where $\mathcal{R}(\boldsymbol{\theta})$ is a spatial rotation with angle vector $\boldsymbol{\theta}$, for $v < c$ form a matrix representation of the Lorentz group. Elements, $\Lambda(\mathbf{v}, \boldsymbol{\theta})$, of the Lorentz group satisfy the defining relation of the Lorentz group:

$$\Lambda(\mathbf{v}, \boldsymbol{\theta})^T G \Lambda(\mathbf{v}, \boldsymbol{\theta}) = G \tag{4-A.10}$$

where the superscript ^{T} specifies the transpose of the matrix.

---

[44] We shall consider only the proper, orthochronous Lorentz group at this point. We assume that the primed and unprimed coordinate systems have parallel axes. So there is no rotation of axes embodied in eq. 15.9.

[45] One can set c = 1 by an appropriate choice of time and spatial distance scales. The demonstration that $\Lambda(\mathbf{v})$ has the form given by eq. 15.9 can be found in many textbooks.

The Lorentz group, with which we are familiar, relates the coordinates of an event in two coordinate systems that differ by a a spatial rotation, and a relative velocity whose magnitude is less than the speed of light. The inhomogenous Lorentz group includes coordinate displacements.[46]

The group elements of the homogeneous Lorentz group can be expressed in terms of the generators **K** of boosts to coordinate systems moving at a constant velocity **v** and the generators **J** of purely spatial rotations by

$$\Lambda(\mathbf{v}, \boldsymbol{\theta}) = \exp[i\omega\hat{\mathbf{u}}\cdot\mathbf{K} + i\boldsymbol{\theta}\cdot\mathbf{J}] \tag{4-A.11}$$

where the vector $\boldsymbol{\theta}$ is a 3-vector specifying the rotation angles, and where $\mathbf{v} = \hat{\mathbf{u}} \tanh\omega$, $\hat{\mathbf{u}}\cdot\hat{\mathbf{u}} = 1$.

The boost transformation $\Lambda(\mathbf{v}) = \Lambda(\mathbf{v}, \mathbf{0})$ has the form

$$\Lambda(\mathbf{v}) = \exp[i\omega\hat{\mathbf{u}}\cdot\mathbf{K}] \tag{4-A.12}$$

Its matrix form is eq. 4-A.9. The matrix form can be expressed in terms of the unit normalized velocity vector $\mathbf{u} = (u_x, u_y, u_z)$ and ω as

$$\Lambda(\omega, \mathbf{u}) = \Lambda(\mathbf{v}) \tag{4-A.13}$$

$$= \begin{bmatrix} \cosh(\omega) & -\sinh(\omega)u_x & -\sinh(\omega)u_y & -\sinh(\omega)u_z \\ -\sinh(\omega)u_x & 1 + (\cosh(\omega) - 1)u_x^2 & (\cosh(\omega) - 1)u_xu_y & (\cosh(\omega) - 1)u_xu_z \\ -\sinh(\omega)u_y & (\cosh(\omega) - 1)u_xu_y & 1 + (\cosh(\omega) - 1)u_y^2 & (\cosh(\omega) - 1)u_yu_z \\ -\sinh(\omega)u_z & (\cosh(\omega) - 1)u_xu_z & (\cosh(\omega) - 1)u_yu_z & 1 + (\cosh(\omega) - 1)u_z^2 \end{bmatrix}$$

where $\Lambda(\omega, \mathbf{u}) = \Lambda(\omega, \mathbf{u}, \boldsymbol{\theta} = \mathbf{0})$ in the previous notation. This definition of the general form of proper, orthochronous, Lorentz boost matrices $\Lambda(\omega, \mathbf{u})$ will be used in subsequent sections to define faster-than-light boost transformations.

The vector form of a Lorentz boost transformation is

$$\mathbf{x}' = \mathbf{x} + (\gamma - 1)\mathbf{x}\cdot\mathbf{v}\,\mathbf{v}/v^2 - \gamma\mathbf{v}t$$
$$t' = \gamma(t - \mathbf{v}\cdot\mathbf{x}/c^2) \tag{4-A.14}$$

where $\gamma = (1 - \beta^2)^{-1/2}$ with $\beta = v/c = v$ (since we set c = 1).

4-A.3 The Nature of Λ(ω, u) for Complex ω

We now turn to the case of complex ω wich includes superluminal (faster-than-light) Lorentz transformations as well as conventional Lorentz transformations. Since, for any complex value z

$$\cosh^2(z) - \sinh^2(z) = 1 \tag{4-A.15}$$

[46] See Weinberg (1995) for a discussion of the inhomogeneous Lorentz group.

it follows that for any complex value of ω, Λ(ω, **u**) is a member of the Lorentz group, and/or of the complex Lorentz group[47] for complex ω:

$$\Lambda(\omega, \mathbf{u})^T G \Lambda(\omega, \mathbf{u}) = G \tag{4-A.16}$$

For certain values of the imaginary part of ω the matrix Λ(ω, **u**) has a particularly simple form, similar to that of Λ(ω, **u**) for real ω, but which generates boosts to relative velocities greater than the speed of light. Among these values are:

$$\omega = \omega_\pm = \omega \pm i\pi/2 \tag{4-A.17}$$

Later we will see that these alternate choices $\omega_\pm$ correspond to specific choices of parity.

4-A.4 Complex Lorentz Group

In the preceding section we saw that the parameter ω can be complex and the boost transformation will still satisfy the Lorentz condition eq. 4-A.10. More generally we can consider complex homogeneous Lorentz transformations Λ(**v**, **θ**) which can be represented by eq. 4-A.11 with complex parameters ω, **û**, and **θ** where **û** and **θ** are complex 3-vectors. **θ** specifies a rotation angle.

In general Λ(**v**, **θ**) is then a transformation between coordinate systems that have complex coordinates. One coordinate system is moving at a constant complex velocity with respect to the other. Coordinate systems do not necessarily have parallel spatial axes in general.

Within the complex Lorentz group, denoted L(C),[48] there are subsets of boosts that play important physical roles in the derivation of the form of The Standard Model. In particular we will see that certain classes of boosts generate faster-than-light transformations. These transformations can be further divided into subclasses of "left-handed" and "right-handed" transformations based on the quantum field theories to which they lead. Further within each subclass there are subclasses of transformations that naturally lead to Dirac-like free field equations that can be described as lepton-like and quark-like.

Thus these boosts are a key ingredient to understanding the form of The Standard Model.

4-A.5 Faster-than-Light Transformations

In this section we will substitute $\omega_\pm$ for ω in Λ(ω, **u**) and then show that we obtain two sets of possible transformations from sublight reference frames to faster-than-light reference frames. One set of transformations, where $\omega_L = \omega + i\pi/2$, will be called *left-handed superluminal boosts*. They eventually lead to the "left-handed" part of The Standard Model. We denote members of this set, $\Lambda_L(\omega, \mathbf{u})$, with the subscript "L" for left-handed.

The other set of boosts where $\omega_R = \omega - i\pi/2$ will be called *right-handed superluminal boosts*. They eventually lead to a right-handed, unphysical,[49] version of The Standard Model. We denote members of this set of boosts, $\Lambda_R(\omega, \mathbf{u})$, with the subscript "R" for right-handed.

---

[47] The complex Lorentz group is defined as the group of all complex transformations that satisfy eq. 15.16.
[48] Streater (2000) points out that the complex Lorentz group is essential to the proof of the CPT theorem.
[49] Currently the case. If a right-handed counterpart to the current Standard Model surfaces at higher energies then the features emerging from right-handed superluminal boosts then become physically important.

Before considering faster-than-light boosts we note the relation between a real-valued ω in a *conventional* Lorentz boost $\Lambda(\omega, \mathbf{u})$, and the magnitude of the relative velocity v for $v < 1$, is

$$\mathbf{v} = \hat{\mathbf{u}} \tanh\omega \qquad \text{with} \qquad \hat{\mathbf{u}}\cdot\hat{\mathbf{u}} = 1$$

$$\cosh(\omega) = \gamma = (1 - v^2)^{-1/2} \qquad (4\text{-A.18})$$
$$\sinh(\omega) = v\gamma = \beta\gamma$$

where $\beta = v = |\mathbf{v}|$.

4-A.6 Left-Handed Superluminal Transformations

Left-handed (proper orthochronous) superluminal boost transformations $\Lambda_L(\mathbf{v})$ have the same form as eq. 4-A.9 for ordinary (proper orthochronous) Lorentz boost transformations. However the magnitude of the relative velocity **v** is greater than the speed of light. Thus $\gamma = (1 - v^2)^{-1/2}$ is pure imaginary and $\Lambda_L(\mathbf{v})$ is complex.

$$\Lambda_L(\mathbf{v}) = \begin{bmatrix} \gamma & -\gamma v_x & -\gamma v_y & -\gamma v_z \\ -\gamma v_x & 1 + (\gamma - 1)v_x^2/v^2 & (\gamma - 1)v_x v_y/v^2 & (\gamma - 1)v_x v_z/v^2 \\ -\gamma v_y & (\gamma - 1)v_x v_y/v^2 & 1 + (\gamma - 1)v_y^2/v^2 & (\gamma - 1)v_y v_z/v^2 \\ -\gamma v_z & (\gamma - 1)v_x v_z/v^2 & (\gamma - 1)v_y v_z/v^2 & 1 + (\gamma - 1)v_z^2/v^2 \end{bmatrix} \qquad (4\text{-A.19})$$

This transformation raises several issues – the most prominent of which is the interpretation of the imaginary coordinates generated by the transformation. Imaginary coordinates would appear at first glance to be unphysical. However we view the measurement of these quantities operationally: an observer measures distances with "rulers", and time with clocks, which both give real numeric values. Thus an observer *in any coordinate system* will always measure real numbers for time and space distances. However an observer *in another coordinate system* that is related to the first coordinate system by a superluminal transformation will view the coordinates in the first system as complex as eq. 4-A.19 indicates.

The reconciliation of these points of view requires the introduction of a new transformation, called a Reality group transformation, in addition to a superluminal Lorentz transformation for the case of faster than light transformations. Reality group transformations maps the complex coordinates generated by a Lorentz transformation to the real coordinates seen by the observer[50] in the "faster than light" reference frame. We describe the Reality group transformations in detail in chapter 7. We show how they imply the Reality group is SU(3)⊗SU(2)⊗U(1)⊗SU(2)⊗U(1).

[50] The linearity of the superluminal transformation makes this secondary transformation physically possible.

4-A.6.1 Cosh-Sinh Representation of Left-Handed Superluminal Boosts

We will now develop the representation of left-handed superluminal boost transformations in terms of cosh(ω) and sinh(ω) for later use in our discussion of tachyons. We find that we must use a complex $\omega_L \equiv \omega + i\pi/2$ to properly describe left-handed superluminal boosts. The relation between ω_L and v is different from eq. 4-A.18 for the case of left-handed superluminal boosts:

$$\begin{aligned} \cosh(\omega_L) &= i\sinh(\omega) = -\gamma = i\gamma_s \\ \sinh(\omega_L) &= i\cosh(\omega) = -\beta\gamma = i\beta\gamma_s \end{aligned} \tag{4-A.20}$$

where $\beta = v > 1$, $\boldsymbol{\omega \geq 0}$, and

$$\gamma_s = (\beta^2 - 1)^{-\frac{1}{2}} \tag{4-A.21}$$

Eq. 4-A.20 implies

$$\begin{aligned} \sinh(\omega) &= \gamma_s \\ \cosh(\omega) &= \beta\gamma_s \end{aligned} \tag{4-A.22}$$

Upon substituting $\boldsymbol{\omega_L}$ for ω in eq. 4-A.13 we obtain another form for a left-handed superluminal transformation (equivalent to that of eq. 4-A.19):

$\Lambda_L(\omega, \mathbf{u}) = \Lambda(\omega + i\pi/2, \mathbf{u})$

$$= \begin{bmatrix} \cosh(\omega_L) & -\sinh(\omega_L)u_x & -\sinh(\omega_L)u_y & -\sinh(\omega_L)u_z \\ -\sinh(\omega_L)u_x & 1+(\cosh(\omega_L)-1)u_x^2 & (\cosh(\omega_L)-1)u_xu_y & (\cosh(\omega_L)-1)u_xu \\ -\sinh(\omega_L)u_y & (\cosh(\omega_L)-1)u_xu_y & 1+(\cosh(\omega_L)-1)u_y^2 & (\cosh(\omega_L)-1)u_yu_z \\ -\sinh(\omega_L)u_z & (\cosh(\omega_L)-1)u_xu_z & (\cosh(\omega_L)-1)u_yu_z & 1+(\cosh(\omega_L)-1)u_z^2 \end{bmatrix}$$

$$= \begin{bmatrix} i\gamma_s & -i\beta\gamma_s u_x & -i\beta\gamma_s u_y & -i\beta\gamma_s u_z \\ -i\beta\gamma_s u_x & 1+(i\gamma_s - 1)u_x^2 & (i\gamma_s - 1)u_xu_y & (i\gamma_s - 1)u_xu_z \\ -i\beta\gamma_s u_y & (i\gamma_s - 1)u_xu_y & 1+(i\gamma_s - 1)u_y^2 & (i\gamma_s - 1)u_yu_z \\ -i\beta\gamma_s u_z & (i\gamma_s - 1)u_xu_z & (i\gamma_s - 1)u_yu_z & 1+(i\gamma_s - 1)u_z^2 \end{bmatrix} = \Lambda_L(\mathbf{v}) \tag{4-A.23}$$

A simple case that illustrates a left-handed superluminal boost is to assume the relative velocity is in the x direction. Then eq. 4-A.23 becomes

$$\Lambda_L(\omega, \mathbf{u} = (1,0,0)) = \begin{bmatrix} i\gamma_s & -i\beta\gamma_s & 0 & 0 \\ -i\beta\gamma_s & i\gamma_s & 0 & 0 \\ 0 & 0 & 1 & 0 \\ 0 & 0 & 0 & 1 \end{bmatrix} \tag{4-A.24}$$

implementing the coordinate transformation:

$$X' = \Lambda_L(\omega, \mathbf{u} = (1,0,0))X$$

or

$$\begin{aligned} t' &= i\gamma_s(t - \beta x) \\ x' &= i\gamma_s(x - \beta t) \\ y' &= y \\ z' &= z \end{aligned} \tag{4-A.25}$$

The addition rule for the x-component of velocity can be computed for infinitesimal displacements in space and time:

$$\begin{aligned} v_x' &= \Delta x' / \Delta t' = (\Delta x\, \gamma_s - \Delta t\, \beta\gamma_s)/(\Delta t\, \gamma_s - \Delta x\, \beta\gamma_s) \\ &= (v_x - \beta)/(1 - \beta v_x) \end{aligned} \tag{4-A.26}$$

in the limit $\Delta t \to 0$ where the x component of a particle's velocity in the unprimed frame is $v_x = \Delta x/\Delta t$. $\Delta t'$ is determined by

$$\Delta t' = i\Delta t\, \gamma_s(1 - \beta v_x) \tag{4-A.27}$$

Note the velocity of light is the same in the primed and unprimed reference frames. (If $v_x = 1$ then $v_x' = 1$.) *Thus left-handed superluminal transformations preserve the constancy of the speed of light in all reference frames.* (Postulate 4-A.2)

Further note that increasing the value of ω in $\Lambda_L(\omega, \mathbf{u})$ corresponds to decreasing the magnitude of the relative velocity v since

$$v = \text{cotanh}(\omega) \tag{4-A.28}$$

by eq. 4-A.22. Thus when $\omega = 0$ then $v = \infty$, and when $\omega = \infty$ then $v = 1$. This is the reverse of the sublight case: by eq. 4-A.18 $v = \tanh(\omega)$. Thus when $\omega = 0$ then $v = 0$, and when $\omega = \infty$ then $v = \infty$.

4-A.6.2 General Velocity Transformation Law – Left-Handed Superluminal Boosts

The general velocity transformation law for a particle moving with velocity **v** in the unprimed reference frame and velocity **v'** in the primed reference frame is

$$\mathbf{v}' = [\mathbf{v} + (\gamma - 1)\mathbf{w}\cdot\mathbf{v}\ \mathbf{w}/w^2 - \gamma\mathbf{w}]/[\ \gamma(1 - \mathbf{w}\cdot\mathbf{v})] \tag{4-A.29}$$

where **w** is the relative velocity of the primed reference frame with respect to the unprimed reference frame, and $\gamma = (1 - w^2)^{-\frac{1}{2}}$. Eq. 4-A.29 is obtained by calculating the derivative dx'/dt' using eqs. 4-A.14. The relative velocity **w** can be greater or less than the speed of light. Eq. 4-A.29 implies

$$v'^2 = 1 + (v^2 - 1)(1 - w^2)/(1 - \mathbf{w}\cdot\mathbf{v})^2 \tag{4-A.30}$$

The relation of the velocities (eq. 4-A.30) will be used to determine the multiplication rules for subluminal and superluminal Lorentz transformations (next subsection).

4-A.6.3 Left-Handed Transformations Multiplication Rules

In this subsection we will determine the multiplication rules of left-handed subluminal and superluminal Lorentz boosts. To do this we will consider three reference frames: an "unprimed" frame, a "primed" frame moving with velocity **w** with respect to the unprimed frame, and a "double-primed" frame moving with velocity **v** with respect to the unprimed frame and velocity **v'** with respect to the primed frame. See Fig.4-A.2.

The velocity **v'** is related to **v** by eqs. 4-A.29 and 4-A.30. Think of the double-primed coordinate system as attached to a particle. In addition note that the transformation law from the unprimed to the double-primed reference frame can be viewed as the product of consecutive transformations (boosts) from the unprimed to the primed reference frames and then from the primed to the double-primed reference frames.

Thus the transformations have the general form:

$$\Lambda_?(\mathbf{v}) = \Lambda_?(\mathbf{v'})\Lambda_?(\mathbf{w}) \tag{4-A.31}$$

where the "?" subscripts indicate subluminal or superluminal transformations (boosts) depending on the magnitude of the relative velocity in the transformation's parentheses.

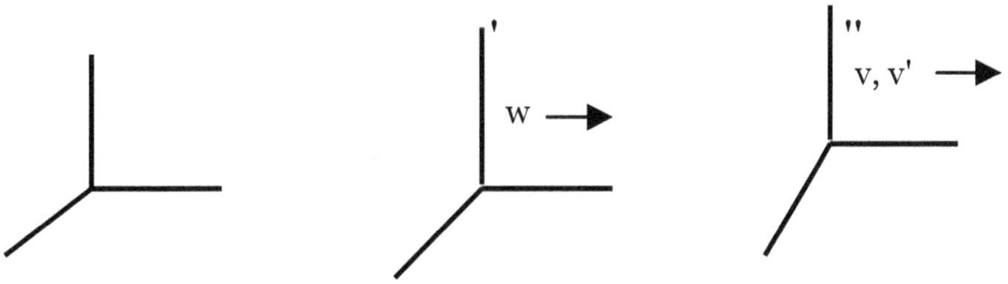

Figure 4-A.2. Three reference frames used to establish transformation multiplication rules.

We now consider the various cases using eq. 4-A.30:

1) If $w > 1$ and $v' > 1$

   then eq. 4-A.30 implies $v < 1$ and thus the left $\Lambda_?(\mathbf{v})$ is a subluminal transformation

$$\Lambda(\mathbf{v}) = \Lambda_L(\mathbf{v'})\Lambda_L(\mathbf{w}) \tag{4-A.32}$$

2) If $w > 1$, $v' < 1$

   then eq. 4-A.30 implies $v > 1$ and thus the left $\Lambda_?(\mathbf{v})$ is a superluminal transformation

$$\Lambda_L(\mathbf{v}) = \Lambda(\mathbf{v'})\Lambda_L(\mathbf{w}) \tag{4-A.33}$$

3) If $w < 1$, $v' > 1$

   then eq. 4-A.30 implies $v > 1$ and thus the left $\Lambda_?(\mathbf{v})$ is a superluminal transformation

$$\Lambda_L(\mathbf{v}) = \Lambda_L(\mathbf{v'})\Lambda(\mathbf{w}) \tag{4-A.34}$$

4) If w < 1, v' < 1

then eq. 4-A.30 implies v < 1 and thus the left Λ?(**v**) is a Lorentz transformation

$$\Lambda(\mathbf{v}) = \Lambda(\mathbf{v'})\Lambda(\mathbf{w}) \qquad (4\text{-A.35})$$

where, in each above case, the transformation on the left side of the equation may be a boost or a combination of a boost and a spatial rotation. Thus we have obtained the multiplication rules for left-handed subluminal and superluminal Lorentz transformations.

4-A.6.4 Inverse of Left-Handed Transformations

The inverse of a Lorentz boost is

$$\Lambda^{-1}(\omega, \hat{\mathbf{u}}) = \exp[-i\omega\hat{\mathbf{u}}\cdot\mathbf{K}] \qquad (4\text{-A.36})$$

where $\omega \geq 0$. Thus the inverse is generated by letting $\omega \to -\omega$. Note that since v = tanhω, the effect of $\omega \to -\omega$ is to let $v \to -v$. In the case of superluminal left-handed boosts, since

$$\Lambda_L(\omega, \mathbf{u}) = \Lambda(\omega + i\pi/2, \mathbf{u}) = \exp[i(\omega + i\pi/2)\hat{\mathbf{u}}\cdot\mathbf{K}] \qquad (4\text{-A.37})$$

we find the inverse is

$$\Lambda_L^{-1}(\omega, \mathbf{u}) = \Lambda(-(\omega + i\pi/2), \mathbf{u}) = \exp[-i(\omega + i\pi/2)\hat{\mathbf{u}}\cdot\mathbf{K}] \qquad (4\text{-A.38})$$

where $\omega \geq 0$. Since $\Lambda_L^{-1}(\omega, \mathbf{u})$ is not the hermitean conjugate of $\Lambda_L(\omega, \mathbf{u})$, superluminal boosts are not unitary. However unitarity is not required since complex Lorentz group elements satisfy the defining relation of the Lorentz group (eq. 4-A.10).

4-A.7 Right-Handed Superluminal Transformations

When we transform between reference frames using a *right-handed*[51] superluminal boost the relation between ω and v is different. The variable ω becomes $\omega_R = \omega - i\pi/2$ and

$$\cosh(\omega_R) = -i\sinh(\omega) = \gamma = -i\gamma_s \qquad (4\text{-A.39})$$
$$\sinh(\omega_R) = -i\cosh(\omega) = \beta\gamma = -i\beta\gamma_s \qquad (4\text{-A.40})$$

where $\beta = v > 1$ and $\omega \geq 0$. Note that $\omega = \mathrm{Re}\,\omega_R$

$$\sinh(\omega) = \gamma_s \qquad (4\text{-A.41})$$
$$\cosh(\omega) = \beta\gamma_s \qquad (4\text{-A.42})$$

with

$$\gamma_s = (\beta^2 - 1)^{-1/2} \qquad (4\text{-A.43})$$

Upon substituting ω_R for ω in eq. 4-A.13 we obtain the form of the right-handed superluminal boost:[52]

---

[51] We call these transformations right-handed because they lead eventually to an alternate right-handed Standard Model This alternate right-handed Standard Model does not appear to correspond to current experimental reality.

$$\Lambda_R(\omega, \mathbf{u}) = \Lambda(\omega - i\pi/2, \mathbf{u}) \tag{4-A.44}$$

$$= \begin{bmatrix} -i\gamma_s & i\beta\gamma_s u_x & i\beta\gamma_s u_y & i\beta\gamma_s u_z \\ i\beta\gamma_s u_x & 1 + (-i\gamma_s - 1)u_x^2 & (-i\gamma_s - 1)u_x u_y & (-i\gamma_s - 1)u_x u_z \\ i\beta\gamma_s u_y & (-i\gamma_s - 1)u_x u_y & 1 + (-i\gamma_s - 1)u_y^2 & (-i\gamma_s - 1)u_y u_z \\ i\beta\gamma_s u_z & (-i\gamma_s - 1)u_x u_z & (-i\gamma_s - 1)u_y u_z & 1 + (-i\gamma_s - 1)u_z^2 \end{bmatrix}$$

A simple case that illustrates right-handed superluminal transformations is to assume a relative velocity in the x direction. Then eq. 4-A.44 becomes

$$\Lambda_R(\omega, \mathbf{u} = (1,0,0)) = \begin{bmatrix} -i\gamma_s & i\beta\gamma_s & 0 & 0 \\ i\beta\gamma_s & -i\gamma_s & 0 & 0 \\ 0 & 0 & 1 & 0 \\ 0 & 0 & 0 & 1 \end{bmatrix} \tag{4-A.45}$$

implementing the coordinate transformation:

$$X' = \Lambda_R(\omega, \mathbf{u})X$$

or

$$\begin{aligned} t' &= -i\gamma_s(t - \beta x) \\ x' &= -i\gamma_s(x - \beta t) \\ y' &= y \\ z' &= z \end{aligned} \tag{4-A.46}$$

Comparing eq. 4-A.45 with eq. 4-A.24 for a left-handed superluminal boost we see that

$$PT\Lambda_L(\omega, \mathbf{u} = (1,0,0)) = \begin{bmatrix} -i\gamma_s & i\beta\gamma_s & 0 & 0 \\ i\beta\gamma_s & -i\gamma_s & 0 & 0 \\ 0 & 0 & -1 & 0 \\ 0 & 0 & 0 & -1 \end{bmatrix}$$

where P is the parity operator and T is the time reversal operator. If we now apply a spatial rotation $\mathfrak{R}$ of π radians around the x axis then we obtain

$$\mathfrak{R}PT\Lambda_L(\omega, \mathbf{u} = (1,0,0))\mathfrak{R}^{-1} = \begin{bmatrix} -i\gamma_s & i\beta\gamma_s & 0 & 0 \\ i\beta\gamma_s & -i\gamma_s & 0 & 0 \\ 0 & 0 & 1 & 0 \\ 0 & 0 & 0 & 1 \end{bmatrix} \tag{4-A.47}$$

$$= \Lambda_R(\omega, \mathbf{u} = (1,0,0))$$

Since P and T commute with spatial rotations we find

---

[52] We note the singularities at $\beta = \pm 1$ or $\omega = \pm\infty$. **As a result we have a branch cut in the complex ω-plane consisting of the entire real ω axis. Therefore three left-handed boosts are not equivalent to a right-handed boost but rather appear on a different Riemann sheet.**

$$\Lambda_R(\omega, \mathbf{u} = (1,0,0)) = PT\mathfrak{R}\Lambda_L(\omega, \mathbf{u} = (1,0,0))\mathfrak{R}^{-1} \quad (4\text{-A.48})$$

or, more generally, performing additional spatial rotations:

$$\Lambda_R(\omega, \mathbf{u}) = PT\mathfrak{R}_u\mathfrak{R}\mathfrak{R}_w\Lambda_L(\omega, \mathbf{w})\mathfrak{R}_w^{-1}\mathfrak{R}^{-1}\mathfrak{R}_u^{-1} \quad (4\text{-A.49})$$

or,

$$\Lambda_R(\omega, \mathbf{u}) = PT\mathfrak{R}_{tot}\Lambda_L(\omega, \mathbf{w})\mathfrak{R}_{tot}^{-1} \quad (4\text{-A.50})$$

where **u** and **w** are unit vectors, and $\mathfrak{R}_{tot} = \mathfrak{R}_u\mathfrak{R}\mathfrak{R}_w$. Alternately,

$$\Lambda_L(\omega, \mathbf{w}) = PT\mathfrak{R}_{tot}^{-1}\Lambda_R(\omega, \mathbf{u})\mathfrak{R}_{tot} \quad (4\text{-A.51})$$

or

$$\Lambda_L(\omega, \mathbf{w}) = PT\Lambda_R(\omega, \mathbf{u}') \quad (4\text{-A.52})$$

for some unit vector **u'**.

Thus we have shown that PT can be used to relate left-handed and right-handed boosts in a one-to-one fashion. *The appearance of the* parity *operator P takes on great significance when we derive features of the Standard Model. The appearance of left-handed form of The Standard Model stems directly from the implicit parity dependence of the left-handed sector of the superluminal part of the complex Lorentz group.*

For a right-handed boost the addition rule for the x-component of velocity can be computed for infinitesimal displacements in space and time:

$$\begin{aligned} v_x' = \Delta x'/\Delta t' &= (\Delta x\,\gamma_s - \Delta t\,\beta\gamma_s)/(\Delta t\,\gamma_s - \Delta x\,\beta\gamma_s) \\ &= (v_x - \beta)/(1 - \beta v_x) \end{aligned} \quad (4\text{-A.53})$$

in the limit $\Delta t \to 0$ where the x component of a particle's velocity in the unprimed frame is $v_x = \Delta x/\Delta t$. Note if $v_x = 1$ then $v_x' = 1$. *Thus right-handed superluminal transformations also preserve the constancy of the speed of light in all reference frames.*

4-A.8 Inhomogeneous Left-Handed Lorentz Group Transformations

The *Left-Handed transformations of the complex Lorentz group* consist of the elements of the real Lorentz group plus left-handed superluminal boost transformations, and combinations of boosts and spatial rotations. Thus the homogeneous left-handed superluminal transformations have the general form:

$$\Lambda_L(\mathbf{v}, \boldsymbol{\theta}) = \exp[i\omega_L\hat{\mathbf{u}}\cdot\mathbf{K} + i\boldsymbol{\theta}\cdot\mathbf{J}] \quad (4\text{-A.54})$$

where $\omega_L' = \omega + i\pi/2$, $\boldsymbol{\theta}$ is the angular vector, and **J** is the angular momentum operator vector. Inhomogeneous left-handed superluminal transformations, which include displacements, can be expressed as

$$\Lambda_L(\mathbf{v}, \boldsymbol{\theta}, \mathbf{d}) = \exp[i\omega_L\hat{\mathbf{u}}\cdot\mathbf{K} + i\boldsymbol{\theta}\cdot\mathbf{J} - i\mathbf{d}\cdot\mathbf{P}] \quad (4\text{-A.55})$$

where **P** is the momentum operator vector and **d** is a displacement vector.

We note

$$\det \Lambda_L(\omega, \mathbf{u}) = \pm 1 \tag{4-A.56}$$

The ordinary Lorentz group is divided into four disjoint subgroups that are often denoted:

$$\begin{aligned}
&L_+^\uparrow: \quad \det \Lambda(\omega, \mathbf{u}) = +1; \quad \text{sgn}\, \Lambda(\omega, \mathbf{u})^0{}_0 = +1 \\
&L_-^\uparrow: \quad \det \Lambda(\omega, \mathbf{u}) = -1; \quad \text{sgn}\, \Lambda(\omega, \mathbf{u})^0{}_0 = +1 \\
&L_+^\downarrow: \quad \det \Lambda(\omega, \mathbf{u}) = +1; \quad \text{sgn}\, \Lambda(\omega, \mathbf{u})^0{}_0 = -1 \\
&L_-^\downarrow: \quad \det \Lambda(\omega, \mathbf{u}) = -1; \quad \text{sgn}\, \Lambda(\omega, \mathbf{u})^0{}_0 = -1
\end{aligned} \tag{4-A.57}$$

where $\text{sgn}\, \Lambda(\omega, \mathbf{u})^0{}_0$ is the sign of the 00 component of the $\Lambda(\omega, \mathbf{u})$ matrix. The various subgroups are related by the discrete transformations of parity P and time reversal T:

$$L_+^\uparrow \xrightarrow{P} L_-^\uparrow$$
$$L_+^\uparrow \xrightarrow{PT} L_+^\downarrow$$
$$L_+^\uparrow \xrightarrow{T} L_-^\downarrow$$

The left-handed superluminal transformations are disjoint in a somewhat different way. By eq. 4-A.56 the determinants are ±1. However the 0-0 matrix element of eq. 4-A.16 gives

$$\Lambda_L{}^0{}_0{}^2 - \Sigma_i\, (\Lambda_L{}^i{}_0)^2 = 1 \tag{4-A.58}$$

The representation of superluminal boosts shows that each factor in eq. 4-A.58 is imaginary. Thus eq. 4-A.58 implies

$$\Sigma_i\, |\Lambda_L{}^i{}_0|^2 \geq 1 \tag{4-A.59}$$

$$|\Lambda_L{}^0{}_0| \geq 0 \qquad\qquad (\text{not } \geq 1) \tag{4-A.60}$$

where || indicates absolute value since the quantities in eq. 4-A.58 are squares – not in absolute value. Thus the magnitude of $\Lambda_L{}^0{}_0$ does not have a gap. Therefore left-handed superluminal transformations can be divided into two categories:

$$\begin{aligned}
{}_L L_+&: \quad \det \Lambda_L(\omega, \mathbf{u}) = +1 \\
{}_L L_-&: \quad \det \Lambda_L(\omega, \mathbf{u}) = -1
\end{aligned} \tag{4-A.61}$$

as one expects for complex Lorentz group transformations.[53]

Earlier we saw that under a PT transformation a left-handed superluminal transformation becomes a right handed superluminal transformation. Again, as in the left-handed case, the various disjoint pieces are related by the discrete transformations of parity P and time reversal T:

---

[53] Streater (2000) p. 13.

$$ {}_{L}L_{+} \xrightarrow{P} {}_{L}L_{-} $$
$$ {}_{L}L_{+} \xrightarrow{T} {}_{L}L_{-} \qquad (4\text{-A}.62) $$

4-A.9 Inhomogeneous Right-Handed Extended Lorentz Group

The inhomogeneous right-handed part of the complex Lorentz group[54] consists of the real Lorentz group plus right-handed superluminal transformations plus rotations and displacements that have the form:

$$ \Lambda_R(\mathbf{v}, \boldsymbol{\theta}, \mathbf{d}) = \exp[i\omega_R\hat{\mathbf{u}}\cdot\mathbf{K} + i\boldsymbol{\theta}\cdot\mathbf{J} - i\mathbf{d}\cdot\mathbf{P}] \qquad (4\text{-A}.63) $$

in general where $\omega_R = \omega - i\pi/2$.

4-A.10 General Forms of Superluminal Boosts

The group elements of the homogeneous complex Lorentz group L(C) can be expressed in terms of the group generators as

$$ \Lambda_C = \exp[i(\omega_r\hat{\mathbf{u}}_r + i\omega_i\hat{\mathbf{u}}_i)\cdot\mathbf{K} + i\boldsymbol{\theta}_c\cdot\mathbf{J}] \qquad (4\text{-A}.64) $$

where the vector $\boldsymbol{\theta}_c$ is a complex 3-vector, $\omega_r \geq 0$ and $\omega_i \geq 0$ are real numbers, and $\hat{\mathbf{u}}_r$ and $\hat{\mathbf{u}}_i$ are real normalized 3-vectors such that $\hat{\mathbf{u}}_r\cdot\hat{\mathbf{u}}_r = 1 = \hat{\mathbf{u}}_i\cdot\hat{\mathbf{u}}_i$. The generators of the homogeneous complex Lorentz group are **K**, and **J** just as for the homogeneous real Lorentz group.

We now focus on boosts because they will be crucial in the determination of the equations of motion of various types of spin ½ particles. A boost has the form

$$ \Lambda_C(\mathbf{v}_c) = \exp[i\omega\hat{\mathbf{w}}\cdot\mathbf{K}] \qquad (4\text{-A}.65) $$

where

$$ \omega = (\omega_r^2 - \omega_i^2 + 2i\omega_r\omega_i\,\hat{\mathbf{u}}_r\cdot\hat{\mathbf{u}}_i)^{1/2} \qquad (4\text{-A}.66) $$

and

$$ \hat{\mathbf{w}} = (\omega_r\hat{\mathbf{u}}_r + i\omega_i\hat{\mathbf{u}}_i)/\omega \qquad (4\text{-A}.67) $$

Since $\hat{\mathbf{u}}_r\cdot\hat{\mathbf{u}}_r = 1 = \hat{\mathbf{u}}_i\cdot\hat{\mathbf{u}}_i$ we see

$$ \hat{\mathbf{w}}\cdot\hat{\mathbf{w}} = 1 \qquad (4\text{-A}.68) $$

The complex relative velocity is

$$ \mathbf{v}_c = \hat{\mathbf{w}}\tanh(\omega) \qquad (4\text{-A}.69) $$

Having placed boost transformations in the form of eq. 4-A.12 we can take advantage of the form of real proper orthchronous Lorentz boost transformations, eq. 4-A.13, and analytically continue to

---

[54] **Since γ_s has branch points at $v = \pm 1$ (which corresponds to $\omega = \pm\infty$ for both the left-handed and right-handed groups) there is a cut along the real ω axis between $-\infty$ and $+\infty$ in the ω complex plane. Therefore, we note, the product of three left-handed Lorentz transformations does not yield a right-handed transformation (as might be supposed from eqs. 2.43 and 2.51) but rather a left-handed transformation on the second sheet. A transformation with $\omega + 3i\pi/2$ is not equivalent to a transformation with $\omega - i\pi/2$.**

complex ω and complex unit vectors $\hat{\mathbf{w}}$ provided eq. 4-A.69 is satisfied. The resulting complex generalization will be the matrix form of proper boosts:

$$\Lambda_C(\mathbf{v_c}) = \exp[i\omega\hat{\mathbf{w}}\cdot\mathbf{K}] \equiv \Lambda_C(\omega, \hat{\mathbf{w}})$$

$$= \begin{bmatrix} \cosh(\omega) & -\sinh(\omega)\hat{w}_x & -\sinh(\omega)\hat{w}_y & -\sinh(\omega)\hat{w}_z \\ -\sinh(\omega)\hat{w}_x & 1+(\cosh(\omega)-1)\hat{w}_x^2 & (\cosh(\omega)-1)\hat{w}_x\hat{w}_y & (\cosh(\omega)-1)\hat{w}_x\hat{w}_z \\ -\sinh(\omega)\hat{w}_y & (\cosh(\omega)-1)\hat{w}_x\hat{w}_y & 1+(\cosh(\omega)-1)\hat{w}_y^2 & (\cosh(\omega)-1)\hat{w}_y\hat{w}_z \\ -\sinh(\omega)\hat{w}_z & (\cosh(\omega)-1)\hat{w}_x\hat{w}_z & (\cosh(\omega)-1)\hat{w}_y\hat{w}_z & 1+(\cosh(\omega)-1)\hat{w}_z^2 \end{bmatrix} \quad (4\text{-A}.70)$$

Since analytic continuations are unique, the above form for $\Lambda_C(\mathbf{v_c})$ is well-defined and unique. It spans the complete set of proper complex Lorentz boosts.

We now will study six classes of boosts that have the property that they boost from a coordinate system with real time and space coordinates to a coordinate system with either a purely real or purely imaginary time, and real, imaginary or complex spatial coordinates. These boosts produce left-handed lepton-like and "quark-like" free Dirac-like equations. They also produce right-handed lepton-like and "quark-like" free Dirac-like equations. We will discuss these Dirac-like equations in detail later. First we describe the four categories of boosts that have the property that they transform the reference frame of a particle at rest to a reference frame where the energy is either purely real or purely imaginary – the distinguishing feature of these four sets of transformations.

4-A.11.1 "Lepton-like" Left-Handed Boosts

If we let

$$\hat{\mathbf{u}}_i = \hat{\mathbf{u}}_r \equiv \hat{\mathbf{u}} \quad (4\text{-A}.71)$$

so that the vector $\hat{\mathbf{u}}_i$ is parallel to $\hat{\mathbf{u}}_r$, and let

$$\omega_i = \pi/2 \quad (4\text{-A}.72)$$

then $\Lambda_C(\mathbf{v_c})$ becomes a lepton-like left-handed boost:[55]

$$\Lambda_C = \exp[i(\omega_r + i\,\pi/2)\hat{\mathbf{u}}_r\cdot\mathbf{K}] \quad (4\text{-A}.73)$$

4-A.11.2 "Lepton-like" Right-Handed Boosts

If we let

$$\hat{\mathbf{u}}_i = -\hat{\mathbf{u}}_r \equiv -\hat{\mathbf{u}} \quad (4\text{-A}.74)$$

so that the vector $\hat{\mathbf{u}}_i$ is anti-parallel to $\hat{\mathbf{u}}_r$, and

$$\omega_i = -\pi/2 \quad (4\text{-A}.75)$$

---

[55] We say "lepton-like" because we obtain a lepton-like Dirac-like equation using these boosts later. Similarly for "quark-like.'

then $\Lambda_C(\mathbf{v}_c)$ becomes a right-handed boost:

$$\Lambda_C = \exp[i(\omega_r - i\,\pi/2)\hat{\mathbf{u}}_r\cdot\mathbf{K}] \tag{4-A.76}$$

4-A.11.3 "Quark-like" Left-Handed Boosts

If the real and imaginary relative vectors parts of $\hat{\mathbf{w}}$, namely $\hat{\mathbf{u}}_r$ and $\hat{\mathbf{u}}_i$, are perpendicular, $\hat{\mathbf{u}}_r\cdot\hat{\mathbf{u}}_i = 0$, then by eq. 4-A.66

$$\omega = (\omega_r^2 - \omega_i^2)^{1/2} \tag{4-A.77}$$

Thus ω is either pure real ($\omega_r \geq \omega_i$) or pure imaginary ($\omega_r < \omega_i$). We choose ω real, and then reset

$$\omega = (\omega_r^2 - \omega_i^2)^{1/2} \rightarrow \omega' = (\omega_r^2 - \omega_i^2)^{1/2} + i\pi/2 = \omega + i\pi/2 \tag{4-A.78}$$

by adding $i\pi/2$ to the ω factor in eq. 4-A.65 since ω is a free parameter. Then the resulting Lorentz transformation then becomes a "quark-like" left-handed boost:[56]

$$\Lambda_C = \exp[i((\omega_r^2 - \omega_i^2)^{1/2} + i\pi/2)(\omega_r\hat{\mathbf{u}}_r + i\omega_i\hat{\mathbf{u}}_i)\cdot\mathbf{K}/\omega] \tag{4-A.79}$$

4-A.11.4 "Quark-like" Right-Handed Boosts

If the real and imaginary relative vectors parts of $\hat{\mathbf{w}}$, namely $\hat{\mathbf{u}}_r$ and $\hat{\mathbf{u}}_i$, are perpendicular, $\hat{\mathbf{u}}_r\cdot\hat{\mathbf{u}}_i = 0$, then by eq. 4-A.66

$$\omega = (\omega_r^2 - \omega_i^2)^{1/2} \tag{4-A.80}$$

Thus ω again starts out either pure real ($\omega_r \geq \omega_i$) or pure imaginary ($\omega_r < \omega_i$). In this case we also choose ω real, and then reset

$$\omega = (\omega_r^2 - \omega_i^2)^{1/2} \rightarrow \omega' = (\omega_r^2 - \omega_i^2)^{1/2} - i\pi/2 \tag{4-A.81}$$

by subtracting $i\pi/2$ from ω in eq. 4-A.65 since ω is a free parameter. The resulting Lorentz boost

$$\Lambda_C = \exp[i((\omega_r^2 - \omega_i^2)^{1/2} - i\pi/2)(\omega_r\hat{\mathbf{u}}_r + i\omega_i\hat{\mathbf{u}}_i)\cdot\mathbf{K}/\omega] \tag{4-A.82}$$

becomes a quark-like right-handed boost.[57]

4-A.11.5 "Quark-like" Boosts

If the real and imaginary relative vectors parts of $\hat{\mathbf{w}}$, namely $\hat{\mathbf{u}}_r$ and $\hat{\mathbf{u}}_i$, are perpendicular, $\hat{\mathbf{u}}_r\cdot\hat{\mathbf{u}}_i = 0$, then by eq. 4-A.66

$$\omega = (\omega_r^2 - \omega_i^2)^{1/2} \tag{4-A.83}$$

---

[56] We say "quark-like" because we will later obtain a quark-like left-handed Dirac-like equation with complex spatial momentum terms using these boosts.

[57] We say "quark-like" because we obtain a quark-like right-handed Dirac-like equation with complex spatial momentum terms using these boosts later.

Thus ω again starts out either pure real ($\omega_r \geq \omega_i$) or pure imaginary ($\omega_r < \omega_i$). In this case choose ω_r real and use ω as defined by eq. 4-A.83.

Then the resulting Lorentz boost

$$\Lambda_C = \exp[i(\omega_r^2 - \omega_i^2)^{1/2}(\omega_r\hat{\mathbf{u}}_r + i\omega_i\hat{\mathbf{u}}_i)\cdot\mathbf{K}/\omega] \quad (4\text{-A.84})$$

becomes a quark-like boost without handedness.[58]

4-A.11.6 Conventional "Dirac" Boosts

If we let

$$\hat{\mathbf{u}}_i = \hat{\mathbf{u}}_r \equiv \hat{\mathbf{u}} \quad (4\text{-A.85})$$

so that the vector $\hat{\mathbf{u}}_i$ is parallel to $\hat{\mathbf{u}}_r$, and let

$$\omega_i = 0 \quad (4\text{-A.86})$$

then $\Lambda_C(\mathbf{v}_c)$ becomes a Dirac boost:[59]

$$\Lambda = \exp[i\omega_r\hat{\mathbf{u}}_r\cdot\mathbf{K}] \quad (4\text{-A.87})$$

This boost can be used to generate the free Dirac equation.

---

[58] We again say "quark-like" because we obtain a quark-like Dirac equation with complex spatial momentum terms using these boosts later.

[59] We say "Dirac" because we obtain a Dirac equation using this boost later.

5. Iota Dirac-like Equations for Particles and Antiparticles

In the previous chapter we found it necessary to add complex 4-dimensional coordinates to iotas to define fermion particles. We then define a Reality group of transformations that mapped complex space to the real-valued 4-dimensional space-time of our experience. The Reality group was then shown to be the origin of the gauge fields necessary for the invariance of dynamical equations, and their consequent physics, under coordinate system transformations. Thus covariant derivatives of the symmetry, and form, of those in the Standard Model were derived.

Having defined coordinates, gauge fields, and covariant driuvatives, we now derive the four species (types) of fermions found in nature by a consideration of *Lorentz group boosts with the important feature that they transform a real-valued energy in one reference frame to a real-valued energy in the transformed reference frame.* The energies of *free* fundamental elementary particles (Ineractions are introduced later.) must be real-valued or they would not be fundamental but would be subject to decay to yet more fundamental particles – contrary to our assumption. The four species of fermions will be identified as charged leptons, neutral leptons – neutrinos, up-type quarks, and down-type quarks.[60] Each lepton species consists of four particles – four generations. Each quark species consists of four generations – each generation consisting of three variants.[61] (We will discuss the four generations of each species, and the three variants in each generation of each quark species in a later chapter.[62])

In developing our theory of fermions (and the previously defined gauge vector bosons associated with the Reality group) we will assume all particles are quantum fields and conform to the rules of canonical quantum field theory including the use of the Faddeev-Popov mechanism.

We begin by defining energy and momentum as Fourier transform variables for functions of coordinates. For any funf f(x) where x is a real or complex 4-vector we define its Fourier transform using an inner product of the coordinates with a 4-vector p that we call the momentum 4-vector consisting of an energy component and a spatial 3-momentum:

$$h(p) = \int d^4x \exp(ip\cdot x)\, g(x) \tag{5.1}$$

where $p\cdot x = g_{\mu\nu}p^{\mu}x^{\nu}$.

BEGINNING OF AMENDED CHAPTERS 2 & 3 EXTRACT FROM BLAHA (2007b)

---

[60] And their anti-particles.

[61] We will also discuss our proposed Dark fermion spectrum later. It is similar in overall form to the normal fermion spectrum except that each quark species generation has only *one* 'variant.'

[62] A lepton species has only one variant in each generation.

5.1 Matrix Representation of Complex Lorentz Group L_C Boosts

We shall now turn our attention to L_C boosts because they will be crucial in the determination of the equations of motion of various types of spin ½ particles. An L_C boost can be expressed in the form

$$\Lambda_C(\mathbf{v}_c) = \exp[i\omega\hat{\mathbf{w}}\cdot\mathbf{K}] \tag{5.2}$$

where

$$\omega = (\omega_r^2 - \omega_i^2 + 2i\omega_r\omega_i\, \hat{\mathbf{u}}_r\cdot\hat{\mathbf{u}}_i)^{½} \tag{5.3}$$

and

$$\hat{\mathbf{w}} = (\omega_r\hat{\mathbf{u}}_r + i\omega_i\hat{\mathbf{u}}_i)/\omega \tag{5.4}$$

Since $\hat{\mathbf{u}}_r\cdot\hat{\mathbf{u}}_r = 1 = \hat{\mathbf{u}}_i\cdot\hat{\mathbf{u}}_i$

$$\hat{\mathbf{w}}\cdot\hat{\mathbf{w}} = 1 \tag{5.5}$$

and the complex relative velocity is

$$\mathbf{v}_c = \hat{\mathbf{w}}\tanh(\omega) \tag{5.6}$$

We now analytically continue to complex ω and complex unit vectors $\hat{\mathbf{w}}$. The resulting complex generalization will be the matrix form of proper L_C boosts:

$$\Lambda_C(\mathbf{v}_c) = \exp[i\omega\hat{\mathbf{w}}\cdot\mathbf{K}] \equiv \Lambda_C(\omega, \hat{\mathbf{w}})$$

$$= \begin{bmatrix} \cosh(\omega) & -\sinh(\omega)\hat{w}_x & -\sinh(\omega)\hat{w}_y & -\sinh(\omega)\hat{w}_z \\ -\sinh(\omega)\hat{w}_x & 1 + (\cosh(\omega) - 1)\hat{w}_x^2 & (\cosh(\omega) - 1)\hat{w}_x\hat{w}_y & (\cosh(\omega) - 1)\hat{w}_x\hat{w}_z \\ -\sinh(\omega)\hat{w}_y & (\cosh(\omega) - 1)\hat{w}_x\hat{w}_y & 1 + (\cosh(\omega) - 1)\hat{w}_y^2 & (\cosh(\omega) - 1)\hat{w}_y\hat{w}_z \\ -\sinh(\omega)\hat{w}_z & (\cosh(\omega) - 1)\hat{w}_x\hat{w}_z & (\cosh(\omega) - 1)\hat{w}_y\hat{w}_z & 1 + (\cosh(\omega) - 1)\hat{w}_z^2 \end{bmatrix} \tag{5.7}$$

Since analytic continuations are unique, the above form for $\Lambda_C(\mathbf{v}_c)$ is well-defined and unique. It spans the complete set of proper L_C boosts.

5.2 Left-handed and Right-handed Parts of L_C

We now describe the Left-handed and Right-handed parts[63] of L_C boosts.

Left-handed Part of L_C

If we let

$$\hat{\mathbf{u}}_i = \hat{\mathbf{u}}_r \equiv \hat{\mathbf{u}} \tag{5.8}$$

so that the vector $\hat{\mathbf{u}}_i$ is parallel to $\hat{\mathbf{u}}_r$, and

[63] The designations Left-handed and Right-handed are chosen to reflect the Left-handed and Right-handed fermion fields that will be used to construct The Standard Model later. See Blaha (2007b) for more detail.

$$\omega_i = \pi/2 \tag{5.9}$$

then $\Lambda_C(\mathbf{v_c})$ becomes a Left-handed L_C boost:

$$\Lambda_C(\mathbf{v_c}) = \Lambda_L(\omega_\mathbf{r}, \mathbf{u}) \tag{5.10}$$

Right-handed part of L_C

If we let

$$\hat{\mathbf{u}}_i = -\hat{\mathbf{u}}_r \equiv -\hat{\mathbf{u}} \tag{5.11}$$

so that the vector $\hat{\mathbf{u}}_i$ is anti-parallel to $\hat{\mathbf{u}}_r$, and

$$\omega_i = -\pi/2 \tag{5.12}$$

then $\Lambda_C(\mathbf{v_c})$ becomes a Right-handed L_C boost:

$$\Lambda_C(\mathbf{v_c}) = \Lambda_R(\omega_\mathbf{r}, \mathbf{u}) \tag{5.13}$$

as described in Blaha (2007b).

5.3 Difference between the Parts of L_C Reduced to Parallelism of $\hat{\mathbf{u}}_r$ and $\hat{\mathbf{u}}_i$

Since the Left-handed L_C part leads to the Standard Model's left-handed features, it seems that the parallel case $\hat{\mathbf{u}}_i = \hat{\mathbf{u}}_r \equiv \hat{\mathbf{u}}$ is more favored by Nature.[64] To some extent this concept of parallel vectors $\hat{\mathbf{u}}_i$ and $\hat{\mathbf{u}}_r$, which leads to the Left-handed L_C, is more intuitively satisfying then the anti-parallel case that leads to the Right-handed L_C part. However, a deeper reason for Nature's choice remains to be found.

5.4 Free Spin ½ Particles – Leptons & Quarks

In this section we begin by developing dynamical equations for spin ½ particles based on the L_C parts. These spin ½ particles are conventional Dirac particles (Majorana particles are also allowed but not discussed), spin ½ tachyons, and "color" versions of both types totalling four species. We will identify leptons and quarks with these fields.

5.4.1 Introduction

Tachyons are particles that move faster than the speed of light. As we saw in earlier books tachyons exist inside Black Holes, and within current theories – particularly SuperString theories. There are also experimental indications that neutrinos are tachyons.

Attempts to create canonical tachyon quantum field theories began in the 1960's. These attempts were made within the framework of the Lorentz group and, consequently, were limited to spin 0 theories

---

[64] It is possible that parity violation might disappear at ultra-high energies. Then we would view the parity symmetric theory as broken to the left-handed Standard Model currently established by experiment with right-handed parts at higher energy.

since there are no finite dimensional representations of the Lorentz group for negative m^2 except for the one-dimensional representation. None of these attempts, or attempts since then, succeeded in creating a canonically quantized spin 0 tachyon quantum field theory.[65]

In this section we will formulate a free spin ½ tachyon Quantum Field Theory. We choose to develop a normal spin ½ theory first. Then we develop a free spin ½ tachyon theory because, as we will see, spin ½ tachyon particles (quarks and leptons) play an extraordinary role in the Standard Model.

We will develop our spin ½ tachyon theory from the "ground up" by applying a Left-Handed L_C boost to the Dirac equation, and its Dirac spinor wave function, for a particle at rest. This procedure will give a tachyon spinor wave function, and the momentum space tachyon equation equivalent of the Dirac equation. Then we will obtain the coordinate space tachyon Dirac equation, define a lagrangian, and proceed to create a canonical quantum field theory for spin ½ tachyons.

The need for dynamical equations arises when we clothe each iota with coordinates to "make" a fermion. Having coordinates leads to describing the motion of particles. Dynamical equations specify the motion of particles. If they have a finite number of terms they can be derived from lagrangians. A lagrangian formalism yields a Hamiltonian (the energy) and the momentum.

5.4.2 First Step - Deriving the Conventional Dirac Equation

In this section we will review a method of obtaining the equation of motion of a particle using a free Dirac equation that is obtained by a Lorentz boost of a spinor wave function[66] of a particle at rest.

In the case of a Lorentz transformation the 4×4 matrix form of a Lorentz transformation of Dirac matrices is

$$S^{-1}(\Lambda(v))\gamma^\nu S(\Lambda(v)) = \Lambda^\nu{}_\mu(v)\gamma^\mu \tag{5.14}$$

where $S(\Lambda(v))$ is

$$\begin{aligned} S(\Lambda(v)) &= \exp(-i\omega\sigma_{0i}v_i/(2|\mathbf{v}|)) = \exp(-\omega\gamma^0\boldsymbol{\gamma}\cdot\mathbf{v}/(2|\mathbf{v}|)) \\ &= \cosh(\omega/2)I + \sinh(\omega/2)\gamma^0\boldsymbol{\gamma}\cdot\mathbf{p}/|\mathbf{p}| \end{aligned} \tag{5.15}$$

with $\omega = \operatorname{arctanh}(|\mathbf{v}|)$, $\cosh(\omega/2) = [(E+m)/(2m)]^{½}$ and $\sinh(\omega/2) = |\mathbf{p}|[2m(E+m)]^{-½}$. Also

$$\begin{aligned} S^{-1}(\Lambda(v)) &= \gamma^0 S^\dagger(\Lambda(v))\gamma^0 = \exp(\omega\gamma^0\boldsymbol{\gamma}\cdot\mathbf{v}/(2|\mathbf{v}|)) \\ &= \cosh(\omega/2)I - \sinh(\omega/2)\gamma^0\boldsymbol{\gamma}\cdot\mathbf{p}/|\mathbf{p}| \end{aligned} \tag{5.16}$$

In constructing fermion dynamical equations *we shall assume that they are linear in derivatives* (although a quadratic form is possible.) We will use the sixteen 4×4 matrices that span the set of transformations of the four values of Asynchronous Logic. Since by theorem[67] all 4 × 4 γ matrices are equivalent up to a unitary transformation we can rotate any constant matrix into a multiple of γ^0 without loss of generality.

We begin by defining a generic positive energy plane wave solution of the Dirac equation for a normal fermion particle at rest with rest energy m as

$$\psi(x) = e^{-imt}w(0) \tag{5.17}$$

---

[65] Except Blaha (2006).

[66] The spinor wave function of a particle at rest is a 4-vector of the 4×4 matrix representation of 4-valued Asynchronous Logic.

[67] R. H. Good, Rev. Mod. Phys., **27**, 187 (1955).

with w(0) a four component logic spinor column vector. *For a free particle at rest, the rest energy $m = m_0$, the iota mass.*[68] The wave function satisfies the momentum space Dirac equation for a fermion at rest:

$$(m\gamma^0 - m)e^{-imt}w(0) = 0 \tag{5.18}$$

Subsequently we will use a similar procedure to construct the free tachyonic Dirac equation.
If we now apply $S(\Lambda(v))$ we find

$$0 = S(\Lambda(v))(m\gamma^0 - m)e^{-imt}w(0) = [mS(\Lambda(v))\gamma^0 S^{-1}(\Lambda(v)) - m]S(\Lambda(v))w(0)$$

A straightforward evaluation shows

$$mS(\Lambda(v))\gamma^0 S^{-1}(\Lambda(v)) = g_{\mu\nu}p^\mu\gamma^\nu = \not{p} \tag{5.19}$$

where $p^0 = (p^2 + m^2)^{1/2}$, $\mathbf{p} = \gamma m\mathbf{v}$, and $p = |\mathbf{p}|$. In addition

$$S(\Lambda(v))w(0) = w(p) \tag{5.20}$$

is a positive energy Dirac spinor. Therefore the Dirac equation for a fermion in motion in momentum space has the form:

$$(\not{p} - m)e^{-ip\cdot x}w(p) = 0 \tag{5.21}$$

where the exponential factor, mt, is also boosted to p·x. Eq. 5.21 implies the well-known free, coordinate space Dirac equation:

$$(i\gamma^\mu\partial/\partial x^\mu - m)\psi(x) = 0 \tag{5.22}$$

5.4.3 Derivation of the Tachyon Dirac Equation

The Left-handed boost has the form:

$$\Lambda_L(\omega, \mathbf{u}) = \Lambda(\omega + i\pi/2, \mathbf{u}) = \exp[i\omega_L\hat{\mathbf{u}}\cdot\mathbf{K}] \tag{5.23}$$

where $\omega_L = \omega + i\pi/2$ and

$$\begin{aligned} \cosh(\omega_L) &= i\sinh(\omega) = -\gamma = i\gamma_s \\ \sinh(\omega_L) &= i\cosh(\omega) = -\beta\gamma = i\beta\gamma_s \end{aligned} \tag{5.24}$$

with, $\beta = v > 1$, $\gamma_s = (\beta^2 - 1)^{-1/2}$, and $\boldsymbol{\omega \geq 0}$. Thus

$$\begin{aligned} \sinh(\omega) &= \gamma_s \\ \cosh(\omega) &= \beta\gamma_s \end{aligned} \tag{5.25}$$

The corresponding spinor transformation is:

---

[68] See section 1.1.3.

$$S_L(\Lambda_L(\omega, \mathbf{u})) = \exp(-i\omega_L\sigma_{0i}v_i/(2|\mathbf{v}|)) = \exp(-\omega_L\gamma^0\boldsymbol{\gamma}\cdot\mathbf{v}/(2|\mathbf{v}|))$$
$$= \cosh(\omega_L/2)I + \sinh(\omega_L/2)\gamma^0\boldsymbol{\gamma}\cdot\mathbf{p}/|\mathbf{p}| \qquad (5.26)$$

The inverse transformation is

$$S_L^{-1}(\Lambda_L(\omega, \mathbf{u})) = \gamma^2\gamma^0 K^{-1}S_L^{\dagger}K\gamma^0\gamma^2 = \gamma^2\gamma^0 S_L^{T}\gamma^0\gamma^2 = \exp(\omega_L\gamma^0\boldsymbol{\gamma}\cdot\mathbf{v}/(2|\mathbf{v}|))$$
$$= \cosh(\omega_L/2)I - \sinh(\omega_L/2)\gamma^0\boldsymbol{\gamma}\cdot\mathbf{p}/|\mathbf{p}| \qquad (5.27)$$

where the superscript T denotes the transpose and K is the complex conjugation operator (that also appears in the time-reversal operator). Note that S_L is not unitary just as the equivalent spinor Lorentz transformation $S(\Lambda(v))$ is not unitary.

We can now apply a left-handed superluminal transformation to the generic positive energy plane wave solution of the Dirac equation for a particle of mass m at rest. The result is

$$0 = S_L(\Lambda_L(\omega, \mathbf{u}))(m\gamma^0 - m)e^{-imt}w(0)$$
$$= [mS_L\gamma^0 S_L^{-1} - m]e^{-imt}S_L w(0)$$

where $S_L = S_L(\Lambda_L(\omega, \mathbf{u}))$. After some algebra

$$mS_L\gamma^0 S_L^{-1} = m[\cosh(\omega_L)\gamma^0 - \sinh(\omega_L)\boldsymbol{\gamma}\cdot\mathbf{p}/|\mathbf{p}|]$$

$$= i\gamma^0 E - i\boldsymbol{\gamma}\cdot\mathbf{p} = i\not{p} \qquad (5.28)$$

using the tachyon energy and momentum expressions

$$\mathbf{p} = m\mathbf{v}\gamma_s \qquad E = m\gamma_s \qquad (5.29)$$

Also

$$S_L w(0) = w_T(p) \qquad (5.30)$$

is a tachyon spinor. See Appendix 5-A (at the end of this section) for a discussion of tachyon spinors.

The momentum space tachyonic Dirac equation is

$$(i\not{p} - m)e^{ip\cdot x}w_T(p) = 0 \qquad (5.31)$$

where $p\cdot x = Et - \mathbf{p}\cdot\mathbf{x}$ after performing a corresponding left-handed superluminal coordinate transformation in the exponential factor. Thus a positive energy wave is transformed into a negative energy wave by the superluminal transformation.

If we apply $i\not{p}$to we find the tachyon mass condition is satisfied

$$-E^2 + \mathbf{p}^2 = m^2 \qquad (5.32)$$

Transforming back to coordinate space we obtain the *tachyon Dirac equation*:

$$(\gamma^\mu\partial/\partial x^\mu - m)\psi_T(x) = 0 \qquad (5.33)$$

The "missing" factor of i in the first term of eq. 5.33 requires the lagrangian to be different from the conventional Dirac lagrangian in order for the lagrangian to be real. The simplest, physically acceptable, free spin ½ tachyon lagrangian density is:

$$\mathcal{L}_T = \psi_T^S(\gamma^\mu\partial/\partial x^\mu - m)\psi_T(x) \tag{5.34}$$

where

$$\psi_T^S = \psi_T^\dagger\, i\gamma^0\gamma^5 \tag{5.35}$$

The corresponding action is

$$I = \int d^4x\, \mathcal{L}_T \tag{5.36}$$

Appendix 3-B of Blaha (2007b) proves I is real. The Hamiltonian density is

$$\mathcal{H} = \pi_T\dot{\psi}_T - \mathcal{L} = i\psi_T^\dagger\gamma^5(\boldsymbol{\alpha}\cdot\nabla + \beta m)\psi_T = -i\psi_T^\dagger\gamma^5\dot{\psi}_T \tag{5.37}$$

using the tachyon Dirac equation to obtain the last equality. The reader will note that the tachyon hamiltonian is hermitean by explicit calculation up to an irrelevant total spatial divergence.

Probability Conservation Law

The tachyon Dirac equation implies a probability conservation law:

$$\partial\rho_5/\partial t = \nabla\cdot\mathbf{j}_5 \tag{5.38}$$

where

$$\rho_5 = \psi_T^\dagger\gamma^5\psi_T \qquad \mathbf{j}_5 = \psi_T^\dagger\gamma^5\boldsymbol{\alpha}\psi_T \tag{5.39}$$

We are thus led to define the conserved axial charge Q_5

$$Q_5 = \int d^3x\, \psi_T^\dagger\gamma^5\psi_T \tag{5.40}$$

Energy-Momentum Tensor

The tachyon energy-momentum tensor is

$$\mathfrak{T}_{T\mu\nu} = -\,g_{\mu\nu}\mathcal{L}_T + \partial\mathcal{L}_T/\partial(\partial\psi_T/\partial x_\mu)\,\partial\psi_T/\partial x^\nu \tag{5.41}$$
$$= i\psi_T^\dagger\gamma^0\gamma^5\gamma_\mu\partial\psi_T/\partial x^\nu \tag{5.42}$$

and thus the conserved energy and momentum are

$$P^0 = H = \int d^3x\, \mathfrak{T}_T^{00} = i\int d^3x\psi_T^\dagger\gamma^5(\boldsymbol{\alpha}\cdot\nabla + \beta m)\psi_T \tag{5.43}$$

and

$$P^i = \int d^3x\, \mathfrak{T}_T^{0i} = -\,i\int d^3x\, \psi_T^\dagger\gamma^5\partial\psi_T/\partial x_i \tag{5.44}$$

Both the energy and momentum differ significantly from the corresponding quantities for conventional Dirac fields.

5.4.4 Tachyon Canonical Quantization

Having defined a suitable tachyon lagrangian we can now proceed to its canonical quantization. The conjugate momentum can be calculated from the above lagrangian density:

$$\pi_{Ta} = \partial\mathcal{L}_T/\partial\dot{\psi}_{Ta} \equiv \partial\mathcal{L}_T/\partial(\partial\psi_{Ta}/\partial t) = -i(\psi_T^{\dagger}\gamma^5)_a \tag{5.45}$$

The resulting non-zero, canonical anti-commutation relations are

$$\{\pi_{Ta}(x), \psi_{Tb}(x')\} = i\,\delta_{ab}\,\delta^3(x - x')$$

or

$$\{\psi_{T\,a}^{\dagger}(x), \psi_{Tb}(x')\} = -[\gamma^5]_{ab}\,\delta^3(x - x') \tag{5.46}$$

At this point we might attempt to complete the canonical quantization procedure in the conventional manner by fourier expanding the quantum field and specifying anti-commutation relations for the fourier component amplitudes. However the incompleteness of the set of plane waves, which are limited by the restriction $|p| \geq m$, *causes the anti-commutator of the fields not to yield a* $\delta^3(x - x')$. *Thus the conventional approach fails to yield the required anti-commutation relations.69*

Other approaches: 1) decompose the tachyon field into left-handed and right-handed parts and then second quantize each part; and 2) second quantize in light-front coordinates ($x^{\pm} = (x^0 \pm x^3)/\sqrt{2}$). These approaches also both fail.[70]

The only approach that does succeed[71] *is to decompose the tachyon field into left-handed and right-handed parts and then second quantize in light-front coordinates. We follow that procedure in the following subsections.*

Separation into Left-Handed and Right-Handed Fields

We will use a transformed set of Dirac matrices to develop our left-handed and right-handed tachyon formulations:

$$\gamma^0 = \begin{bmatrix} 0 & -I \\ -I & 0 \end{bmatrix} \qquad \gamma^i = \begin{bmatrix} 0 & \sigma_i \\ -\sigma_i & 0 \end{bmatrix} \qquad \gamma^5 = \begin{bmatrix} I & 0 \\ 0 & -I \end{bmatrix} \tag{5.47}$$

which are obtained from the usual Dirac matrices by applying the unitary transformation $U = 2^{-\frac{1}{2}}(I + \gamma^5\gamma^0)$. *I is the 4×4 identity matrix in eq. 5.47.* The γ^5 chirality operator's eigenvalues define handedness: +1 corresponds to right-handed; and −1 corresponds to left-handed:

---

[69] See G. Feinberg, Phys. Rev. **159**, 1089 (1967) for example.

[70] See the first edition Blaha (2006) where these possibilities were considered and found to fail.

[71] Blaha (2006) discusses this case in detail.

$$\gamma^5\psi_L = -\psi_L \qquad \gamma^5\psi_R = \psi_R \tag{5.48}$$

Consequently, we can define left-handed and right-handed tachyon fields with the projection operators:

$$\begin{aligned} C^{\pm} &= ½(I \pm \gamma^5) \\ C^{+} + C^{-} &= I \\ C^{\pm\,2} &= C^{\pm} \\ C^{+}C^{-} &= 0 \end{aligned} \tag{5.49}$$

with the result

$$\begin{aligned} \psi_{TL} &= C^{-}\psi_T \\ \psi_{TR} &= C^{+}\psi_T \end{aligned} \tag{5.50}$$

We can calculate the commutation relations of the left-handed and right-handed tachyon fields from eq. 5.46 by pre-multiplying and post-multiplying by $½(1-\gamma^5)$ and $½(1+\gamma^5)$. The results are:

$$\{\psi_{TLa}^{\dagger}(x), \psi_{TLb}(x')\} = ½(1-\gamma^5)_{ab}\,\delta^3(x-x') \tag{5.51}$$

$$\{\psi_{TRa}^{\dagger}(x), \psi_{TRb}(x')\} = -½(1+\gamma^5)_{ab}\,\delta^3(x-x') \tag{5.52}$$

$$\{\psi_{TLa}^{\dagger}(x), \psi_{TRb}(x')\} = \{\psi_{TRa}^{\dagger}(x), \psi_{TLb}(x')\} = 0 \tag{5.53}$$

The lagrangian density above decomposes into left-handed and right-handed parts:

$$\mathcal{L}_T = \psi_{TL}^{\dagger}\gamma^0 i\gamma^\mu\partial_\mu\psi_{TL} - \psi_{TR}^{\dagger}\gamma^0 i\gamma^\mu\partial_\mu\psi_{TR} - im[\psi_{TR}^{\dagger}\gamma^0\psi_{TL} - \psi_{TL}^{\dagger}\gamma^0\psi_{TR}] \tag{5.54}$$

Further Separation into + and – Light-Front Fields

There have been many studies of light-front (infinite momentum frame) physics in the past forty years.[72] Light-front coordinates *cannot* be obtained by a Lorentz transformation, or by a superluminal transformation, from a standard set of coordinate system variables even in a limiting sense. Instead they are a defined set of variables that have been used to develop quantum field theories that have been shown to be equivalent to quantum field theories based on conventional coordinates. In particular, light-front quantum field theories have been shown to yield fully Lorentz covariant S matrix elements that are the same as S matrix elements calculated in the conventional way.

Light-front variables can be defined by:

$$\begin{aligned} x^{\pm} &= (x^0 \pm x^3)/\sqrt{2} \\ \partial/\partial x^{\pm} &\equiv \partial^{\mp} \equiv (\partial/\partial x^0 \pm \partial/\partial x^3)/\sqrt{2} \end{aligned} \tag{5.55}$$

---

[72] L. Susskind, Phys. Rev. **165**, 1535 (1968); K. Bardakci and M. B. Halpern Phys. Rev. **176**, 1686 (1968), S. Weinberg, Phys. Rev. **150**, 1313 (1966); J. Kogut and D. Soper, Phys. Rev. **D1**, 2901 (1970); J. D. Bjorken, J. Kogut, and D. Soper, Phys. Rev. **D3**, 1382 (1971); R. A. Neville and F. Rohrlich, Nuov. Cim. **A1**, 625 (1971); F. Rohrlich, Acta Phys Austr. Suppl. **8**, 277 (1971); S-J Chang, R. Root, and T-M Yan, Phys. Rev. **D7**, 1133 (1973); S-J Chang, and T-M Yan, Phys. Rev. **D7**, 1147 (1973); T-M Yan, Phys. Rev. **D7**, 1761 (1973); T-M Yan, Phys. Rev. **D7**, 1780 (1973); C. Thorn, Phys. Rev. **D19**, 639 (1979); and references therein.

with the "transverse" coordinate variables, x^1 and x^2, unchanged.

The inner product of two 4-vectors has the form

$$x \cdot y = x^+y^- + y^+x^- - x^1y^1 - x^2y^2 \tag{5.56}$$

and the light-front definition of Dirac matrices is:

$$\gamma^{\pm} = (\gamma^0 \pm \gamma^3)/\surd 2 \tag{5.57}$$

with transverse matrices γ^1 and γ^2 defined as usual. Note the useful identity:

$$\gamma^{\pm\,2} = 0$$

We define "+" and "–" tachyon fields with the projection operators:

$$R^{\pm} = ½(I \pm \gamma^0\gamma^3) \tag{5.58}$$

They are:

$$\begin{aligned} &\text{Left-handed, } \pm \text{ light-front fields:} \quad && \psi_{TL}{}^{\pm} = R^{\pm}C^-\psi_T \\ &\text{Right-handed, } \pm \text{ light-front fields:} \quad && \psi_{TR}{}^{\pm} = R^{\pm}C^+\psi_T \end{aligned} \tag{5.59}$$

Now if we transform to light-front variables and fields as above we obtain the light-front free tachyon lagrangian:

$$\begin{aligned} \mathcal{L}_T = {} & 2^{½}\psi_{TL}{}^{+\dagger}i\partial^-\psi_{TL}{}^+ + 2^{½}\psi_{TL}{}^{-\dagger}i\partial^+\psi_{TL}{}^- - \psi_{TL}{}^{+\dagger}\gamma^0 i\gamma^j\partial^j\psi_{TL}{}^- - \psi_{TL}{}^{-\dagger}\gamma^0 i\gamma^j\partial^j\psi_{TL}{}^+ - \\ & - 2^{½}\psi_{TR}{}^{+\dagger}i\partial^-\psi_{TR}{}^+ - 2^{½}\psi_{TR}{}^{-\dagger}i\partial^+\psi_{TR}{}^- + \psi_{TR}{}^{+\dagger}\gamma^0 i\gamma^j\partial^j\psi_{TR}{}^- + \psi_{TR}{}^{-\dagger}\gamma^0 i\gamma^j\partial^j\psi_{TR}{}^+ - \\ & - im[\psi_{TR}{}^{+\dagger}\gamma^0\psi_{TL}{}^- - \psi_{TL}{}^{+\dagger}\gamma^0\psi_{TR}{}^- + \psi_{TR}{}^{-\dagger}\gamma^0\psi_{TL}{}^+ - \psi_{TL}{}^{-\dagger}\gamma^0\psi_{TR}{}^+] \end{aligned} \tag{5.60}$$

with implied sums over j = 1,2. In contrast to the light-front tachyon lagrangian we note the corresponding light-front "normal" Dirac fermion lagrangian is

$$\begin{aligned} \mathcal{L}_{Dirac} = {} & 2^{½}\psi_L{}^{+\dagger}i\partial^-\psi_L{}^+ + 2^{½}\psi_L{}^{-\dagger}i\partial^+\psi_L{}^- - \psi_L{}^{+\dagger}\gamma^0 i\gamma^j\partial^j\psi_L{}^- - \psi_L{}^{-\dagger}\gamma^0 i\gamma^j\partial^j\psi_L{}^+ - \\ & - 2^{½}\psi_R{}^{+\dagger}i\partial^-\psi_R{}^+ + 2^{½}\psi_R{}^{-\dagger}i\partial^+\psi_R{}^- - \psi_R{}^{+\dagger}\gamma^0 i\gamma^j\partial^j\psi_R{}^- - \psi_R{}^{-\dagger}\gamma^0 i\gamma^j\partial^j\psi_R{}^+ - \\ & - im[\psi_R{}^{+\dagger}\gamma^0\psi_L{}^- + \psi_L{}^{+\dagger}\gamma^0\psi_R{}^- + \psi_R{}^{-\dagger}\gamma^0\psi_L{}^+ + \psi_L{}^{-\dagger}\gamma^0\psi_R{}^+] \end{aligned} \tag{5.61}$$

The difference in signs between these lagrangians will turn out to be a crucial factor in the derivation of features of the Standard Model later.

Returning to the tachyon lagrangian eq. 5.60 we obtain equations of motion through the standard variational techniques:

$$\begin{aligned} 2^{½}i\partial^-\psi_{TL}{}^+ - \gamma^0 i\gamma^j\partial^j\psi_{TL}{}^- + im\gamma^0\psi_{TR}{}^- &= 0 \\ 2^{½}i\partial^-\psi_{TR}{}^+ - \gamma^0 i\gamma^j\partial^j\psi_{TR}{}^- + im\gamma^0\psi_{TL}{}^- &= 0 \\ 2^{½}i\partial^+\psi_{TL}{}^- - \gamma^0 i\gamma^j\partial^j\psi_{TL}{}^+ + im\gamma^0\psi_{TR}{}^+ &= 0 \end{aligned} \tag{5.62}$$

$$2^{½}i\partial^{+}\psi_{TR}^{\;-} - \gamma^{0}i\gamma^{j}\partial^{j}\psi_{TR}^{\;+} + im\gamma^{0}\psi_{TL}^{\;+} = 0$$

Eqs. 5.62 show that $\psi_{TL}^{\;-}$ and $\psi_{TR}^{\;-}$ are dependent fields that are functions of $\psi_{TL}^{\;+}$ and $\psi_{TR}^{\;+}$ on the light-front where x^{+} equals a constant. They can be expressed in an integral form as well. (The independent fields $\psi_{TL}^{\;+}$ and $\psi_{TR}^{\;+}$ play a fundamental role in tachyon theory and are used to define "in" and "out" tachyon states in perturbation theory.)

The conjugate momenta are

$$\pi_{TL}^{\;+} = \partial\mathcal{L}/\partial(\partial^{-}\psi_{TL}^{\;+}) = 2^{½}i\psi_{TL}^{\;+\dagger} \tag{5.63}$$
$$\pi_{TL}^{\;-} = \partial\mathcal{L}/\partial(\partial^{-}\psi_{TL}^{\;-}) = 0$$
$$\pi_{TR}^{\;+} = \partial\mathcal{L}/\partial(\partial^{-}\psi_{TR}^{\;+}) = -2^{½}i\psi_{TR}^{\;+\dagger} \tag{5.64}$$
$$\pi_{TR}^{\;-} = \partial\mathcal{L}/\partial(\partial^{-}\psi_{TR}^{\;-}) = 0$$

Quantization on surfaces of constant x^{+} (light-front surfaces) has been shown to support satisfactory formulations of Quantum Electrodynamics and other quantum field theories. Thus x^{+} plays the role of the "time" variable in light-front quantized theories. So we will define canonical equal x^{+} anti-commutation relations for spin ½ tachyons.

The resulting canonical equal-light-front ($x^{+} = y^{+}$) anti-commutation relations of the independent fields are:

$$\{\psi_{TL}^{\;+\dagger}{}_{a}(x), \psi_{TL}^{\;+}{}_{b}(y)\} = 2^{-1}[C^{-}R^{+}]_{ab}\,\delta(x^{-} - y^{-})\delta^{2}(x - y) \tag{5.65}$$
$$\{\psi_{TR}^{\;+\dagger}{}_{a}(x), \psi_{TR}^{\;+}{}_{b}(y)\} = -2^{-1}[C^{+}R^{+}]_{ab}\,\delta(x^{-} - y^{-})\delta^{2}(x - y) \tag{5.66}$$
$$\{\psi_{TL}^{\;+\dagger}{}_{a}(x), \psi_{TR}^{\;+}{}_{b}(y)\} = \{\psi_{TR}^{\;+\dagger}{}_{a}(x), \psi_{TL}^{\;+}{}_{b}(y)\} = 0 \tag{5.67}$$
$$\{\psi_{TL}^{\;+}{}_{a}(x), \psi_{TR}^{\;+}{}_{b}(y)\} = \{\psi_{TR}^{\;+\dagger}{}_{a}(x), \psi_{TL}^{\;+\dagger}{}_{b}(y)\} = 0 \tag{5.68}$$

where the factors of 2^{-1} are the result of the $2^{½}$ factor in eqs. 5.63 and 5.64, and the factor of $2^{-½}$ in the definition of x^{-} above.

If we compare eqs. 5.65 and 5.66 with the corresponding anti-commutation relations of *conventional* Dirac quantum fields:

$$\{\psi_{L}^{\;+\dagger}{}_{a}(x), \psi_{L}^{\;+}{}_{b}(y)\} = 2^{-1}[C^{-}R^{+}]_{ab}\,\delta(x^{-} - y^{-})\delta^{2}(x - y) \tag{5.69}$$
$$\{\psi_{R}^{\;+\dagger}{}_{a}(x), \psi_{R}^{\;+}{}_{b}(y)\} = 2^{-1}[C^{+}R^{+}]_{ab}\,\delta(x^{-} - y^{-})\delta^{2}(x - y) \tag{5.70}$$

we see that the right-handed tachyon anti-commutation relation has a minus sign relative to the corresponding right-handed conventional anti-commutation relation. The right-handed tachyon anti-commutation relation with its minus sign will require compensating minus signs in its creation and annihilation Fourier component operators' anti-commutation relations.

The sign differences between the lagrangian terms in eqs. 5.63 and 5.64 ultimately lead to parity violating features in the Standard Model lagrangian and thus resolve the long-standing question:

Why parity violation? Answer: Nature preferentially chooses the Left-handed part of the complex Lorentz group.. This choice is not a consequence of Ockham's Razor. But it does conform to

Leibniz's Minimax Principle – a minor differentiation based on parity results in "maximal" physical consequences.

Left-Handed Tachyons

The free, "+" light-front, left-handed tachyon wave function Fourier expansion is:

$$\psi_{TL}^{+}(x) = \sum_{\pm s}\int d^2p dp^+ N_{TL}^{+}(p)\theta(p^+)[b_{TL}^{+}(p, s)u_{TL}^{+}(p, s)e^{-ip\cdot x} + d_{TL}^{+\dagger}(p, s)v_{TL}^{+}(p, s)e^{+ip\cdot x}] \quad (5.71)$$

and its hermitean conjugate is

$$\psi_{TL}^{+\dagger}(x) = \sum_{\pm s}\int d^2p dp^+ N_{TL}^{+}(p)\theta(p^+) [b_{TL}^{+\dagger}(p, s)u_{TL}^{+\dagger}(p,s)e^{+ip\cdot x} + d_{TL}^{+}(p, s)v_{TL}^{+\dagger}(p, s)e^{-ip\cdot x}] \quad (5.72)$$

where † indicates hermitean conjugate, where

$$N_{TL}^{+}(p) = [2m|\mathbf{p}|/((2\pi)^3(p^+(p^+ - p^-) + p_\perp^{2}))]^{1/2} \quad (5.73)$$

where the anti-commutation relations of the Fourier coefficient operators are

$$\begin{aligned}
\{b_{TL}^{+}(q,s), b_{TL}^{+\dagger}(p,s')\} &= \delta_{ss'}\delta^2(\mathbf{q} - \mathbf{p})\delta(q^+ - p^+)\\
\{d_{TL}^{+}(q,s), d_{TL}^{+\dagger}(p,s')\} &= \delta_{ss'}\delta^2(\mathbf{q} - \mathbf{p})\delta(q^+ - p^+)\\
\{b_{TL}^{+}(q,s), b_{TL}^{+}(p,s')\} &= \{d_{TL}^{+}(q,s), d_{TL}^{+}(p,s')\} = 0\\
\{b_{TL}^{+\dagger}(q,s), b_{TL}^{+\dagger}(p,s')\} &= \{d_{TL}^{+\dagger}(q,s), d_{TL}^{+\dagger}(p,s')\} = 0\\
\{b_{TL}^{+}(q,s), d_{TL}^{+\dagger}(p,s')\} &= \{d_{TL}^{+}(q,s), b_{TL}^{+\dagger}(p,s')\} = 0\\
\{b_{TL}^{+\dagger}(q,s), d_{TL}^{+\dagger}(p,s')\} &= \{d_{TL}^{+}(q,s), b_{TL}^{+}(p,s')\} = 0
\end{aligned} \quad (5.74)$$

and where the spinors are

$$\begin{aligned}
u_{TL}^{+}(p, s) &= C^- R^+ S_L(\Lambda_L(\mathbf{p}))w^1(0)\\
u_{TL}^{+}(p, -s) &= C^- R^+ S_L(\Lambda_L(\mathbf{p}))w^2(0)\\
v_{TL}^{+}(p, s) &= C^- R^+ S_L(\Lambda_L(\mathbf{p}))w^3(0)\\
v_{TL}^{+}(p, -s) &= C^- R^+ S_L(\Lambda_L(\mathbf{p}))w^4(0)\\
\\
u_{TL}^{+\dagger}(p, s) &= w^{1T}(0)S_L^{\dagger}(\Lambda_L(\mathbf{p}))R^+C^-\\
u_{TL}^{+\dagger}(p, -s) &= w^{2T}(0)S_L^{\dagger}(\Lambda_L(\mathbf{p}))R^+C^-\\
v_{TL}^{+\dagger}(p, s) &= w^{3T}(0)S_L^{\dagger}(\Lambda_L(\mathbf{p}))R^+C^-\\
v_{TL}^{+\dagger}(p, -s) &= w^{4T}(0)S_L^{\dagger}(\Lambda_L(\mathbf{p}))R^+C^-
\end{aligned} \quad (5.75)$$

where the superscript "T" indicates the transpose. (These spinors are described in Appendix 5-A.)

The canonical left-handed, light-front anti-commutation relation results in:

$$\{\psi_{TL}{}^{+}{}_{a}(x), \psi_{TL}{}^{+\dagger}{}_{b}(y)\} = \sum_{\pm s,s'} \int d^2p dp^+ \int d^2p' dp'^+ N_{TL}{}^{+}(p) N_{TL}{}^{+}(p') \theta(p^+)\theta(p'^+) \cdot$$

$$\cdot[\{b_{TL}{}^{+\dagger}(p',s'), b_{TL}{}^{+}(p,s)\} u_{TL}{}^{+}{}_{a}(p,s) u_{TL}{}^{+\dagger}{}_{b}(p',s') e^{+ip'\cdot y - ip\cdot x} +$$

$$+ \{d_{TL}{}^{+}(p',s'), d_{TL}{}^{+\dagger}(p,s)\} v_{TL}{}^{+}{}_{a}(p,s) v_{TL}{}^{+\dagger}{}_{b}(p',s') e^{-ip'\cdot y + ip\cdot x}]$$

$$= \sum_{\pm s} \int d^2p dp^+ N_{TL}{}^{+2}(p) \theta(p^+)[u_{TL}{}^{+}{}_{a}(p,s) u_{TL}{}^{+\dagger}{}_{b}(p,s) e^{+ip\cdot(y-x)} +$$

$$+ v_{TL}{}^{+}{}_{a}(p,s) v_{TL}{}^{+\dagger}{}_{b}(p,s) e^{-ip\cdot(y-x)}]$$

$$= -i\int d^2p dp^+ \theta(p^+) N_{TL}{}^{+2}(p)(2m|\mathbf{p}|)^{-1}\{[C^-R^+(i\not{p} - m)\gamma\cdot\mathbf{p}R^+C^-]_{ab} e^{+ip\cdot(y-x)} +$$

$$+ [C^-R^+(i\not{p} + m)\gamma\cdot\mathbf{p}R^+C^-]_{ab} e^{-ip\cdot(y-x)}\}$$

$$= -i\int d^2p_\perp \int_0^\infty dp^+ N_{TL}{}^{+2}(p)\{[C^-R^+(ip^+(p^+ - p^-) + ip_\perp{}^2 - mp_\perp\cdot\gamma_\perp)C^-]_{ab} e^{+ip^+(y^- - x^-) - ip_\perp\cdot(y_\perp - x_\perp)} -$$

$$- [C^-R^+(-ip^+(p^+ - p^-) - ip_\perp{}^2 - mp_\perp\cdot\gamma_\perp)C^-]_{ab} e^{-ip^+(y^- - x^-) + ip_\perp\cdot(y_\perp - x_\perp)}\}/(2m|\mathbf{p}|)$$

$$= \int d^2p_\perp \int_{-\infty}^\infty dp^+ N_{TL}{}^{+2}(p)[C^-R^+(p^+(p^+ - p^-) + p_\perp{}^2)]_{ab}\, e^{+ip^+(y^- - x^-) - ip_\perp\cdot(y_\perp - x_\perp)}/(2m|\mathbf{p}|)$$

upon letting $p^+ \rightarrow -p^+$ and $\mathbf{p}_\perp \rightarrow -\mathbf{p}_\perp$ in the second term after using $N_{TL}{}^{+2}(p)(p^+(p^+ - p^-) + p_\perp{}^2) = 1$. The result

$$= \tfrac{1}{2}\int d^2p_\perp \int_{-\infty}^\infty dp^+ (2\pi)^{-3}[C^-R^+]_{ab} e^{+ip^+(y^- - x^-) - ip_\perp\cdot(y_\perp - x_\perp)}$$

$$= 2^{-1}[C^-R^+]_{ab}\,\delta(y^- - x^-)\delta^2(\mathbf{y} - \mathbf{x}) \tag{5.76}$$

Therefore we have left-handed, light-front quantized tachyons with canonical commutation relations and localized tachyons. As a result we have a canonical Tachyon Quantum Field Theory unlike previous efforts.

Right-Handed Tachyons

The case of right-handed tachyons is similar to the left-handed case with only two differences: a minus sign in the creation and annihilation operator anti-commutation relations, and the use of right-handed projection operators. The right-handed tachyon wave function light-front Fourier expansion is:

$$\psi_{TR}{}^{+}(x) = \sum_{\pm s} \int d^2p dp^+ N_{TR}{}^{+}(p)\theta(p^+)[b_{TR}{}^{+}(p, s) u_{TR}{}^{+}(p, s) e^{-ip\cdot x} + d_{TR}{}^{+\dagger}(p, s) v_{TR}{}^{+}(p, s) e^{+ip\cdot x}] \tag{5.77}$$

and its hermitean conjugate is

$$\psi_{TR}{}^{+\dagger}(x) = \sum_{\pm s}\int d^2p dp^+ N_{TR}{}^{+}(p)\theta(p^+)\,[b_{TR}{}^{+\dagger}(p, s)u_{TR}{}^{+\dagger}(p, s)e^{+ip\cdot x} + d_{TR}{}^{+}(p, s)v_{TR}{}^{+\dagger}(p, s)e^{-ip\cdot x}] \qquad (5.78)$$

where $N_{TR}{}^{+}(p) = N_{TL}{}^{+}(p)$, where the anti-commutation relations of the Fourier coefficient operators are

$$\begin{aligned}
&\{b_{TR}{}^{+}(q,s), b_{TR}{}^{+\dagger}(p,s')\} = -\delta_{ss'}\delta^2(\mathbf{q}-\mathbf{p})\delta(q^+ - p^+) \\
&\{d_{TR}{}^{+}(q,s), d_{TR}{}^{+\dagger}(p,s')\} = -\delta_{ss'}\delta^2(\mathbf{q}-\mathbf{p})\delta(q^+ - p^+) \\
&\{b_{TR}{}^{+}(q,s), b_{TR}{}^{+}(p,s')\} = \{d_{TR}{}^{+}(q,s), d_{TR}{}^{+}(p,s')\} = 0 \\
&\{b_{TR}{}^{+\dagger}(q,s), b_{TR}{}^{+\dagger}(p,s')\} = \{d_{TR}{}^{+\dagger}(q,s), d_{TR}{}^{+\dagger}(p,s')\} = 0 \\
&\{b_{TR}{}^{+}(q,s), d_{TR}{}^{+\dagger}(p,s')\} = \{d_{TR}{}^{+}(q,s), b_{TR}{}^{+\dagger}(p,s')\} = 0 \\
&\{b_{TR}{}^{+\dagger}(q,s), d_{TR}{}^{+\dagger}(p,s')\} = \{d_{TR}{}^{+}(q,s), b_{TR}{}^{+}(p,s')\} = 0
\end{aligned} \qquad (5.79)$$

and where the spinors are

$$u_{TR}{}^{+}(p, s) = C^+R^+u_T(p,s) \qquad (5.80)$$

$$v_{TR}{}^{+}(p, s) = C^+R^+v_T(p,s) \qquad (5.81)$$

by Appendix 5-A (eq. 5-A.7).

The right-handed anti-commutation relation with the minus sign follows in particular because of the minus signs found earlier.

5.4.5 Interpretation of Tachyon Creation and Annihilation Operators

To properly discuss the physical interpretation of tachyon creation and annihilation operators we must first determine the Hamiltonian and momentum operators in terms of creation and annihilation operators.

The energy-momentum tensor density is the symmetrized version of

$$\mathfrak{T}^{\mu\nu} = \sum_i \partial\mathcal{L}/\partial(\partial\chi_i/\partial x_\mu)\,\partial\chi_i/\partial x_\nu - g^{\mu\nu}\mathcal{L} \qquad (5.82)$$

where the sum over i is over the fields. The light-front hamiltonian is

$$H \equiv P^- = T^{+-} = \int dx^- d^2x\,\mathfrak{T}^{+-} \qquad (5.83)$$

and the "momenta" are

$$P^+ = T^{++} = \int dx^- d^2x\,\mathfrak{T}^{++} \qquad (5.84)$$

$$P^i = T^{+i} = \int dx^- d^2x\,\mathfrak{T}^{+i} \qquad (5.85)$$

for i = 1,2.

The light-front, left-handed and right-handed tachyon lagrangian $\mathcal{L}_T$ and its equations of motion imply

$$\begin{aligned}
H = i2^{-1/2}\int dx^- d^2x\,[&\psi_{TL}{}^{+\dagger}\partial^-\psi_{TL}{}^{+} - \partial^-\psi_{TL}{}^{+\dagger}\psi_{TL}{}^{+} + \psi_{TL}{}^{-\dagger}\partial^+\psi_{TL}{}^{-} - \partial^+\psi_{TL}{}^{-\dagger}\psi_{TL}{}^{-} - \\
&- \psi_{TR}{}^{+\dagger}\partial^-\psi_{TR}{}^{+} + \partial^-\psi_{TR}{}^{+\dagger}\psi_{TR}{}^{+} - \psi_{TR}{}^{-\dagger}\partial^+\psi_{TR}{}^{-} + \partial^+\psi_{TR}{}^{-\dagger}\psi_{TR}{}^{-} + \text{mass terms}]
\end{aligned} \qquad (5.86)$$

After substituting for the various fields we find the *independent fields* (which create the in and out particle states) have the hamiltonian terms:

$$H = \sum_{\pm s} \int d^2p dp^+ \, p^- [b_{TL}{}^{+\dagger}(p,s) b_{TL}{}^{+}(p,s) - d_{TL}{}^{+}(p,s) d_{TL}{}^{+\dagger}(p,s) - b_{TR}{}^{+\dagger}(p,s) b_{TR}{}^{+}(p,s) + d_{TR}{}^{+}(p,s) d_{TR}{}^{+\dagger}(p,s)] \tag{5.87}$$

$$= \sum_{\pm s} \int d^2p dp^+ \, p^- [b_{TL}{}^{+\dagger}(p,s) b_{TL}{}^{+}(p,s) + d_{TL}{}^{+\dagger}(p,s) d_{TL}{}^{+}(p,s) - b_{TR}{}^{+\dagger}(p,s) b_{TR}{}^{+}(p,s) - d_{TR}{}^{+\dagger}(p,s) d_{TR}{}^{+}(p,s)] \tag{5.88}$$

up to the usual infinite constants due to left-handed operator rearrangement and right-handed operator rearrangement that are discarded. Eq. 5.88 is the basis for our particle interpretation of tachyon creation and annihilation operators based on Dirac's hole theory. Dirac hole theory as applied in light-front coordinates assumes all negative p^- ("energy") states are filled.

Left-Handed Tachyon Creation and Annihilation Operators

1. We identify $b_{TL}{}^{+\dagger}(p,s)$ and $d_{TL}{}^{+}(p,s)$ as creation operators for left-handed tachyons. $b_{TL}{}^{+\dagger}(p,s)$ creates a positive p^- ("energy") state and $d_{TL}{}^{+}(p,s)$ creates a negative p^- ("energy") state.

2. $b_{TL}{}^{+}(p,s)$ and $d_{TL}{}^{+\dagger}(p,s)$ are the corresponding annihilation operators for left-handed tachyons. $b_{TL}{}^{+}(p,s)$ annihilates a positive p^- ("energy") state and $d_{TL}{}^{+\dagger}(p,s)$ annihilates a negative p^- ("energy") state.

3. We assume Dirac hole theory holds for the left-handed tachyon vacuum with all negative energy states filled. There is no tachyon energy gap as there is for Dirac fermions. There is also the problem that the left-handed tachyon vacuum is not invariant under ordinary Lorentz transformations or Superluminal transformations. *However if we confine ourselves to light-front coordinates for computations no ambiguity can result and the Lorentz covariant quantities that we calculate, such as the S matrix, are well-defined.*

4. Using tachyon hole theory we identify $b_{TL}{}^{+}(p,s)$ and $d_{TL}{}^{+\dagger}(p,s)$ as annihilation operators for left-handed tachyons. $b_{TL}{}^{+}(p,s)$ annihilates a positive p^- ("energy") state and $d_{TL}{}^{+\dagger}(p,s)$ annihilates a negative p^- ("energy") state – thus creating a hole in the tachyon sea that we view as the creation of a positive p^- ("energy"), left-handed antitachyon. $d_{TL}{}^{+}(p,s)$ annihilates a positive p^- ("energy"), left-handed antitachyon.

Right-Handed Tachyon Creation and Annihilation Operators

The anti-commutation relations of right-handed tachyon creation and annhilation operators and the right-handed Hamiltonian terms have the "wrong" sign compared to corresponding Dirac operators and left-handed tachyon operators. This situation is completely analogous to the situation of time-like photons in the covariant formulation of quantum Electrodynamics.[73] In the case of time-like photons it

[73] Bogoliubov (1959) pp. 130-136.

was possible to introduce an indefinite metric (Gupta-Bleuler formulation), and then to use the subsidiary condition $\partial A^\nu/\partial x^\nu = 0$ to reduce the dynamics of QED to the transverse components. Thus the time-like photons were intermediate artifacts needed to have a manifestly covariant formulation while QED observables depended solely on the transverse components of the electromagnetic field.

In the present case of free tachyons, and in leptonic ElectroWeak Theory there is no evident "subsidiary condition" to eliminate the right-handed tachyon fields. But since the only manner in which the right-handed leptonic tachyon fields[74] interact is through mass terms, which can be easily 'integrated out", right-handed leptonic tachyon fields are removed from the observable part of the leptonic ElectroWeak Theory by their "lack of interaction" with left-handed fields.

In the case of quark ElectroWeak Theory right-handed tachyon quark fields have charge (–1/3) and thus experience an electromagnetic interaction as well as a Z interaction. However, since quarks are totally confined, right-handed tachyon quarks will not be able to continuously emit photons or Z's due to energy conservation and their confinement to bound states of fixed positive energy. Earlier, when we consider complex Lorentz group boosts, we will suggest that quarks may not consist of Dirac particles or tachyons of the type considered up to this point in this chapter. Rather they may be variants on Dirac particles and tachyons satisfying different dynamical equations. However, the preceding comments on quarks would still apply.

Thus right-handed tachyons are analogous to time-like photons – necessary theoretically but prevented from causing a negative energy disaster by the forms of their interactions. We discuss this subject in more detail in the following chapters.

5.4.6 Tachyon Feynman Propagator

In this section we develop the light-front propagator for tachyons. We begin with a subsection describing the light-front propagators of Dirac fields.

Dirac Field Light-Front Propagators

The light-front Feynman propagator for the ψ^+ field of a Dirac fermion is

$$iS^+_F(x,y)\gamma^0 = \theta(x^+ - y^+)<0|\psi^+(x)\psi^{+\dagger}(y)|0> - \theta(y^+ - x^+)<0|\psi^{+\dagger}(y)\psi^+(x)|0> \quad (5.89)$$

and does not contain a non-covariant piece due to the projection operators:

$$iS^+_F(x,y) = \int d^2p dp^+\theta(p^+)[1/(2(2\pi)^3p^+)]\{\theta(x^+- y^+)[R^+(\not{p} +m)R^-]\, e^{-ip\cdot(x-y)} + \theta(y^+-x^+)[R^+(-\not{p}+m)R^-]e^{+ip\cdot(x-y)}\}$$
$$= R^+iS_F(x,y)R^- \quad (5.90)$$

where $S_F(x,y)$ is the usual Feynman propagator.

The light-front Feynman propagator for a *left-handed* <u>Dirac</u> field ψ^+ is

$$iS^+_{LF}(x,y) = \int d^2p dp^+\theta(p^+)[1/(2(2\pi)^3p^+)]\{\theta(x^+-y^+)[C^-R^+(\not{p}+m)R^-C^-]e^{-ip\cdot(x-y)} + $$
$$+ \theta(y^+- x^+)[C^-R^+(-\not{p} + m)R^-C^-]e^{+ip\cdot(x-y)}\}$$

$$= C^-R^+iS_F(x,y)R^-C^- \quad (5.91)$$

---

[74] The tachyon fields are provisionally assumed to be neutrino fields in the leptonic sector, and d, s and b quarks in the quark sector.

Tachyon Field Light Front Propagators

Turning now to tachyons, the light-front Feynman propagator for the left-handed ψ_{TL}^{+} *tachyon* field is (using the previous Fourier expansion of the left-handed tachyon field):

$$iS^{+}_{TLF}(x,y) = \theta(x^+ - y^+)<0|\psi_{TL}^{+}(x)\psi_{TL}^{++}(y)\gamma^0|0> - \theta(y^+ - x^+)<0|\psi_{TL}^{++}(y)\gamma^0\psi_{TL}^{+}(x)|0>$$
$$= -i\int d^2p dp^+ \theta(p^+) N_{TL}^{+2}(2m|\mathbf{p}|)^{-1} C^- R^+ \{\theta(x^+ - y^+)[(i\not p - m)\gamma\cdot\mathbf{p}]e^{-ip\cdot(x-y)} +$$
$$+ \theta(y^+ - x^+)[(i\not p + m)\gamma\cdot\mathbf{p}]e^{+ip\cdot(x-y)}\} R^+ C^- \gamma^0$$

If we define the on-shell momentum variable

$$p_0^- = (p_0^1 p_0^1 + p_0^2 p_0^2 - m^2)/(2p_0^+),\ p_0^+ = p^+,\ p_0^j = p^j \text{ (for } j = 1, 2),\ p_{\perp 0}^2 = p_0^j p_0^j \text{ and } \not p_0 = p_0\cdot\gamma$$

then the above equation can be rewritten as

$$iS^{+}_{TLF}(x,y) = -iC^- R^+ \int d^4p[32\pi^4(p_0^+(p_0^+ - p_0^-) + p_{0\perp}^2)]^{-1} e^{-ip\cdot(x-y)} \{\theta(p^+)(i\not p_0 - m)\gamma\cdot\mathbf{p}_0]/[p^- - p_0^- + i\varepsilon] +$$
$$+ \theta(-p^+)(i\not p_0 + m)\gamma\cdot\mathbf{p}_0]/[p^- + p_0^- - i\varepsilon]\} R^+ C^- \gamma^0$$
$$= -\tfrac{1}{2}\, i\int d^4p(2\pi)^{-4}[C^- R^+(i\not p - m)\gamma\cdot\mathbf{p}R^+ C^- \gamma^0]e^{-ip\cdot(x-y)}[(p^2 + m^2 + i\varepsilon)(p^+(p^+ - p^-) + p_\perp^2))]^{-1}$$

and using $C^- R^+(i\not p - m)\gamma\cdot\mathbf{p}R^+ C^- = i\, C^- R^+(p^+(p^+ - p^-) + p_\perp^2)$ we find

$$iS^{+}_{TLF}(x,y) = \tfrac{1}{2} C^- R^+ \gamma^0 \int d^4p(2\pi)^{-4}\, p^+ e^{-ip\cdot(x-y)}/(p^2 + m^2 + i\varepsilon) \tag{5.92}$$

Similarly the light-front Feynman propagator for the right-handed ψ_{TR}^{+} tachyon field is

$$iS^{+}_{TRF}(x,y) = \theta(x^+ - y^+)<0|\psi_{TR}^{+}(x)\psi_{TR}^{++}(y)\gamma^0|0> - \theta(y^+ - x^+)<0|\psi_{TR}^{++}(y)\gamma^0\psi_{TR}^{+}(x)|0>$$
$$= -\tfrac{1}{2} C^+ R^+ \gamma^0 \int d^4p(2\pi)^{-4}\, p^+ e^{-ip\cdot(x-y)}/(p^2 + m^2 + i\varepsilon) \tag{5.93}$$

where the relative minus sign between eqs. 5.92 and 5.93 is due to the relative minus signs of the Fouier component operator anti-commutation relations.

Thus we find *tachyon* pole terms in the tachyon propagators as one would expect.

5.5 Complex Space and 3-Momentum & Real-Valued Energy Fermions (Quarks)

In this section we will use L_C boosts to develop a wider set of dynamical equations for free spin ½ fermions with real-valued energy and complex-valued 3-momentum.[75] Earlier we defined L_C boosts with

---

[75] The complexon theory that we develop and use for quark dynamics in the Standard Model is not required. Our Standard Model could use Dirac fermion dynamics for the up-type quarks and tachyon dynamics for down-type quarks. We choose to use complexon dynamics for all quark types because they have an internal SU(3)-like structure

$$\Lambda_C(\mathbf{v}_c) = \exp[i\omega\hat{\mathbf{w}}\cdot\mathbf{K}] \tag{5.94}$$
$$\omega = (\omega_r^2 - \omega_i^2 + 2i\omega_r\omega_i\, \hat{\mathbf{u}}_r\cdot\hat{\mathbf{u}}_i)^{1/2} \tag{5.95}$$
$$\hat{\mathbf{w}} = (\omega_r\hat{\mathbf{u}}_r + i\omega_i\hat{\mathbf{u}}_i)/\omega \tag{5.96}$$
$$\hat{\mathbf{w}}\cdot\hat{\mathbf{w}} = \hat{\mathbf{u}}_r\cdot\hat{\mathbf{u}}_r = \hat{\mathbf{u}}_i\cdot\hat{\mathbf{u}}_i = 1 \tag{5.97}$$
$$\mathbf{v}_c = \hat{\mathbf{w}}\tanh(\omega) \tag{5.98}$$

5.5.1 L_C Spinor "Normal" Lorentz Boosts & More Spin ½ Particle Types

Spinor boost transformations were used in previous sections to develop the dynamical equations for Dirac fields and tachyon fields. In this section we will use L_C spinor boosts to generate additional fermion field dynamical equations.

The form of the L_C spinor boost transformation corresponding to the coordinate transformation is:

$$\begin{aligned} S_C(\omega, \mathbf{v}_c) &= \exp(-i\omega\sigma_{0k}\hat{w}_k/2) = \exp(-\omega\gamma^0\boldsymbol{\gamma}\cdot\hat{\mathbf{w}}/2) \\ &= \cosh(\omega/2)I + \sinh(\omega/2)\gamma^0\boldsymbol{\gamma}\cdot\hat{\mathbf{w}} \end{aligned} \tag{5.99}$$

The inverse transformation is

$$\begin{aligned} S_C^{-1}(\omega, \mathbf{v}_c) &= \gamma^2\gamma^0 K^{-1}S_C^{\dagger}K\gamma^0\gamma^2 = \gamma^2\gamma^0 S_C^{T}\gamma^0\gamma^2 = \exp(\omega\gamma^0\boldsymbol{\gamma}\cdot\hat{\mathbf{w}}/2) \\ &= \cosh(\omega/2)I - \sinh(\omega/2)\gamma^0\boldsymbol{\gamma}\cdot\hat{\mathbf{w}} \end{aligned} \tag{5.100}$$

where the superscript T denotes the transpose and K is the complex conjugation operator (that also appears in the time-reversal operator). Note that S_C is not unitary just as in previous cases considered in this chapter.

We now redo the development of spin ½ dynamical equations of motion of earlier sections for this more general case of complex ω and $\hat{\mathbf{w}}$. Again we apply a boost to a Dirac equation for a positive energy plane wave particle of mass m at rest:

$$\begin{aligned} 0 &= S_C(\omega, \mathbf{v}_c))(m\gamma^0 - m)e^{-imt}w(0) \\ &= [mS_C\gamma^0 S_C^{-1} - m]e^{-imt}S_C w(0) \end{aligned} \tag{5.101}$$

where $S_C = S_C(\omega, \hat{\mathbf{w}})$. After some algebra

$$mS_C\gamma^0 S_C^{-1} = m[\cosh(\omega)\gamma^0 - \sinh(\omega)\boldsymbol{\gamma}\cdot\hat{\mathbf{w}}] \tag{5.102}$$

Case 1: Parallel Real and Imaginary Relative Vectors

If the real and imaginary relative vectors parts of $\hat{\mathbf{w}}$, namely $\hat{\mathbf{u}}_r$ and $\hat{\mathbf{u}}_i$, are parallel, then $\hat{\mathbf{u}}_r\cdot\hat{\mathbf{u}}_i = 1$ and

$$\omega = \omega_r + i\omega_i \tag{5.103}$$

Eq. 5.102 can be re-expressed as

$$mS_C\gamma^0 S_C^{-1} = m[\cosh(\omega_r)\cos(\omega_i) + i\sinh(\omega_r)\sin(\omega_i)]\gamma^0 - m[\sinh(\omega_r)\cos(\omega_i) + i\cosh(\omega_r)\sin(\omega_i)]\boldsymbol{\gamma}\cdot\hat{\mathbf{u}}_r \tag{5.104}$$

or equivalently

$$mS_C\gamma^0 S_C^{-1} = \cos(\omega_i)\gamma\cdot p_r + i\sin(\omega_i)\gamma\cdot p_i \tag{5.105}$$

---

suggestive of color SU(3). More importantly, their spin dynamics is different and thus may resolve the differences between theory and experiment – particularly for the deep inelastic parton spin-dependent structure functions.

where

$$p_r^0 = m\cosh(\omega_r) \qquad p_i^0 = m\sinh(\omega_r) \tag{5.106}$$

and

$$\mathbf{p}_r = m\hat{\mathbf{u}}_r \sinh(\omega_r) \qquad \mathbf{p}_i = m\hat{\mathbf{u}}_r \cosh(\omega_r) \tag{5.107}$$

If $\omega_i = 0$, then we recover the momentum space Dirac equation. If $\omega_i = \pi/2$, then we obtain the left-handed momentum space tachyon equation. Since the range of ω_i is $[0, \infty>$ (due to the cut along the real ω-plane axis) eq. 5.105 corresponds to the results of the Left-Handed Lorentz boost part discussed earlier.

Case 2: Anti-Parallel Real and Imaginary Relative Vectors

If the real and imaginary relative vectors parts of $\hat{\mathbf{w}}$, $\hat{\mathbf{u}}_r$ and $\hat{\mathbf{u}}_i$, are anti-parallel $\hat{\mathbf{u}}_r = -\hat{\mathbf{u}}_i$, then $\hat{\mathbf{u}}_r \cdot \hat{\mathbf{u}}_i = -1$ and

$$\omega = \omega_r - i\omega_i \tag{5.108}$$

We can then express eq. 5.105 as

$$mS_C\gamma^0 S_C^{-1} = m[\cosh(\omega_r)\cos(\omega_i) - i\sinh(\omega_r)\sin(\omega_i)]\gamma^0 - m[\sinh(\omega_r)\cos(\omega_i) - i\cosh(\omega_r)\sin(\omega_i)]\gamma\cdot\hat{\mathbf{u}}_r \tag{5.109}$$

or

$$mS_C\gamma^0 S_C^{-1} = \cos(\omega_i)\gamma\cdot p_r - i\sin(\omega_i)\gamma\cdot p_i \tag{5.110}$$

where

$$p_r^0 = m\cosh(\omega_r) \qquad p_i^0 = m\sinh(\omega_r) \tag{5.111}$$

and

$$\mathbf{p}_r = m\hat{\mathbf{u}}_r \sinh(\omega_r) \qquad \mathbf{p}_i = m\hat{\mathbf{u}}_r \cosh(\omega_r) \tag{5.112}$$

If $\omega_i = 0$, then we again recover the momentum space Dirac equation, If $\omega_i = \pi/2$, then we obtain the right-handed momentum space tachyon equation. (The range of ω_i is again $[0, \infty>$.)

Note: Since the matrix elements in the boost depend on $\gamma = (1 - \beta^2)^{-1/2}$ with a singularities at $\beta = \pm 1$, which in turn corresponds to $\omega = \pm\infty$, there is a branch cut along the ω axis in the complex ω-plane. Therefore we point out again the product of three Left-handed transformations is not equivalent to a Right-handed transformation.

Case 3: Complexons: A New Type of Particle with Perpendicular Real and Imaginary 3-Momenta

If the real and imaginary relative vectors parts of $\hat{\mathbf{w}}$, namely $\hat{\mathbf{u}}_r$ and $\hat{\mathbf{u}}_i$, are perpendicular, $\hat{\mathbf{u}}_r \cdot \hat{\mathbf{u}}_i = 0$, then

$$\omega = (\omega_r^2 - \omega_i^2)^{1/2} \tag{5.113}$$

Thus ω is either pure real ($\omega_r \geq \omega_i$) or pure imaginary ($\omega_r < \omega_i$).

The momentum space equation generated by the corresponding L_C spinor boost is

$$\{m\cosh(\omega)\gamma^0 - m\sinh(\omega)\gamma\cdot(\omega_r\hat{\mathbf{u}}_r + i\omega_i\hat{\mathbf{u}}_i)/\omega - m\}e^{-ip\cdot x}w_c(p) = 0 \tag{5.114}$$

Defining the momentum 4-vector

$$p = (p^0, \mathbf{p}) \tag{5.115}$$

where

$$p^0 = m\cosh(\omega) \qquad \mathbf{p} = \mathbf{p_r} + i\mathbf{p_i} \qquad (5.116)$$
$$\mathbf{p_r} = m\omega_r\hat{\mathbf{u}}_r \sinh(\omega)/\omega \qquad \mathbf{p_i} = m\omega_i\hat{\mathbf{u}}_i \sinh(\omega)/\omega \qquad (5.117)$$

and

$$\mathbf{p_r}\cdot\mathbf{p_i} = 0 \qquad (5.118)$$

then we obtain a positive energy Dirac-like equation with complex 3-momentum

$$[p\cdot\gamma - m]e^{-ip\cdot x}w_c(p) = 0$$

or, explicitly, (5.119)

$$[p^0\gamma^0 - (\mathbf{p_r} + i\mathbf{p_i})\cdot\boldsymbol{\gamma} - m]e^{-ip\cdot x}w_c(p) = 0$$

with a complex 3-momentum **p** and the 4-momentum mass shell condition:

$$p^2 = p^{0\,2} - \mathbf{p_r}\cdot\mathbf{p_r} + \mathbf{p_i}\cdot\mathbf{p_i} = m^2 \qquad (5.120)$$

Note

$$|\mathbf{v}| = |\mathbf{p}|/p^0 = [(\mathbf{p_r} + i\mathbf{p_i})\cdot(\mathbf{p_r} + i\mathbf{p_i})]^{1/2}/p^0 = \tanh(\omega) \qquad (5.121)$$

and thus the Lorentz factor

$$\gamma = \cosh(\omega) \qquad (5.122)$$

Eq. 5.119 is the momentum space equivalent of the wave equation

$$[i\gamma^0\partial/\partial t + i\boldsymbol{\gamma}\cdot(\nabla_r + i\nabla_i) - m]\psi_C(t, \mathbf{x_r}, \mathbf{x_i}) = 0 \qquad (5.123)$$

where

$$x_c = (t, \mathbf{x_r} - i\mathbf{x_i}) \qquad (5.123a)$$

and where the grad operators ∇_r and ∇_i are with respect to $\mathbf{x_r}$ and $\mathbf{x_i}$ respectively. Since $\hat{\mathbf{u}}_r\cdot\hat{\mathbf{u}}_i = 0$, we see that there is a subsidiary condition on the wave function

$$\nabla_r\cdot\nabla_i\,\psi_C(t, \mathbf{x_r}, \mathbf{x_i}) = 0 \qquad (5.124)$$

We will call the particles satisfying eqs.5.123 and 5.124 *complexons*. In addition eq. 5.118 implies the anti-commutation relation

$$\{\boldsymbol{\gamma}\cdot\mathbf{p_r}, \boldsymbol{\gamma}\cdot\mathbf{p_i}\} = 0 \qquad (5.125)$$

which in turn implies

$$\boldsymbol{\gamma}\cdot\nabla_r\boldsymbol{\gamma}\cdot\nabla_i\psi_C(t, \mathbf{x_r}, \mathbf{x_i}) = \boldsymbol{\gamma}\cdot\nabla_i\boldsymbol{\gamma}\cdot\nabla_r\psi_C(t, \mathbf{x_r}, \mathbf{x_i}) = 0 \qquad (5.126)$$

We note that eq. 5.125 is covariant under the real Lorentz group and eq. 5.126 can be easily put into covariant form since the difference of these 4-vectors squared is a real Lorentz group invariant: $[\gamma^0\partial/\partial t + \boldsymbol{\gamma}\cdot(\nabla_r + i\nabla_i)]^2 - [\gamma^0\partial/\partial t + i\boldsymbol{\gamma}\cdot(\nabla_r - i\nabla_i)]^2 = 4\nabla_r\cdot\nabla_i$.

Before considering a lagrangian formulation and the Fourier operator representation of $\psi_C(t, \mathbf{x_r}, \mathbf{x_i})$ we will define the spinors and associated real and imaginary spin operators.

The spinor generated from a spin up Dirac spinor at rest by a complex boost is

$$w_c(p) = S_C(p)w(0) = [\cosh(\omega/2)I + \sinh(\omega/2)\gamma^0\boldsymbol{\gamma}\cdot\hat{\mathbf{w}}]w(0) \quad (5.127)$$

Following a procedure similar to Appendix 5-A (which the reader may wish to examine first) we define four spinors for Dirac particles at rest:

$$w^k(0) = \begin{bmatrix} \delta_{1k} \\ \delta_{2k} \\ \delta_{3k} \\ \delta_{4k} \end{bmatrix} \quad (5\text{-A.2})$$

where Kronecker deltas appear in the brackets. Then by applying eq. 5.127 to the spinors defined by eq. 5-A.2 we find the L_C spinors

$$S_C w^k(0) = w_{Cr}{}^k(p) + i w_{Ci}{}^k(p) \quad (5.128)$$

where

$$\begin{aligned} S_{Cr} &= \cosh(\omega/2)I + (\omega_r/\omega)\sinh(\omega/2)\gamma^0\boldsymbol{\gamma}\cdot\hat{\mathbf{u}}_r \\ &= [(m + E)/(2m)]^{1/2}I + [m(m + E)]^{-1/2}\gamma^0\boldsymbol{\gamma}\cdot\mathbf{p}_r = aI + b\gamma^0\boldsymbol{\gamma}\cdot\mathbf{p}_r \end{aligned} \quad (5.129)$$

Thus the "real" spinors $w_{Cr}{}^k(p)$ are the columns of

$$S_{Cr} = \begin{bmatrix} \underline{w_{Cr}{}^1(p)} & \underline{w_{Cr}{}^2(p)} & \underline{w_{Cr}{}^3(p)} & \underline{w_{Cr}{}^4(p)} \\ a & 0 & bp_{r\,z} & bp_{r-} \\ 0 & a & bp_{r\,+} & -bp_{r\,z} \\ bp_{r\,z} & bp_{r-} & a & 0 \\ bp_{r\,+} & -bp_{r\,z} & 0 & a \end{bmatrix} \quad (5.130)$$

where $p_{r\,\pm} = p_{r\,x} \pm ip_{r\,y}$. The "imaginary" spinors are the columns of

$$S_{Ci} = (\omega_i/\omega)\sinh(\omega/2)\gamma^0\boldsymbol{\gamma}\cdot\hat{\mathbf{u}}_i = [m(m + E)]^{-1/2}\gamma^0\boldsymbol{\gamma}\cdot\mathbf{p}_i = b\gamma^0\boldsymbol{\gamma}\cdot\mathbf{p}_i \quad (5.131)$$

$$S_{Ci} = \begin{bmatrix} \underline{w_{Ci}{}^1(p)} & \underline{w_{Ci}{}^2(p)} & \underline{w_{Ci}{}^3(p)} & \underline{w_{Ci}{}^4(p)} \\ 0 & 0 & bp_{i\,z} & bp_{i-} \\ 0 & 0 & bp_{i\,+} & -bp_{i\,z} \\ bp_{i\,z} & bp_{i-} & 0 & 0 \\ bp_{i\,+} & -bp_{i\,z} & 0 & 0 \end{bmatrix} \quad (5.132)$$

where $p_{i\,\pm} = p_{i\,x} \pm ip_{i\,y}$.

Eqs. 5.127 through 5.132 imply that the wave function solution of eq. 5.123, subject to the subsidiary condition eq. 5.124, is[76, 77]

$$\psi_C(x_r, x_i) = \sum_{\pm s}\int d^3p_r d^3p_i\, N_C(p)\delta(\mathbf{p}_r\cdot\mathbf{p}_i/m^2)[b_C(p,s)u_C(p, s)e^{-i(p\cdot x + p^*\cdot x^*)/2} + d_C^\dagger(p,s)v_C(p, s)e^{+i(p\cdot x + p^*\cdot x^*)/2}] \quad (5.133)$$

where $\mathbf{p} = \mathbf{p}_r + i\mathbf{p}_i$ (eq. 3.95), $\mathbf{x} = \mathbf{x}_r - i\mathbf{x}_i$, $p\cdot x = p^0x^0 - \mathbf{p}\cdot\mathbf{x}$, and where we use

$$(p\cdot x + p^*\cdot x^*)/2 = p^0x^0 - \mathbf{p}_r\cdot\mathbf{x}_r - \mathbf{p}_i\cdot\mathbf{x}_i \quad (5.134)$$

in the exponentials in order to avoid divergences that would appear in the calculation of the equal-time commutator, the Feynman propagator and other quantities of interest after second quantization. Note that

$$(\nabla_r + i\nabla_i)e^{-i(p\cdot x + p^*\cdot x^*)/2} = i(\mathbf{p}_r + i\mathbf{p}_i)e^{-i(p\cdot x + p^*\cdot x^*)/2} \quad (5.135)$$

and

$$(\nabla_r + i\nabla_i)e^{-ip^*\cdot x^*} = 0 \quad (5.136)$$

for all p.

The wave function's conjugate (the hermitean conjugate modified by letting $\mathbf{x}_i \rightarrow -\mathbf{x}_i$ in addition to hermitean conjugation) is

$$\psi_C^\dagger(x) = \psi_C^\dagger(x_r, -x_i) = \sum_{\pm s}\int d^3p_r d^3p_i\, \delta(\mathbf{p}_r\cdot\mathbf{p}_i/m^2)N_C(p^*)\cdot$$
$$\cdot[b_C^\dagger(p^*,s)u_C^\dagger(p^*,s)e^{+i(p\cdot x^* + p^*\cdot x)/2} + d_C(p^*,s)v_C^\dagger(p^*,s)e^{-i(p\cdot x^* + p^*\cdot x)/2}] \quad (5.137)$$

where $\mathbf{p} = \mathbf{p}_r + i\mathbf{p}_i$, $\mathbf{x} = \mathbf{x}_r - i\mathbf{x}_i$, $p\cdot x = p^0x^0 - \mathbf{p}\cdot\mathbf{x}$, and † indicates hermitean hermitean conjugation.

The spinors are

$$\begin{aligned}
u_C(p, s) &= S_C(p)w^1(0)\\
u_C(p, -s) &= S_C(p)w^2(0)\\
v_C(p, s) &= S_C(p)w^3(0)\\
v_C(p, -s) &= S_C(p)w^4(0)\\
u_C^\dagger(p^*, s) &= w^{1T}(0)S_C^\dagger(p^*)= w^{1T}(0)S_C(p)\\
u_C^\dagger(p^*, -s) &= w^{2T}(0)S_C^\dagger(p^*) = w^{2T}(0)S_C(p)\\
v_C^\dagger(p^*, s) &= w^{3T}(0)S_C^\dagger(p^*) = w^{3T}(0)S_C(p)\\
v_C^\dagger(p^*, -s) &= w^{4T}(0)S_C^\dagger(p^*) = w^{4T}(0)S_C(p)
\end{aligned} \quad (5.138)$$

---

76 Note that when $|\mathbf{p}_i| \geq |\mathbf{p}_r|$ (for imaginary $\omega = (\omega_r^2 - \omega_i^2)^{1/2}$) the 3-momentum becomes imaginary $\mathbf{p}\cdot\mathbf{p} < 0$. However, since we will be identifying confined quarks with this type of particle – much modified by a confining color quark interaction – the issue of an imaginary 3-momentum in the hypothetical free quark case becomes moot. We note the energy gap between positive and negative energy states disappears so E = 0 is possible. Thus real Lorentz transformations can mix positive and negative energy states. The solution is to do all calculations in the light-front frame as we do for tachyons. Then the mixing issue is resolved. In the present case we second quantize on the "time-front" for illustrative purposes.

77 We scale $\mathbf{p}_r\cdot\mathbf{p}_i$ with m^2 in the delta function for convenience. All fermions have at least a minimal mass – the mass of the iota.

with the superscript "T" indicating the transpose. Note that

$$S_C^\dagger(p^*) = [S_C(p^*)]^\dagger = S_C(p) \quad (5.139)$$

The normalization factor $N_C(p)$ is

$$N_C(p) = [2m/((2\pi)^6 p^0)]^{1/2} \quad (5.140)$$

Since $\mathbf{p}_r = \mathbf{p}_i = 0$ in the particle rest frame prior to the complex group boost, the boosted particle spin 4-vector s^μ satisfies

$$s^\mu p_r^{\ \mu} = s^\mu p_i^{\ \mu} = 0 \quad (5.141)$$

Note that s^μ is itself complex[78] and, if the spin points in the z-direction prior to the complex boost, then the boosted s^μ has the form

$$s^\mu = (-\sinh(\omega)\hat{w}_z, (0,0,1) + (\cosh(\omega) - 1)\hat{w}_z\hat{\mathbf{w}}) \quad (5.142)$$

with $\hat{\mathbf{w}}$ defined earlier: $\hat{\mathbf{w}} = (\omega_r\hat{\mathbf{u}}_r + i\omega_i\hat{\mathbf{u}}_i)/\omega = \mathbf{p}/(m\sinh(\omega))$.

A Global SU(3) Symmetry Revealed

Before proceding to consider the second quantization of this case, we will consider a global SU(3) symmetry implicit in the previous equations. The defining property of the group SU(3) is that it preserves the invariance of inner products of complex 3-vectors of the form:

$$u^*\cdot v = u^{1*}v^1 + u^{2*}v^2 + u^{3*}v^3 \quad (5.143)$$

If we examine the dynamical equation eq. 5.123 we see that the differential operator is invariant under an SU(3) transformation U (using $\nabla_c = (\nabla_c^*)^* = \mathbf{D}_c^*$)

$$[i\gamma^0\partial/\partial t + i\mathbf{D}_c^*\cdot\gamma - m] = [i\gamma^0\partial/\partial t + i\mathbf{D}_c'^*\cdot\gamma' - m] \quad (5.144)$$

where

$$\mathbf{D}_c^* = \nabla_c = \nabla_r + i\nabla_i$$

and

$$\gamma'^a = U^{ab}\gamma'^b$$
$$D_c'^{*a} = D_c'^{*b}U^{\dagger ab}$$

where U is a global SU(3) transformation and $U^\dagger = U^{-1}$. By theorem[79] all 4×4 γ matrices such as γ' are equivalent up to a unitary transformation V. Thus $V^\dagger\gamma' V = \gamma$ and eq. 5.144 is equivalent to

$$[i\gamma^0\partial/\partial t + i\mathbf{D}_c^*\cdot\gamma - m] = [i\gamma^0\partial/\partial t + i\mathbf{D}_c'^*\cdot\gamma - m] \quad (5/145)$$
$$= [i\gamma^0\partial/\partial t + i\nabla_c'\cdot\gamma - m]$$

---

[78] This feature of partons, which is not present in ordinary Dirac particles, might be the source of the discrepancies between theory and experiment in deep inelstic parton spin physics which is based on conventional real parton spins.
[79] R. H. Good, Rev. Mod. Phys., **27**, 187 (1955).

where $\nabla_c{}'_a = U^{ab}\nabla_{cb}$, This demonstrates that eq. 5.123 is invariant under an SU(3) transformation if

$$\psi_C(t, \mathbf{x}_c) = \psi_C(t, U\mathbf{x}_c) = \psi_C{}'(t, \mathbf{x}_c') \tag{5.146}$$

where $\psi_C(t, \mathbf{x}_c) \equiv \psi_C(t, \mathbf{x}_r, \mathbf{x}_i)$.

The subsidiary condition eq. 5.124 can be seen to transform as

$$\nabla_r\cdot\nabla_i\, \psi_C(t, \mathbf{x}_c) = \nabla_r{}^*\cdot\nabla_i\, \psi_C(t, \mathbf{x}_c) = \nabla_r{}'^*\cdot\nabla_i{}'\psi_C{}'(t, \mathbf{x}_c') = 0 \tag{5.147}$$

under an SU(3) rotation. The invariance of the orthogonality condition is preserved.

The wave function (eq. 5.123) transforms in the following way under the SU(3) transformation U. If we define

$$q^{*\mu} = (q^0, \mathbf{q}^*) = (p^0, \mathbf{p}_r + i\mathbf{p}_i) = (p^0, \mathbf{p}) = p^\mu \tag{5.148}$$

then eq. 5.133 can be rewritten in an invariant form under a SU(3) transformation:

$$\psi_C(x) = \sum_{\pm s}\int d^3q_r d^3q_i\, N_C(p^0)\delta(\mathbf{q}_r{}^*\cdot\mathbf{q}_i/m^2)[b_C(q^*,s)u_C(q^*,s)e^{-i(q^*\cdot x\, +\, q\cdot x^*)/2} + d_C{}^\dagger(q^*,s)v_C(q^*,s)e^{+i(q^*\cdot x\, +\, q\cdot x^*)/2}] \tag{5.149}$$

where $x = x_c$ subject to an examination of the transformation properties of the fourier coefficients and spinors. Note both terms in each exponential are separately invariant under global SU(3). (Note also $\mathbf{q}_r{}^* = \mathbf{q}_r$ since $\mathbf{q}_r$ is real.)

From the form of S_C above it is clear that an argument similar to that for the dynamical equations shows S_C is invariant under an SU(3) transformation and thus their spinors are also invariant under SU(3) transformations. The fourier coefficients, if second quantized in a direct generalization of the usual manner, have covariant anti-commutation relations under an SU(3) transformation. For example

$$\{b_C(q,s), b_C{}^\dagger(q'^*,s')\} = \delta_{ss'}\delta^3(q_r - q'_{r'})\delta^3(q_i - q'_{i'}) \tag{5.150}$$

Under an SU(3) transformation, z = Uq and z' = Uq', the right side of eq. 5.150 transforms to

$$\delta^3(q_r - q'_{r'})\delta^3(q_i - q'_{i'}) \rightarrow \delta^3(z_r - z'_{r'})\delta^3(z_i - z'_{i'})/|\partial(q)/\partial(z)| = \delta^3(z_r - z'_{r'})\delta^3(z_i - z'_{i'}) \tag{5.151}$$

where

$$|\partial(q)/\partial(z)| = |\partial(q_r{}^1,q_r{}^2,q_r{}^3,q_i{}^1, q_i{}^2, q_i{}^3)/\partial(z_r{}^1,z_r{}^2,z_r{}^3,z_i{}^1, z_i{}^2, z_i{}^3)| = 1 \tag{5.152}$$

is the Jacobian of the transformation U. Thus the fourier coefficients transform trivially under SU(3). For example,

$$b_C(q^*,s) \rightarrow b_C(z^*,s) \tag{5.153}$$

Since the integrand transforms as

$$\int d^3q_r d^3q_i \;\rightarrow\; \int d^3z_r d^3z_i\, |\partial(q)/\partial(z)| = \int d^3z_r d^3z_i \tag{5.154}$$

the wave function $\psi_C(t, \mathbf{x})$ transforms as an SU(3) scalar up to an inessential unitary transformation V of γ matrices: $\psi_C(t, \mathbf{x}) \rightarrow V\psi_C(t, \mathbf{x})$.[80]

Global SU(3) Spin ½ Complexon Fields

Having uncovered an SU(3) symmetry in the scalar field equations of Case 3A the generalization of the scalar field equations to the 3 representation of SU(3) is direct:

$$\psi_C{}^a(x) = \sum_{\pm s}\int d^3p_r d^3p_i\, N_C(p)\delta(\mathbf{p}_r\cdot\mathbf{p}_i/m^2)[b_C(p,a,s)u_C{}^a(p, s)e^{-i(p\cdot x + p^*\cdot x^*)/2} + d_C{}^\dagger(p,a,s)v_C{}^a(p, s)e^{+i(p\cdot x + p^*\cdot x^*)/2}] \tag{5.155}$$

where $x = x_c$ for a = 1,2, 3 with $u_C{}^a(p, s)$ and $v_C{}^a(p, s)$ being the product a spinor of type eq. 5.138 and a 3 element column vector c^a with b^{th} element

$$c^a(b) = \delta^{ab} \tag{5.156}$$

Under a global SU(3) transformation U the 3 complexon wave functions transform as

$$\psi_C{}'^a(x) = U^{ab}\psi_C{}^b(x) \tag{5.157}$$

In a subsequent discussion we will extend the global SU(3) symmetry described in these subsections to be color local SU(3) upon the introduction of the Yang-Mills color gluon interaction.

Lagrangian Formulation and Second Quantization of Complexons

In this subsection we will outline the canonical quantization of SU(3) singlet complexons with the quantum field equation

$$[i\gamma^0\partial/\partial t + i\boldsymbol{\gamma}\cdot(\boldsymbol{\nabla}_r + i\boldsymbol{\nabla}_i) - m]\psi_C(t, \mathbf{x}_r, \mathbf{x}_i) = 0 \tag{5.158}$$

and subsidiary condition

$$\boldsymbol{\nabla}_r\cdot\boldsymbol{\nabla}_i\, \psi_C(t, \mathbf{x}_r, \mathbf{x}_i) = 0 \tag{5.159}$$

We begin with the Lagrangian density

$$\mathcal{L} = \bar{\psi}_C(i\gamma^\mu D_\mu - m)\psi_C(x) \tag{5.160}$$

where $\bar{\psi}_C = \psi_C{}^\dagger\gamma^0$:

$$\psi_C{}^\dagger = [\psi_C(\mathbf{x}_r, \mathbf{x}_i)]^\dagger\,\big|_{\mathbf{x}_i = -\mathbf{x}_i} \tag{5.161}$$

$$\begin{aligned} D_0 &= \partial/\partial x^0 \\ D_k &= \partial/\partial x^k + i\,\partial/\partial x_i{}^k \end{aligned} \tag{5.162}$$

with $x^k = x_r{}^k$ for k = 1, 2, 3. The invariant action (under real Lorentz transformations) is

---

[80] The spinors $u_C(q^*,s)$ and $v_C(q^*,s)$ are unchanged up to a unitary transformation of the γ matrices ($V^\dagger\gamma' V = \gamma$). Thus the term $(U\mathbf{w})^*\cdot\gamma' = \mathbf{w}^*\cdot V\gamma V^\dagger \equiv \mathbf{w}^*\cdot\gamma$ in the expressions for the $u_C(q^*,s)$ and $v_C(q^*,s)$ spinors.

$$I = \int d^7x\mathcal{L} \tag{5.163}$$

It is easy to show that the action is real

$$I^* = I \tag{5.164}$$

in a manner similar to the case considered in Appendix 5-A due to the form of $\psi_C^\dagger$ in eq. 5.161. (One has to change the integration over $\mathbf{x}_i$ to $-\mathbf{x}_i$ after taking the complex conjugate of I and performing manipulations similar to those in Appendix 5-A.)

The conjugate momentum is

$$\pi_{Ca} = \partial\mathcal{L}/\partial\dot{\psi}_{Ca} \equiv \partial\mathcal{L}/\partial(\partial\psi_{Ca}/\partial x^0) = i\psi^\dagger_{C\,a} \tag{5.165}$$

where a is a spinor index. It yields the non-zero anti-commutation relation

$$\{\psi^\dagger_{C\,a}(x), \psi_{Cb}(y)\} = \delta_{ab}\,\delta^3(x_r - y_r)\delta^3(x_i - y_i) \tag{5.166}$$

where x and y are complex. However we will see that the constraint eq. 5.159 is required. So the correct anti-commutator turns out to be

$$\{\psi^\dagger_{C\,a}(x), \psi_{Cb}(y)\} = -\delta_{ab}\delta'(\nabla_r\cdot\nabla_i/m^2)[\delta^3(x_r - y_r)\delta^3(x_i - y_i)] \tag{5.167}$$

where all ∇_r and ∇_i are ∇ derivatives with respect to x, and where $\delta'(\nabla_r\cdot\nabla_i)$ is the derivative of a delta function with the argument being differential operators such as those in eq. 5.159. The minus sign is due to the presence of a *derivative* of a delta-function and is not an issue.

The hamiltonian density is

$$\mathcal{H} = \pi_C\dot{\psi}_C - \mathcal{L} = \psi_C^\dagger(-i\boldsymbol{\alpha}\cdot\mathbf{D} + \beta m)\psi_C \tag{5.168}$$

and the (unsymmetrized) energy-momentum tensor is

$$\mathcal{T}_{\mu\nu} = -\,g_{\mu\nu}\,\mathcal{L} + \partial\mathcal{L}/\partial(D^\mu\psi_C)D_\nu\psi_C \tag{5.169}$$

The conserved energy and momentum are

$$P^0 = H = \int d^3x_r d^3x_i\, \mathcal{T}^{00} = \int d^3x_r d^3x_i\, \mathcal{H} \tag{5.170}$$

and

$$P^i = \int d^3x_r d^3x_i\, \mathcal{T}^{0i} \tag{5.171}$$

We now proceed to establish the canonical anti-commutation relations. First, the second quantization of the complexon field uses the above fourier coefficient anti-commutation relations (suitably rewritten):

$$\begin{aligned}
&\{b_C(p,s), b_C^\dagger(p'^*,s')\} = \delta_{ss'}\delta^3(\mathbf{p}_r - \mathbf{p}'_{r'})\delta^3(\mathbf{p}_i + \mathbf{p}'_{i'})\\
&\{d_C(p,s), d_C^\dagger(p'^*,s')\} = \delta_{ss'}\,\delta^3(\mathbf{p}_r - \mathbf{p}'_{r'})\delta^3(\mathbf{p}_i + \mathbf{p}'_{i'})\\
&\{b_C(p,s), b_C(p'^*,s')\} = \{d_C(p,s), d_C(p'^*,s')\} = 0\\
&\{b_C^\dagger(p,s), b_C^\dagger(p'^*,s')\} = \{d_C^\dagger(p,s), d_C^\dagger(p'^*,s')\} = 0
\end{aligned} \tag{5.172}$$

$$\{b_C(p,s), d_C^\dagger(p'^*,s')\} = \{d_C(p,s), b_C^\dagger(p'^*,s')\} = 0$$
$$\{b_C^\dagger(p,s), d_C^\dagger(p'^*,s')\} = \{d_C(p,s), b_C(p'^*,s')\} = 0$$

The delta-function arguments $\delta^3(\mathbf{p}_i + \mathbf{p'}_{i'})$ above have a positive sign in order to obtain $\delta^3(\mathbf{x}_i - \mathbf{y}_i)$ in the field anti-commutator eq. 5.167.

The spinors, eq. 5.138, satisfy

$$\sum_{\pm s} u_\alpha(p, s)\bar{u}_\beta(p^*, s) = (2m)^{-1}(\not{p} + m)_{\alpha\beta} \tag{5.173}$$
$$\sum_{\pm s} v_\alpha(p, s)\bar{v}_\beta(p^*, s) = (2m)^{-1}(\not{p} - m)_{\alpha\beta}$$

remembering

$$\bar{u}_C(p^*,s) = w^{1T}(0)S_C(p)\gamma^0 = w^{1T}(0)[\cosh(\omega/2)I + \sinh(\omega/2)\gamma^0\boldsymbol{\gamma}\cdot\hat{\mathbf{w}}]\gamma^0 \tag{5.174}$$

by eqs. 5.137 since $\hat{\mathbf{w}}^{**} = \hat{\mathbf{w}}$.

We will now evaluate the equal-time anti-commutation relation using eqs. 5.136 and 5.137:

$$\{\psi_{C\,a}^{\dagger}(x), \psi_{Cb}(y)\} = \sum_{\pm s,\, s'} \int d^3p_r d^3p_i\, d^3p'_r d^3p'_i\, \delta(\mathbf{p}_r\cdot\mathbf{p}_i/m^2)\delta(\mathbf{p'}_r\cdot\mathbf{p'}_i/m^2)\, N_C(p')N_C(p)\cdot$$
$$\cdot[\{b_C^\dagger(p^*,s)u_{Ca}^\dagger(p^*,s)e^{+i(p\cdot x^* + p^*\cdot x)/2}, b_C(p',s')u_{Cb}(p', s')e^{-i(p'\cdot y + p'^*\cdot y^*)/2}\} +$$
$$+ \{d_C(p^*,s)v_{Ca}^\dagger(p^*,s)e^{-i(p\cdot x^* + p^*\cdot x)/2}, d_C^\dagger(p',s')v_{Cb}(p', s')e^{+i(p'\cdot y + p'^*\cdot y^*)/2}\}]$$

$$= \int d^3p_r d^3p_i\, N_C^2(p)[\delta(\mathbf{p}_r\cdot\mathbf{p}_i/m^2)]^2[((\not{p} + m)\gamma^0)_{ba} e^{+i(p\cdot x^* + p^*\cdot x)/2 - i(p^*\cdot y + p\cdot y^*)/2} +$$
$$+((\not{p} - m)\gamma^0)_{ba} e^{-i(p\cdot x^* + p^*\cdot x)/2 + i(p^*\cdot y + p\cdot y^*)/2}]/(2m)$$

Next we use eq. 5.140 and the identity

$$[\delta(x - y)]^2 = -\tfrac{1}{2}\,\delta'(x - y) \equiv -\tfrac{1}{2}\, d\delta(x - y)/dx \tag{5.175}$$

which can be derived from the step function identity $\theta(x - y) = [\theta(x - y)]^2$ to obtain

$$\{\psi_{C\,a}^{\dagger}(x),\psi_{Cb}(y)\} = -\tfrac{1}{2}\int d^3p_r d^3p_i N_C^2(p)\delta'(\mathbf{p}_r\cdot\mathbf{p}_i/m^2)[((\not{p}+m)\gamma^0)_{ba} e^{-ipr\cdot(xr - yr) + ipi\cdot(xi - yi)} +$$
$$+ ((\not{p} - m)\gamma^0)_{ba} e^{+ipr\cdot(xr - yr) - ipi\cdot(xi - yi)}]/(2m)$$
$$= -\tfrac{1}{2}\delta_{ba}\int d^3p_r d^3p_i N_C^2(p)\delta'(\mathbf{p}_r\cdot\mathbf{p}_i/m^2)p^0 e^{-ipr\cdot(xr - yr) + ipi\cdot(xi - yi)}/m$$

$$= -\,\delta_{ab}\,\delta'(\nabla_r\cdot\nabla_i/m^2)[\delta^3(x_r - y_r)\delta^3(x_i - y_i)] \tag{5.176}$$

The grad operators, ∇_r and ∇_i, are derivatives are with respect to x in the Dirac delta functions. The factor[81] $\delta'(\nabla_r\cdot\nabla_i)$ expresses the orthogonality constraint in coordinate space on the momenta. It is analogous to the transversality constraint on the electromagnetic vector potential commutator:

---

[81] A derivative of a delta function containing grad operators.

$$[\pi_A{}^j(x), A_k(y)] = -i\,\delta^{tr}{}_{jk}(x-y) \quad (5.177)$$

$$\delta^{tr}{}_{jk}(x-y) = (\delta_{jk} - \partial_j\partial_k/\nabla^2)\,\delta^3(x-y) \quad (5.178)$$

where $\partial_k = \partial/\partial x_k$.

Complexon Feynman Propagator

The complexon Feynman propagator for ψ_C is[82]

$$\begin{aligned} iS_C(x, y) &= \theta(x^0 - y^0)\langle 0|\psi_C(x)\psi_C{}^\dagger(y)\gamma^0|0\rangle - \theta(y^0 - x^0)\langle 0|\psi_C{}^\dagger(y)\gamma^0\psi_C(x)|0\rangle \quad (5.179)\\ &= \int d^3p_r d^3p_i N_C{}^2(p)[\delta(\mathbf{p_r}\cdot\mathbf{p_i}/m^2)]^2\{\theta(x^0 - y^0)(\not p + m)e^{-i(p^*\cdot(x-y) + p\cdot(x^*-y^*))/2} - \\ &\qquad - \theta(y^0 - x^0)(\not p - m)e^{+i(p^*\cdot(x-y) + p\cdot(x^*-y^*))/2}\}/(2m) \end{aligned}$$

$$= -(4\pi)^{-1}\int dp^0 d^3p_r d^3p_i (2\pi)^{-6}\delta'(\mathbf{p_r}\cdot\mathbf{p_i}/m^2)(\not p+m)e^{-i(p^*\cdot(x-y) + p\cdot(x^*-y^*))/2}/(p^2-m^2+i\varepsilon)$$

$$= -\tfrac{1}{2}\int dp^0 d^3p_r d^3p_i\, \delta'(\mathbf{p_r}\cdot\mathbf{p_i}/m^2)(\not p + m)(2\pi)^{-7}\exp[-ip^0(x^0 - y^0) + i\mathbf{p_r}\cdot(\mathbf{x_r} - \mathbf{y_r}) - i\mathbf{p_i}\cdot(\mathbf{x_i} - \mathbf{y_i})]/(p^2 - m^2 + i\varepsilon) \quad (5.180)$$

The integral can be written in the form:

$$\begin{aligned} I &= \int dp^0 d^3p_r d^3p_i \delta'(\mathbf{p_r}\cdot\mathbf{p_i}/m^2)(\not p+m)\exp[-ip^0(x^0-y^0)+i\mathbf{p_r}\cdot(\mathbf{x_r}-\mathbf{y_r})-i\mathbf{p_i}\cdot(\mathbf{x_i}- \mathbf{y_i})]/(p^2 - m^2 + i\varepsilon)\\ &= \int d^4p_r dM^2\delta'(\nabla_r\cdot\nabla_i/m^2)(p^0\gamma^0-(\mathbf{p_r}-\nabla_i)\cdot\gamma+m)\exp[-ip^0(x^0-y^0)+i\mathbf{p_r}\cdot(\mathbf{x_r}-\mathbf{y_r})]J(\mathbf{x_i} - \mathbf{y_i}, M^2)/(p_r{}^2-M^2 + i\varepsilon) \end{aligned} \quad (5.181)$$

where $p_r{}^2 = p^{0\,2} - \mathbf{p_r}\cdot \mathbf{p_r}$ and

$$\begin{aligned} J(\mathbf{x_i} - \mathbf{y_i}, M^2) &= (2\pi)^{-3}\int d^3p_i\,\delta(M^2 + \mathbf{p_i}{}^2 - m^2)\exp[-i\mathbf{p_i}\cdot(\mathbf{x_i}- \mathbf{y_i})] \quad (5.182)\\ &= (2\pi)^{-2}|\mathbf{x_i}- \mathbf{y_i}|^{-1}\theta(m^2 - M^2)\sin((m^2 - M^2)^{1/2}|\mathbf{x_i}- \mathbf{y_i}|) \end{aligned}$$

The complexon Feynman propagator can be rearranged into the form of a spectral integral:

$$iS_C(x, y) = -\int dM\,(i\gamma^0\partial/\partial x^0 - i(\nabla_r - i\nabla_i)\cdot\gamma + m)\delta'(\nabla_r\cdot\nabla_i/m^2)J(\mathbf{x_i} - \mathbf{y_i}, M^2)\Delta_F(x-y, M) \quad (5.183)$$

where

$$\Delta_F(x-y, M) = (2\pi)^{-4}\int d^4p_r\,\exp[-ip^0(x^0 - y^0) + i\mathbf{p_r}\cdot(\mathbf{x_r} - \mathbf{y_r})]/(p_r{}^2 - M^2 + i\varepsilon) \quad (5.184)$$

---

[82] The reader, upon seeing the additional integrations $\int d^3p_i$ might suspect that they would ultimately lead to divergence issues in perturbation theory calculations. However the $\delta'(\mathbf{p_r}\cdot\mathbf{p_i}/m^2)$ term compensates in part for the additional integrations by four powers of momentum since $\delta'(\mathbf{p_r}\cdot\mathbf{p_i}/m^2) = (|\mathbf{p_r}||\mathbf{p_i}|/m^2)^{-2}\delta'(\cos\theta_{ri})$ where θ_{ri} is the angle between the momenta. As a result only 2 fermion and 3 fermion loop integrations would potentially have difficulties if one uses the conventional approach to perturbation theory. If one uses the approach of Blaha (2003) and (2005a) then there are no divergences.

Case 4: Left-handed Tachyon Complexons

In this case $\hat{\mathbf{u}}_r \cdot \hat{\mathbf{u}}_i = 0$ again. However we add an imaginary term to ω to obtain a manifest Left-handed L_C boost[83]

$$\Lambda_{CL}(\mathbf{v}_c) = \exp[i(\omega + i\pi/2)\hat{\mathbf{w}}\cdot\mathbf{K}] \tag{5.185}$$

where ω remains

$$\omega = (\omega_r^2 - \omega_i^2)^{1/2} \tag{5.186}$$

and

$$\hat{\mathbf{w}} = (\omega_r \hat{\mathbf{u}}_r + i\omega_i \hat{\mathbf{u}}_i)/\omega \tag{5.187}$$

$$\hat{\mathbf{w}}\cdot\hat{\mathbf{w}} = \hat{\mathbf{u}}_r\cdot\hat{\mathbf{u}}_r = \hat{\mathbf{u}}_i\cdot\hat{\mathbf{u}}_i = 1 \tag{5.188}$$

$$\mathbf{v}_c = \hat{\mathbf{w}} \tanh(\omega + i\pi/2) = \hat{\mathbf{w}} \operatorname{cotanh}(\omega) \tag{5.189}$$

Letting $\omega_L = \omega + i\pi/2$ we find, as before,

$$\begin{aligned} \cosh(\omega_L) &= i \sinh(\omega) = -\gamma = i\,\gamma_s \\ \sinh(\omega_L) &= i \cosh(\omega) = -\beta\gamma = i\beta\gamma_s \end{aligned} \tag{5.190}$$

with, $\beta = v_c = |\mathbf{v}_c| > 1$, $\gamma_s = (\beta^2 - 1)^{-1/2}$, and

$$\begin{aligned} \sinh(\omega) &= \gamma_s \\ \cosh(\omega) &= \beta\gamma_s \end{aligned} \tag{5.191}$$

Thus we denote $\Lambda_{CL}(\mathbf{v}_c)$ by

$$\Lambda_{CL}(\mathbf{v}_c) \equiv \Lambda_{CL}(\omega, \hat{\mathbf{w}}) \tag{5.192}$$

The corresponding spinor boost transformation is:

$$\begin{aligned} S_{CL}(\Lambda_{CL}(\omega, \hat{\mathbf{w}})) &= \exp(-i\omega_L \sigma_{0i} \hat{w}_i/2) = \exp(-\omega_L \gamma^0 \boldsymbol{\gamma}\cdot\hat{\mathbf{w}}/2) \\ &= \cosh(\omega_L/2)I + \sinh(\omega_L/2)\gamma^0 \boldsymbol{\gamma}\cdot\hat{\mathbf{w}} \end{aligned} \tag{5.193}$$

The momentum space equation generated by $S_{CL}(\Lambda_{CL}(\omega, \hat{\mathbf{w}}))$ is

$$\{m \cosh(\omega_L)\gamma^0 - m \sinh(\omega_L)\boldsymbol{\gamma}\cdot(\omega_r \hat{\mathbf{u}}_r + i\omega_i \hat{\mathbf{u}}_i)/\omega - m\}e^{+ip\cdot x} w_{cL}(p) = 0 \tag{5.194}$$

or

$$\{im \sinh(\omega)\gamma^0 - im \cosh(\omega)\boldsymbol{\gamma}\cdot(\omega_r \hat{\mathbf{u}}_r + i\omega_i \hat{\mathbf{u}}_i)/\omega - m\}e^{+ip\cdot x} w_{cL}(p) = 0 \tag{5.195}$$

where $p\cdot x = Et - \mathbf{p}\cdot\mathbf{x}$ after performing a corresponding left-handed superluminal coordinate transformation in the exponential factor. Thus the positive energy wave is transformed into a negative energy wave by the transformation.

The momentum 4-vector is defined by

$$p = (p^0, \mathbf{p}) \tag{5.196}$$

where

[83] The reader can readily verify the form is consistent that generated by an L_C boost transformation.

$$p^0 = m \sinh(\omega) \qquad \mathbf{p} = \mathbf{p}_r + i\mathbf{p}_i \tag{5.197}$$

with

$$\mathbf{p}_r = m\omega_r\hat{\mathbf{u}}_r \cosh(\omega)/\omega \qquad \mathbf{p}_i = m\omega_i\hat{\mathbf{u}}_i \cosh(\omega)/\omega \tag{5.198}$$

and

$$\mathbf{p}_r \cdot \mathbf{p}_i = 0 \tag{5.199}$$

then eq. 5.195 becomes the complexon tachyon equation

$$[ip\cdot\gamma - m]e^{+ip\cdot x}w_{cL}(p) = 0 \tag{5.200}$$

with a complex 3-momentum **p** and the tachyon 4-momentum mass shell condition:[84]

$$p^2 = p^{0\,2} - \mathbf{p}_r^{\,2} + \mathbf{p}_i^{\,2} = -m^2 \tag{5.201}$$

Eq. 5.200 is the momentum space equivalent of the wave equation

$$[\gamma^0\partial/\partial t + \boldsymbol{\gamma}\cdot(\boldsymbol{\nabla}_r + i\boldsymbol{\nabla}_i) - m]\psi_{CL}(t, \mathbf{x}_r, \mathbf{x}_i) = 0 \tag{5.202}$$

or

$$[\gamma\cdot\nabla - m]\psi_{CL}(t, \mathbf{x}_r, \mathbf{x}_i) = 0 \tag{5.203}$$

with the subsidiary condition on the wave function

$$\boldsymbol{\nabla}_r\cdot\boldsymbol{\nabla}_i\, \psi_{CL}(t, \mathbf{x}_r, \mathbf{x}_i) = 0 \tag{5.204}$$

also holds. We note that eq. 5.202 is covariant under the real Lorentz group and eq. 5.204 can be easily put into (real Lorentz group) covariant form.

Before considering a lagrangian formulation and the Fourier operator representation of $\psi_{CL}(t, \mathbf{x}_r, \mathbf{x}_i)$ we will define the tachyon spinors, and its associated real and imaginary spin operators.

The spinor generated from a spin up Dirac spinor at rest by the L_C spinor boost eq. 5.193 is

$$w_{cL}(p) = S_{CL}w(0) = [\cosh(\omega_L/2)I + \sinh(\omega_L/2)\gamma^0\boldsymbol{\gamma}\cdot\hat{\mathbf{w}}]w(0) \tag{5.205}$$

Following a procedure similar to Appendix 5-A (which the reader may wish to examine first) we define four spinors for Dirac particles at rest with eq. 5-A.2. Then by applying a boost to these rest spinors we find the L_C tachyon spinors:

$$S_{CL}w^k(0) = w_{cL}{}^k(p) \tag{5.206}$$

and from these tachyon spinors we generalize to tachyon spinors $u_{CL}(p, s)$ and $v_{CL}(p, s)$ in a manner similar tothat of the previous case.

Eqs. 5.200 through 5.204 imply that the wave function solution of eq. 5.200, subject to the subsidiary condition eq. 5.204, has the form

---

[84] Note that the presence of the $\mathbf{p}_i^2$ term does not change the tachyon requirement that $\mathbf{p}_r^2 \geq m^2$ as seen in the previous cases.

$$\psi_{CL}(x) = \sum_{\pm s} \int_{\mathbf{p}_r^2 \geq m^2} d^3p_r d^3p_i\, N_{CL}(p)\delta(\mathbf{p}_r \cdot \mathbf{p}_i/m^2)[b_{CL}(p,s)u_{CL}(p, s)e^{-i(p\cdot x + p^*\cdot x^*)/2} + d_{CL}^\dagger(p,s)v_{CL}(p, s)e^{+i(p\cdot x + p^*\cdot x^*)/2}] \quad (5.207)$$

where $\mathbf{p} = \mathbf{p}_r + i\mathbf{p}_i$, $\mathbf{x} = \mathbf{x}_r - i\mathbf{x}_i$, $p\cdot x = p^0x^0 - \mathbf{p}\cdot\mathbf{x}$, and $b_{CL}(p, s)$ and $d_{CL}(p,s)$ are tachyon fourier coefficients.

Global SU(3) Symmetry

We can show that there is also a global SU(3) symmetry present here as shown in the previous case. The demonstration is similar to that of eqs. 5.143 – 5.156.

Light-Front Quantization of Tachyonic Complexons

Because of the momentum constraint $\mathbf{p}_r^2 \geq m^2$ the set of solutions of the form of eq. 5.207 is incomplete and the result of second quantization would not be an equal time anti-commutator expression consisting of derivatives of delta functions (eq. 5.176) but rather an analogue to previous unsuccessful attempts to create a second quantized tachyon theory.[85]

Therefore we will use light-front coordinates, and left and right handed field operators (as previously) to obtain a successful second quantization of this new type of tachyon.

The "missing" factor of i in the first term of eq. 5.203 requires the lagrangian to be different from the conventional Dirac lagrangian in order for the lagrangian to be real. The simplest, physically acceptable, free spin ½ tachyon lagrangian density for ψ_{CL} is:

$$\mathcal{L}_{CL} = \psi_{CL}{}^C(x)(\gamma\cdot\nabla - m)\psi_{CL}(x) \quad (5.208)$$

where

$$\psi_{CL}{}^C(x) = [\psi_{CL}(x)]^\dagger|_{\mathbf{x}_i = -\mathbf{x}_i}\, i\gamma^0\gamma^5 \quad (5.209)$$

is similar to eq. 5.161. In words, eq. 5.209 states: take the hermitean conjugate of $\psi_{CL}(x)$; change $\mathbf{x}_i$ to $-\mathbf{x}_i$; and then post-multiply by the indicated factors.

The free complexon invariant action (under real Lorentz transformations) is

$$I = \int d^7x \mathcal{L}_{CL} \quad (5.210)$$

The action can be shown to be real

$$I^* = I \quad (5.211)$$

in a manner similar to the case considered in Appendix 5-A. The tachyonic complexon's energy-momentum tensor is

$$\begin{aligned}\mathfrak{T}_{CL\mu\nu} &= -\, g_{\mu\nu}\mathcal{L}_{CL} + \partial\mathcal{L}_{CL}/\partial(D^\mu\psi_{CL})\, D_\nu\psi_{CL} \quad (5.212)\\ &= i\psi_{CL}{}^C\gamma^0\gamma^5\gamma_\mu D_\nu\psi_{CL}\end{aligned}$$

where

$$D_0 = \partial/\partial x^0$$

[85] Such as G. Feinberg, Phys. Rev. **159**, 1089 (1967).

$$D_k = \partial/\partial x_r^{\ k} + i\ \partial/\partial x_i^{\ k} \tag{5.213}$$

and thus the conserved energy and momentum are

$$P^0 = H = \int d^3x_r d^3x_i\ \mathfrak{T}_{CL}^{\ 00} = i\int d^3x_r d^3x_i \psi_{CL}^{\ \ C}\gamma^5(\boldsymbol{\alpha}\cdot\mathbf{D} + \beta m)\psi_{CL} \tag{5.214}$$

$$P^k = \int d^3x_r d^3x_i\ \mathfrak{T}_{CL}^{\ 0k} = -\,i\int d^3x_r d^3x_i\ \psi_{CL}^{\ \ C}\gamma^5 D^k\psi_{CL} \tag{5.215}$$

Having defined a suitable tachyon lagrangian we can now proceed to its canonical quantization. The conjugate momentum can be calculated from the lagrangian density eq. 5.212:

$$\pi_{CLa} = \partial\mathfrak{L}_{CL}/\partial\dot{\psi}_{CLa} \equiv \partial\mathfrak{L}_{CL}/\partial(\partial\psi_{CLa}/\partial t) = -i([\psi_{CL}(x)]^\dagger|_{\mathbf{x}_i = -\mathbf{x}_i}\gamma^5)_a \tag{5.216}$$

The resulting non-zero, canonical anti-commutation relations are

$$\{\pi_{CL_a}(x), \psi_{CLb}(y)\} = i\ \delta_{ab}\ \delta^3(x_r - y_r)\delta^3(x_i - y_i)$$

based on locality in both real and imaginary coordinates:

$$\{\psi_{CL}{}^{\dagger}{}_{a}(x)|_{\mathbf{x}_i = -\mathbf{x}_i}, \psi_{Tb}(y)\} = -\,[\gamma^5]_{ab}\ \delta^3(x_r - y_r)\delta^3(x_i - y_i) \tag{5.217}$$

At this point we might attempt to complete the canonical quantization procedure in the conventional manner by Fourier expanding the field and specifying anti-commutation relations for the fourier component amplitudes. However the incompleteness of the set of plane waves, which are limited by the restriction $\mathbf{p}_r^{\,2} \geq m^2$, causes the equal time anti-commutator of the fields *not* to yield a δ-functions.

Therefore we turn to the previous successful approach to tachyon quantization[86] and decompose the tachyonic complexon field into left-handed and right-handed parts and then second quantize in light-front coordinates.

Separation into Left-Handed and Right-Handed Fields

As before we will use a transformed set of Dirac matrices to develop our left-handed and right-handed tachyon formulations. The γ^5 chirality operator's eigenvalues define handedness: +1 corresponds to right-handed; and −1 corresponds to left-handed:

$$\gamma^5\psi_{CLL} = -\,\psi_{CLL} \qquad\qquad \gamma^5\psi_{CLR} = \psi_{CLR} \tag{5.218}$$

We define left-handed and right-handed tachyon fields with the projection operators:

$$\begin{aligned} C^{\pm} &= \tfrac{1}{2}(I \pm \gamma^5) \\ C^{+} + C^{-} &= I \\ C^{\pm\,2} &= C^{\pm} \end{aligned} \tag{5.219}$$

---

[86] Blaha (2006) discusses this case in detail.

$$C^+C^- = 0$$

with the result

$$\psi_{CLL} = C^-\psi_{CL}$$
$$\psi_{CLR} = C^+\psi_{CL} \tag{5.220}$$

We can calculate the commutation relations of the left-handed and right-handed tachyonic complexon fields from eq. 5.217 by pre-multiplying and post-multiplying by ½(1 – γ^5) and ½(1 + γ^5). The results are:

$$\{\psi_{CLLa}{}^\dagger(x)|_{\mathbf{x_i} = -\mathbf{x_i}}, \psi_{CLLb}(y)\} = C^-{}_{ab}\,\delta^6(x - y) \tag{5.221}$$

$$\{\psi_{CLRa}{}^\dagger(x)|_{\mathbf{x_i} = -\mathbf{x_i}}, \psi_{CLRb}(y)\} = -C^+{}_{ab}\,\delta^6(x - y) \tag{5.222}$$

$$\{\psi_{CLLa}{}^\dagger(x)|_{\mathbf{x_i} = -\mathbf{x_i}}, \psi_{CLRb}(y)\} = \{\psi_{CLRa}{}^\dagger(x)|_{\mathbf{x_i} = -\mathbf{x_i}}, \psi_{CLLb}(x')\} = 0 \tag{5.223}$$

where

$$\delta^6(x - y) = \delta^3(x_r - y_r)\delta^3(x_i - y_i) \tag{5.224}$$

The lagrangian density of eq. 5.208 decomposes into left-handed and right-handed parts: (The change $\mathbf{x_i}$ to $-\mathbf{x_i}$ will be understood in $\psi_{CLL}{}^\dagger(x)$ and $\psi_{CLR}{}^\dagger(x)$ in the following.)

$$\mathcal{L}_{CL} = \psi_{CLL}{}^\dagger\gamma^0 i\gamma^\mu\partial_\mu\psi_{CLL} - \psi_{CLR}{}^\dagger\gamma^0 i\gamma^\mu\partial_\mu\psi_{CLR} - im[\psi_{CLR}{}^\dagger\gamma^0\psi_{CLL} - \psi_{CLL}{}^\dagger\gamma^0\psi_{CLR}] \tag{5.225}$$

Further Separation into + and – Light-Front Complexon Fields

As previously, we now use light-front coordinates and quantization to obtain a successful second quantization of this form of tachyon field. Light-front variables, in the present case where we have to contend with complex 3-vectors, are defined by real coordinates and derivatives:

$$x^\pm = (x^0 \pm x_r{}^3)/\surd 2$$
$$\partial/\partial x^\pm \equiv \partial^\mp \equiv (\partial/\partial x^0 \pm \partial/\partial x_r{}^3)/\surd 2 \tag{5.226}$$

with the "transverse" real coordinate variables, $x_r{}^1$ and $x_r{}^2$, and imaginary coordinate variables $x_i{}^1$, $x_i{}^2$, and $x_i{}^3$.

The inner product of two 4-vectors has the form

$$x\cdot y = x^+y^- + y^+x^- + i[y_i{}^3(x^+ - x^-) + x_i{}^3(y^+ - y^-)]/\surd 2 + x_i{}^3y_i{}^3 - (\mathbf{x}_{r\perp} - i\mathbf{x}_{i\perp})\cdot(\mathbf{y}_{r\perp} - i\mathbf{y}_{i\perp}) \tag{5.227}$$

with

$$\mathbf{x}_{r\perp} = (x_r{}^1, x_r{}^2) \qquad \mathbf{x}_{i\perp} = (x_i{}^1, x_i{}^2)$$
$$\mathbf{y}_{r\perp} = (y_r{}^1, y_r{}^2) \qquad \mathbf{y}_{i\perp} = (y_i{}^1, y_i{}^2) \tag{5.228}$$

where $x = (x^0, \mathbf{x} = \mathbf{x_r} - i\mathbf{x_i})$ and $y = (y^0, \mathbf{y} = \mathbf{y_r} - i\mathbf{y_i})$. Momenta are always defined as $p = (p^0, \mathbf{p} = \mathbf{p_r} + i\mathbf{p_i})$.

The light-front definition of Dirac matrices is:

$$\gamma^{\pm} = (\gamma^0 \pm \gamma^3)/\sqrt{2} \tag{5.229}$$

with transverse matrices γ^1 and γ^2 defined as usual. Note:

$$\gamma^{\pm\,2} = 0$$

We define "+" and "–" tachyon fields with the projection operators:

$$R^{\pm} = \tfrac{1}{2}(I \pm \gamma^0\gamma^3) \tag{5.230}$$

$$\begin{aligned} &\text{Left-handed, } \pm \text{ light-front fields:} && \psi_{CLL}{}^{\pm} = R^{\pm}C^{-}\psi_{CL} \\ &\text{Right-handed, } \pm \text{ light-front fields:} && \psi_{CLR}{}^{\pm} = R^{\pm}C^{+}\psi_{CL} \end{aligned} \tag{5.231}$$

Transforming to light-front variables and fields as above we obtain the light-front free tachyon lagrangian:

$$\begin{aligned} \mathcal{L}_{CL} = {} & 2^{½}\psi_{CLL}{}^{+\dagger} i\partial^{-}\psi_{CLL}{}^{+} + 2^{½}\psi_{CLL}{}^{-\dagger} i\partial^{+}\psi_{CLL}{}^{-} - \psi_{CLL}{}^{+\dagger}\gamma^0[i\gamma_\perp\cdot\nabla_{r\perp} - \gamma\cdot\nabla_i]\psi_{CLL}{}^{-} - \\ & - \psi_{CLL}{}^{-\dagger}\gamma^0[i\gamma_\perp\cdot\nabla_{r\perp} - \gamma\cdot\nabla_i]\psi_{CLL}{}^{+} - 2^{½}\psi_{CLR}{}^{+\dagger} i\partial^{-}\psi_{CLR}{}^{+} - 2^{½}\psi_{CLR}{}^{-\dagger} i\partial^{+}\psi_{CLR}{}^{-} + \\ & + \psi_{CLR}{}^{+\dagger}\gamma^0[i\gamma_\perp\cdot\nabla_{r\perp} - \gamma\cdot\nabla_i]\psi_{CLR}{}^{-} + \psi_{CLR}{}^{-\dagger}\gamma^0[i\gamma_\perp\cdot\nabla_{r\perp} - \gamma\cdot\nabla_i]\psi_{CLR}{}^{+} - \\ & - im[\psi_{CLR}{}^{+\dagger}\gamma^0\psi_{CLL}{}^{-} - \psi_{CLL}{}^{+\dagger}\gamma^0\psi_{CLR}{}^{-} + \psi_{CLR}{}^{-\dagger}\gamma^0\psi_{CLL}{}^{+} - \psi_{CLL}{}^{-\dagger}\gamma^0\psi_{CLR}{}^{+}] \end{aligned} \tag{5.232}$$

(Note the similarity to the previous tachyon case.) Again the difference in signs between the left-handed and right-handed terms will be a crucial factor in the derivation of the left-handed features of the Standard Model.

Eq. 5.232 generates the equations of motion:

$$\begin{aligned} &2^{½}i\partial^{-}\psi_{CLL}{}^{+} - \gamma^0[i\gamma_\perp\cdot\nabla_{r\perp} - \gamma\cdot\nabla_i]\psi_{CLL}{}^{-} + im\gamma^0\psi_{CLR}{}^{-} = 0 \\ &2^{½}i\partial^{-}\psi_{CLR}{}^{+} - \gamma^0[i\gamma_\perp\cdot\nabla_{r\perp} - \gamma\cdot\nabla_i]\psi_{CLR}{}^{-} + im\gamma^0\psi_{CLL}{}^{-} = 0 \\ &2^{½}i\partial^{+}\psi_{CLL}{}^{-} - \gamma^0[i\gamma_\perp\cdot\nabla_{r\perp} - \gamma\cdot\nabla_i]\psi_{CLL}{}^{+} + im\gamma^0\psi_{CLR}{}^{+} = 0 \\ &2^{½}i\partial^{+}\psi_{CLR}{}^{-} - \gamma^0[i\gamma_\perp\cdot\nabla_{r\perp} - \gamma\cdot\nabla_i]\psi_{CLR}{}^{+} + im\gamma^0\psi_{CLL}{}^{+} = 0 \end{aligned} \tag{5.233}$$

Eqs. 5.233 show that $\psi_{CLL}{}^{-}$ and $\psi_{CLR}{}^{-}$ are dependent fields that are functions of $\psi_{CLL}{}^{+}$ and $\psi_{CLR}{}^{+}$ on the light-front where x^+ equals a constant. They can be expressed in an integral form as well. (The independent fields $\psi_{CLL}{}^{+}$ and $\psi_{CLR}{}^{+}$ play a fundamental role in tachyonic complexon theory and are used to define "in" and "out" tachyon states in perturbation theory.)

The conjugate momenta implied by eq. 5.232 are

$$\begin{aligned} &\pi_{CLL}{}^{+} = \partial\mathcal{L}/\partial(\partial^{-}\psi_{CLL}{}^{+}) = 2^{½}i\psi_{CLL}{}^{+\dagger} \\ &\pi_{CLL}{}^{-} = \partial\mathcal{L}/\partial(\partial^{-}\psi_{CLL}{}^{-}) = 0 \end{aligned} \tag{5.234}$$

$$\begin{aligned} &\pi_{CLR}{}^{+} = \partial\mathcal{L}/\partial(\partial^{-}\psi_{CLR}{}^{+}) = -2^{½}i\psi_{CLR}{}^{+\dagger} \\ &\pi_{CLR}{}^{-} = \partial\mathcal{L}/\partial(\partial^{-}\psi_{CLR}{}^{-}) = 0 \end{aligned} \tag{5.235}$$

x^+ plays the role of the "time" variable in light-front quantized theories. So we define canonical equal x^+ anti-commutation relations for spin ½ tachyonic complexons also.

The canonical equal-light-front ($x^+ = y^+$) anti-commutation relations of the independent fields would normally be:

$$\{\psi_{CLL}{}^{+\dagger}{}_a(x), \psi_{CLL}{}^{+}{}_b(y)\} = 2^{-1}[C^-R^+]_{ab}\delta(x^- - y^-)\delta^2(x_r - y_r)\delta^3(x_I - y_i) \tag{5.236}$$

$$\{\psi_{CLR}{}^{+\dagger}{}_a(x), \psi_{CLR}{}^{+}{}_b(y)\} = -2^{-1}[C^+R^+]_{ab}\,\delta(x^- - y^-)\delta^2(x_r - y_r)\delta^3(x_I - y_i) \tag{5.237}$$

$$\{\psi_{CLL}{}^{+\dagger}{}_a(x), \psi_{CLR}{}^{+}{}_b(y)\} = \{\psi_{CLR}{}^{+\dagger}{}_a(x), \psi_{CLL}{}^{+}{}_b(y)\} = 0 \tag{5.238}$$

$$\{\psi_{CLL}{}^{+}{}_a(x), \psi_{CLR}{}^{+}{}_b(y)\} = \{\psi_{CLR}{}^{+\dagger}{}_a(x), \psi_{CLL}{}^{+\dagger}{}_b(y)\} = 0 \tag{5.239}$$

But as in the previous case they will be modified.

Again we see that the right-handed tachyon anti-commutation relation (eq. 5.237) has a minus sign relative to the corresponding conventional right-handed anti-commutation relation.

The sign differences between the left-handed and right-handed lagrangian terms ultimately lead to parity violating features in the Standard Model lagrangian.

Left-Handed Tachyonic Complexons

The free, "+" light-front, left-handed tachyonic complexon Fourier expansion is:

$$\psi_{CLL}{}^{+}(x_r, x_i) = \sum_{\pm s}\int d^2p_r dp^+ d^3p_i\, N_{CLL}{}^{+}(p)\theta(p^+)\delta((p_i{}^3(p^+ - p^-)/\surd 2 + \mathbf{p}_{r\perp}\cdot\mathbf{p}_{i\perp})/m^2)\cdot$$
$$\cdot[b_{CLL}{}^{+}(p, s)u_{CLL}{}^{+}(p, s)e^{-i(p\cdot x + p^*\cdot x^*)/2} + d_{CLL}{}^{+\dagger}(p, s)v_{CLL}{}^{+}(p, s)e^{+i(p\cdot x + p^*\cdot x^*)/2}] \tag{5.240}$$

Its hermitean conjugate is

$$\psi_{CLL}{}^{+\dagger}(x_r, x_i) = \sum_{\pm s}\int d^2p_r dp^+ d^3p_i\, N_{CLL}{}^{+}(p)\theta(p^+)\delta((p_i{}^3(p^+ - p^-)/\surd 2 + \mathbf{p}_{r\perp}\cdot\mathbf{p}_{i\perp})/m^2)\cdot$$
$$\cdot[b_{CLL}{}^{\dagger}(p^*,s)u_{CLL}{}^{\dagger}(p^*,s)e^{+i(p^*\cdot x + p\cdot x^*)/2} + d_{CLL}(p^*,s)v_{CLL}{}^{\dagger}(p^*,s)e^{-i(p^*\cdot x + p\cdot x^*)/2}] \tag{5.241}$$

where $\mathbf{p} = \mathbf{p}_r + i\mathbf{p}_i$, $\mathbf{x} = \mathbf{x}_r - i\mathbf{x}_i$, $p\cdot x = p^0x^0 - \mathbf{p}\cdot\mathbf{x}$, and † indicates hermitean conjugate. The spinors are

$$\begin{aligned}
u_{CLL}{}^{+}(p, s) &= C^- R^+ S_{CL}w^1(0)\\
u_{CLL}{}^{+}(p, -s) &= C^- R^+ S_{CL}w^2(0)\\
v_{CLL}{}^{+}(p, s) &= C^- R^+ S_{CL}w^3(0)\\
v_{CLL}{}^{+}(p, -s) &= C^- R^+ S_{CL}w^4(0)\\
u_{CLL}{}^{+\dagger}(p^*, s) &= w^{1T}(0)S_{CL}R^+C^-
\end{aligned} \tag{5.242}$$

$$u_{CLL}^{+\dagger}(p^*, -s) = w^{2T}(0)S_{CL}R^+C^-$$
$$v_{CLL}^{+\dagger}(p^*, s) = w^{3T}(0)S_{CL}R^+C^-$$
$$v_{CLL}^{+\dagger}(p^*, -s) = w^{4T}(0)S_{CL}R^+C^-$$

where the superscript "T" indicates the transpose (These spinors are described in Appendix 5-A.) and

$$N_{CLL}^{+}(p) = (2\pi)^{-3}(2m/p^+)^{1/2} \tag{5.243}$$

The anti-commutation relations of the Fourier coefficient operators are

$$\begin{aligned}
&\{b_{CLL}(p,s), b_{CLL}^{\dagger}(p'^*,s')\} = 2^{-1/2}\delta_{ss'}\delta(p^+ - p'^+)\delta^2(\mathbf{p_r} - \mathbf{p'_{r'}})\delta^3(\mathbf{p_i} + \mathbf{p'_{i'}})\\
&\{d_{CLL}(p,s), d_{CLL}^{\dagger}(p'^*,s')\} = 2^{-1/2}\delta_{ss'}\,\delta(p^+ - p'^+)\delta^2(\mathbf{p_r} - \mathbf{p'_{r'}})\delta^3(\mathbf{p_i} + \mathbf{p'_{i'}})\\
&\{b_{CLL}(p,s), b_{CLL}(p'^*,s')\} = \{d_{CLL}(p,s), d_{CLL}(p'^*,s')\} = 0\\
&\{b_{CLL}^{\dagger}(p,s), b_{CLL}^{\dagger}(p'^*,s')\} = \{d_{CLL}^{\dagger}(p,s), d_{CLL}^{\dagger}(p'^*,s')\} = 0\\
&\{b_{CLL}(p,s), d_{CLL}^{\dagger}(p'^*,s')\} = \{d_{CLL}(p,s), b_{CLL}^{\dagger}(p'^*,s')\} = 0\\
&\{b_{CLL}^{\dagger}(p,s), d_{CLL}^{\dagger}(p'^*,s')\} = \{d_{CLL}(p,s), b_{CLL}(p'^*,s')\} = 0
\end{aligned} \tag{5.244}$$

The delta-function arguments $\delta^3(\mathbf{p_i} + \mathbf{p'_{i'}})$ above have a positive sign in order to obtain $\delta^3(\mathbf{x_i} - \mathbf{y_i})$ in the field anti-commutators.

The spinors, eq. 5.242, satisfy

$$\begin{aligned}
&\sum_{\pm s} u_{CLL\,\alpha}^{+}(p, s)\bar{u}_{CLL\,\beta}^{+}(p^*, s) = (2m)^{-1}[C^-R^+(i\not{p} + m)R^-C^+]_{\alpha\beta}\\
&\sum_{\pm s} v_{CLL\,\alpha}^{+}(p, s)\bar{v}_{CLL\,\beta}^{+}(p^*, s) = (2m)^{-1}[C^-R^+(i\not{p} - m)R^-C^+]_{\alpha\beta}
\end{aligned} \tag{5.245}$$

where $\bar{u}_{CLL}^{+} = u_{CLL}^{+\dagger}\gamma^0$ and $\bar{v}_{CLL}^{+} = v_{CLL}^{+\dagger}\gamma^0$.

We now evaluate the canonical left-handed, light-front anti-commutation relation:

$$\begin{aligned}
\{\psi_{CLL\,a}^{+}(x), \psi_{CLL\,b}^{+\dagger}(y)\} = &\sum_{\pm s,s'} \int d^3p_i d^2p dp^+ \int d^3p_i' d^2p' dp'^+ N_{CLL}^{+}(p)\, N_{CLL}^{+}(p')\cdot\\
&\cdot\theta(p^+)\theta(p'^+)\delta((p_i^3(p^+ - p^-)/\surd 2 + \mathbf{p_{r\perp}}\cdot\mathbf{p_{i\perp}})/m^2)\, \delta((p_i'^3(p'^+ - p'^-)/\surd 2 + \mathbf{p'_{r\perp}}\cdot\mathbf{p'_{i\perp}})/m^2)\cdot\\
&\cdot[\{b_{CLL}^{+\dagger}(p'^*,s'),b_{CLL}^{+}(p,s)\}u_{CLL\,a}^{+}(p,s)u_{CLL\,b}^{+\dagger}(p'^*,s')e^{+i(p'^*\cdot y+p'\cdot y^*)2 - i(p\cdot x+p^*\cdot x^*)/2} +\\
&+\{d_{CLL}^{+}(p'^*,s'),d_{CLL}^{+\dagger}(p,s)\}v_{CLL\,a}^{+}(p,s)v_{CLL\,b}^{+\dagger}(p'^*,s')e^{-i(p'^*\cdot y+p'\cdot y^*)/2 + i(p\cdot x + p^*\cdot x^*)/2}]
\end{aligned}$$

$$\begin{aligned}
= &\,2^{-1/2}\sum_{\pm s}\int d^3p_i d^2p_r dp^+ [N_{CLL}^{+}(p)]^2\theta(p^+)[\delta((p_i^3(p^+ - p^-)/\surd 2 + \mathbf{p_{r\perp}}\cdot\mathbf{p_{i\perp}})/m^2)]^2\cdot\\
&\cdot[u_{CLL\,a}^{+}(p,s)u_{CLL\,b}^{+\dagger}(p^*,s)e^{+i(p^*\cdot(y-x)+p\cdot(y^*-x^*))/2} + v_{CLL\,a}^{+}(p,s)v_{CLL\,b}^{+\dagger}(p^*,s)e^{-i(p^*\cdot(y-x)+p\cdot(y^*-x^*))/2}]
\end{aligned}$$

$$\begin{aligned}
= &-2^{-3/2}\int d^3p_i d^2p dp^+\theta(p^+)[N_{CLL}^{+}(p)]^2\delta'((p_i^3(p^+ - p^-)/\surd 2 + \mathbf{p_{r\perp}}\cdot\mathbf{p_{i\perp}})/m^2)(2m)^{-1}\cdot\\
&\cdot\{[C^-R^+(i\not{p} + m)\gamma^0R^+C^-]_{ab}e^{+i(p^*\cdot(y-x)+p\cdot(y^*-x^*))/2} + [C^-R^+(i\not{p} - m)\gamma^0R^+C^-]_{ab}e^{-i(p^*\cdot(y-x)+p\cdot(y^*-x^*))/2}\}
\end{aligned}$$

∞

$$= -(1/2)C^-R^+\delta_{ab} \int_0 d^3p_i\, d^2p_\perp \int dp^+ \delta'((p_i^3(p^+ - p^-)/\surd 2 + \mathbf{p}_\perp\cdot\mathbf{p}_{i\perp})/m^2)(2\pi)^{-6}\cdot$$

$$\cdot\{ e^{+i\{p^+(y^- - x^-) - \mathbf{p}_{r\perp}\cdot(\mathbf{y}_{r\perp} - \mathbf{x}_{r\perp}) + \mathbf{p}_i\cdot(\mathbf{y}_i - \mathbf{x}_i)\}} + e^{-i\{p^+(y^- - x^-) - \mathbf{p}_{r\perp}\cdot(\mathbf{y}_{r\perp} - \mathbf{x}_{r\perp}) + \mathbf{p}_i\cdot(\mathbf{y}_i - \mathbf{x}_i)\}}\}$$

$$= -C^-R^+\delta_{ab}(4\pi)^{-1}\int_0^\infty dp^+\delta'(\nabla_r\cdot\nabla_i/m^2)\delta^3(y_i - x_i)\,\delta^2(y_r - x_r)\{e^{+ip^+(y^- - x^-)} + e^{-ip^+(y^- - x^-)}\}$$

whereupon we revert back to the original form of the constraint: $\delta(\nabla_r\cdot\nabla_i/m^2)$

$$\{\psi_{CLL}{}^{+}{}_a(x), \psi_{CLL}{}^{++}{}_b(y)\} = -(1/2)C^-R^+\delta_{ab}\,\delta'(\nabla_r\cdot\nabla_i/m^2)\delta(y^- - x^-)\delta^2(y_r - x_r)\delta^3(y_i - x_i) \quad (5.246)$$

The result is the left-handed, light-front equivalent of the earlier non-tachyon result. Again the constraint is apparent in the anti-commutator. (The factor of 2 difference is due to light-front coordinate definitions.)

Therefore we have left-handed, light-front quantized tachyonic complexons with the equivalent of canonical anti-commutation relations, and with localized tachyonic complexons. As a result we have a canonical tachyonic complexon Quantum Field Theory.

Left-handed Case 4: Tachyonic Complexon Feynman Propagator

The light-front Feynman propagator for the left-handed $\psi_{CLL}{}^{+}$ *tachyonic* complexon field is

$$\begin{aligned} iS^+{}_{CLLF}(x,y) &= \theta(x^+ - y^+)\langle 0|\psi_{CLL}{}^+(x)\psi_{CLL}{}^{++}(y)\gamma^0|0\rangle - \theta(y^+ - x^+)\langle 0|\psi_{CLL}{}^{++}(y)\gamma^0\psi_{CLL}{}^+(x)|0\rangle \quad (5.247)\\ &= -\tfrac{1}{2}\int d^3p_i d^2p_r dp^+\theta(p^+)N_{CLL}{}^{+2}\delta'((p_i^3(p^+ - p^-)/\surd 2 + \mathbf{p}_{r\perp}\cdot\mathbf{p}_{i\perp})/m^2)(2m)^{-1}C^-R^+\cdot\\ &\quad\cdot\{\theta(x^+ - y^+)[(i\not{p} + m)\gamma^0]e^{+i(p^*\cdot(y - x) + p\cdot(y^* - x^*))/2} +\\ &\quad + \theta(y^+ - x^+)[(i\not{p} - m)\gamma^0]e^{-i(p^*\cdot(y - x) + p\cdot(y^* - x^*))/2}\}R^+C^-\gamma^0 \end{aligned}$$

If we define the on-shell momentum variables

$$\begin{aligned} &p_0{}^- = (p_{r0}{}^1 p_{r0}{}^1 + p_{r0}{}^2 p_{r0}{}^2 - \mathbf{p}_{i0}\cdot\mathbf{p}_{i0} - m^2)/(2p_0{}^+)\\ &p_0{}^+ = p^+,\ p_{r0}{}^j = p_r{}^j \qquad (\text{for } j = 1, 2),\\ &\mathbf{p}_{i0} = \mathbf{p}_i,\ p_{r\perp 0}{}^2 = p_{r0}{}^j p_{r0}{}^j\\ &\not{p}_0 = p_0\cdot\gamma \end{aligned}$$

with $p_0 = (p^0, \mathbf{p}_{r0} + i\mathbf{p}_{r0})$ then the above equation can be rewritten as

$$= -\tfrac{1}{2}C^-R^+\int d^4p d^3p_i N_{CLL}{}^{+2}\delta'((p_{i0}{}^3(p_0{}^+ - p_0{}^-)/\surd 2 + \mathbf{p}_{r\perp 0}\cdot\mathbf{p}_{i\perp 0})/m^2)(4\pi m)^{-1}e^{+i(p^*\cdot(y - x) + p\cdot(y^* - x^*))/2}\,.$$

$$\cdot\{\theta(p^+)(i\not{p}_0 + m)\gamma^0]/[p^- - p_0{}^- + i\varepsilon] + \theta(-p^+)(i\not{p}_0 - m)\gamma^0]/[p^- + p_0{}^- - i\varepsilon]\}R^+C\gamma^0$$

$$\begin{aligned} &= -\tfrac{1}{2}\int d^4p_r d^3p_i\, N_{CLL}{}^{+2}\delta'((p_{i0}{}^3(p^+ - p^-)/\surd 2 + \mathbf{p}_{r\perp}\cdot\mathbf{p}_{i\perp})/m^2)(p^+/4\pi m)\, e^{+i(p^*\cdot(y - x) + p\cdot(y^* - x^*))/2}\,.\\ &\quad\cdot[C^-R^+(i\not{p} + m)\gamma^0R^+C^-\gamma^0][(p^2 + m^2 + i\varepsilon)]^{-1} \end{aligned}$$

with $p_r = (p^0, \mathbf{p_r})$ and $p = (p^0, \mathbf{p_r} + i\mathbf{p_r})$. Substituting for N_{CLL} and using $x\delta'(x) = -\delta(x)$ we obtain

$$= -\tfrac{1}{2}\int d^4p_r d^3p_i (2\pi)^{-7}\delta'(\mathbf{p_r}\cdot\mathbf{p_i}/m^2)\exp[ip^0(y^0 - x^0) - i\mathbf{p_r}\cdot(\mathbf{y_r} - \mathbf{x_r}) + i\mathbf{p_i}\cdot(\mathbf{y_i} - \mathbf{x_i})]\cdot$$
$$\cdot [C^-R^+(i\not{p} + m)R^-C^+]/(p^2 + m^2 + i\varepsilon)$$

since $C^-R^+(i\not{p} + m)\gamma^0R^+C^-\gamma^0 = C^-R^+(i\not{p} + m)R^-C^+$. The integral can be written:

$$= \int d^4p_r d^3p_i \delta'(\mathbf{p_r}\cdot\mathbf{p_i}/m^2)C^-R^+(i\not{p} + m)R^-C^+\cdot$$
$$\cdot\exp[-ip^0(x^0 - y^0) + i\mathbf{p_r}\cdot(\mathbf{x_r} - \mathbf{y_r}) - i\mathbf{p_i}\cdot(\mathbf{x_i} - \mathbf{y_i})]/(p^2 + m^2 + i\varepsilon)$$
$$= \int d^4p_r dM^2\delta'(\nabla_r\cdot\nabla_i/m^2)C^-R^+(ip^0\gamma^0 - (\nabla_r - i\nabla_i)\cdot\boldsymbol{\gamma} + m)R^-C^+\cdot$$
$$\cdot\exp[-ip^0(x^0 - y^0) + i\mathbf{p_r}\cdot(\mathbf{x_r} - \mathbf{y_r})]J_2(\mathbf{x_i} - \mathbf{y_i}, M^2)/(p_r^2 + M^2 + i\varepsilon)$$

where

$$J_2(\mathbf{x_i} - \mathbf{y_i}, M^2) = (2\pi)^{-3}\int d^3p_i\,\delta(M^2 - \mathbf{p_i}^2 - m^2)\exp[-i\mathbf{p_i}\cdot(\mathbf{x_i} - \mathbf{y_i})] \tag{5.248}$$
$$= (2\pi)^{-2}|\mathbf{x_i} - \mathbf{y_i}|^{-1}\theta(M^2 - m^2)\sin((M^2 - m^2)^{1/2}|\mathbf{x_i} - \mathbf{y_i}|)$$

This tachyonic complexon Feynman propagator can be rearranged into the form of a spectral integral:

$$iS^+_{CLLF}(x, y) = -\int dM\, C^-R^+(\gamma^0\partial/\partial x^0 + (\nabla_r - i\nabla_i)\cdot\boldsymbol{\gamma} - m)R^-C^+\delta'(\nabla_r\cdot\nabla_i/m^2)J_2(\mathbf{x_i} - \mathbf{y_i}, M^2)\Delta_{FT}(x - y, M) \tag{5.249}$$

with ∇_r and ∇_i derivatives with respect to $\mathbf{x_r}$ and $\mathbf{x_i}$ and where

$$\Delta_{FT}(x - y, M) = (2\pi)^{-4}\int d^4p_r \exp[-ip^0(x^0 - y^0) + i\mathbf{p_r}\cdot(\mathbf{x_r} - \mathbf{y_r})]/(p_r^2 + M^2 + i\varepsilon) \tag{5.250}$$

Case 5: Right-Handed Tachyonic Complexons

The case of right-handed tachyonic complexons is similar to left-handed complexons with only one difference: a minus sign in the canonical right-handed equal-time commutation relations resulting in a minus sign in the creation and annihilation operator anti-commutation relations. The right-handed tachyonic complexon wave function light-front Fourier expansion is:

$$\psi_{CLR}^+(x_r, x_i) = \sum_{\pm s}\int d^2p_r dp^+ d^3p_i\, N_{CLR}^+(p)\theta(p^+)\delta((p_i^3(p^+ - p^-)/\surd 2 + \mathbf{p_{r\perp}}\cdot\mathbf{p_{i\perp}})/m^2)\cdot$$
$$\cdot[b_{CLR}^+(p, s)u_{CLR}^+(p, s)e^{-i(p\cdot x + p^*\cdot x^*)/2} + d_{CLR}^{+\dagger}(p, s)v_{CLR}^+(p, s)e^{+i(p\cdot x + p^*\cdot x^*)/2}] \tag{5.251}$$

where

$$N_{CLR}^+(p) = (2\pi)^{-3}(2m/p^+)^{1/2} \tag{5.252}$$

Its hermitean conjugate is

$$\psi_{CLR}^{+\dagger}(x_r, x_i) = \sum_{\pm s}\int d^2p_r dp^+ d^3p_i\, N_{CLR}^+(p)\theta(p^+)\delta((p_i^3(p^+ - p^-)/\surd 2 + \mathbf{p_{r\perp}}\cdot\mathbf{p_{i\perp}})/m^2)\cdot$$
$$\cdot[b_{CLR}^\dagger(p^*, s)u_{CLR}^\dagger(p^*, s)e^{+i(p^*\cdot x + p\cdot x^*)/2} + d_{CLR}(p^*, s)v_{CLR}^\dagger(p^*, s)e^{-i(p^*\cdot x + p\cdot x^*)/2}] \tag{5.253}$$

where $\mathbf{p} = \mathbf{p}_r + i\mathbf{p}_i$, $\mathbf{x} = \mathbf{x}_r - i\mathbf{x}_i$, $p \cdot x = p^0x^0 - \mathbf{p} \cdot \mathbf{x}$, and † indicates hermitean conjugate. The right-handed spinors are

$$\begin{aligned}
u_{CLR}{}^{+}(p, s) &= C^+ R^+ S_{CR} w^1(0) \\
u_{CLR}{}^{+}(p, -s) &= C^+ R^+ S_{CR} w^2(0) \\
v_{CLR}{}^{+}(p, s) &= C^+ R^+ S_{CR} w^3(0) \\
v_{CLR}{}^{+}(p, -s) &= C^+ R^+ S_{CR} w^4(0) \\
u_{CLR}{}^{+\dagger}(p^*, s) &= w^{1T}(0) S_{CR} R^+ C^+ \\
u_{CLR}{}^{+\dagger}(p^*, -s) &= w^{2T}(0) S_{CR} R^+ C^+ \\
v_{CLR}{}^{+\dagger}(p^*, s) &= w^{3T}(0) S_{CR} R^+ C^+ \\
v_{CLR}{}^{+\dagger}(p^*, -s) &= w^{4T}(0) S_{CR} R^+ C^+
\end{aligned} \qquad (5.254)$$

where the superscript "T" indicates the transpose. The anti-commutation relations of the Fourier coefficient operators are

$$\begin{aligned}
&\{b_{CLR}(p,s), b_{CLR}{}^\dagger(p'^*,s')\} = -2^{-1/2}\delta_{ss'}\delta(p^+ - p'^+)\delta^2(\mathbf{p}_r - \mathbf{p}'_{r'})\delta^3(\mathbf{p}_i + \mathbf{p}'_{i'}) \\
&\{d_{CLR}(p,s), d_{CLR}{}^\dagger(p'^*,s')\} = -2^{-1/2}\delta_{ss'}\delta(p^+ - p'^+)\delta^2(\mathbf{p}_r - \mathbf{p}'_{r'})\delta^3(\mathbf{p}_i + \mathbf{p}'_{i'}) \\
&\{b_{CLR}(p,s), b_{CLR}(p'^*,s')\} = \{d_{CLR}(p,s), d_{CLR}(p'^*,s')\} = 0 \\
&\{b_{CLR}{}^\dagger(p,s), b_{CLR}{}^\dagger(p'^*,s')\} = \{d_{CLR}{}^\dagger(p,s), d_{CLR}{}^\dagger(p'^*,s')\} = 0 \\
&\{b_{CLR}(p,s), d_{CLR}{}^\dagger(p'^*,s')\} = \{d_{CLR}(p,s), b_{CLR}{}^\dagger(p'^*,s')\} = 0 \\
&\{b_{CLR}{}^\dagger(p,s), d_{CLR}{}^\dagger(p'^*,s')\} = \{d_{CLR}(p,s), b_{CRR}(p'^*,s')\} = 0
\end{aligned} \qquad (5.255)$$

The spinors satisfy

$$\sum_{\pm s} u_{CLR}{}^{+}{}_{\alpha}(p, s)\bar{u}_{CLR}{}^{+}{}_{\beta}(p^*, s) = (2m)^{-1}[C^+R^+(-i\not{p} + m)R^-C^-]_{\alpha\beta} \qquad (5.256)$$

$$\sum_{\pm s} v_{CLR}{}^{+}{}_{\alpha}(p, s)\bar{v}_{CLR}{}^{+}{}_{\beta}(p^*, s) = (2m)^{-1}[C^+R^+(-i\not{p} - m)R^-C^-]_{\alpha\beta}$$

where $\bar{u}_{CLR}{}^{+} = u_{CLR}{}^{+\dagger}\gamma^0$ and $\bar{v}_{CLR}{}^{+} = v_{CLR}{}^{+\dagger}\gamma^0$.

The right-handed anti-commutation relation with a minus sign follows in particular because of the minus signs in eqs. 5.255.

Right-handed Case 5: Tachyonic Complexon Feynman Propagator

The Feynman propagator for right-handed tachyonic complexons can be obtained from eqs. 5.249 and 5.250 by changing the parity projection operator and some numerator signs in the integral (basically $p \to -p$) resulting in

$$iS^+_{CLRF}(x, y) = \int dM\, C^+R^+(\gamma^0\partial/\partial x^0 + (\nabla_r - i\nabla_i)\cdot\gamma - m)R^-C^-\delta'(\nabla_r\cdot\nabla_i/m^2)J_2(\mathbf{x}_i - \mathbf{y}_i, M^2)\Delta_{FT}(x - y, M) \qquad (5.257)$$

with $\nabla_r + i\nabla_i$ derivatives with respect to $\mathbf{x}_r$ and $\mathbf{x}_i$ and where

$$\Delta_{FT}(x - y, M) = (2\pi)^{-4}\int d^4p_r \exp[-ip^0(x^0 - y^0) + i\mathbf{p_r}\cdot(\mathbf{x_r} - \mathbf{y_r})]/(p_r^2 + M^2 + i\varepsilon) \tag{5.258}$$

Other Cases? No

The four cases considered above are the only cases having symmetry under the real Lorentz group L and a single real energy (with a corresponding single real time parameter) that is independent of the direction of the boost thus preserving (real) spatial rotation invariance. The reality of the time variable survives the breakdown to conventional Lorentz invariance.

One might think that using the other type of spinor boost operator.

$$\begin{aligned} S_{CR}(\Lambda_{CR}(\omega, \hat{\mathbf{w}})) &= \exp(-i\omega_R\sigma_{0i}w_i/2) = \exp(-\omega_R\gamma^0\boldsymbol{\gamma}\cdot\hat{\mathbf{w}}/2) \\ &= \cosh(\omega_R/2)I + \sinh(\omega_R/2)\gamma^0\boldsymbol{\gamma}\cdot\hat{\mathbf{w}} \end{aligned} \tag{5.259}$$

where $\omega_R = \omega - i\pi/2$ might lead to more possible forms of spin ½ wave equations and particles. In fact it merely leads to the same particle types but with the role of the left-handed and right-handed fields reversed. The result would be a "right-handed" Standard Model contrary to experiment.

5.6 Spinor Boosts Generate 4 Species of Particles: Leptons and Quarks

In this chapter we have found four types of fermions using complex Lorentz boosts that correspond in a natural way with the four general species (types) of known fermions: charged leptons, neutrinos, up-type color quarks and down-type color quarks.[87]

Charged lepton fermions

The conventional Dirac equation and solutions.

Neutrinos

Simple tachyons with real energy and 3-momentum. Their free field equation is:

$$(\gamma^\mu\partial/\partial x^\mu - m)\psi_T(x) = 0 \tag{5.260}$$

and their left-handed $\psi_{TL}^{\ +}$ Feynman propagator is:

$$iS^+_{TLF}(x, y) = \tfrac{1}{2}C^-R^+\gamma^0\int d^4p(2\pi)^{-4}\, p^+e^{-ip\cdot(x-y)}/(p^2 + m^2 + i\varepsilon) \tag{5.261}$$

Similarly the light-front Feynman propagator for the right-handed $\psi_{TR}^{\ +}$ tachyon field is

$$iS^+_{TRF}(x,y) = -\tfrac{1}{2}C^+R^+\gamma^0\int d^4p(2\pi)^{-4}\, p^+e^{-ip\cdot(x-y)}/(p^2 + m^2 + i\varepsilon) \tag{5.262}$$

---

[87] We call each type of fermion a *species*. Each species has three known generations.

Up-type Color Quarks

Up-type quarks are assumed[88] to be fermions with complex 3-momenta - complexons, and an internal color SU(3) symmetry, that satisfy $p^2 = m^2$. Their field equation with a color SU(3) index, denoted a, inserted is

$$[i\gamma^0\partial/\partial t + i\gamma\cdot(\nabla_r + i\nabla_i) - m]\psi_C{}^a(t, \mathbf{x}_r, \mathbf{x}_i) = 0 \tag{5.263}$$

with the subsidiary condition

$$\nabla_r\cdot\nabla_i\, \psi_C{}^a(t, \mathbf{x}_r, \mathbf{x}_i) = 0 \tag{5.264}$$

The free field solution is:

$$\psi_C{}^a(x) = \sum_{\pm s}\int d^3p_r d^3p_i\, N_C(p)\delta(\mathbf{p}_r\cdot\mathbf{p}_i/m^2)[b_C(p,a,s)u_C{}^a(p, s)e^{-i(p\cdot x + p^*\cdot x^*)/2} + d_C{}^\dagger(p,a,s)v_C{}^a(p, s)e^{+i(p\cdot x + p^*\cdot x^*)/2}] \tag{5.265}$$

The free Feynman propagator arranged into the form of a spectral integral is

$$iS_C{}^{ab}(x,y) = -\delta^{ab}\int dM\, (i\gamma^0\partial/\partial x^0 - i(\nabla_r - i\nabla_i)\cdot\gamma + m)\delta'(\nabla_r\cdot\nabla_i/m^2)J(\mathbf{x}_i - \mathbf{y}_i, M^2)\Delta_F(x - y, M) \tag{5.266}$$

where

$$\Delta_F(x - y, M) = (2\pi)^{-4}\int d^4p_r \exp[-ip^0(x^0 - y^0) + i\mathbf{p}_r\cdot(\mathbf{x}_r - \mathbf{y}_r)]/(p_r{}^2 - M^2 + i\varepsilon) \tag{5.267}$$

and

$$\begin{aligned} J(\mathbf{x}_i, M^2) &= (2\pi)^{-3}\int d^3p_i\, \delta(M^2 + \mathbf{p}_i{}^2 - m^2) \exp[-i\mathbf{p}_i\cdot(\mathbf{x}_i - \mathbf{y}_i)] \\ &= (2\pi)^{-2}|\mathbf{x}_i - \mathbf{y}_i|^{-1}\theta(m^2 - M^2)\sin((m^2 - M^2)^{1/2}|\mathbf{x}_i - \mathbf{y}_i|) \end{aligned} \tag{5.268}$$

Down-type Color Quarks

Tachyonic complexons with complex 3-momenta, and an internal global SU(3) symmetry, that have mass shell condition $p^2 = -m^2$. Their field equation with a color SU(3) index, denoted a, inserted is

$$[\gamma^0\partial/\partial t + \gamma\cdot(\nabla_r + i\nabla_i) - m]\psi_{CL}{}^a(t, \mathbf{x}_r, \mathbf{x}_i) = 0 \tag{5.269}$$

with the subsidiary condition on the wave function

$$\nabla_r\cdot\nabla_i\, \psi_{CL}{}^a(t, \mathbf{x}_r, \mathbf{x}_i) = 0 \tag{5.270}$$

Its free field left-handed solution is:

---

[88] The complexon theory that we develop and use for quark dynamics in the Standard Model is not required. Our Standard Model could use Dirac fermion dynamics for the up-type quarks and tachyon dynamics for down-type quarks. Then the (broken) Left-handed complex Lorentz boosts would have the basic space-time group rather than L_C. We choose to use complexon dynamics for quarks because they have an internal SU(3)-like structure suggestive of color SU(3). More importantly, their spin dynamics is different and thus may resolve the differences between theory and experiment for the deep inelastic parton spin-dependent structure functions.

$$\psi_{CLL}{}^{+a}(x_r, x_i) = \sum_{\pm s} \int d^2p_r dp^+ d^3p_i\, N_{CLL}{}^{+}(p)\theta(p^+)\delta((p_i{}^3(p^+ - p^-)/\surd 2 + \mathbf{p}_{r\perp}\cdot\mathbf{p}_{i\perp})/m^2)\cdot$$
$$\cdot[b_{CLL}{}^{+}(p,a,s)u_{CLL}{}^{a}(p,a,s)e^{-i(p\cdot x + p^*\cdot x^*)/2} + d_{CLL}{}^{+\dagger}(p,a,s)v_{CLL}{}^{+a}(p,a,s)e^{+i(p\cdot x + p^*\cdot x^*)/2}] \quad (5.271)$$

and its right-handed solution is

$$\psi_{CLR}{}^{+a}(x_r, x_i) = \sum_{\pm s} \int d^2p_r dp^+ d^3p_i\, N_{CLR}{}^{+}(p)\theta(p^+)\delta((p_i{}^3(p^+ - p^-)/\surd 2 + \mathbf{p}_{r\perp}\cdot\mathbf{p}_{i\perp})/m^2)\cdot$$

$$\cdot[b_{CLR}{}^{+}(p,a,s)u_{CLR}{}^{+a}(p,a,s)e^{-i(p\cdot x+p^*\cdot x^*)/2} + d_{CLR}{}^{+\dagger}(p,a,s)v_{CLR}{}^{+a}(p,a,s)e^{+i(p\cdot x+p^*\cdot x^*)/2}] \quad (5.272)$$

The free left-handed Feynman propagator arranged into the form of a spectral integral is

$$iS^+_{CLLF}{}^{ab}(x,y) = -\delta^{ab}\int dM\, C^- R^+(\gamma^0\partial/\partial x^0 + (\nabla_r - i\nabla_i)\cdot\gamma - m)R^- C^+ \delta'(\nabla_r\cdot\nabla_i/m^2)J_2(\mathbf{x}_i - \mathbf{y}_i, M^2)\Delta_{FT}(x - y,M) \quad (5.273)$$

with ∇_r and ∇_i derivatives with respect to $\mathbf{x}_r$ and $\mathbf{x}_i$ and where

$$\Delta_{FT}(x - y, M) = (2\pi)^{-4}\int d^4p_r \exp[-ip^0(x^0 - y^0) + i\mathbf{p}_r\cdot(\mathbf{x}_r - \mathbf{y}_r)]/(p_r{}^2 + M^2 + i\varepsilon) \quad (5.274)$$

and

$$J_2(\mathbf{x}_i, M^2) = (2\pi)^{-3}\int d^3p_i\, \delta(M^2 - \mathbf{p}_i{}^2 - m^2) \exp[-i\mathbf{p}_i\cdot(\mathbf{x}_i - \mathbf{y}_i)] \quad (5.275)$$

$$= (2\pi)^{-2}|\mathbf{x}_i - \mathbf{y}_i|^{-1}\theta(M^2 - m^2)\sin((M^2 - m^2)^{1/2}|\mathbf{x}_i - \mathbf{y}_i|)$$

The free right-handed Feynman propagator arranged into the form of a spectral integral is

$$iS^+_{CLRF}{}^{ab}(x, y) = \delta^{ab}\int dM\, C^+ R^+(\gamma^0\partial/\partial x^0 + (\nabla_r - i\nabla_i)\cdot\gamma - m)R^- C^- \delta'(\nabla_r\cdot\nabla_i/m^2)J_2(\mathbf{x}_i - \mathbf{y}_i, M^2)\Delta_{FT}(x - y,M) \quad (5.276)$$

with ∇_r and ∇_i derivatives with respect to $\mathbf{x}_r$ and $\mathbf{x}_i$, and where

$$\Delta_{FT}(x - y, M) = (2\pi)^{-4}\int d^4p_r \exp[-ip^0(x^0 - y^0) + i\mathbf{p}_r\cdot(\mathbf{x}_r - \mathbf{y}_r)]/(p_r{}^2 + M^2 + i\varepsilon) \quad (5.277)$$

5.7 First Step towards The Standard Model

Thus we have found a set of four fermion species that corresponds to the known fermions of one fermion generation. In subsequent chapters we will derive the one generation model in detail. Then we will introduce three generations with mixing to complete the derivation of the form of the Standard Model. Then the only remaining major issue will be the values of the coupling constants and other numerical parameters.

The overall pattern that begins to emerge from the developments in this chapter divides particles and interactions into two categories (as seen in Nature):

Particles with real 4-Momenta	Complexons (Complex 3-Momenta)
Leptons	color quarks
$SU(2)\otimes U(1)$ Vector Bosons	Color SU(3) gluons
Higgs Particles	Possibly Higgs Particles

We will explore these issues in detail in the following chapters. But basically the leptons, $SU(2)\otimes U(1)$ Vector Bosons and a set of Higgs particles appear to be based on the Left-handed boosts. These particles have real energies and momenta although some are "normal" and some are tachyons.

Another category of particles, complexons, emerges from our study of L_C. These particles have real energies and complex 3-momenta. In perturbation theory the loop integrations of loops of these particles would consist of a 7-fold integration over energy and complex 3-momenta with corresponding 7-fold delta functions to enforce energy-momentum conservation. As pointed out earlier the complex 3-momenta of these types of fermions has an SU(3) symmetry that it is natural to generalize to local color SU(3). (The other category of fermions lacks this global SU(3) symmetry – as leptons lack color SU(3).) Thus we see the beginnings of the structure of the Standard Model in this chapter on spin ½ particles. The following chapters lead to a detailed derivation of the form of the Standard Model.

5.8 Dirac-like Equations of Matter from 4-Valued Logic

In Blaha's derivation every truly fundamental particle of matter, whether quark or lepton, has spin ½. We have seen in chapter 10 of Blaha (2011c) that the basic algebra of Operator Logic eigenvalue operators, and that of its raising and lowering operators, is the same as the algebra of creation and annihilarion operators for free spin ½ particles. Our goal is to build our theory on the scaffolding of Operator Logic. We view a fermion particle as an iota core which is dressed in spatial coordinates:

$$\text{Iota core} + \text{coordinates} = \text{fermion particle} \tag{5.278}$$

The creation and annihilation operators b(p,s) and $d^\dagger(p,s)$ (and their hermitean conjugates $b^\dagger(p,s)$ and d(p,s)) are mathematically similar to the raising and lowering operators of Operator (Matrix) Logic. They satisfy the anticommutation relations

$$\begin{aligned} \{b(q,s), b^\dagger(p,s')\} &= \delta_{ss'}\delta^3(\mathbf{q}-\mathbf{p}) \\ \{d(q,s), d^\dagger(p,s')\} &= \delta_{ss'}\delta^3(\mathbf{q}-\mathbf{p}) \end{aligned} \tag{5.279}$$

Thus we see spin ½ particle wave functions originating from the Dirac spinors, and raising and lowering operators of the spinor formulation of Operator Logic.

When particles interact the quantum field theory interaction terms use fermion creation operators, b(q,s) and $d^\dagger(q,s)$, *and annihilation operators,* $b^\dagger(p,s')$ and d(q,s), *to implement the transformations between the Iotas of the interacting particles. Thus the mathematics of the embedded iotas' logic values is automatically implemented within quantum field theoretic calculations.*

An interesting point that emerges from this discussion is the nature of spin ½ particle states such as

$$|p, s\rangle = b^\dagger(p, s)|0\rangle \tag{5.280}$$

This state is interpreted as a one particle state. It also has an analogous interpretation in Operator Logic as creating a one term universe of discourse – a construct which is in part linguistic and in part logic. Thus particles are embodiments of Logic values and particle interactions change the logic values of the initial particles to those of the emergent particles. All in all, our universe can be viewed as an extraordinarily intricate logic machine. Serendipitously we are now seeing the use of particles to create quantum computers, which, in a sense, is bringing us full circle. Particles are Logic; Logic machines emerge from particle interactions.

5.9 Why Second Quantization of Fields?

One might have argued that the fermion field types that we have found could be treated as ordinary c-number fields and not be second quantized. However, particles are discrete entities that can be enumerated with integers. Second quantization implements the discrete particle concept in the most direct way and thus by Leibiz's Principle as well as Ockham's Razor second quantization is the best solution to obtain particle discreteness.

Appendix 5-A. Leptonic Tachyon Spinors

The general form of the solutions of the free tachyon Dirac equation eq. 3.18 can be written

$$\psi_T{}^r(x) = e^{-i\chi_r p\cdot x} w^r(p) \qquad (5\text{-A.1})$$

where $\chi_r = +1$ for r = 1, 2 and $\chi_r = -1$ for r = 3, 4. Denoting the spinors $w^r(p) = w^r(0)$ for a particle is at rest in a frame (E = m) we see they can take the form

$$w^r(0) = \begin{bmatrix} \delta_{1r} \\ \delta_{2r} \\ \delta_{3r} \\ \delta_{4r} \end{bmatrix} \qquad (5\text{-A.2})$$

where Kronecker deltas appear in the brackets. From eq. 5.30 we find

$$S_L(\Lambda_L(\omega, \mathbf{u}))w^r(0) = w_T{}^r(p) \qquad (5\text{-A.3})$$

Using eq. 3.11 for $S_L(\Lambda_L(\omega, \mathbf{u}))$ and

$$\mathbf{p} = m\mathbf{v}\gamma_s \qquad E = m\gamma_s \qquad (5\text{-A.4})$$

we see that eq. 5-A.3 implies the columns of the resulting $S_L(\Lambda_L(\omega, \mathbf{u}))$ matrix are

$$S_L(\Lambda_L(\omega, \mathbf{u})) = \begin{bmatrix} \underline{w_T{}^3(p)} & \underline{w_T{}^4(p)} & \underline{w_T{}^1(p)} & \underline{w_T{}^2(p)} \\ \cosh(\omega_L/2) & 0 & \sinh(\omega_L/2)p_z/p & \sinh(\omega_L/2)p_-/p \\ 0 & \cosh(\omega_L/2) & \sinh(\omega_L/2)p_+/p & -\sinh(\omega_L/2)p_z/p \\ \sinh(\omega_L/2)p_z/p & \sinh(\omega_L/2)p_-/p & \cosh(\omega_L/2) & 0 \\ \sinh(\omega_L/2)p_+/p & -\sinh(\omega_L/2)p_z/p & 0 & \cosh(\omega_L/2) \end{bmatrix} \qquad (5\text{-A.5})$$

based on the superluminal transformation of positive energy states to negative energy states (eqs. 5.30 and 5.31) with $p_\pm = p_x \pm ip_y$ and where $p = |\mathbf{p}|$. It is easy to verify

$$(i\not{p} - \chi_r m)w_T{}^r(p) = 0 \tag{5-A.6}$$

where $\chi_r = -1$ for r = 1, 2 and $\chi_r = +1$ for r = 3, 4.

The spinors that we defined earlier can be generalized in a manner similar to Dirac spinors. We will use a similar notation to the Dirac spinor notation:

$$\begin{aligned} u_T(p, s) &= w_T{}^1(p) \\ u_T(p, -s) &= w_T{}^2(p) \\ v_T(p, s) &= w_T{}^3(p) \\ v_T(p, -s) &= w_T{}^4(p) \end{aligned} \tag{5-A.7}$$

We define "double dagger" spinors:

$$\begin{aligned} u_T{}^{\ddagger}(p, s) &= u_T{}^{\dagger}(p, s)i\gamma\cdot\mathbf{p}/|\mathbf{p}| \\ u_T{}^{\ddagger}(p, -s) &= u_T{}^{\dagger}(p, -s)i\gamma\cdot\mathbf{p}/|\mathbf{p}| \\ v_T{}^{\ddagger}(p, s) &= v_T{}^{\dagger}(p, s)i\gamma\cdot\mathbf{p}/|\mathbf{p}| \\ v_T{}^{\ddagger}(p, -s) &= v_T{}^{\dagger}(p, -s)i\gamma\cdot\mathbf{p}/|\mathbf{p}| \end{aligned} \tag{5-A.8}$$

.

where † indicates hermitean conjugate, which appear in important spinor "completeness" sums:

$$\sum_{\pm s} u_{T\alpha}(p, s)u_{T\beta}{}^{\ddagger}(p, s) = (2m)^{-1}(i\not{p} - m)_{\alpha\beta} \tag{5-A.9}$$

$$\sum_{\pm s} v_{T\alpha}(p, s)v_{T\beta}{}^{\ddagger}(p, s) = (2m)^{-1}(i\not{p} + m)_{\alpha\beta} \tag{5-A.10}$$

or

$$\sum_{\pm s} u_{T\alpha}(p, s)u_{T\beta}{}^{\dagger}(p, s) = -i(2m)^{-1}[(i\not{p} - m)\gamma\cdot\mathbf{p}/|\mathbf{p}|]_{\alpha\beta} \tag{5-A.11}$$

$$\sum_{\pm s} v_{T\alpha}(p, s)v_{T\beta}{}^{\dagger}(p, s) = -i(2m)^{-1}[(i\not{p} + m)\gamma\cdot\mathbf{p}/|\mathbf{p}|]_{\alpha\beta} \tag{5-A.12}$$

Lastly we define light-front, left-handed tachyon spinors by

$$\begin{aligned} u_{TL}{}^{+}(p, s) &= C^- R^+ S_L(\Lambda_L(\omega, \mathbf{u}))w^1(0) \\ u_{TL}{}^{+}(p, -s) &= C^- R^+ S_L(\Lambda_L(\omega, \mathbf{u}))w^2(0) \\ v_{TL}{}^{+}(p, s) &= C^- R^+ S_L(\Lambda_L(\omega, \mathbf{u}))w^3(0) \\ v_{TL}{}^{+}(p, -s) &= C^- R^+ S_L(\Lambda_L(\omega, \mathbf{u}))w^4(0) \end{aligned} \tag{5-A.13}$$

$$\begin{aligned} u_{TL}{}^{+\dagger}(p, s) &= w^{1T}(0)\, S_L{}^{\dagger}(\Lambda_L(\omega, \mathbf{u}))\, R^+C^- \\ u_{TL}{}^{+\dagger}(p, -s) &= w^{2T}(0)\, S_L{}^{\dagger}(\Lambda_L(\omega, \mathbf{u}))R^+C^- \\ v_{TL}{}^{+\dagger}(p, s) &= w^{3T}(0)\, S_L{}^{\dagger}(\Lambda_L(\omega, \mathbf{u}))R^+C^- \\ v_{TL}{}^{+\dagger}(p, -s) &= w^{4T}(0)\, S_L{}^{\dagger}(\Lambda_L(\omega, \mathbf{u}))R^+C^- \end{aligned} \tag{5-A.14}$$

where the superscript "T" indicates the transpose and † indicates hermitean conjugate.

END OF AMENDED CHAPTERS 2 & 3 EXTRACT FROM BLAHA (2007b)

Appendix 5-B. Experimental Evidence for Faster-Than-Light Particles & Physics

BEGINNING OF CHAPTER 3 EXTRACT FROM BLAHA (2013b)

Among the key assumptions of our extended Standard Models are 1) that the speed of light is the same in all inertial reference frames and 2) that some fundamental particles (neutrinos and down-type quarks) travel faster than the speed of light.

In this chapter we describe convincing evidence for faster than light physics. In chapter 4 we will show the need for a complex space-time that, suitably formulated, has a transformation between inertial reference frames that preserves the constancy of the speed of light in all inertial reference frames.

Until 1907 physicists thought that there was no limit on the speed of a particle or lump of matter. In 1907 Einstein and Poincaré showed that there was an inherent limit on the speed of a massive object – the speed of light. For the past 100 odd years physicists have generally accepted the speed of light as the limiting speed for particles with mass. Several theoretical physicists in the 1960's (E. C. Sudarshan and Gerald Feinberg) investigated the possibility of faster than light particles. They found that faster than light particles were theoretically possible but their theories – particularly their quantum field theories – had numerous discrepancies from canonical quantum field theory. These differences were taken by many to indicate that faster than light particles (called tachyons) were not present in nature. This belief was further supported by the happenings at particle accelerators where it was impossible to accelerate normal charged particles such as protons faster than the speed of light.

In the past ten years this author[89] developed a satisfactory quantum field theory of faster than light particles and found that if neutrinos and down-type quarks were faster than light particles he could derive the form of The Standard Model of Elementary Particles in detail. This theoretical development seems to have stimulated experimental groups at the new Linear Hadron Collider (LHC) at the CERN laboratory in Switzerland and the Gran Sasso Laboratory in Italy to measure the speed of neutrinos emitted in LHC particle collisions. The results, described below, were mixed and one can fairly say they neither proved nor disproved that neutrinos were tachyons.

However there is other experimental data that strongly indicate that neutrinos are tachyons, and that quantum mechanics requires – not just faster than light behavior – but in some circumstances instantaneous effects at a distance – infinite speed of transmission!

In this chapter we will look at experimentally proven instantaneous Quantum Mechanical effects, at tritium decay experiments over the past 20 years that imply faster than light neutrinos, at neutrino speed measurements at the CERN LHC and Gran Sasso, at tachyonic particle behavior inside of Black Holes, and at the tachyonic behavior of Higgs particles, the "so-called God particle." *The cumulative result of these considerations is that faster than light particles, and physics, are a part of nature.*

---

[89] See Blaha (2012b) and earlier books extending back nine years.

5-B.1 Instantaneous Quantum Mechanical Effects

Quantum entanglement is a quantum phenomenon wherein parts of a physical system are in a certain quantum state but are separated by a space-like distance. If a change is made in part of a quantum entangled system then it is known theoretically, and experimentally, that other parts of the system change instantaneously.[90] Many experiments have shown that the change in other parts of a system is instantaneous and thus can be viewed as taking place at infinite speed – obviously beyond the speed of light.[91] The most recent experiment by Juan Yin et al[92] has shown directly that quantum mechanical effects travel faster than 10,000 times the speed of light. These experimental results are consistent with the instantaneous speed predicted by quantum mechanics. Thus faster than light behavior is implicit in quantum theory and is experimentally verified.

5-B.2 Tritium Decay Experiments Yielding Neutrinos

Fact: Particles with negative values for the square of their mass are tachyons – particles moving faster than light.

A series of experiments by various groups over recent years imply that electron neutrinos produced in tritium decay have negative mass squared despite the best efforts of experimenters to obtain positive values for the neutrino mass squared.

Experiment	measured mass squared	Year
Mainz	$-1.6 \pm 2.5 \pm 2.1$	2000
Troitsk	$-1.0 \pm 3.0 \pm 2.1$	2000
Zürich	$-24 \pm 48 \pm 61$	1992
Tokyo INS	$- 65 \pm 85 \pm 65$	1991
Los Alamos	$- 147 \pm 68 \pm 41$	1991
Livermore	$- 130 \pm 20 \pm 15$	1995
China	$- 31 \pm 75 \pm 48$	1995
1998 Average	-27 ± 20	1998

Table 5-B.1 Electron neutrino mass squared values found in various tritium decay experiments. (Masses are in units of eV.) The average mass squared is negative suggesting electron neutrinos are tachyons.

Table 5-B.1 summarizes the measured electron mass squared in these experiments. These experiments strongly suggest that neutrinos have negative mass squared and are thus faster-than-light particles - tachyons. However their small masses indicate that they only exceed the speed of light by a small amount.

[90] Matson, John, "Quantum Teleportation Achieved Over Record Distances" *Nature* **13**, August 2012.

[91] Francis, Matthew, "Quantum Entanglement Shows that Reality Can't be Local", *Ars Technica*, 30 October 2012.

[92] Juan Yin et al, arXiv[quant-ph]: 1303.0614V1 (March 4, 2013).

5-B.3 LHC/Gran Sasso Direct Measurements of Neutrino Speeds

Two groups performed experiments at Gran Sasso Laboratory in Italy. They detected neutrinos emitted in interactions at the CERN LHC in Switzerland. The LVD collaboration in an exhaustive study of neutrino velocities found that the question was still open according to their data. Their refereed Physical Review Letter Abstract stated:

> We report the measurement of the time of flight of ν_μ on the CNGS baseline (732 km) with the Large Volume Detector (LVD) at the Gran Sasso Laboratory. The CERN-SPS accelerator has been operated from May 10th to May 24th 2012, with a tightly bunched-beam structure to allow the velocity of neutrinos to be accurately measured on an event-by-event basis. LVD has detected 48 neutrino events, associated with the beam, with a high absolute time accuracy. These events allow us to establish the following limit on the difference between the neutrino speed and the light velocity: $-3.8\times10^{-6} < (v_\nu - c)/c < 3.1\times10^{-6}$ (at 99% C.L.). This value is an order of magnitude lower than previous direct measurements.[93]

These results (involving at least 35 neutrino detections) slightly favor, and do not rule out, faster-than-light neutrinos. Another experiment at the same locations by the ATLAS group stated that they found neutrino velocities (Five neutrinos were measured.) were below c. This group has not published their results as yet. We conclude that the published data appears to support faster than light neutrinos – consistent with our theory of The Standard Model.

A new project is in the planning stages to measure neutrino beams at larger distances. The hope is that the masses of the various neutrinos will be determined by the experiment. If the neutrino mass squared values turn out to be negative then it will constitute additional proof that neutrinos are tachyons (confirming tritium decay data), and thus support this author's formulation of The Standard Model of Elementary Particles.

5-B.4 Tachyonic Behavior Within Black Holes

Inside a black hole (such as the Schwarzschild solution of General Relativity) the time coordinate effectively becomes a spatial coordinate and the radius coordinate effectively becomes a time coordinate. An in-falling particle has a constantly decreasing radial distance from the center of the black hole just as time always increases outside a black hole.

As a result of the interchange of the roles of time and radius the velocity of a particle descending radially inside a Black Hole has a speed faster than light and is tachyonic.

5-B.5 Higgs Fields are Tachyons

Recently groups at the LHC CERN laboratory have announced the discovery of Higgs particles. The dynamic equations for Higgs bosons in The Standard Model have a negative mass squared. The mass squared must be negative or the Higgs Mechanism could not generate particle masses. Having negative mass terms implies that Higgs fields are tachyonic – faster than light particles. Their tachyonic nature is masked by a quartic self-interaction that generates a condensate and thereby the masses of other particles.

---

[93] N. Yu. Agafonova et al. (LVD Collaboration), "Measurement of the Velocity of Neutrinos from the CNGS Beam with the Large Volume Detector" Phys. Rev. Lett. **109**, 070801 (15 August 2012).

5-B.6 Conclusion: Faster-Than-Light Particles – Tachyons Exist in Nature

The bulk of the experimental and theoretical evidence presented in previous sections strongly favors the existence of faster-than-light particles such as neutrinos. Tachyonic neutrinos are an important part of our form of The Standard Model. This form of the theory also strongly suggests that quarks are tachyonic in parallel with tachyonic neutrinos in order to obtain the symmetries of The Standard Model.

END OF AMENDED CHAPTER 3 EXTRACT FROM BLAHA (2013b)

Appendix 5-C. Phenomena Beyond the Light Barrier

BEGINNING OF APPENDIX 1-A EXTRACT FROM BLAHA (2007b) AND EARLIER WORK

5-C.1 Superluminal (Faster-than-Light) Transformations

In this Appendix we will briefly survey some of the very different features of faster-than light physical phenomena. We will frame our discussion in terms of the two simple reference frames depicted in Fig. 5-C.1. The prime frame is moving at a speed $v > c$ (the speed of light) in the positive x direction with respect to the unprimed reference frame.

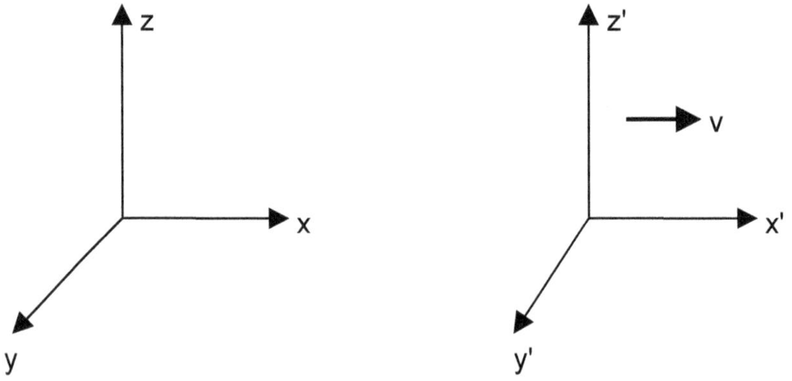

Figure 5-C.1. Two coordinate systems having a relative speed v in the x direction.

As shown later in the text we define a superluminal (faster-than-light) transformation between coordinates in these reference frames with (Eqd. 2.16 and 2.13 are in Blaha (2007b))

$$\begin{aligned} t' &= \gamma_s(t - \beta x/c) \\ x' &= \gamma_s(x - \beta ct) \\ y' &= iy \\ z' &= iz \end{aligned} \tag{2.16}$$

where

$$\gamma_s = (\beta^2 - 1)^{-\frac{1}{2}} \tag{2.13}$$

and $\beta = v/c > 1$. The appearance of imaginary values for y' and z' is not a cause for alarm. An observer resident in the prime coordinate system will measure real y and z distances with a ruler. The only purpose of the factors of i is to relate the y and z coordinates to y' and z'. An observer in either coordinate system will view his/her coordinates as real.

The energy and momentum of a tachyon (faster-than-light) particle of mass m traveling at a speed v > c is

$$E = \gamma_s mc^2 \tag{5-C.1}$$

and

$$\mathbf{p} = m\gamma_s \mathbf{v} \tag{5-C.2}$$

Note that the tachyon defining condition is satisfied:

$$E^2 - c^2\mathbf{p}^2 = -m^2c^4 \tag{5-C.3}$$

Also note that in the limit $\beta \to \infty$ that

$$E = 0 \tag{5-C.4}$$

and

$$p = mc \tag{5-C.5}$$

where $p = |\mathbf{p}|$. Tachyons are always in motion. The minimal momentum of a tachyon is given by eq. 5-C.5. It corresponds to zero energy. It is the tachyon equivalent of Einstein's famous $E = mc^2$.

5-C.2 Length Dilations and Time Contractions

In ordinary Lorentz transformations a moving ruler will appear to be shorter in the direction of its motion when measured in another reference frame. This phenomenon is called *Lorentz contraction.*

Superluminal Length Dilation/Contraction

In the case of a superluminal transformation we find precisely the opposite effect, *superluminal length dilation*, is a possibility. Consider the case of the transformation of eq. 2.16 above (coresponding to Fig. 5-C.1), which relates the prime reference frame traveling at speed v in the positive x direction to the unprimed reference frame. A ruler perpendicular to the x-axis will have the same length in both reference frames if its endpoints are simultaneously measured – perhaps by photographing it. The y and z equations in eqs. 2.16 specify this fact up to an extraneous factor of i.

If the ruler is at rest in the prime reference frame and parallel to the x' axis, then a simultaneous measurement of its endpoints at the same time t_0 by an observer in the unprimed reference frame (perhaps by photographing it) will reveal both *length contraction and dilation* depending on the value of β. If the length is $L' = x'_2 - x'_1$ in the prime frame and $L = x_2 - x_1$ in the unprimed frame, then the equations:

$$x'_1 = \gamma_s(x_1 - \beta ct_0) \tag{5-C.7}$$
$$x'_2 = \gamma_s(x_2 - \beta ct_0) \tag{5-C.8}$$

imply

$$L' = \gamma_s L = (\beta^2 - 1)^{-\frac{1}{2}} L \tag{5-C.9}$$

Thus we have three cases:

Case 1: $\beta \in <1, \sqrt{2}>$:	$L < L'$	Contraction	(5-C.10)
Case 2: $\beta = \sqrt{2}$:	$L = L'$	Equality	(5-C.11)

Case 3: $\beta \in <\sqrt{2}, \infty>$: $L > L'$ Dilation (5-C.12)

Superluminal Time Contraction/Dilation

In the case of a superluminal transformation we find *superluminal time contraction* is a possibility. Consider again the case of the transformation of eq. 2.16 above coresponding to Fig. 5-C.1 relating the prime reference frame traveling at speed v in the positive x direction to the unprimed reference frame. Consider the time interval between two events occurring at the same point x'_0 in the prime reference frame. From the viewpoint of an observer in the unprimed frame the events take place at different points x_1 and x_2. If the time interval is $T' = t'_2 - t'_1$ in the prime frame and $T = t_2 - t_1$ in the unprimed frame, then the inverse of eqs. 2.16 give:

$$t_1 = \gamma_s(t'_1 + \beta x'_0/c) \quad (5\text{-C.13})$$
$$t_2 = \gamma_s(t'_2 + \beta x'_0/c) \quad (5\text{-C.14})$$

and imply

$$T = \gamma_s T' = (\beta^2 - 1)^{-1/2} T' \quad (5\text{-C.15})$$

Again we have three cases:

Case 1: $\beta \in <1, \sqrt{2}>$: $T > T'$ Dilation (5-C.16)

Case 2: $\beta = \sqrt{2}$: $T = T'$ Equality (5-C.17)

Case 3: $\beta \in <\sqrt{2}, \infty>$: $T < T'$ Contraction (5-C.18)

The time interval in the unprimed frame can be less than, equal to, or greater than the time interval in the frame where the events take place at the same spatial point.

Thus superluminal transformations are more complex than Lorentz transformations with respect to space and time, dilation and contraction.

5-C.3 Tachyon Fission to More Massive Particles – Reverse Fission

Another way in which faster-than-light phenomena differ from sublight phenomena is particle fission. Normally when a particle or nucleus decays or fissions the masses of the particles produced by the decay are smaller than the mass of the original particle or nucleus. And energy is released. We are familiar with fission as the source of nuclear energy.

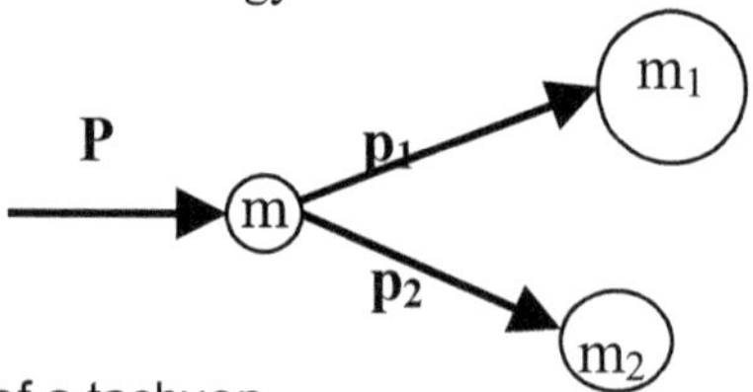

Figure 5-C.2. Two particle decay of a tachyon.

In the case of faster-than-light particles, tachyons, a much different possibility is present: a tachyon can decay into heavier tachyons: *a particle's spatial 3-momentum can be transformed into mass.*

We will consider the specific case of a tachyon decaying into two particles to illustrate this possibility. (See Fig. 5-C.2.)

We will assume the initial tachyon has zero energy[94] and thus the tachyons emerging from the decay also have zero energy. The analysis is based on conservation of total energy and momentum.

Momentum conservation implies

$$\mathbf{P} = \mathbf{p}_1 + \mathbf{p}_2 \tag{5-C.19}$$

Since all energies are zero

$$\begin{aligned}(cP)^2 &= (c\mathbf{P})^2 = m^2 \\ (cp_1)^2 &= (c\mathbf{p}_1)^2 = m_1^2 \\ (cp_2)^2 &= (c\mathbf{p}_2)^2 = m_2^2\end{aligned} \tag{5-C.20}$$

where $P = |\mathbf{P}|$, $p_1 = |\mathbf{p}_1|$, and $p_2 = |\mathbf{p}_2|$. If we now square eq. 5-C.19 and use eqs. 5-C.20 we obtain

$$m^2 = m_1^2 + m_2^2 + 2m_1m_2 \cos\theta \tag{5-C.21}$$

where θ is the angle between the emerging particles momenta $\mathbf{p}_1$ and $\mathbf{p}_2$.

Eq. 5-C.21 has a number of interesting cases:

Case $\theta = 0$:

$$m = m_1 + m_2 \tag{5-C.22}$$

The masses of the outgoing tachyons sum to the mass of the original tachyon.

Case $\theta = \pi/2$:

$$m^2 = m_1^2 + m_2^2 \tag{5-C.23}$$

The masses of each outgoing tachyon is less than the mass of the original tachyon.

Case $\theta = \pi$:

$$m^2 = (m_1 - m_2)^2 \tag{5-C.24}$$

In this case either $m_1 > m$ or $m_2 > m$. Thus one of the outgoing tachyons has a greater mass than the original tachyon. Mass is effectively created from the spatial momentum of the particle. This process is the inverse of normal particle decay or fission where the sum of the outgoing masses is always less than the original particle's mass and the difference is mass converted into energy in the form of additional photons via $E = mc^2$.

This last case, where one of the outgoing particles is more massive than the original particle, is not just for $\theta = \pi$. Since

$$\cos\theta = (m^2 - m_1^2 - m_2^2)/(2m_1m_2) \tag{5-C.25}$$

---

[94] If a particle has zero energy its velocity is infinite so the case considered is somewhat artificial. However the results would still be approximately true for very large velocities. The simplicity of the kinematics led us to consider this case.

we see that *the sum of the outgoing tachyon masses is always greater than the original tachyon mass (except when $\theta = 0$)* since

$$\cos\theta = 1 + [m^2 - (m_1 + m_2)^2]/(2m_1m_2) \le 1 \qquad (5\text{-}C.26)$$

and thus

$$[m^2 - (m_1 + m_2)^2]/(2m_1m_2) \le 0 \qquad (5\text{-}C.27)$$

Note $m = m_1 + m_2$ only if $\theta = 0$.

Since we can transform the above discussion to the case of tachyons with a non-zero energy using an ordinary Lorentz transformation the above discussion in this subsection is general.

We therefore conclude that when a tachyon decays into two tachyons the sum of the masses of the produced tachyons is greater than the mass of the original tachyon except if the angle between the momenta of the produced tachyons is zero. In that case the sum of the masses of the produced tachyon equals the mass of the original tachyon.

*Thus tachyons can engage in **reverse fission** in which **momentum is converted into mass so the outgoing particles have a total mass greater than the incoming particle**.* In the case of "normal" fission part of the mass of a particle can be converted to energy and the sum of the masses of the decay product particles is less than the mass of the original particle.

5-C.4 Light Chasing Faster-than-Light Particles?

Einstein told a story that he imagined positioning himself in a (Galilean) reference frame moving at the speed of light and seeing electromagnetic waves "frozen" in time so that they were no longer vibrating. This vision inspired him to reconsider the transformation laws between coordinate systems and to derive the theory of Special Relativity. In Special Relativity the speed of light is the same in all reference frames.

In this subsection we will consider a light pulse from the points of view of two reference frames whose relative speed v is greater than the speed of light. We will use the example considered earlier and add a pulse of light traveling in the positive x direction. (See Fig. 5-C.3.)

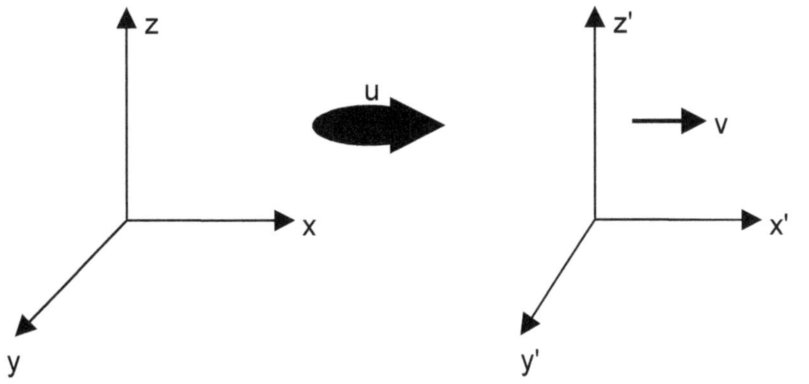

Figure 5-C.3. Two coordinate systems having a relative speed v in the x direction. A pulse of light is displayed as a thick arrow.

The general law for the addition of velocities in a situation such as depicted in Fig. 5-C.3 is well known. If we adapt it to the present example and let u be the speed of the pulse in the unprimed frame (temporarily forgetting it is a light pulse) we find it implies

$$u' = (u - \beta c)/(1 - \beta u/c) \tag{2.17a}$$

where $\beta = v/c > 1$, and u' is the speed of the pulse in the prime frame. Then if we set u = c we see that u' = c as well. *Thus our superluminal transformations preserve the constancy of the speed of light just like Lorentz transformations.*

As a result the pulse of light will intersect the z' axis eventually. However if superluminal transformations did not preserve the speed of light in all frames the pulse might never reach the z' axis. For example under a Galilean transformation the speed of the pulse would be u' = u – v = c – v and the pulse would actually be falling further and further behind the z' axis.

5-C.5 Electromagnetic Field of a Charged Tachyon – A Pancake Effect?

The electric field of a charge q at rest in a reference frame is:

$$\mathbf{E} = (q/(4\pi\varepsilon_0)\check{\mathbf{r}}/r^2 \tag{5-C.28}$$

in spherical coordinates where $\check{\mathbf{r}}$ is a unit vector in the radial direction.

Sublight Charged Particle

The electric and magnetic fields of a charge q moving in the positive x direction with speed v < c are

$$\mathbf{E} = (q/(4\pi\varepsilon_0)\check{\mathbf{r}}(1 - \beta^2)/[r^2(1 - \beta^2\sin^2\theta)^{\frac{3}{2}}] \tag{5-C.29}$$
$$\mathbf{B} = (q/(4\pi\varepsilon_0)\check{\mathbf{r}}\beta(1 - \beta^2)\sin\theta/[r^2(1 - \beta^2\sin^2\theta)^{\frac{3}{2}}] \tag{5-C.30}$$

where $\check{\mathbf{r}}$ is the radial unit vector, $\beta = v/c$, and θ is measured with respect to the polar axis which is taken to be the x axis. As $\beta \rightarrow 1$ the electric and magnetic fields develop a "pancake" form with large field strengths in the directions perpendicular to the direction of motion similar to the transverse fields of electromagnetic quanta. This feature is the basis of the Weizsäcker-Williams method of virtual quanta.

Charged Tachyon

The electric and magnetic fields of a tachyon of charge q moving in the positive x direction with speed $v > c$ are

$$\mathbf{E} = (q/(4\pi\varepsilon_0)\check{\mathbf{r}}(\beta^2 - 1)/[r^2(\beta^2\sin^2\theta - 1)^{\frac{3}{2}}] \quad (5\text{-C.31})$$

$$\mathbf{B} = (q/(4\pi\varepsilon_0)\check{\mathbf{r}}\beta(\beta^2 - 1)\sin\theta/[r^2(\beta^2\sin^2\theta - 1)^{\frac{3}{2}}] \quad (5\text{-C.32})$$

where $\beta = v/c > 1$, and θ is again measured with respect to the polar axis which is taken to be the x axis. In the case of tachyons there are three cases of interest.

Case $\beta^2\sin^2\theta - 1 < 0$:
The electric and magnetic fields are pure imaginary and are excluded from the forward and backward cones surrounding the x axis defined by $|\sin\theta| < \beta^{-1}$.
Case $\beta^2\sin^2\theta - 1 = 0$:
The electric and magnetic fields are infinite. Thus the field strengths are infinite on a cone at the angle θ with respect to the x-axis. By comparison, a magnetic monopole only has a one-dimensional, singularity line extending from the monopole to infinity.
Case $\beta^2\sin^2\theta - 1 > 0$:
The electric and magnetic fields decrease in strength as $\sin^2\theta$ increases. Thus the region of maximum field strength are the forward and backward cones where $|\sin\theta|$ is greater than but near β^{-1} in value. The pancake picture of the sublight charged particle does not hold for charged tachyons.

Are Tachyonic Cones in the Au-Au Scattering Quark-Gluon Plasma?

Cones have been observed in high energy Au-Au scattering in which a quark-gluon plasma is created. These cones have been attributed to a variety of causes such as hydrodynamically generated Mach cones, and Cherenkov radiation. The possibility exists that tachyonic excitations may transiently exist in the quark-gluon plasma and may, in part, explain the observed cones and dips. The above described cones in the case of a moving charged tachyon are remarkably similar in character. See the CERES collaboration paper arXiv:nucl-ex/0701023, and references therein, for experimental findings.

5-C.6 Superluminal (Tachyon) Physics is Different

The simple classical examples presented in this appendix demonstrate that superluminal physics has many interesting new features that are worthy of interest. Since tachyons exist in Black Holes, and, perhaps, in other contexts, their study is a worthwhile endeavor.

END OF APPENDIX 1-A EXTRACT FROM BLAHA (2007b)

Appendix 5-D. Superluminal (Faster Than Light) Kinetic Theory and Thermodynamics

BEGINNING OF CHAPTER 40 EXTRACT FROM BLAHA (2012a)

This chapter changes the flow of topics of these volumes from gravitation and particle theory to superluminal many particle dynamics. We will progress from superluminal Kinetic theory to Thermodynamics. We will see that there are strong similarities with non-relativistic Thermodynamics.

5-D.1 Superluminal Kinetic Theory

Assemblages of large numbers of particles embody the Maxwell-Boltzmann distribution. The Boltzmann H theorem is the beginning point for derivations of the non-relativistic Maxwell-Boltzmann distribution. The non-relativistic Maxwell-Boltzmann distribution has the form

$$f(\mathbf{v}, \mathbf{r}) = n(m/(2\pi kT))^{3/2}\exp\{-[m(\mathbf{v} - \mathbf{v}_0)^2/2 + V(r)]/(kT)\} \tag{5-D.1}$$

where n is the particle density, T is the temperature, $\mathbf{v}_0$ is the average velocity, m is the particle mass, V(r) is an external conservative force, and k is Boltzmann's constant. In terms of a hamiltonian

$$H(\mathbf{v}, \mathbf{r}) = m\mathbf{v}^2/2 + V(r) \tag{5-D.2}$$

we can express the Maxwell-Boltzmann distribution as

$$f(\mathbf{v}, \mathbf{r}) = n(m/(2\pi kT))^{3/2}\exp\{-H(\mathbf{v} - \mathbf{v}_0, \mathbf{r})/(kT)\} \tag{5-D.3}$$

5-D.1.1 Relativistic Form of the Maxwell-Boltzmann Distribution

If we assume that we have a container containing a distribution of relativistic (sublight) particles with an average velocity $\mathbf{v}_0 = 0$, and no external force, then the form of eq. 5-D.3 generalizes to the relativistic Maxwell-Boltzmann distribution

$$f_R(\mathbf{v}) = C_R \exp\{-H/(kT)\} \tag{5-D.4}$$

where C_R is a normalization constant and H is the relativistic hamiltonian for a free particle:

$$H = c(m^2c^2 + \mathbf{p}^2)^{1/2} \tag{5-D.5}$$

with $\mathbf{p} = \gamma m\mathbf{v}$ and $\gamma = (1 - v^2/c^2)^{-1/2}$. C_R is determined by the condition

$$\int d^3v f_R(\mathbf{v}) = 1 \quad (5\text{-D.6})$$

5-D.1.2 Superluminal Form of the Maxwell-Boltzmann Distribution

The superluminal form of Maxwell-Boltzman distribution is based on the form of the mass shell condition for superluminal particles:

$$E^2 - c^2\mathbf{p}^2 = m^2c^4 \quad (5\text{-D.7})$$

which implies a free hamiltonian

$$H_S = c(\mathbf{p}^2 - m^2c^2)^{1/2} \quad (5\text{-D.8})$$

where

$$\mathbf{p} = \gamma_s m\mathbf{v} \quad (5\text{-D.9})$$

and

$$\gamma_s = (v^2/c^2 - 1)^{-1/2}$$

The seemingly slight difference between eqs. 5-D.8 and 5-D.9, and eq. 5-D.5 causes major differences between superluminal and relativistic kinetic theory and thermodynamics. On the other hand relativistic kinetic theory and thermodynamics are qualitatively similar in many ways with their non-relativistic counterparts.

One major difference is the behavior of kinematic variables near the speed of light:

As $v \to c$ Below the Speed of Light

$$p \to \infty$$
$$H \to \infty$$

As $v \to c$ From Above the Speed of Light

$$p \to \infty$$
$$H_S \to \infty$$

As $v \to \infty$

$$p \to mc$$
$$H_S \to 0$$

Thus as v ranges from c to ∞, H_S decreases monotonically from ∞ to zero and p decreases from ∞ to mc. This behavior contrasts with H in eq. 5-D.5, which increases monotonically with p as v increases from 0 to c. Thus the sublight Maxwell-Boltzmann distribution decreases with v as v increases from 0 to c.

The superluminal Maxwell-Boltzmann distribution *increases* with v as v increases from c to ∞ as we see below. The superluminal Maxwell-Boltzmann distribution decreases with p as p increases from mc to ∞. *As a result the natural physical parametrization of the Maxwell-Boltzmann distribution should be in terms of the momentum rather than the velocity.* Thus Boltzmann's H function which normally is

$$H_B(t) = \int d^3v\, f(\mathbf{v}, t) \log f(\mathbf{v}, t)$$

must be replaced with[95]

$$H_{BS}(t) = \int d^3p\, f_S(\mathbf{p}, t) \log f_S(\mathbf{p}, t)$$

The equilibrium superluminal Maxwell-Boltzman distribution can be derived from $H_S(t)$. It has the same general form as the relativistic distribution

$$f_S(\mathbf{p}) = C_S \exp\{-H_S/(kT)\} \tag{5-D.10}$$

where C_S is a normalization constant and H_S is the superluminal hamiltonian for a free particle.

We now apply the normalization condition[96]

$$n = N/V = \int d^3p f_S(\mathbf{p}) = C_S \int d^3p \exp\{-H_S/(kT)\} \tag{5-D.11}$$

where n is the particle density, N is the number of particles in the system, and V is the volume of the system. We calculate C_S by evaluating the integral:

$$n = 4\pi C_S \int_m^\infty dp\, p^2 \exp\{-H_S/(kT)\} \tag{5-D.12}$$

Letting $x = p/(mc)$ and $\alpha = mc^2/(kT)$ we see eq. 5-D.12 becomes

$$n = 4\pi m^3c^3C_S \int_1^\infty dx\, x^2 \exp\{-\alpha(x^2 - 1)^{1/2}\} \tag{5-D.13}$$

Then letting $y^2 = x^2 - 1$ we find

$$n = 4\pi m^3c^3C_S \int_0^\infty dy\, y(y^2 + 1)^{1/2} \exp(-\alpha y)$$

$$= -m^3c^3C_S G^{31}{}_{13}(\alpha^2/4 \mid {}^{0}_{-3/2,0,\,1/2}) \tag{5-D.14}$$

where $G^{31}{}_{13}(\ldots)$ is Meijer's G-Function.[97] Therefore

$$C_S = -[m^3c^3G^{31}{}_{13}((mc^2/(2kT))^2 \mid {}^{0}_{-3/2,0,\,1/2})/n]^{-1} \tag{5-D.15}$$

The most probable momentum of a particle p_p is the maximum of

$$p_p = \text{Max}\{p^2\exp[-H_S/(kT)]\}$$

$$= \{(2(kT)^2/c^2)[1 + (1 - m^2c^4/(kT)^2)^{1/2}]\}^{1/2} \tag{5-D.16}$$

For large T or small T the maximum is

---

[95] Note the additional factor of m^3 in $\int d^3p$ will be absorbed in the normalization (eq. 5-D.11).

[96] We note that using $\int d^3v$ rather than $\int d^3p$ in eq. 5-D.11 would result in a divergence – another reason for our choice of integration parameter.

[97] See Gradshteyn (1965) integral 3.389.2 and p. 1068 for the properties of Meijer's G-Function.

$$p_p \approx 2kT/c \ > mc$$

The velocity v_p corresponding to the maximum in the momentum is

$$v_p = cp_p/(p_p{}^2 - m^2c^2)^{1/2}$$

For large T or small T, the velocity v_p corresponding to the maximum in the momentum is approximately

$$v_p \approx c + {1/2}\, m^2c^5/(2kT)^2$$

5-D.2 Superluminal Thermodynamics

Turning now to the thermodynamics of a dilute superluminal gas implied by the superluminal Maxwell-Boltzman distribution we begin by calculating the average energy per particle

$$\varepsilon = C_S \int d^3p\, H_S \exp[-H_S/(2kT)] \Big/ \int d^3p\, C_S \exp[-H_S/(kT)] \tag{5-D.17}$$

$$= (C_S/n) \int d^3p\, H_S \exp[-H_S/(kT)]$$

$$= (C_S/n)2kT\alpha\, 4\pi m^2c^3 \int_0^\infty dy\, y^2(y^2 + 1)^{1/2} \exp(-\alpha y)$$

$$= -(C_S/n)m^3c^5G^{31}{}_{13}(\alpha^2/4\, |^{-1/2}{}_{-2,0,\, 1/2})$$

$$= mc^2\, G^{31}{}_{13}((mc^2/(2kT))^2|^{-1/2}{}_{-2,0,\, 1/2}) \Big/ G^{31}{}_{13}((mc^2/(2kT))^2|^{0}{}_{-3/2,0,\, 1/2})$$

(5-D.18)

The Maxwell-Boltzman normalization factor is related to the energy per particle by

$$C_S = -n\varepsilon/(m^3c^5G^{31}{}_{13}(\alpha^2/4\, |^{-1/2}{}_{-2,0,\, 1/2})) \tag{5-D.19}$$

Note that C_S is proportional to the energy in contrast to the non-relativistic case where the Maxwell-Boltzman normalization factor $C = (3m/(4\pi\varepsilon))^{3/2}$.

We now calculate the superluminal pressure for the case of a distribution of superluminal particles bouncing on a wall perpendicular to the z-axis. The wall is assumed to be a perfectly reflecting plane. The pressure is the average force per unit area due to the gas of superluminal particles. The number of particles bombarding the wall per second is with $v_z > 0$ is $v_z f_S(\mathbf{p})d^3p$. Thus the pressure is

$$P = \int d^3p\, 2p_z v_z f_S(\mathbf{p}) \tag{5-D.20}$$

where the particle momentum changes by $2p_z$ due to reflection. Due to spherical symmetry one expects the average values for the various components of **v** to be equal. Consequently we can re-express eq. 5-D.20 as

$$P = 1/3 \int d^3p\, 2m\gamma_s v^2 f_S(\mathbf{p}) \tag{5-D.21}$$

$$= 1/3 \int d^3p2pvf_S(\mathbf{p})$$

Since

$$v = cp/(p^2 - m^2c^2)^{½} \quad (5\text{-D.22})$$

we see

$$P = 8\pi c/3 \int_m^\infty dpp^4 f_S(\mathbf{p})/(p^2 - m^2c^2)^{½}$$

Following steps similar to eqs. 5-D.12 – 5-D.15 leads to

$$P = m^4c^4\, C_S G^{31}{}_{13}((mc^2/(2kT))^2|\,^{½}{}_{-2,0,\,½}) \quad (5\text{-D.23})$$

The *equation of state* relating the pressure and energy is

$$P = -(m/c)\{G^{31}{}_{13}((mc^2/(4kT))^2|\,^{½}{}_{-2,0,\,½})/G^{31}{}_{13}(\alpha^2/4\,|^{-½}{}_{-2,0,\,½}))\}n\varepsilon \quad (5\text{-D.24})$$

Substituting for ε we find

$$P = -(nm^2c)\{G^{31}{}_{13}(\rho\,|\,^{½}{}_{-2,0,\,½})/\,G^{31}{}_{13}(\rho\,|^{0}{}_{-3/2,0,\,½})\} \quad (5\text{-D.25})$$

where

$$\rho = (mc^2/(2kT))^2 \quad (5\text{-D.26})$$

Turning now to the consideration of a dilute gas the internal energy of the gas can be defined to be[98]

$$U(t) = N\varepsilon \quad (5\text{-D.27})$$

We note that the work done by the superluminal gas if its volume increases by dV is PdV. Then the superluminal (and usual) form of the first law of thermodynamics is

$$dQ = dU + PdV \quad (5\text{-D.28})$$

where Q is the heat absorbed. The heat capacity of the system for constant volume is

$$C_V = (\partial U/\partial T)_V \quad (5\text{-D.29})$$

The second law of thermodynamics, Boltzmann's H theorem, is based on

$$H = -S/kV \quad (5\text{-D.30})$$

[98] The internal energy of a gas of non-interacting non-relativistic particles is U(t) = 3NkT/2. In the superluminal case it appears that it is eq. 5-D.18.

where H is the negative of the entropy divided k times the volume V. In systems where there are no superluminal particles, the H theorem states that the entropy never decreases for an isolated gas of fixed volume.

We can calculate H for a superluminal system under equilibrium conditions, H_e, from[99]

$$H_e = \int d^3p f_S(\mathbf{p}) \ln(f_S(\mathbf{p})) \tag{5-D.31}$$

$$= \int d^3p f_S(\mathbf{p})[\ln C_S - H_S/(kT)] \tag{5-D.32}$$

$$= n \ln C_S - \int d^3p f_S(\mathbf{p}) H_S/(kT)$$

$$= n \ln C_S - n\varepsilon/(kT) \tag{5-D.33}$$

by eqs. 5-D.11 and 5-D.17. Therefore

$$S = -kVH_{Se} = -kN \ln C_S + N\varepsilon/T \tag{5-D.34}$$

Consequently we obtain the superluminal *and* standard non-relativistic result

$$1/T = (\partial S/\partial U)_x \tag{5-D.35}$$

where x represents all other extensive variables.

5-D.3 Approximate Calculation of Kinetic and Thermodynamic Quantities

We can obtain more tractable expressions for kinetic and thermodynamic quantities by assuming $\mathbf{p}^2 \gg m^2c^2$ and approximating the hamiltonian (eq. 5-D.8) with

$$H_{Sa} = cp \tag{5-D.36}$$

The approximate normalization condition is

$$n = N/V = \int d^3p f_{Sa}(\mathbf{p}) = C_{Sa} \int d^3p \exp\{-H_{Sa}/(kT)\} \tag{5-D.37}$$

where n is the particle density, N is the number of particles in the system, and V is the volume of the system. C_S is determined by

$$n = 4\pi C_{Sa} \int_{mc}^{\infty} dp\, p^2 \exp\{-cp/(kT)\} \tag{5-D.38}$$

Letting $\alpha = c/(kT)$ we see eq. 5-D.38 becomes

---

[99] We consistently assume that integrals over the momentum $\int d^3p$ are the proper integration (rather than integrations over velocity $\int d^3v$) because, for example, the calculation of the normalization constant eq. 5-D.11 would diverge if the integration were over $\int d^3v$.

$$n = 4\pi C_{Sa}\, d^2/d\alpha^2 \int_{mc}^{\infty} dp\, \exp(-\alpha p)$$

$$= 4\pi C_{Sa}\, d^2/d\alpha^2\, [(1/\alpha) \exp(-\alpha mc)] \qquad (5\text{-D.39})$$

Therefore the normalization factor is

$$C_{Sa} = n/\{4\pi\, d^2/d\alpha^2\, [(1/\alpha)\exp(-\alpha mc)]\} \qquad (5\text{-D.40})$$

The most probable momentum of a particle p_p is the maximum of

$$p_{pa} = \text{Max}\{p^2 \exp[-H_{Sa}/(kT)]\}$$
$$= 2kT/c \qquad (5\text{-D.41})$$

The velocity v_{pa} corresponding to the maximum in the momentum is

$$v_{pa} = cp_{pa}/(p_{pa}^{\;2} - m^2c^2)^{1/2}$$

For large T or small T, the velocity v_{pa} corresponding to the maximum in the momentum is approximately

$$v_{pa} \approx c + \tfrac{1}{2}\, m^2c^5/(2kT)^2$$

Turning now to the thermodynamics implied by the superluminal Maxwell-Boltzman distribution we begin by calculating the average energy per particle

$$\varepsilon_a = \int d^3p\, H_{Sa} \exp[-H_{Sa}/(kT)] \Big/ \int d^3p\, \exp[-H_{Sa}/(kT)] \qquad (5\text{-D.42})$$

$$= (C_{Sa}/n) \int d^3p\, H_{Sa} \exp[-H_{Sa}/(kT)]$$

$$= -(4\pi c C_{Sa}/n)\, d^3/d\alpha^3[(1/\alpha)\exp(-\alpha mc)] \rightarrow 3kT \text{ for } T \gg mc$$

where $\alpha = c/(kT)$.[100]

The Maxwell-Boltzman normalization factor is related to the energy per particle by

$$C_{Sa} = -n\varepsilon_a/\{4\pi c\, d^3/d\alpha^3[(1/\alpha)\exp(-\alpha mc)]\} \qquad (5\text{-D.43})$$

Note that C_{Sa} is proportional to the energy ε_a in contrast to the non-relativistic case where the Maxwell-Boltzman normalization factor $C = (3m/(4\pi\varepsilon))^{3/2}$.

We now calculate the superluminal pressure for the case of a distribution of superluminal particles bouncing on a wall perpendicular to the z-axis. The wall is assumed to be a perfectly reflecting plane. The pressure is the average force per unit area due to the gas of superluminal particles. The number of particles bombarding the wall per second is with $v_z > 0$ is $v_z f_{Sa}(\mathbf{p})d^3p$. Thus the pressure is

---

[100] The Superluminal case differs from the non-relativistic case: $\varepsilon_a = 3kT/2$. An example of $\varepsilon_a = 3kT$ is a crystal with a potential energy of compression. See p. 192 Morse (1964).

$$P_a = \int d^3p\, 2p_z v_z f_{Sa}(\mathbf{p}) \tag{5-D.44}$$

where the particle momentum changes by $2p_z$ due to reflection. Due to spherical symmetry one expects the average values for the various components of **v** to be equal. Consequently we can re-express eq. 5-D.44 as

$$P_a = 1/3 \int d^3p\, 2m\gamma_s v^2 f_{Sa}(\mathbf{p}) \tag{5-D.45}$$
$$= 1/3 \int d^3p\, 2pv f_{Sa}(\mathbf{p})$$

Since

$$v = cp/(p^2 - m^2c^2)^{1/2} \tag{5-D.46}$$

we see

$$P_a = 8\pi c/3 \int_{mc}^{\infty} dp\, p^4 f_{Sa}(\mathbf{p})/(p^2 - m^2c^2)^{1/2} \tag{5-D.47}$$

$$\cong 8\pi c/3 \int_{mc}^{\infty} dp\, p^3 f_{Sa}(\mathbf{p})$$

Evaluating eq. 5-D.47 yields

$$P_a = -(8\pi c/3)\, C_{Sa}\, d^3/d\alpha^3[(1/\alpha)\exp(-\alpha mc)] \tag{5-D.48}$$

The *equation of state* relating the pressure and energy is[101]

$$P_a = 2/3\; n\varepsilon_a \tag{5-D.49}$$

For $T \gg mc$ we found[102]

$$\varepsilon_a \rightarrow 3kT \tag{5-D.50}$$

then, contrary to non-relativistic kinetic theory, we find ($T \gg mc$)

$$P_a = 2nkT \tag{5-D.51}$$

Turning now to the consideration of a dilute gas the internal energy of the gas for $T \gg mc$ is

$$U(t) = N\varepsilon \rightarrow 3NkT \tag{5-D.52}$$

We note again that the work done by the superluminal gas if its volume increases by dV is PdV. Then the superluminal (and usual) form of the first law of thermodynamics is

$$dQ = dU + PdV \tag{5-D.53}$$

---

[101] The same equation of state as non-relativistic kinetic theory. See p. 72 Huang (1965).

[102] Later we will define temperature in terms of the entropy S as $1/T = (\partial S/\partial U)_x$ where x is all other extensive variables.

where Q is the heat absorbed. The heat capacity of the system for constant volume is ($T \gg mc$)

$$C_V \rightarrow 3Nk \tag{5-D.54}$$

The second law of thermodynamics, Boltzmann's H theorem, is based on

$$H = -S/kV \tag{5-D.55}$$

where H is the negative of the entropy divided k times the volume V. In systems where there are no superluminal particles, the H theorem states that the entropy never decreases for an isolated gas of fixed volume.

We can calculate H_{BS} for a superluminal system under equilibrium conditions, H_{BSea}, from[103]

$$H_{BSea} = \int d^3p f_{Sa}(\mathbf{p}) \ln(f_{Sa}(\mathbf{p})) \tag{5-D.56}$$

$$= \int d^3p f_{Sa}(\mathbf{p})[\ln C_{Sa} - H_{Sa}/(kT)] \tag{5-D.57}$$

$$= n \ln C_{Sa} - \int d^3p f_{Sa}(\mathbf{p}) H_{Sa}/(kT)$$

$$= n \ln C_{Sa} - n\varepsilon_a/(kT) \tag{5-D.58}$$

by eqs. 5-D.11 and 5-D.17. Therefore

$$S_a = -kVH_{BSea} = -kN \ln C_{Sa} + N\varepsilon_a/T \tag{5-D.59}$$

The superluminal *and* standard non-relativistic result still holds

$$1/T = (\partial S/\partial U)_x \tag{5-D.60}$$

where x represents all other extensive variables.

5-D.4 Superluminal Kinetics and Themodynamics Are Similar to the Non-Relativistic Case

In the previous sections we have shown that kinetic theory and the laws of themodynamics are usually similar in the superluminal and non-relativistic cases modulo detail differences in the values of the various quantities due to differences between superluminal kinematics and non-relativistic kinematics.

END OF CHAPTER 40 EXTRACT FROM BLAHA (2012a)

---

[103] We consistently assume that integrals over the momentum $\int d^3p$ are the proper integration (rather than integrations over velocity $\int d^3v$) because, for example, the calculation of the normalization constant eq. 5-D.11 would diverge if the integration were over $\int d^3v$.

6. ElectroWeak Doublets from Rotations between Tachyon and Normal Particles

BEGINNING OF AMENDED CHAPTER 6 FROM BLAHA (2007b)

6.0 Introduction

In this chapter we begin by generalizing the free Dirac equation to a 2×2 matrix of Dirac-like equations that have a larger group covariance. This matrix of equations is applied to a doublet consisting of a normal Dirac particle wave function and a tachyon wave function. We will identify these doublets with ElectroWeak lepton doublets initially.

Then starting in section 6.5 we consider a generalized 2×2 matrix of equations (covariant under the L_C group) for doublets of complexon particles (quarks) with complex 3-momenta consisting of an up-type complexon and a down-type tachyonic complexon. Because of an inherent SU(3) symmetry we will identify these doublets with quark ElectroWeak doublets. SU(3) symmetry leads us to identify each complexon quark in a doublet as a color SU(3) triplet, and leads thence to SU(3) color quark confinement (described in a subsequent chapter).

6.1 Transformations of Dirac and Tachyon Equations

A Left-handed boost of the Dirac equation transforms the Dirac equation into the spin ½ tachyon equation, and vice versa:

$$S_L(\Lambda_L(\omega, \mathbf{u}))\psi(x) \rightarrow \psi_T'(x') \tag{6.1a}$$
$$S_L(\Lambda_L(\omega, \mathbf{u}))\psi_T(x) \rightarrow \psi'(x')$$

and (noting the appearance of a γ^5)

$$S_L(\Lambda_L(\omega, \mathbf{u}))(\gamma^\mu\partial/\partial x^\mu - m)S_L^{-1}(\Lambda_L(\omega, \mathbf{u})) = (i\gamma^\mu\partial/\partial x'^\mu - m) \tag{6.1b}$$
$$S_L(\Lambda_L(\omega, \mathbf{u}))\gamma^5(i\gamma^\mu\partial/\partial x^\mu - m)\gamma^5 S_L^{-1}(\Lambda_L(\omega, \mathbf{u})) = (\gamma^\mu\partial/\partial x'^\mu - m)$$

where

$$x'^\mu = i\Lambda_L{}^\mu{}_\nu(\omega, \mathbf{u})x^\nu \tag{6.1c}$$
$$\partial/\partial x'^\mu = -i\Lambda_L{}^\nu{}_\mu(\omega, \mathbf{u})\partial/\partial x^\nu$$

by the matrix equation is

$$x' = E(\mathbf{v})x = i\Lambda_L(\mathbf{v})x$$

Eqs. 6.1a – 6.1c imply

$$S_L(\Lambda_L(\omega, \mathbf{u}))(\gamma^\mu\partial/\partial x^\mu - m)\psi_T(x) = (i\gamma^\mu\partial/\partial x'^\mu - m)S_L(\Lambda_L(\omega, \mathbf{u}))\psi_T(x)$$
$$= (i\gamma^\mu\partial/\partial x'^\mu - m)\psi'(x') \tag{6.1d}$$

and

$$S_L(\Lambda_L(\omega, \mathbf{u}))\gamma^5(i\gamma^\mu\partial/\partial x^\mu - m)\psi(x) = (\gamma^\mu\partial/\partial x'^\mu - m)S_L(\Lambda_L(\omega, \mathbf{u}))\gamma^5\psi(x)$$
$$= (\gamma^\mu\partial/\partial x'^\mu - m)\psi_T'(x') \qquad (6.1e)$$

where

$$\psi'(x') = S_L(\Lambda_L(\omega, \mathbf{u}))\psi_T(x) \qquad (6.1f)$$

and

$$\psi_T'(x') = S_L(\Lambda_L(\omega, \mathbf{u}))\gamma^5\psi(x) \qquad (6.1g)$$

Note the Dirac equation is not left-handed complex Lorentz covariant.

6.2 Doublet Extended Dirac Equations

We will now consider the issue of generalizing the Dirac equation so that the extended equation is covariant under both Lorentz transformations and Left-handed complex Lorentz transformations.

The only obvious method to obtain an extended Dirac equation that is covariant under complex Lorentz transformations is to define an 8×8 matrix generalization. Let

$$đ(x) = \begin{bmatrix} (\gamma^\mu\partial/\partial x^\mu - m) & 0 \\ 0 & (i\gamma^\mu\partial/\partial x^\mu - m) \end{bmatrix} \qquad (6.2)$$

be an 8×8 matrix operator with the 4×4 matrix elements shown, and let

$$\Psi(x) = \begin{bmatrix} \psi_T(x) \\ \psi(x) \end{bmatrix} \qquad (6.3)$$

be an 8 component column vector composed of a Dirac field and a tachyon field. Then the extended Dirac equation is

$$đ(x)\Psi(x) = 0 \qquad (6.4)$$

We now define the 8×8 Left-handed complex Lorentz transformation

$$S_{L8}(\Lambda_L(v)) = \begin{bmatrix} 0 & S_L(\Lambda_L(v))\gamma^5 \\ S_L(\Lambda_L(v)) & 0 \end{bmatrix} \qquad (6.5)$$

with inverse transformation

$$S_{L8}^{-1}(\Lambda_L(v)) = \begin{bmatrix} 0 & S_L^{-1}(\Lambda_L(v)) \\ \gamma^5 S_L^{-1}(\Lambda_L(v)) & 0 \end{bmatrix} \qquad (6.6)$$

Note: we use the notations $S_L(\Lambda_L(v))$ and $S_L(\Lambda_L(\omega, \mathbf{u}))$ interchangeably. Applying S_{L8} to eq. 6.4 yields

$$0 = S_{L8}(\Lambda_L(v))đ(x)\Psi(x) = đ(x')\Psi'(x') \tag{6.7}$$

where

$$\Psi'(x') = \begin{bmatrix} S_L\gamma^5\psi(x) \\ S_L\psi_T(x) \end{bmatrix} = \begin{bmatrix} \psi_T'(x') \\ \psi'(x') \end{bmatrix} \tag{6.8}$$

Thus the extended Dirac equation is covariant under generalized Left-handed complex Lorentz transformations such as eqs. 6.5-6.6. Covariance requires tachyon and the Dirac particles must have the same absolute value for the mass which is the iota mass in the free fermion case.

It is easy to show that the extended Dirac equation eq. 6.4 is also covariant under conventional Lorentz transformations in the 8×8 representation:

$$S_8(\Lambda(v)) = \begin{bmatrix} S(\Lambda(v)) & 0 \\ 0 & S(\Lambda(v)) \end{bmatrix} \tag{6.9}$$

with inverse

$$S_8^{-1}(\Lambda(v)) = \begin{bmatrix} S^{-1}(\Lambda(v)) & 0 \\ 0 & S^{-1}(\Lambda(v)) \end{bmatrix} \tag{6.10}$$

and non-diagonal Lorentz transformations:

$$S_{8A}(\Lambda(v)) = \begin{bmatrix} 0 & S(\Lambda(v)) \\ S(\Lambda(v)) & 0 \end{bmatrix} \tag{6.11}$$

with inverse transformation

$$S_{8A}^{-1}(\Lambda(v)) = \begin{bmatrix} 0 & S^{-1}(\Lambda(v)) \\ S^{-1}(\Lambda(v)) & 0 \end{bmatrix} \tag{6.12}$$

Under a conventional Lorentz transformation we find

$$0 = S_8(\Lambda(v))đ(x)\Psi(x) = đ(x')\Psi'(x') \tag{6.13}$$

$$0 = S_{8A}(\Lambda(v))đ(x)\Psi(x) = đ(x')\Psi'(x')$$

The lagrangian density that corresponds to our 8-dimensional construction is

$$\mathcal{L}_8 = \overline{\Psi}(x)\not{d}(x)\Psi(x) \tag{6.14}$$

where

$$\overline{\Psi}(x) = \Psi^{\dagger}\Gamma^0 \tag{6.15}$$

and

$$\Gamma^0 = \begin{bmatrix} i\gamma^0\gamma^5 & 0 \\ 0 & \gamma^0 \end{bmatrix} \tag{6.16}$$

The action is

$$I = \int d^4x \mathcal{L}_8 \tag{6.17}$$

is invariant under Lorentz transformations S_8 and S_{8A}.

The Hamiltonian density for the 8-dimensional theory is

$$\mathcal{H}_8(x) = \begin{bmatrix} i\psi_T^{\dagger}\gamma^5(\boldsymbol{\alpha}\cdot\nabla + \beta m)\psi_T & 0 \\ 0 & \psi^{\dagger}(-i\boldsymbol{\alpha}\cdot\nabla + \beta m)\psi \end{bmatrix} \tag{6.18}$$

6.3 Non-Invariance of the Extended Free Action under a Left-handed Extended Lorentz Transformation

The action 6.17 is not invariant under Left-handed complex Lorentz transformations. The fundamental cause of this non-invariance is the three dimensional nature of space. In the case of Dirac particles one can define a Lorentz invariant action because time is one- dimensional. Thus one can use $\psi^{\dagger}\gamma^0 = \bar{\psi}$ to form the Dirac field lagrangian and action. A key factor in Lorentz invariance is the relation between the inverse and hermitean conjugate of the spinor boost operator

$$\gamma^0 S^{-1}\gamma^0 = S^{\dagger} \tag{6.19}$$

In the case of the tachyon lagrangian and action Left-handed complex Lorentz invariance is not possible because the tachyonic equivalent to eq. 6.19 is[104]

$$S_L^{-1}(\Lambda(\mathbf{v}))\boldsymbol{\gamma}\cdot\mathbf{p}/|\mathbf{p}| = i\gamma^0 S_L^{\dagger}(\Lambda(\mathbf{v})) \tag{6.20}$$

where $\mathbf{p} = m\gamma_s\mathbf{v}$. The appearance of $\boldsymbol{\gamma}\cdot\mathbf{p}/|\mathbf{p}|$ in eq. 6.20 precludes the invariance of the free tachyon action.

We will now show the effect of a Left-handed complex Lorentz transformation (eqs. 6.5 and 6.6) on the lagrangian density eq. 6.14. The two non-zero parts of the lagrangian density $\mathcal{L}_8$ (eq. 6.14) are

---

[104] This relation is derivable from eqs. 3.11 and 3.12.

$$\mathcal{L}_1 = \psi_T^{\dagger} i\gamma^0\gamma^5(\gamma^\mu\partial/\partial x^\mu - m)\psi_T(x) \tag{6.21}$$

and

$$\mathcal{L}_2 = \psi^{\dagger}\gamma^0(i\gamma^\mu\partial/\partial x^\mu - m)\psi(x) \tag{6.22}$$

The effect of the transformation, eqs. 6.5-6.6, on these terms is

$$\begin{aligned}\mathcal{L}_1' &= \psi_T^{\dagger} i\gamma^0\gamma^5 S_L^{-1}S_L(\gamma^\mu\partial/\partial x^\mu - m)\, S_L^{-1}S_L\psi_T(x)\\ &= \psi_T^{\dagger} i\gamma^0\gamma^5 S_L^{-1}(i\gamma^\mu\partial/\partial x'^\mu - m)S_L\psi_T(x)\\ &= -\psi_T^{\dagger}S_L^{\dagger}\gamma^5(\boldsymbol{\gamma}\cdot\mathbf{p}/|\mathbf{p}|)(i\gamma^\mu\partial/\partial x'^\mu - m)S_L\psi_T(x)\\ &= \psi'^{\dagger}(x')(\boldsymbol{\gamma}\cdot\mathbf{p}/|\mathbf{p}|)\gamma^5(i\gamma^\mu\partial/\partial x'^\mu - m)\psi'(x')\end{aligned} \tag{6.23}$$

and

$$\begin{aligned}\mathcal{L}_2' &= \psi^{\dagger}\gamma^0\gamma^5 S_L^{-1}S_L\gamma^5(i\gamma^\mu\partial/\partial x^\mu - m)\gamma^5 S_L^{-1}S_L\gamma^5\psi(x)\\ &= \psi^{\dagger}\gamma^0\gamma^5 S_L^{-1}(\gamma^\mu\partial/\partial x'^\mu - m)S_L\gamma^5\psi(x)\\ &= i\psi^{\dagger}\gamma^5 S_L^{\dagger}(\boldsymbol{\gamma}\cdot\mathbf{p}/|\mathbf{p}|)(\gamma^\mu\partial/\partial x'^\mu - m)S_L\gamma^5\psi(x)\\ &= i\psi_T'^{\dagger}(x')(\boldsymbol{\gamma}\cdot\mathbf{p}/|\mathbf{p}|)(\gamma^\mu\partial/\partial x'^\mu - m)\psi_T'(x')\end{aligned} \tag{6.24}$$

using eqs. 6.20, 6.1f and 6.1g, where $\psi'(x')$ is a solution of the Dirac equation obtained by Left-handed complex Lorentz boosting (by $\mathbf{v} = \mathbf{p}/(\gamma m)$) of a tachyon field and where $\psi_T'(x')$ is a solution of the tachyon equation obtained by Left-handed complex Lorentz boosting (by $\mathbf{v} = \mathbf{p}/(\gamma m)$) of a Dirac field. Eqs. 6.23-6.24 clearly show that $\mathcal{L}_8$ is *not* invariant under Left-handed complex Lorentz transformations.

Consequently the action of eq. 6.17 is only invariant under inhomogeneous Lorentz transformations. *This state of affairs is actually an advantage when we derive features of the Standard Model because it will be seen to prevent any interplay between unbroken ElectroWeak SU(2) rotations and Left-handed complex Lorentz transformations.*

6.4 The Diracian Dilemma – To what do Left-handed Extended Lorentz Boost Particles Correspond? Answer: Leptons

The development of this 8-dimensional formalism, and in particular, the "bi-spinor" wave function consisting of a Dirac spinor and and a tachyon spinor, raises the question, "Is there a particle interpretation for the "bi-spinor" wave function?" Dirac faced a similar issue in 1928-1930 with the negative energy states of the Dirac equation. He developed "hole theory" which eventually led to the interpretation of holes in the sea of filled negative energy states as *positrons*. We now face the same problem: with what pairs of particles do we identify the doublets consisting of a Dirac particle and a tachyon?

The obvious natural interpretation of these 8-spinors is ElectroWeak isodoublets such as:

$$\Psi_\ell(x) = \begin{bmatrix} \psi_{\ell T} \\ \psi_\ell \end{bmatrix} \sim \begin{bmatrix} \nu \\ e \end{bmatrix} \tag{6.25}$$

for leptons where "e" represents a charged lepton and ν represents a neutrino. With this interpretation we can introduce SU(2)⊗U(1) gauge interactions and develop one-generation, leptonic ElectroWeak theory naturally.

6.5 To what do Complexons Correspond? Quarks

We have identified two of the four types of spin ½ fermions as leptons. The remaining two types of spin ½ fermions – complexons – ψ_C and ψ_{CT} seem to naturally correspond to quarks since their equations of motion and wave functions have a natural SU(3) symmetry as we pointed out earlier. We therefore associate a color SU(3) symmetry with these two types of spin ½ complexons. The Electroweak doublet of quarks then is

$$\Psi_q^{\ a}(x) = \begin{bmatrix} \psi_C^{\ a} \\ \\ \psi_{CT}^{\ a} \end{bmatrix} \sim \begin{bmatrix} u^a \\ \\ d^a \end{bmatrix} \tag{6.26}$$

where u is an "up" type quark and d is a "down" type quark.[105]

The rationale for constructing quark doublets is the same as in the leptonic case: We wish to define a generalization of the "Dirac-like" equations of motion that is covariant under L_C boosts.

6.6 Quark Doublets

We assume that quark doublets consist of a complexon[106] and a tachyonic complexon[107] and to this extent they mirror lepton doublets. In this section we will develop a generalized free complexon equation and describe its features.

---

[105] While the lepton situation is clear in the sense that charged leptons cannot be tachyons since their masses are known (Thus only tachyonic neutrinos are the only currently allowed possibility.), the quark situation is somewhat unclear. We have provisionally chosen the "down" type of quark (d, s, and b) as tachyonic. The association of bound states of these quarks such as the K^0 and B^0 systems which are known to have CP violation, and the CP violation engendered by tachyons, encourages this interpretation.

In addition, $W^\pm$ charge asymmetry in $p\bar{p}$ collisions indicate the d sea in a proton is greater than the u sea (K. Abe et al, PRL **74**, 850 (1995)) as does the asymmetry of Drell-Yan production in deep inelastic scattering on p and n targets (A. Baldit et al, Phys. Lett. **B332**, 244 (1994)). These results are to be expected since there is no mass gap for a d tachyon sea while there is a mass gap for a u Dirac particle sea. Complexon quarks may explicate the discrepancies between theory and experiment in the spin structure functions of the parton model for nucleons.

[106] **An "ordinary" complexon can "exceed the speed of light" just like a tachyonic complexon because a complexon has a complex valued velocity enabling it to evade the real-valued singularity at v = c.**

[107] The global SU(3) symmetry of complexons makes their identification with quarks reasonable. However, the complexon theory that we develop and use for quark dynamics in the Standard Model is not required. Our Standard Model could use Dirac fermion dynamics for the up-type quarks and tachyon dynamics for down-type quarks. Then the (broken) Left-handed Extended Lorentz group would be the basic space-time group rather than L_C. We choose to use complexon dynamics for quarks because they have an internal SU(3)-like structure suggestive of color SU(3). More importantly, their spin dynamics is different and thus may resolve the differences between theory and experiment for the deep inelastic parton spin-dependent structure functions. Nevertheless, quarks could be similar to leptons in this regard and form a doublet of a Dirac fermion and an ordinary tachyon. Whether quarks are

Summary of L_C Boosts to Generate Spin ½ Equations

We begin by recapitulating L_C boost features for coordinates and spinors:[108]

$$\Lambda_C(\mathbf{v_c}) = \exp[i\omega\hat{\mathbf{w}}\cdot\mathbf{K}] \tag{2.61}$$
$$\omega = (\omega_r^2 - \omega_i^2 + 2i\omega_r\omega_i\,\hat{\mathbf{u}}_r\cdot\hat{\mathbf{u}}_i)^{1/2} \tag{2.62}$$
$$\hat{\mathbf{w}} = (\omega_r\hat{\mathbf{u}}_r + i\omega_i\hat{\mathbf{u}}_i)/\omega \tag{2.63}$$
$$\hat{\mathbf{w}}\cdot\hat{\mathbf{w}} = \hat{\mathbf{u}}_r\cdot\hat{\mathbf{u}}_r = \hat{\mathbf{u}}_i\cdot\hat{\mathbf{u}}_i = 1 \tag{2.64a}$$
$$\mathbf{v_c} = \hat{\mathbf{w}}\tanh(\omega) \tag{2.64b}$$

The corresponding L_C spinor boost for $m^2 > 0$ particles with complex 3-momenta is

$$S_C(\omega, \mathbf{v_c}) = \exp(-i\omega\sigma_{0k}\hat{w}_k/2) = \exp(-\omega\gamma^0\boldsymbol{\gamma}\cdot\hat{\mathbf{w}}/2)$$
$$= \cosh(\omega/2)I + \sinh(\omega/2)\gamma^0\boldsymbol{\gamma}\cdot\hat{\mathbf{w}} \tag{3.78}$$

with inverse transformation

$$S_C^{-1}(\omega, \mathbf{v_c}) = \gamma^2\gamma^0 K^{-1}S_C^{\dagger}K\gamma^0\gamma^2 = \gamma^2\gamma^0 S_C^{\;T}\gamma^0\gamma^2 = \exp(\omega\gamma^0\boldsymbol{\gamma}\cdot\hat{\mathbf{w}}/2)$$
$$= \cosh(\omega/2)I - \sinh(\omega/2)\gamma^0\boldsymbol{\gamma}\cdot\hat{\mathbf{w}} \tag{3.79}$$

The Dirac-like complexon equation resulting from the boost is

$$[i\gamma^0\partial/\partial t + i\boldsymbol{\gamma}\cdot(\nabla_r + i\nabla_i) - m]\psi_C(t, \mathbf{x}_r, \mathbf{x}_i) = 0 \tag{3.101}$$

where $\mathbf{x} = \mathbf{x}_r - i\mathbf{x}_i$. The subsidiary condition is

$$\nabla_r\cdot\nabla_i\,\psi_C(t, \mathbf{x}_r, \mathbf{x}_i) = 0 \tag{3.102a}$$

The L_C coordinate boost that leads to $m^2 < 0$ tachyonic complexons with complex 3-momenta is

$$\Lambda_{CL}(\mathbf{v_c}) \equiv \Lambda_{CL}(\omega, \hat{\mathbf{w}}) = \exp[i(\omega + i\pi/2)\hat{\mathbf{w}}\cdot\mathbf{K}] \tag{3.151}$$

where

$$\omega = (\omega_r^2 - \omega_i^2)^{1/2} \tag{2.62}$$
$$\hat{\mathbf{w}} = (\omega_r\hat{\mathbf{u}}_r + i\omega_i\hat{\mathbf{u}}_i)/\omega \tag{2.63}$$
$$\hat{\mathbf{w}}\cdot\hat{\mathbf{w}} = \hat{\mathbf{u}}_r\cdot\hat{\mathbf{u}}_r = \hat{\mathbf{u}}_i\cdot\hat{\mathbf{u}}_i = 1 \tag{2.64a}$$
$$\mathbf{v_c} = \hat{\mathbf{w}}\tanh(\omega + i\pi/2) = \hat{\mathbf{w}}\,\mathrm{cotanh}(\omega) \tag{3.152}$$
$$\omega_L = \omega + i\pi/2$$

The L_C spinor boost for tachyonic complexons is

$$S_{CL}(\Lambda_{CL}(\omega, \hat{\mathbf{w}})) = \exp(-i\omega_L\sigma_{0i}\hat{w}_i/2) = \exp(-\omega_L\gamma^0\boldsymbol{\gamma}\cdot\hat{\mathbf{w}}/2)$$

---

complexons or not is an experimental question! From a Leibniz Minimax Principle viewpoint the use of complexon quarks shows the maximum benefit from the underlying theoretical development we are pursuing.

[108] The equation numbering of this subsection follows that of Blaha (2007b).

$$= \cosh(\omega_L/2)I + \sinh(\omega_L/2)\gamma^0\boldsymbol{\gamma}\cdot\hat{\mathbf{w}} \tag{3.154}$$

The resulting Dirac-like tachyonic complexon equation is

$$[\gamma^0\partial/\partial t + \boldsymbol{\gamma}\cdot(\boldsymbol{\nabla}_r + i\boldsymbol{\nabla}_i) - m]\psi_{CL}(t, \mathbf{x_r}, \mathbf{x_i}) = 0 \tag{3.163a}$$

with the subsidiary condition

$$\boldsymbol{\nabla}_r\cdot\boldsymbol{\nabla}_i\ \psi_{CL}(t, \mathbf{x_r}, \mathbf{x_i}) = 0 \tag{3.164}$$

L_C Boosts between Complexons and Tachyonic Complexons

An L_C spinor boost of a complexon can change it into a tachyonic complexon and vice versa:

$$\begin{aligned} S_{CL}(\Lambda_{CL}(\omega, \hat{\mathbf{w}}))\psi_C(x) &\rightarrow \psi_{CT}'(x') \\ S_{CL}(\Lambda_{CL}(\omega, \hat{\mathbf{w}}))\psi_{CT}(x) &\rightarrow \psi_C'(x') \end{aligned} \tag{6.27}$$

Similarly the differential operator used in the equations of motion can also be transformed.

$$\begin{aligned} S_{CL}(\Lambda_{CL}(\omega, \hat{\mathbf{w}}))(\gamma^\mu D_\mu - m)S_{CL}^{-1}(\Lambda_{CL}(\omega, \hat{\mathbf{w}})) &= (i\gamma^\mu D'_\mu - m) \\ S_{CL}(\Lambda_{CL}(\omega, \hat{\mathbf{w}}))\gamma^5(i\gamma^\mu D_\mu - m)\gamma^5 S_{CL}^{-1}(\Lambda_{CL}(\omega, \hat{\mathbf{w}})) &= (\gamma^\mu D'_\mu - m) \end{aligned} \tag{6.28}$$

where

$$\begin{aligned} x'^\mu &= i\Lambda_{CL}{}^\mu{}_\nu(\omega, \mathbf{u})x^\nu \\ D'_\mu &= -i\Lambda_{CL}{}^\nu{}_\mu(\omega, \mathbf{u})D_\nu \end{aligned} \tag{6.29}$$

or in matrix form

$$X' = E_{CL}(\omega, \hat{\mathbf{w}})X \equiv i\Lambda_{CL}(\omega, \hat{\mathbf{w}})X \tag{6.30}$$

Eqs. 6.27 – 6.29 imply

$$\begin{aligned} S_{CL}(\Lambda_{CL}(\omega, \hat{\mathbf{w}}))(\gamma^\mu D_\mu - m)\psi_{CT}(x) &= (i\gamma^\mu D'_\mu - m)S_{CL}(\Lambda_{CL}(\omega, \hat{\mathbf{w}}))\psi_{CT}(x) \\ &= (i\gamma^\mu D'_\mu - m)\psi_C'(x') \end{aligned} \tag{6.31}$$

and

$$\begin{aligned} S_{CL}(\Lambda_{CL}(\omega,\hat{\mathbf{w}}))\gamma^5(i\gamma^\mu D_\mu - m)\psi_C(x) &= (\gamma^\mu D'_\mu - m)S_{CL}(\Lambda_{CL}(\omega,\hat{\mathbf{w}}))\gamma^5\psi_C(x) \\ &= (\gamma^\mu D'_\mu - m)\psi_{CT}'(x') \end{aligned} \tag{6.32}$$

where

$$\psi_C'(x') = S_{CL}(\Lambda_{CL}(\omega, \hat{\mathbf{w}}))\psi_{CT}(x) \tag{6.33}$$

and

$$\psi_{CT}'(x') = S_{CL}(\Lambda_{CL}(\omega, \hat{\mathbf{w}}))\gamma^5\psi_C(x) \tag{6.34}$$

Thus neither complexon dynamical equation is L_C covariant.

Doublet Dynamical Equation for Complexons

We will now consider the issue of generalizing the complexon dynamical equations so that the generalized equation is covariant under both Lorentz transformations and L_C boosts.

The only obvious method to obtain a generalized equation that is covariant under L_C boosts is to define an 8×8 matrix generalization. Let

$$đ_C(x) = \begin{bmatrix} (i\gamma^\mu D_\mu - m) & 0 \\ 0 & (\gamma^\mu D_\mu - m) \end{bmatrix} \tag{6.35}$$

be an 8×8 matrix operator with the 4×4 matrix elements shown, and let

$$\Psi_C(x) = \begin{bmatrix} \psi_C(x) \\ \psi_{CT}(x) \end{bmatrix} \tag{6.36}$$

be an 8 component column vector composed of a complexon field and a tachyonic complexon field. Then the generalized complexon equation is

$$đ_C(x)\Psi_C(x) = 0 \tag{6.37}$$

We now define the 8×8 Left-handed L_C boost transformation

$$S_{CL8} \equiv S_{CL8}(\Lambda_{CL}(\omega, \hat{\mathbf{w}})) = \begin{bmatrix} 0 & S_{CL}(\Lambda_{CL}(\omega, \hat{\mathbf{w}})) \\ S_{CL}(\Lambda_{CL}(\omega, \hat{\mathbf{w}}))\gamma^5 & 0 \end{bmatrix} \tag{6.38}$$

with inverse transformation

$$S_{CL8}^{-1} \equiv S_{CL8}^{-1}(\Lambda_{CL}(\omega, \hat{\mathbf{w}})) = \begin{bmatrix} 0 & \gamma^5 S_{CL}^{-1}(\Lambda_{CL}(\omega, \hat{\mathbf{w}})) \\ S_{CL}^{-1}(\Lambda_{CL}(\omega, \hat{\mathbf{w}})) & 0 \end{bmatrix} \tag{6.39}$$

Applying S_{CL8} to eq. 6.37 yields

$$0 = S_{CL8}đ_C(x)\Psi_C(x) = đ_C(x')\Psi_C'(x') \tag{6.40}$$

where

$$\Psi_C'(x') = \begin{bmatrix} S_{CL8}\psi_{CT}(x) \\ S_{CL8}\gamma^5\psi_C(x) \end{bmatrix} = \begin{bmatrix} \psi_C'(x') \\ \psi_{CT}'(x') \end{bmatrix} \tag{6.41}$$

Thus the generalized complexon equation is covariant under L_C boosts. Covariance requires the complexon, and tachyonic complexon, must have the same absolute value for the (iota) mass.

It is easy to show that the generalized complexon equation is also covariant under conventional Lorentz transformations represented as 4×4 diagonal blocks in an 8×8 matrix representation. (The demonstration is analogous to eqs. 6.9 – 6.13.)

The lagrangian density that corresponds to our 8-dimensional construction is

$$\mathcal{L}_{C8} = \overline{\Psi}_C(x)\not{d}_C(x)\Psi_C(x) \tag{6.42}$$

where

$$\overline{\Psi}_C(x) = \Psi_C^{\dagger}|_{\mathbf{x_i} = -\mathbf{x_i}}\Gamma_C^{\ 0} \tag{6.43}$$

and

$$\Gamma_C^{\ 0} = \begin{bmatrix} \gamma^0 & 0 \\ 0 & i\gamma^0\gamma^5 \end{bmatrix} \tag{6.44}$$

The action

$$I = \int d^4x \mathcal{L}_{C8} \tag{6.45}$$

is invariant under Lorentz transformations S_8 and S_{8A} (eqs. 6.9 – 6.12).

The Hamiltonian density for the 8-dimensional theory is

$$\mathcal{H}_{C8}(x) = \begin{bmatrix} \psi_C^{\dagger}(-i\boldsymbol{\alpha}\cdot\nabla_C + \beta m)\psi_C & 0 \\ 0 & i\psi_{CT}^{\dagger}\gamma^5(\boldsymbol{\alpha}\cdot\nabla_C + \beta m)\psi_{CT} \end{bmatrix} \tag{6.46}$$

where the spatial vector part of D^{μ} is

$$\nabla_C = \mathbf{D} \tag{6.47}$$

Non-Invariance of the Generalized Free Complexon Action under an L_C Boost

The action 6.45 is not invariant under L_C boosts. The reason is similar to that of section 6.3 for the "leptonic" type of particle: there is no simple relation between the hermitean conjugate of an L_C spinor boost and its inverse (similar to eq. 6.19 for the Dirac boost case).

Consequently the action of eq. 6.45 is only invariant under inhomogeneous Lorentz transformations. *This state of affairs is again an advantage when we derive features of the Standard Model because it prevents any interplay between unbroken ElectroWeak SU(2) rotations and* L_C *transformations in the complexon (quark) sector.*

END OF AMENDED CHAPTER 6 EXTRACT FROM BLAHA (2007b)

7. ElectroWeak SU(2)⊗U(1) from Real Superluminal Velocities

In the discussions up to this point we have not considered imaginary (and more generally complex) coordinates resulting from a superluminal Lorentz transformation. In this section we show that they require us to introduce another transformation that maps complex coordinates to real coordinates. This transformation will be of significance because it leads to an SU(2)⊗U(1) symmetry. It emerges when we consider superluminal transformations but is not required for ordinary sublight Lorentz transformations. This new SU(2)⊗U(1) symmetry is identified with the SU(2)⊗U(1) symmetry of the ElectroWeak sector of The Standard Model.

We introduce this new transformation by reconsidering the previous simple example wherein one coordinate system is traveling at a speed v in the x direction with respect to the "laboratory" system.

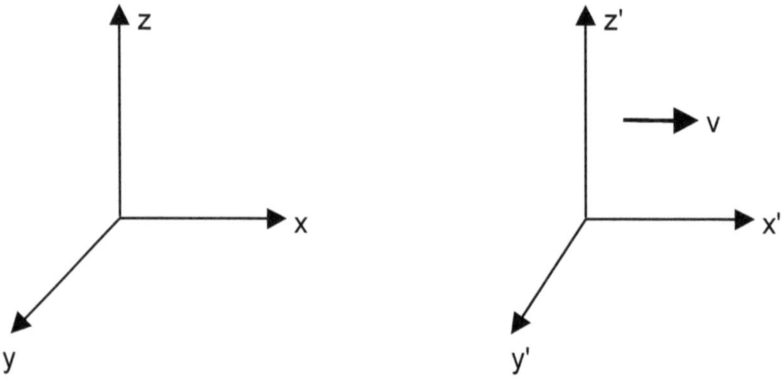

Figure 7.1. Depiction of two coordinate systems. The "primed" coordinate system is moving with velocity v in the positive x direction with respect to the "unprimed" coordinate system. We choose parallel axes for convenience.

The (left-handed[109]) Lorentz transformation is given by eq. 7.1 and the coordinates in the two reference frames are related by eq. 7.2.

$$\Lambda_L(\omega, \mathbf{u} = (1,0,0)) = \begin{bmatrix} i\gamma_s & -i\beta\gamma_s & 0 & 0 \\ -i\beta\gamma_s & i\gamma_s & 0 & 0 \\ 0 & 0 & 1 & 0 \\ 0 & 0 & 0 & 1 \end{bmatrix} \qquad (7.1)$$

[109] The right-handed Lorentz transformation case is analogous.

implementing the coordinate transformation:

$$X' = \Lambda_L(\omega, \mathbf{u} = (1,0,0))X$$

or

$$\begin{aligned} t' &= i\gamma_s(t - \beta x) \\ x' &= i\gamma_s(x - \beta t) \\ y' &= y \\ z' &= z \end{aligned} \tag{7.2}$$

We now define a transformation that maps the real coordinates of the unprimed reference frame to real coordinates in the primed reference frame.

$$\Pi_L(\mathbf{u}) = \begin{bmatrix} -i & 0 & 0 & 0 \\ 0 & -i & 0 & 0 \\ 0 & 0 & 1 & 0 \\ 0 & 0 & 0 & 1 \end{bmatrix} \tag{7.3}$$

where **u** is the unit vector corresponding to the direction of **v** (the positive x direction in this example). Using $\Pi_L(\mathbf{u})$ we obtain an overall transformation from real coordinates to real coordinates:

$$X'' = \Pi_L(\mathbf{u})\Lambda_L(\omega, \mathbf{u} = (1,0,0))X$$

or

$$\begin{aligned} t'' &= \gamma_s(t - \beta x) \\ x'' &= \gamma_s(x - \beta t) \\ y'' &= y \\ z'' &= z \end{aligned} \tag{7.4}$$

where $\gamma_s = (\beta^2 - 1)^{-½}$. An observer in the primed reference frame would consider his/her time to be real when measured on a clock, and distances along the x axis to be real when measured with a ruler. Thus eq. 7.4 makes good sense physically because in any reference frame, observers measure real distances and real times. For this reason we will call transformations of the type of eq. 7.4 – from real coordinates to real coordinates – *physical* superluminal transformations for real-valued velocities.

It is important to note that $\Pi_L(\mathbf{u})$ is position dependent in general for more complicated Λ_L's and so the Reality group is a local group of the Yang-Mills type. This is clear from eqs. 7.1 and 7.2. We will see the Reality group contains the local (Yang-Mills) ElectroWeak SU(2)⊗U(1) symmetry as it should.

This simple example generalizes to arbitrary relative real velocities **v**. First we note that the Lorentz transformation for a velocity **v** that is a rotation of the velocity in the x-direction ($\mathbf{v} = |\mathbf{v}|R\mathbf{u}$ where R is the relevant rotation matrix) has the form

$$\Lambda_L(\omega, \mathbf{v}) = \mathfrak{R}(\mathbf{v}/v, \mathbf{u})\Lambda_L(\omega, \mathbf{u} = (1,0,0))\mathfrak{R}^{-1}(\mathbf{v}/v, \mathbf{u}) \tag{7.5}$$

where $\mathfrak{R}(\mathbf{v}/v, \mathbf{u})$ is a rotation from the velocity direction **u** to direction **v**/v.

The original transformation (eq. 7.2) can be written as

$$\Pi_L(\mathbf{u})\Lambda_L(\omega, \mathbf{u} = (1,0,0)) = \Pi_L(\mathbf{u})\mathfrak{R}^{-1}(\mathbf{v}/v, \mathbf{u})\Lambda_L(\omega, \mathbf{v})\mathfrak{R}(\mathbf{v}/v, \mathbf{u}) \tag{7.6}$$

Consequently the combined transformation for velocity **v** is

$$\begin{aligned}\mathfrak{R}(\mathbf{v}/\text{v},\, \mathbf{u})\Pi_L(\mathbf{u})\Lambda_L(\omega,\, \mathbf{u} = (1,0,0))\mathfrak{R}^{-1}(\mathbf{v}/\text{v},\, \mathbf{u}) \\ = \mathfrak{R}(\mathbf{v}/\text{v},\, \mathbf{u})\Pi_L(\mathbf{u})\mathfrak{R}^{-1}(\mathbf{v}/\text{v},\, \mathbf{u})\Lambda_L(\omega,\, \mathbf{v}) \\ = \Pi_L(\mathbf{v}/\text{v})\Lambda_L(\omega,\, \mathbf{v})\end{aligned} \tag{7.7}$$

Thus for a Lorentz transformation $\Lambda_L(\omega, \mathbf{v})$ for velocity **v** we see that we can define a subsidiary transformation $\Pi_L(\mathbf{v}/\text{v})$ of the form

$$\Pi_L(\mathbf{v}/\text{v}) = \mathfrak{R}(\mathbf{v}/\text{v},\, \mathbf{u})\Pi_L(\mathbf{u})\mathfrak{R}^{-1}(\mathbf{v}/\text{v},\, \mathbf{u}) \tag{7.8}$$

The general form of $\mathfrak{R}(\mathbf{v}/\text{v}, \mathbf{u})$, is

$$\mathfrak{R}(\mathbf{v}/\text{v},\, \mathbf{u}) = \begin{bmatrix} 1 & 0 \quad 0 \quad 0 \\ 0 & \\ 0 & \mathfrak{R}_3(\mathbf{v}/\text{v},\, \mathbf{u}) \\ 0 & \end{bmatrix} \tag{7.9}$$

where $\mathfrak{R}_3(\mathbf{v}/\text{v}, \mathbf{u})$ is a 3 × 3 rotation matrix that can be expressed in terms of the generators of the 3-dimensional rotation group as

$$\mathfrak{R}_3(\mathbf{v}/\text{v},\, \mathbf{u}) = \exp(i\boldsymbol{\theta}\cdot\mathbf{J}) \tag{7.10}$$

The rotation angles $\boldsymbol{\theta}$ are real numbers since we are rotating the real vector **u** to the real vector **v**/v. Given the form of eq. 7.10 then we see that the form of $\Pi_L(\mathbf{v}//\text{v})$ is

$$\Pi_L(\mathbf{v}/\text{v}) = \begin{bmatrix} -i & 0 \quad 0 \quad 0 \\ 0 & \\ 0 & \mathfrak{R}_3(\mathbf{v}/\text{v},\, \mathbf{u})\Pi_{L3}(\mathbf{u})\mathfrak{R}_3^{-1}(\mathbf{v}/\text{v},\, \mathbf{u}) \\ 0 & \end{bmatrix} \tag{7.11}$$

where

$$\Pi_{L3}(\mathbf{u}) = \begin{bmatrix} -i & 0 & 0 \\ 0 & 1 & 0 \\ 0 & 0 & 1 \end{bmatrix} \tag{7.12}$$

If we consider the case of an infinitesimal rotation $\boldsymbol{\theta}$ to first order in $\boldsymbol{\theta}$

$$\mathfrak{R}_3(\mathbf{v}/\text{v},\, \mathbf{u}) \simeq I + i\boldsymbol{\theta}\cdot\mathbf{J} \tag{7.13}$$

then

$$\begin{aligned}\Pi_{L3}(\mathbf{v}/\text{v}) = \mathfrak{R}_3(\mathbf{v}/\text{v},\, \mathbf{u})\Pi_{L3}(\mathbf{u})\mathfrak{R}_3^{-1}(\mathbf{v}/\text{v},\, \mathbf{u}) \simeq \Pi_{L3}(\mathbf{u}) + i\boldsymbol{\theta}\cdot\mathbf{J}\Pi_{L3}(\mathbf{u}) - i\Pi_{L3}(\mathbf{u})\boldsymbol{\theta}\cdot\mathbf{J} \\ \simeq \Pi_{L3}(\mathbf{u})[I + i\Pi_{L3}^{-1}(\mathbf{u})[\boldsymbol{\theta}\cdot\mathbf{J},\, \Pi_{L3}(\mathbf{u})]\end{aligned} \tag{7.14}$$

where $\Pi_{L3}^{-1}(\mathbf{u})$ is the inverse of $\Pi_{L3}(\mathbf{u})$ and [...] represents the commutator. Thus for arbitrary rotations eq. 7.14 implies

$$\Pi_{L3}(\mathbf{v}/\mathrm{v}) = \mathfrak{R}_3(\mathbf{v}/\mathrm{v}, \mathbf{u})\Pi_{L3}(\mathbf{u})\mathfrak{R}_3^{-1}(\mathbf{v}/\mathrm{v}, \mathbf{u}) = \Pi_{L3}(\mathbf{u})\exp\{i\Pi_{L3}^{-1}(\mathbf{u})[\boldsymbol{\theta}\cdot\mathbf{J}, \Pi_{L3}(\mathbf{u})]\} \tag{7.15}$$

We can find the general form of $\Pi_{L3}(\mathbf{v}/\mathrm{v})$ by considering the case of eq. 7.6 in more detail. The exponentiated matrix expression in 7.15 can written

$$\Pi_{L3}^{-1}(\mathbf{u})[\boldsymbol{\theta}\cdot\mathbf{J}, \Pi_{L3}(\mathbf{u})] = \Pi_{L3}^{-1}(\mathbf{u})\boldsymbol{\theta}\cdot\mathbf{J}\Pi_{L3}(\mathbf{u}) - \boldsymbol{\theta}\cdot\mathbf{J} = \boldsymbol{\theta}\cdot\mathbf{Q} \tag{7.16}$$

where

$$\mathbf{Q} = \Pi_{L3}^{-1}(\mathbf{u})\mathbf{J}\Pi_{L3}(\mathbf{u}) - \mathbf{J} = \mathbf{Q}' - \mathbf{J} \tag{7.17}$$

The matrices Q_i can be evaluated using eq. 7.12 and the matrix representations of rotation generators J_i: which are equivalent in form to the SU(2) generators T_i:

$$J_1 = \begin{bmatrix} 0 & 0 & 0 \\ 0 & 0 & -i \\ 0 & i & 0 \end{bmatrix} = T_1 \tag{7.18}$$

$$J_2 = \begin{bmatrix} 0 & 0 & i \\ 0 & 0 & 0 \\ -i & 0 & 0 \end{bmatrix} = T_2 \tag{7.19}$$

$$J_3 = \begin{bmatrix} 0 & -i & 0 \\ i & 0 & 0 \\ 0 & 0 & 0 \end{bmatrix} = T_3 \tag{7.20}$$

The rotation generators satisfy the commutation relations

$$[J_i, J_j] = i\epsilon_{ijk}J_k \tag{7.21}$$

as do the SU(2) generators:

$$[T_i, T_j] = i\epsilon_{ijk}T_k \tag{7.22}$$

We can calculate Q' and obtain

$$Q'_1 = \begin{bmatrix} 0 & 0 & 0 \\ 0 & 0 & -i \\ 0 & i & 0 \end{bmatrix} \tag{7.23}$$

$$Q'_2 = \begin{bmatrix} 0 & 0 & -1 \\ 0 & 0 & 0 \\ -1 & 0 & 0 \end{bmatrix} \tag{7.24}$$

$$Q'_3 = \begin{bmatrix} 0 & 1 & 0 \\ 1 & 0 & 0 \\ 0 & 0 & 0 \end{bmatrix} \tag{7.25}$$

We note that each Q'_i is hermitean and the Q'_i satisfy the commutation relations:

$$[Q'_i, Q'_j] = i\epsilon_{ijk}Q'_k \tag{7.26}$$

Consequently the set of Q'_i are also equivalent to SU(2) generators. As a result the exponential factor

$$\Pi_{L3}(\mathbf{v}/v) = \Pi_{L3}(\mathbf{u})\exp\{i\boldsymbol{\theta}\cdot(\mathbf{Q}' - \mathbf{J})\} \tag{7.27}$$

is equivalent to a combination of SU(2) rotations not only in this case but in general for superluminal transformations. The factor $\Pi_{L3}(\mathbf{u})$ is not an SU(2) matrix since its determinant is not +1 but

$$\Pi'_{L3}(\mathbf{u}) = -i\Pi_{L3}(\mathbf{u}) \tag{7.28}$$

is an SU(2) matrix since

$$\Pi'^{-1}_{L3}(\mathbf{u}) = \Pi'^{\dagger}_{L3}(\mathbf{u}) \tag{7.29}$$

$$\det \Pi'_{L3}(\mathbf{u}) = 1 \tag{7.30}$$

and

$$\Pi'_{L3}(\mathbf{v}/v) = \Pi'_{L3}(\mathbf{u})\exp\{i\boldsymbol{\theta}\cdot(\mathbf{Q}' - \mathbf{J})\} \tag{7.31}$$

is similarly an SU(2) rotation.

Thus the general form of superluminal, real velocity, transformation from a real set of coordinates to a real set of coordinates is[110]

$$\Pi_L(\mathbf{v}/v)\Lambda_L(\omega, \mathbf{v}) \tag{7.32}$$

where

$$\Pi_L(\mathbf{v}/v) = \begin{bmatrix} -i & 0 & 0 & 0 \\ 0 & & & \\ 0 & & \Pi_{L3}(\mathbf{u})\exp\{i\,\boldsymbol{\theta}\cdot(\mathbf{Q}' - \mathbf{J})\} & \\ 0 & & & \end{bmatrix} \tag{7.33}$$

The Lorentz condition for real to real physical transformations generalizes to

$$\Lambda(\mathbf{v})^T\Pi_L(\mathbf{v}/v)^\dagger G\, \Pi_L(\mathbf{v})\Lambda(\mathbf{v}/v) = G \tag{7.34}$$

Since superluminal transformations $\Lambda_L(\omega, \mathbf{v})$ transform real coordinates to complex coordinates in general, we can generalize the form of a superluminal transformation to

$$e^{i\varphi}\Pi_L(\mathbf{v}'/v')\Lambda_L(\omega, \mathbf{v}) \tag{7.35}$$

where φ is a constant phase and $\mathbf{v}'$ is an arbitrary velocity. This generalization will satisfy the generalized Lorentz condition

$$\Lambda(\mathbf{v})^T\Pi_L(\mathbf{v}'/v')^\dagger e^{-i\varphi}G\, e^{i\varphi}\Pi_L(\mathbf{v}'/v')\Lambda(\mathbf{v}) = G \tag{7.36}$$

but the transformation will, in general, yield a complex set of coordinates when applied to a set of real coordinates.

---

[110] The choice of the unit vector **u** and the angle vector **θ** must be such that applying the transformation to a real set of coordinates yields a real set of coordinates.

These considerations imply:

1. Any observer in a coordinate system will treat a complex 4-dimensional coordinate system as if it were a real 4-dimensional coordinate system with complex-valued straight lines along each dimension (assuming rectangular coordinates).
2. The transformation $e^{i\varphi}\Pi'_{L3}(\mathbf{v}/v)$ is a SU(2)⊗U(1) transformation that takes complex 3-dimensional spatial coordinates to complex 3-dimensional spatial coordinates. In particular straight lines map to straight lines.
3. Physical observations in the observer's coordinate system are invariant under SU(2)⊗U(1) rotations of the spatial coordinates and the multiplication of the time component by an arbitrary phase.
4. The matrix

$$\Pi'_L(\mathbf{v}/v, \chi, \varphi) = \begin{bmatrix} e^{i\chi} & 0 & 0 & 0 \\ 0 & & & \\ 0 & & e^{i\varphi}\Pi'_{L3}(\mathbf{u})\exp\{i\,\boldsymbol{\theta}\cdot(\mathbf{Q}' - \mathbf{J})\} & \\ 0 & & & \end{bmatrix} \tag{7.37}$$

(where χ and φ are real numbers and $\mathbf{u}$ is a unit vector along any convenient coordinate axis) is an SU(2)⊗U(1) transformation that transforms complex 4-dimensional coordinates to complex 4-dimensional coordinates. Note, $\Pi_L(\mathbf{v}/v) = \Pi'_L(\mathbf{v}/v, 3\pi/2, \pi/2)$ is a special case of $\Pi'_L(\mathbf{v}/v, \chi, \varphi)$. Due to the manifest form of 7.37 we see

$$\Pi'_L{}^{\mu}{}_{\alpha}{}^{*}\Pi'_L{}^{\mu}{}_{\beta} = [\Pi'_L{}^{\dagger}\Pi'_L]_{\alpha\beta} = I_{\alpha\beta} \tag{7.38}$$

(with an implied sum over μ) or, in matrix form,

$$\Pi'_L{}^{\dagger}\,\Pi'_L = I \tag{7.39}$$

and also[111]

$$\Pi'_L{}^{\dagger}G\Pi'_L = G \tag{7.40}$$

5. Complex coordinate values of the type generated by superluminal transformations with real-valued velocities are transformable to real coordinates. The complex coordinates are thus physically equivalent to corresponding real coordinate values in the sense that an observer in that frame would automatically use the real coordinates so obtained since rulers and clocks always measure real spatial coordinates and times. *Therefore physical theory is invariant under global SU(2)⊗U(1) coordinate transformations since complex coordinates, so generated, can be rotated back to real coordinates.*
6. The complex coordinates of any point obtained through a superluminal transformation can be transformed to a real set of coordinates by the above SU(2)⊗U(1) transformation. This SU(2)⊗U(1) invariance is the SU(2)⊗U(1) symmetry of the ElectroWeak interactions.

---

[111] Eq.7.40 is close to the defining condition for a Lorentz group element but the presence of complex conjugation rather than a transpose means Π'_L is outside the real and complex Lorentz groups.

8. Dark ElectroWeak SU(2)⊗U(1) from Complex Superluminal Velocities

In this section[112] we will consider superluminal transformations based on complex-valued velocities. These transformations will require us to introduce another group transformation that maps complex coordinates to real coordinates. This transformation will be of significance because it leads to a hitherto unstated SU(2)⊗U(1) symmetry. We identify this SU(2)⊗U(1) symmetry with an SU(2)⊗U(1) symmetry of a Dark ElectroWeak sector of an extended The Standard Model. This "Dark ElectroWeak sector" remains to be found experimentally but there is some preliminary suggestive data from the CERN LHC.

We introduce this new transformation by extending the previous simple example to Fig. 8.1 wherein one coordinate system is traveling at a speed u in the x direction with respect to the "laboratory" system. In the new transformation the relative velocity is complex and has two components: a real-valued component in the x direction and an imaginary-valued component in the y direction. $\mathbf{u} = u_x\mathbf{i} + iu_y\mathbf{j}$. In the complex case $\beta = \tanh(\omega_L)$ is real-valued by eq. 4-A.21 where $\omega_L = \omega + i\pi/2$ and ω is real.

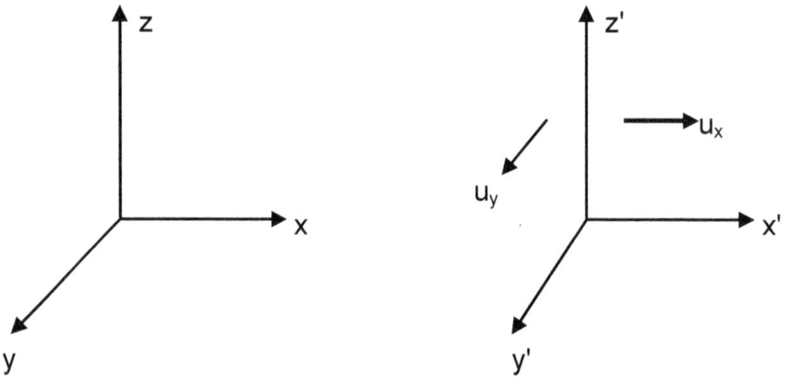

Figure 8.1. Depiction of two coordinate systems. The "primed" coordinate system is moving with velocity $u = u_x i + iu_y j$ with respect to the "unprimed" coordinate system. We choose parallel axes for convenience.

The (left-handed[113]) Lorentz transformation is given by eq. 4-A.23:

[112] Most of the material in this chapter appeared in Blaha (2011c) originally.

[113] The right-handed Lorentz transformation case is analogous.

$$\Lambda_L(\omega, \mathbf{u}) = \begin{bmatrix} i\gamma_s & -i\beta\gamma_s u_x & \beta\gamma_s u_y & 0 \\ -i\beta\gamma_s u_x & 1 + (i\gamma_s - 1)u_x^2 & i(i\gamma_s - 1)u_x u_y & 0 \\ \beta\gamma_s u_y & i(i\gamma_s - 1)u_x u_y & 1 - (i\gamma_s - 1)u_y^2 & 0 \\ 0 & 0 & 0 & 1 \end{bmatrix} \tag{8.1}$$

$$= \Lambda(\omega + i\pi/2, \mathbf{u})$$

implementing the coordinate transformation:

$$X' = \Lambda_L(\omega, \mathbf{u} = (u_x, iu_y, 0))X$$

or

$$\begin{aligned} t' &= i\gamma_s(t - \beta u_x - i\beta u_y) \\ x' &= -i\gamma_s\beta u_x t + i\gamma_s x + u_x x - u_x^2 x + i(i\gamma_s - 1)u_x u_y y \\ y' &= \gamma_s\beta u_y t + i(i\gamma_s - 1)u_x u_y x + [1 - (i\gamma_s - 1)u_y^2]y \\ z' &= z \end{aligned} \tag{8.2}$$

We now define a transformation that maps the real coordinates of the unprimed reference frame to real coordinates in the primed reference frame.

$$\Pi_L(\mathbf{u}, \mathbf{X}) = \begin{bmatrix} e^{ia} & e^{ib} & e^{ic} & 0 \\ e^{id} & e^{ie} & e^{if} & 0 \\ e^{ig} & e^{ih} & e^{ij} & 0 \\ 0 & 0 & 0 & 1 \end{bmatrix} \tag{8.3}$$

where **u** is the unit vector corresponding to the direction of the relative velocity. It is important to note that $\Pi_L(\mathbf{u}, \mathbf{X})$ is position dependent and so the Reality group is a local group of the Yang-Mills type. This is clear from eqs. 8.1 and 8.2. The Reality group which contains the ElectroWeak SU(2)⊗U(1) in the previous chapter is also a local Yang-Mills theory, as it should be.

Earlier we defined complex boosts. We summarize the definition below:

$$\Lambda_L(\omega, \mathbf{u}) = \Lambda_L(\mathbf{v}_c) = \exp[i(\omega + i\pi/2)\mathbf{u}\cdot\mathbf{K}] \tag{8.4}$$

where ω remains

$$\omega = (\omega_r^2 - \omega_i^2)^{1/2}$$

and

$$\mathbf{u} = (\omega_r \mathbf{u}_r + i\omega_i \mathbf{u}_i)/\omega$$
$$\mathbf{u}\cdot\mathbf{u} = \mathbf{u}_r\cdot\mathbf{u}_r = \mathbf{u}_i\cdot\mathbf{u}_i = 1$$
$$\mathbf{v}_c = \mathbf{u}\tanh(\omega + i\pi/2) = \mathbf{u}\,\mathrm{cotanh}(\omega)$$

In the example we are considering we set $\mathbf{u}_x = \omega_r \mathbf{u}_r$ and $\mathbf{u}_y = \omega_i \mathbf{u}_i$.

Using $\Pi_L(\mathbf{u}, \mathbf{X})$ we obtain an overall transformation from real coordinates to real coordinates:

$$X'' = \Pi_L(\mathbf{u}, \mathbf{X})\Lambda_L(\omega, \mathbf{u} = (u_x, u_y, 0))X$$

with the coordinates of X and X'' real-valued. An observer in the double primed reference frame would consider his/her time to be real when measured on a clock, and distances along the x and y axes to be real when measured with a ruler.

The velocity vectors: $u_x\mathbf{i}$ and $iu_y\mathbf{j}$ in our example define a plane in space. There are two types of rotations that are possible. 1) An angular rotation in the plane defined by the vectors. This is a U(1) transformation. 2) a spatial rotation of the plane that is an SU(2) rotation. Thus the joint rotations of **u** have an SU(2)⊗U(1) symmetry group. The R group – the Reality group for 4-dimensions – has two SU(2)⊗U(1) factors.[114] We see that this "newly found" group can be *assumed* to be the Dark ElectroWeak symmetry group. This group of transformations has associated Dark interactions. Very recent data from the CERN LHC suggests that Dark Matter has interactions. We suggest that these interactions are analogous to the known ElectroWeak interactions.

We discuss Dark Matter and its interactions in detail in chapter 13 in sections 13.2.1 and 13.2.2.

---

[114] Note that the ElectroWeak SU(2)⊗U(1) symmetry was found in chapter 7 to associate its SU(2) transformations with the spatial part of coordinates only, and the SU(2)⊗U(1) group found in this chapter associates its transformations with both spatial and time coordinates we see the two subgroups are distinctively different. Thus we find the Reality group R = SU(2)⊗U(1)⊗SU(2)⊗U(1)⊗SU(3). The SU(3) part which yields the strong interactions will be derived in the next chapter.

9. Color SU(3)

9.1 Two Possible Approaches to Color SU(3)

There are two approaches to obtaining the Strong interaction and Color SU(3) symmetry:

1. Assume up-type and down-type quarks are in 3 representations of Color SU(3). This assumption sheds no light on a deeper origin of the Strong interaction and Color SU(3). It simply assumes the color SU(3) of the Strong interaction sector of the Standard Model. Thus our understanding is not deepened. A postulate corresponding to this assumption is:

Postulate: Quarks are in the 3 representation of Color SU(3). The SU(3) symmetry is gauged with local Yang-Mills SU(3) fields called gluons that constitute the Strong interaction of the quark sector. Quarks are minimally coupled to the gluons in a gauge covariant fashion.

2. In the preceding chapter, and appendix, the ElectroWeak sector of the Standard Model (modulo generations and their mixing) was shown to essentially follow from the L(C) covariance of the doublet equations of motion for leptons and quarks, minimal coupling to gauge fields and stability requirements for tachyons, and from a previously unrecognized SU(2)⊗U(1)⊕U(1) symmetry. Thus we have a significant geometrical basis for the form of the ElectroWeak sector. We would like to establish a similar geometrical basis for the Strong interaction and Color SU(3). *We now suggest that the subset of SU(3) generators of U(4) is the source of Color SU(3). If we now extend the parameters to be real functions of the space-time coordinates (i.e. local SU(3) transformations), then we obtain a geometrical basis for color SU(3). A key factor in this interpretation is the global covariance of complexon equations of motion under global SU(3).* We thus establish a basis for ElectroWeak symmetry and for color SU(3) symmetry in geometry. This line of thought leads to the following alternative postulate **which we accept**.

Postulate Quarks are complexons.

The following sections are taken from chapter 17 of Blaha (2011c) with some changes.

9.2 A Global SU(3) Symmetry of Complexon Quarks

We will now consider a global SU(3) covariance implicit in eqs. 5.123 – 5.127. The defining property of the group SU(3) is that it preserves the invariance of inner products of complex 3-vectors of the form:

$$u^*\cdot v = u^{1*}v^1 + u^{2*}v^2 + u^{3*}v^3 \tag{9.1}$$

If we examine the dynamical equation eq. 5.123 we see that the differential operator is covariant under a global SU(3) transformation U of the complex spatial 3-coordinates:

$$[i\gamma^0\partial/\partial t + i\mathbf{D}_c{}^*\cdot\gamma - m] = [i\gamma^0\partial/\partial t + i\mathbf{D}_c{}'^*\cdot\gamma' - m] \tag{9.2}$$

where

$$\mathbf{D}_c{}^* = \nabla_c = \nabla_r + i\nabla_i$$

and

$$\gamma^a = U^{ab}\gamma'^b \tag{9.3a}$$
$$D_c{}^{*a} = D_c{}'^{*b}U^{ab*} \tag{9.3b}$$

where $U^\dagger = U^{-1}$. We now wish to exhibit the covariance of eq. 5.123. Since we can view the three spatial γ-matrices as SU(3) 3-vectors, we can express eq. 9.3 as the result of an SU(3) rotation V of the γ-matrices (on the spinor indices)

$$V\gamma^a V^{-1} = U^{ab}\gamma'^b \tag{9.4}$$

where V is a 4×4 reducible representation[115] of SU(3), namely, $\underline{3} + \underline{1}$. Since V commutes with γ^0 in the Pauli matrix representation of the γ matrices we see that V can have the form

$$V = \begin{bmatrix} A\exp(i\alpha_i\sigma_i) & 0 \\ 0 & B\exp(i\beta_i\sigma_i) \end{bmatrix}$$

where A, B, α_i and β_i are constants, and the zeroes represent 2 by 2 zero matrices. The inverse of V is $V^\dagger$. Eq. 9.4 becomes

$$V\gamma^a V^{-1} = \begin{bmatrix} 0 & AB^*\exp(i\alpha_i\sigma_i)\sigma_a\exp(-i\beta_i\sigma_i) \\ -A^*B\exp(i\beta_i\sigma_i)\sigma_a\exp(-i\alpha_i\sigma_i) & 0(-i\beta_i\sigma_i) \end{bmatrix}$$

We now note the generators of the global SU(3) symmetry under discussion have a 4×4 matrix reducible representation ($\underline{3} + \underline{1}$). The generators of this reducible representation are F_i and F_0 (a diagonal matrix diag(0,0,0,0,0,0,0,0,1).with F_i being the Gell-Mann SU(3) generators for i = 1, 2, …, 8.

Projection operators can be defined to project out the $\underline{3}$ representation piece P_3 and the $\underline{1}$ representations piece P_1. of the complexon spinor fields:

Thus the $\underline{3}$ complexon field is

$$\psi_{C3}(t, \mathbf{x}_r, \mathbf{x}_i) = P_3\psi_C(t, \mathbf{x}_r, \mathbf{x}_i) \tag{9.5}$$

while the $\underline{1}$ complexon field is

$$\psi_{C1}(t, \mathbf{x}_r, \mathbf{x}_i) = P_1\psi_C(t, \mathbf{x}_r, \mathbf{x}_i) \tag{9.6}$$

---

[115] Eqs. 16.50, and 16.53 furnish a reducible 4×4 matrix representation described later in section 17.6.1. The generators of this reducible representation are $_4F_i = F_i + \Omega_{+00} - F_9$ with F_i being the Gell-Mann SU(3) generators for i = 1, 2, …, 8.

Since P_1 and P_3 do not commute with Lorentz transformations, a Lorentz transformation mixes ψ_{C1} and ψ_{C3}.[116] Since P_1 and P_3 do not commute with γ5, left-handed and right-handed complexons would also be mixed by these projection operators. The matrix V has a $\underline{3} + \underline{1}$ reducible representation.

In a manner similar to the covariance proof of the Dirac equation[117] we see that eq. 9.2 is covariant under SU(3) transformations:

$$V[i\gamma^0\partial/\partial t + i\boldsymbol{\gamma}\cdot\mathbf{D}_c{}^* - m]V^{-1}V\psi_C(t, \mathbf{x}_r, \mathbf{x}_i) = 0$$

or

$$[i\gamma^{0\prime}\partial/\partial t' + i\mathbf{D}_c{}'^*\cdot\boldsymbol{\gamma}' - m]V\psi_C(t, \mathbf{x}_r, \mathbf{x}_i) = 0 \tag{9.7}$$

(Note $\gamma^{0\prime} = V\gamma^0V^{-1}$ and $t' = t$.) The SU(3) transformed wave function $\psi_C'(t, \mathbf{x}')$ is

$$\psi_C'(t', \mathbf{x}') = V\psi_C(t, \mathbf{x}) = V\psi_C(t', U\mathbf{x}') \tag{9.8}$$

Thus the complexon Dirac equation is covariant under global SU(3).

The subsidiary condition,

$$\nabla_r\cdot\nabla_i\, \psi_{Cu}(t, \mathbf{x}_r, \mathbf{x}_i) = 0 \tag{9.9}$$

is also covariant under an SU(3) rotation:

$$\nabla_r{}'^*\cdot\nabla_i{}'\psi_C'(t, \mathbf{x}') = \nabla_r\cdot\nabla_i\, V\psi_C(t, \mathbf{x}) = V\nabla_r{}^*\cdot\nabla_i\, \psi_C(t, \mathbf{x}) = 0 \tag{9.10}$$

We now examine the transformation of the wave function eq. 9.8 under the SU(3) transformation U. If we define

$$q^{*\mu} = (q^0, \mathbf{q}^*) = (p^0, \mathbf{p}_r + i\mathbf{p}_i) = (p^0, \mathbf{p}) = p^\mu \tag{9.11}$$

then $\psi_C(t, \mathbf{x})$ will be seen to be covariant form under an SU(3) transformation:

$$\psi_C(t, x) = \sum_{\pm s} \int d^3q_r d^3q_i\, N_C(p^0)\delta(\mathbf{q}_r{}^*\cdot\mathbf{q}_i/m^2)[b_C(q^*,s)u_C(q^*,s)e^{-i(q^*\cdot x + q\cdot x^*)/2} + d_C{}^\dagger(q^*,s)v_C(q^*,s)e^{+i(q^*\cdot x + q\cdot x^*)/2}] \tag{9.12}$$

Note both terms in each exponential are separately invariant under global SU(3). ($\mathbf{q}_r{}^* = \mathbf{q}_r$ since $\mathbf{q}_r$ is real.)

Eq. 9.8 implies that the spinors appearing in eq. 9.12 are covariant under SU(3) transformations

$$u_C'(q'^*,s') = Vu_C(q^*,s) \tag{9.13}$$

---

[116] At this point it is worth noting that the construction of complexon fields, based on a boost from a particle rest state, guarantees that a reference frame exists in which any complexon particle has a single real time variable. Similarly a reference frame exists for a set of complexon particles (that is within a Lorentz of the center of momentum frame) with a single real time variable. The time variables of the individual complexon particles in the set are complex in general but are functions of the center of momentum real time variable. So there is only one real time variable for each complexon in the set although the time variable of an individual particle may be a complex function of the real center of momentum time variable.

[117] For example see Bjorken (1964) pp. 18 – 20.

$$v_C'(q'^*,s') = Vv_C(q^*,s) \tag{9.14}$$

The fourier coefficients, if second quantized in a complex spatial coordinate generalization of the usual manner, also have covariant anti-commutation relations under an SU(3) transformation:

$$\{b_C(q,s),\ b_C^\dagger(q'^*,s')\} = \delta_{ss'}\delta^3(q_r - q'_{r'})\delta^3(q_i - q'_{i'}) \tag{9.15}$$

Under an SU(3) transformation, z = Uq and z' = Uq', the right side of eq. 9.15 transforms to

$$\delta^3(q_r - q'_{r'})\delta^3(q_i - q'_{i'}) \rightarrow \delta^3(z_r - z'_{r'})\delta^3(z_i - z'_{i'})/|\partial(q)/\partial(z)| = \delta^3(z_r - z'_{r'})\delta^3(z_i - z'_{i'}) \tag{9.16}$$

where

$$|\partial(q)/\partial(z)| = |\partial(q_r^{\,1},q_r^{\,2},q_r^{\,3},q_i^{\,1},\ q_i^{\,2},\ q_i^{\,3})/\partial(z_r^{\,1},z_r^{\,2},z_r^{\,3},z_i^{\,1},\ z_i^{\,2},\ z_i^{\,3})| = 1 \tag{9.17}$$

is the Jacobian of the transformation U. The fourier coefficients transform trivially under SU(3):

$$b_C(q^*,s) \rightarrow b_C(z^*,s) \tag{9.18}$$

Since the integrand transforms as

$$\int d^3q_r d^3q_i \rightarrow \int d^3z_r d^3z_i\, |\partial(q)/\partial(z)| = \int d^3z_r d^3z_i \tag{9.19}$$

we see that the wave function $\psi_C(t, \mathbf{x})$ transforms covariantly.

9.3 Local Color SU(3) and the Strong Interactions

In the previous section we showed that the equations of motion of free Dirac-like, complexon, up-type quarks are covariant under global SU(3).. The free, tachyon, complexon, down-type quark equations of motion are also easily seen to be covariant under this SU(3) subgroup. In this section we will show this covariance is the "source" of local Color SU(3) symmetry of quarks, and then we will introduce the Strong interaction via minimal coupling to SU(3) Yang-Mills gluons in gauge covariant derivatives.

We now introduce a complexon field with a global SU(3) index a which takes values from 1 to 3 making the field a member of the 3 representation of global SU(3):

$$\psi_C^{\,a}(t, \mathbf{x}) \tag{9.20}$$

Due to the SU(3) index the transformation property of $\psi_C^{\,a}(t, \mathbf{x})$ changes from eq. 9.8 to

$$\psi_C''^{a}(t, \mathbf{x}') = U^{ab}V\psi_C^{\,b}(t, \mathbf{x}) = U^{ab}V\psi_C^{\,b}(t, U\mathbf{x}') \tag{9.21}$$

where U^{ab} is an SU(3) rotation of 3 representation "vectors" such as $\psi_C^{\,b}$ and **x.** V is the corresponding rotation of the spinor indices of $\psi_C^{\,b}(t, \mathbf{x})$.

Note that the SU(3) rotation of the field factorizes into an SU(3) rotation of the three fields $\psi_C{}^b$ by U^{ab} and an SU(3) rotation of the four spinor components of each individual field $\psi_C{}^b$ by V.

This factorization enables us to consider a global SU(3) rotation of the $\psi_C{}^b$ fields while holding the coordinates fixed:

$$\psi_C{}'^a(t, \mathbf{x}) = U^{ab}\psi_C{}^b(t, \mathbf{x}) \tag{9.22}$$

The equations of motion are covariant under this global transformation

$$0 = U^{ab}[i\gamma^0\partial/\partial t + i\boldsymbol{\gamma}\cdot\mathbf{D_c}^* - m]\psi_C{}^b(t, \mathbf{x_r}, \mathbf{x_i})$$

$$= [i\gamma^0\partial/\partial t + i\boldsymbol{\gamma}\cdot\mathbf{D_c}^* - m]\psi'_C{}^a(t, \mathbf{x_r}, \mathbf{x_i}) \tag{9.23}$$

We now note the form of eq. 9.22 is the same as that of a *local* Yang-Mills rotation:

$$\psi_C{}'^a(t, \mathbf{x}) = \Theta^{ab}(t, \mathbf{x})\psi_C{}^b(t, \mathbf{x}) \tag{9.24}$$

where $\mathbf{x} = \mathbf{x_r} + i\mathbf{x_i}$. Therefore if we introduce a local SU(3) Yang-Mills field $A_{C\nu}(t, \mathbf{x_r}, \mathbf{x_i})$ and define a covariant derivative we can generalize eq. 9.21 to the case of local, color SU(3) if we do <u>not</u> perform the spinor rotation V.[118]

$$\mathcal{D}_\nu = D_\nu - igA_{C\nu} \tag{9.25}$$

where

$$A_{C\nu} = A_C{}^a{}_\nu t^a \tag{9.26}$$

and where $D_\nu = D_{q\nu}$ is given by

$$D_0 = \partial/\partial x^0$$
$$D_k = \partial/\partial x_r{}^k + i\,\partial/\partial x_i{}^k \tag{9.27}$$

The SU(3) 3×3 matrix generators satisfy

$$[t^a, t^b] = if^{abc}t^c \tag{9.28}$$

We can represent $\Theta_{ab}(x)$ in the form:

$$\Theta_{ab}(x) = [\exp(-i\varphi_c(x)t^c)]_{ab} \tag{9.29}$$

where $\varphi_c(x)$ is a local parameter dependent on $x = (x^0, \mathbf{x} = \mathbf{x_r} + i\mathbf{x_i})$, and t^c is an SU(3) generator.

Applying a gauge transformation to the gauge covariant derivative of a complexon fermion field $\mathcal{D}_\nu\psi_C(x)$:

$$\Theta\mathcal{D}_\nu\psi_C(x) = \Theta D_\nu\psi_C(x) - ig\Theta A_{C\nu}\Theta^{-1}\Theta\psi_C(x) \tag{9.30}$$

---

[118] This approach enables us to avoid the dilemmas associated with mixing coordinate and internal symmetries as described by Coleman, S., Phys. Rev. **138** B1262 (1965) and others in the case of SU(6) in the 1960's. Note that the spinor rotation V is expressed in terms of numerical matrices while, in the second quantized formulation, the U^{ab} rotation is expressed in terms of second quantized fields as well as numeric matrices. Thus the factorization is reflected in the form of the transformation.

$$= D_\nu\psi_C'(x) - igA_C{}'_\nu\psi_C'(x) = (\mathcal{D}_\nu\psi_C(x))'$$

where

$$\psi_C'(x) = \Theta(x)\psi_C(x) \tag{9.31}$$

we find

$$A_C{}'_\nu = (-i/g)(D_\nu\Theta(x))\Theta^{-1}(x) + \Theta(x)A_{C\nu}(x)\Theta^{-1}(x) \tag{9.32}$$

The reader will note that the form of eqs. 9.25 – 9.31 is identical to those associated with a conventional non-abelian gauge interaction with the replacement:

$$\partial/\partial x^\nu \rightarrow D_\nu \tag{9.33}$$

with D_ν given by eq. 9.27. Note that $\varphi_c(x)$, the local parameter in eq. 9.29 is dependent in general, on time, and the real and imaginary parts of the complex spatial 3-vector.

Introducing the SU(3) gauge covariant derivative transforms eq. 9.23 to

$$0 = [i\gamma^\nu\mathcal{D}_\nu - m]\psi_C{}^a(t, \mathbf{x}_r, \mathbf{x}_i) \tag{9.34}$$

The preceding argumentation supports the following postulates:

Postulate: Quarks are in a 3̲ representation of a global SU(3) subgroup of GL(4). Their transformation law under SU(3) transformations is eq. 19.6 wherein the space-time coordinates are not transformed.

Postulate: The covariance of the quark equations of motion under global SU(3) transformations becomes covariance under local color SU(3) transformations when the equations of motion are generalized to gauge covariant derivatives. We assume the equations of motion are so generalized. The interaction terms introduced constitute the Strong interaction.

We note the case of tachyon complexon quarks differs only in small details from the above discussion of Dirac-type complexon quarks.

9.4 Interactions Resulting from Complex Space-Time Projected to Real Physical Space-Time

This chapter and chapter 7 have shown that the complex Lorentz group and the Reality group generate the familiar interactions of The Standard Model: SU(3)⊗SU(2)⊗U(1) plus an additional set of SU(2)⊗U(1) interactions that we take to be the interactions of Dark Matter. In the next chapter we will display the lagrangian for the Extended Standard Model that results. Thus we have realized the goal of reducing Elementary Particle Physics to a fundamental theory based ultimately on Logic supplemented by the principles described in chapter 1.[119]

We will introduce General Relativity and an additional set of interactions (and their broken symmetry) in chapters that follow.

---

[119] Some of the postulates embedded in the text are not independent assumptions but actually follow from the mathematics of prior constructs.

10. Derivation of One Generation Leptonic Standard Model Features

BEGINNING OF AMENDED CHAPTERS 7-11 EXTRACT FROM BLAHA (2007b)

10.1 The Standard Model Leptonic Sector as a Consequence of the Complex Lorentz Group Coordinates Projected to Real Values

The development of the Standard Model in the 1960's and 1970's was a major step forward in our understanding of the major forces of nature. However the strange, and, in many physicists' opinions, unattractive form of the theory led particle theorists to conclude that it was a provisional theory that would eventually be replaced by a more elegant fundamental theory. Work in these directions has focussed upon 1) embedding the Standard Model within a larger ("elegant") symmetry group; 2) embedding the Standard Model within a space-time with extra dimensions that generate the Standard Model through some mechanism such as the Kaluza-Klein mechanism; and 3) viewing the Standard Model as somehow a "low energy" phenomenology that emerges from a Superstring Theory.

In this chapter we take an alternate view. We will derive the leptonic sector of the Standard Model based on the L_c and Reality group covariance of the equations of motion. Thus, unlike previous efforts, we view the Standard Model as in a natural form dictated by L_c and Reality group covariance and certain other fundamental physical requirements. Beauty being in the eye of the beholder we shall endeavor to show the attractiveness of the derivation of the Standard Model based on this covariance.

The naturalness of the derivation, and its close connection to the left-handed L_c *group, strongly suggest the Standard Model, which was grown by theorists from experiment, has an undeniable quality of genuineness that will likely survive the passage of time. The basis of the derivation, in a more fundamental theoretical framework, raises the hope that we have found a new, deeper level of understanding of elementary particle dynamics.*

10.2 Assumptions for the Leptonic Sector of the Standard Model

We will make certain assumptions, consistent with our construction approach, that provide a basis for the derivation of most aspects of the leptonic sector the Standard Model (The quark sector is derived in a subsequent chapter.). These assumptions, some of which will be derived later, connect our theory with physical reality.

1. One generation of leptons is assumed in this chapter. (The three generation case is treated later.)
2. The form of the equations of motion of the unbroken leptonic sector of the Standard Model is determined by covariance under the L_c and Reality groups.

3. Leptonic matter is composed of spin ½ Dirac particles (electrons) with charge -1 and tachyons (neutrinos) of charge zero as well as their anti-particles.[120]
4. A neutrino is a tachyon with a non-zero bare mass that includes the iota mass, and that may be changed by symmetry breaking effects.
5. Gauge fields are massless "before" spontaneous symmetry breaking and are thus conventional gauge fields without a tachyon equivalent in the theory.
6. L_c and Reality group covariance of the dynamical equations of motion, are spontaneously broken through the appearance of additional fermion and vector boson mass terms generated by a mechanism such as the Higgs mechanism.
7. Spatial coordinates have been rotated to real values using the Reality group.

10.3 Derivation of the Leptonic Sector of the Standard Model

The steps of the construction/derivation are:

A. L_c and Reality group covariance of the dynamical equations of motion requires that spin ½ particles be described by a generalization of the Dirac equation to an 8×8 matrix form in eqs 6.2 – 6.18 based on a doublet consisting of a Dirac particle and a tachyon.

B. We identify the Dirac particle with a charged lepton and the tachyon with a neutrino. The bare (iota) masses of these particles have the same numeric value (before symmetry breaking).

C. The leptonic sector free field lagrangian (without gauge fields introduced yet) is explicitly

$$\mathcal{L}_{\text{freelep}} = \Psi^\dagger(x) \begin{bmatrix} \gamma^0\gamma^5 i(\gamma^\mu\partial/\partial x^\mu - m) & 0 \\ 0 & \gamma^0(i\gamma^\mu\partial/\partial x^\mu - m) \end{bmatrix} \Psi(x) \tag{10.1}$$

Focussing on the derivative term we see that it can be put in the form

$$\Psi^\dagger(x)\gamma^0 \begin{bmatrix} \gamma^5 & 0 \\ 0 & I_4 \end{bmatrix} i\gamma^\mu\partial/\partial x^\mu\, \Psi(x) = \Psi^\dagger(x)\gamma^0[C^+ I - C^- \sigma_3] i\gamma^\mu\partial/\partial x^\mu \Psi(x) \tag{10.2}$$

where I_4 is a 4×4 identity matrix, where I and σ_3 are 2×2 matrices, and where C^+ and C^- are defined in chapter 5. The Pauli matrices σ_i are

$$\sigma_1 = \begin{bmatrix} 0 & 1 \\ 1 & 0 \end{bmatrix} \qquad \sigma_2 = \begin{bmatrix} 0 & -i \\ i & 0 \end{bmatrix} \qquad \sigma_3 = \begin{bmatrix} 1 & 0 \\ 0 & -1 \end{bmatrix} \tag{10.3}$$

[120] If neutrinos are Majorana particles then the derivation must be modified.

Expression 10.2 can be re-expressed in terms of left-handed and right-handed fields as

$$\Psi_L^\dagger(x)\gamma^0 i\gamma^\mu\partial/\partial x^\mu\Psi_L(x) - \Psi_R^\dagger(x)\gamma^0 i\gamma^\mu\partial/\partial x^\mu\sigma_3\Psi_R(x) \tag{10.4}$$

D. At this point we are in a position to introduce couplings to gauge fields. In view of the doublet nature of the fields $\Psi_L(x)$ and $\Psi_R(x)$ it would appear, at first glance, that the symmetry group of the gauge fields would be $SU(2)_L \otimes SU(2)_R$. However the right-handed tachyon field in expression 10.4 has the wrong sign in the lagrangian, as has been noted in the previous discussion of the free tachyon lagrangian and anti-commutator. Consequently the right-handed tachyon field *cannot* have trilinear or higher order couplings. If it did have such interactions then it would rapidly degrade to lower and lower energy by the emission of particles since right-handed leptonic tachyons can exist in principle as free particles (modulo possible Higgs terms). (In this regard the situation is similar to that of time-like photons, except that the set of tachyon physical states cannot be defined in a manner analogous to Gupta-Bleuler electrodynamics where the timelike and longitudinal photons "cancel" each other so that only transverse photons have physical effects.) *Thus there must be no right-handed leptonic tachyon interactions.*[121]

The doublet nature of the left-handed sector implies at least local SU(2) symmetries implemented with a covariant derivative.

The restricted nature of the right-handed leptonic sector indicates that *only* the Dirac particle in the "right-handed doublet" can have an interaction. Also the appearance of σ_3 in the right-handed term in expression 10.4 breaks SU(2) invariance <u>if</u> the left-handed covariant derivative (eq. 10.5 below) were substituted for $\partial/\partial x^\mu$ in the right-handed term. Thus a U(1) local gauge field interaction, restricted to the Dirac field member of the right-handed doublet and coupling to both members of the left-handed doublet, is the only allowed possibility. Without the U(1) interaction, the "right-handed doublet" would have no trilinear or higher order elementary particle interactions and would be physically irrelevant (except gravitationally) in unbroken gauge theory before spontaneous breakdown.

Putting these symmetries together we obtain a left-handed covariant derivative implementing local SU(2)⊗U(1) invariance found earlier:

$$D_{L\mu} = \partial/\partial x^\mu + \tfrac{1}{2} i g_2 \boldsymbol{\sigma}\cdot\mathbf{W}_\mu + \tfrac{1}{2} i g' B_\mu \tag{10.5}$$

and a right-handed covariant derivative[122]

$$\begin{aligned} D_{R\mu} &= \partial/\partial x^\mu\sigma_3 + \tfrac{1}{2} i g' B_\mu |Q|\sigma_3 \\ &= \partial/\partial x^\mu\sigma_3 + \tfrac{1}{2} i g' B_\mu |Q| \end{aligned} \tag{10.6}$$

where Q is the charge operator using the relation $|Q|\sigma_3 = |Q|$ for leptons. We use the absolute value of Q in order to achieve consistency in form with the right-handed quark sector described in the next chapter. As a result expression 10.4 becomes

---

[121] Right-handed neutrinos must interact with gravitons due to their mass-energy and the universality of the gravitational interaction. The extreme weakness of the gravitational interaction mitigates this effect.

[122] The coupling constants are defined by $e = g'\cos\theta_W = g_2\sin\theta_W$.

$$\Psi_L^\dagger(x)\gamma^0 i\gamma^\mu D_{L\mu}\Psi_L(x) - \Psi_R^\dagger(x)\gamma^0 i\gamma^\mu D_{R\mu}\Psi_R(x) \quad (10.7)$$

Thus the leptonic sector of the lagrangian[123] (modulo mass/Higgs terms) is

$$\begin{aligned}\mathcal{L}_{lep1} &= \Psi_L^\dagger\gamma^0 i\gamma^\mu D_{L\mu}\Psi_L - \Psi_R^\dagger\gamma^0 i\gamma^\mu D_{R\mu}\Psi_R \quad (10.8)\\ &= \Psi_L^\dagger\gamma^0 i\gamma^\mu D_{L\mu}\Psi_L + \bar{\psi}_{eR}\gamma^0 i\gamma^\mu D_{R\mu}\psi_{eR} - \bar{\psi}_{\upsilon R}\gamma^0 i\gamma^\mu \partial/\partial x^\mu \psi_{\upsilon R}\\ &= \Psi_L^\dagger\gamma^0 i\gamma^\mu D_{L\mu}\Psi_L + \bar{\psi}_{eR}\gamma^0 i\gamma^\mu(\partial/\partial x^\mu + ½ig'B_\mu)\psi_{eR} - \bar{\psi}_{\upsilon R}\gamma^0 i\gamma^\mu \partial/\partial x^\mu \psi_{\upsilon R}\end{aligned}$$

where we identify the tachyon as a neutrino and the Dirac particle as a charged lepton such as the electron. Our leptonic sector lagrangian is now the usual leptonic sector ElectroWeak lagrangian with a tachyonic neutrino (neglecting mass related terms).

E. We have local gauge invariance prior to symmetry breaking: The gauge field sector has the usual Yang-Mills lagrangian terms and the B field has lagrangian terms similar to that of the QED lagrangian.

F. Spontaneous symmetry breaking of gauge symmetry, and of L_c group covariance, via the Higgs mechanism (or an alternative mechanism) can be implemented in such a way as to give the electron its known mass as well as the massive vector bosons. Since spontaneous symmetry breaking breaks L_c covariance to Lorentz covariance it is a moot point whether the Higgs sector exhibits a similar covariance.

That concludes the derivation of leptonic sector of the Standard Model (except for mass/Higgs terms. We have shown that the form of the leptonic sector of the Standard Model is fundamental in nature and based on L_c and Reality group covariance of the equations of motion. Thus it is not correct to view the Standard Model as the result of the breakdown of a larger internal symmetry group to SU(2)⊗U(1). On the contrary, the SU(2)⊗U(1) interaction form is a consequence of the L_c and Reality group covariance of the free dynamical equations.

[123] Note that the gauge fields do not have a tachyon equivalent since they are initially massless prior to spontaneous symmetry breaking.

11. Can Neutrinos Be Tachyons?

Recently the three species of neutrinos have been found to have masses[124] in neutrino oscillation experiments although the masses are very small. If we denote the three neutrinos as m_1, m_2, and m_3; and let $\Delta m_{ij}^2 = m_j^2 - m_i^2$, then[125]

$$\Delta m_{12}^2 \cong 8 \times 10^{-5}\ \text{eV}^2$$

$$\Delta m_{23}^2 \cong 2.8 \times 10^{-3}\ \text{eV}^2$$

It is claimed that the sign of each Δm^2 is unambiguously determined.[126] However the mass difference values so obtained can be interpreted as differences in tachyon masses as well as differences in "normal" particle masses. The dependence of neutrino oscillations on masses squared results from the time evolution of mixed neutrino states. We consider mixtures of two neutrino states in the vacuum (for the sake of simplicity):

$$|\nu_a\rangle = \cos\theta\, |\nu_e\rangle - \sin\theta\, |\nu_\mu\rangle$$

$$|\nu_b\rangle = \sin\theta\, |\nu_e\rangle + \cos\theta\, |\nu_\mu\rangle$$

with differing masses m_a and m_b.[127] The phase factors determining the time dependence of $|\nu_a\rangle$ and $|\nu_b\rangle$ generate the neutrino oscillations from which the mass relations were obtained. The phase of the kth state is

$$|\nu_k(t)\rangle \sim e^{-im_k^2 t/(2p)}$$

for normal neutrinos. In the case of tachyons the factor in the exponential changes sign:

Tachyonic Neutrinos

$$E = (\mathbf{p}^2 - m^2)^{1/2} \cong p - m^2/(2p)$$

"Normal" Neutrinos

$$E = (\mathbf{p}^2 + m^2)^{1/2} \cong p + m^2/(2p)$$

Thus tachyonic neutrinos would exhibit the time dependence

---

[124] Note the neutrino mass differences are approximately of the order of the mass squared estimate of an iota: $m_0^2 \approx 4 \times 10^{-4}\ \text{eV}^2$ (section 1.1.3).

[125] S. M. Bilenky, "Neutrino Masses, Mixing and Oscillations", arXiv:hep-ph/050175 (October 13, 2005).

[126] A. B. McDonald, "Evidence for Neutrino Oscillations I", p. 8 arXiv:nucl-ex/0412005 (December, 2005).

[127] L. Wolfenstein, Phys. Rev. **D17**, 2369 (1978).

$$|\nu_k(t)> \sim e^{+im_{Tk}^2 t/(2p)}$$

where we use the mass subscript "T" to indicate a tachyon.

As a result, if neutrinos are "normal" the experimental results above would suggest the neutrino masses satisfy

$$m_3^2 > m_2^2 > m_1^2$$

On the other hand, if neutrinos are tachyons the experimental results above would suggest the value of the tachyon neutrino masses satisfies

$$m_1^2 > m_2^2 > m_3^2$$

an inverted spectrum compared to hypothetical "normal" neutrinos. When one considers the fact that tachyon masses squared are negative we see that an ordering of the tachyon neutrino mass spectrum, consistent with experimental neutrino results, is:

A Negative Neutrino m^2 Spectrum

$m^2 = 0$ ----------

m_1^2----------

m_2^2----------

m_3^2-----------

12. Derivation of One Generation Quark Sector Standard Model Features

12.1 Quark Sector of the Standard Model

The derivation of the form of the quark sector of the Standard Model is very similar to the preceding derivation of the form of the leptonic sector – but with some important points of difference. The primary difference is that we identify quarks with complexons. (This is a plausible assumption in view of the appearance of global SU(3) symmetry within complexon fields. Since complexons have very different spin characteristics compared to Dirac fields, complexon quarks may resolve difficulties in parton spin studies.) Complexon quarks have complex 3-momentum as we showed earlier. Complexon gluons also have complex 4-momentum (required by consistency with quark terms in dynamic equations.

12.2 Quark Sector Assumptions

We will make assumptions that will provide a basis for a construction/derivation of most aspects of single generation, quark sector Standard Model. These assumptions, based primarily on earlier discussions, connect our theory with physical reality.

1. One generation of quarks is assumed in this chapter. The three generation Standard Model is developed in a subsequent chapter.
2. The equations of motion of the unbroken form of the Standard Model are determined by covariance under the complex group L_C, and the Reality group (the source of the gauge fields.)
3. Quarks are composed of spin ½ "normal" complexons and tachyonic complexons.
4. The L_C and Reality group covariance of the dynamical equations are spontaneously broken through the appearance of mass terms generated by a mechanism such as the Higgs mechanism.

12.3 Derivation of the Form of the Standard Model Quark Sector

The construction/derivation:

A. L(C) covariance of the dynamical equations of motion requires that spin ½ particles be described by equations[128] generalized to an 8×8 matrix form (eqs 6.35 – 6.37) based on a doublet consisting of a

---

[128] The global SU(3) symmetry of complexons makes their identification with quarks reasonable. However, the complexon theory that we develop and use for quark dynamics in the Standard Model is not required. Our Standard Model could use Dirac fermion dynamics for the up-type quarks and tachyon dynamics for down-type quarks. We choose to use complexon dynamics for quarks because they have an SU(3)-like structure suggestive of color SU(3). More importantly, their spin dynamics is different and this difference may resolve the differences between theory and experiment for the deep inelastic parton spin-dependent structure functions. Nevertheless, quarks could be similar to leptons in this regard and form a doublet of a Dirac fermion and an ordinary tachyon. Whether quarks are complexons or not is an experimental question!

Dirac-type particle and a tachyon. (They may or may not be complexons.) In in the case of quarks the Dirac-type particle is the top component in the doublet and the tachyon is the bottom component.

B. Thus the 8 × 8 quark matrix formalism is:

$$đ_q(x)\Psi_C(x) = 0 \tag{12.1}$$

where[129]

$$đ_q(x) = \begin{bmatrix} (i\gamma^\mu D_{q\mu} - m_0) & 0 \\ 0 & (\gamma^\mu D_{q\mu} - m_0) \end{bmatrix} \tag{12.2}$$

with $D_{q\mu}$ given by eq. 9.27 and

$$\Psi_C(x) = \begin{bmatrix} \psi_{Cu}(x) \\ \psi_{Cd}(x) \end{bmatrix} \tag{12.3}$$

The upper 4-component field is a u-type field and the lower 4-component field is a d-type tachyonic field. The generalized equation eq. 12.1 is covariant under 8 × 8 L(C) transformations similar to eqs. 6.38 – 6.41. The generalized quark fermion equation eq. 12.1 is also covariant under conventional Lorentz transformations in the 8 × 8 representation.

The free quark sector lagrangian density that corresponds to the 8-dimensional fermion equation eq. 12.1 is

$$\mathcal{L}_{freeQuark} = \overline{\Psi}_C(x)đ_q(x)\Psi_C(x) \tag{12.4}$$

where

$$\overline{\Psi}_C(x) = \Psi_C^{\dagger}\Gamma_C^{\ 0} \tag{12.5}$$

with

$$\Psi_C^{\dagger} = (\Psi_C^{\dagger})|_{\mathbf{x_i} = -\mathbf{x_i}} \tag{12.6}$$

for complexon quarks, and with

$$\Psi_C^{\dagger} = \text{the hermitean conjugate of } \Psi_C \tag{12.7}$$

in the case of non-complexon quarks. The † in the parentheses on the right side of eq. 12.6 indicates hermitean conjugation. Also

---

129 If we wish to obtain the Standard Model without complexon quarks (which is the conventional Standard Model) then $D_{q\mu} = \partial/\partial x^\mu$. If we wish to obtain the Standard Model with complexon quarks (which is a new form of the Standard Model) then $D_{q\mu}$ is given by eq. 9.27.

$$\Gamma_C{}^0 = \begin{bmatrix} \gamma^0 & 0 \\ 0 & i\gamma^0\gamma^5 \end{bmatrix} \qquad (12.8)$$

C. The free quark field lagrangian is explicitly

$$\mathcal{L}_{freeQuark} = \Psi_C{}^\dagger \begin{bmatrix} \gamma^0(i\gamma^\mu D_{q\mu} - m_0) & 0 \\ 0 & \gamma^0\gamma^5 i(\gamma^\mu D_{q\mu} - m_0) \end{bmatrix} \Psi_C \qquad (12.9)$$

Focussing on the derivative term we see that it can be put in the form

$$\Psi_C{}^\dagger\gamma^0 \begin{bmatrix} I_4 & 0 \\ 0 & \gamma^5 \end{bmatrix} i\gamma^\mu D_{q\mu}\Psi_C = \Psi_C{}^\dagger\gamma^0[C^+ I + C^- \sigma_3]i\gamma^\mu D_{q\mu}\Psi_C \qquad (12.10)$$

where I_4 is a 4×4 identity matrix, where I, and the Pauli matrix σ_3, are 2×2 matrices, and C^+ and C^- are defined earlier. Expression 12.10 can be expressed in terms of left-handed and right-handed fields as

$$\Psi_{CL}{}^\dagger\gamma^0 i\gamma^\mu D_{q\mu}\Psi_{CL} + \Psi_{CR}{}^\dagger\gamma^0 i\gamma^\mu D_{q\mu}\sigma_3\Psi_{CR} \qquad (12.11)$$

D. At this point we are in a position to introduce couplings to gauge fields. In view of the doublet nature of the fields $\Psi_{CL}(x)$ and $\Psi_{CR}(x)$ it would again appear, at first glance, that the symmetry group of the gauge fields would be $SU(2)_L \otimes SU(2)_R$. However the right-handed tachyonic field has the wrong sign in the lagrangian, as has been noted in the earlier discussion of the free tachyon lagrangian and anti-commutator. Consequently, **if quarks were *not* confined,** the right-handed quark tachyon field could not have trilinear or higher order couplings. **If free tachyon quarks existed, and had interactions, then they would rapidly degrade to lower and lower energy by the emission of particles.**

However, because of quark confinement in bound states with discrete energy levels, a bound tachyon quark can only emit particles if a lower energy bound state exists. As a result right-handed tachyon quarks can have interactions, such as the electromagnetic interaction, because quark confinement "tames" their propensity to emit particles due to the "wrong sign" in the lagrangian.

Again there is an analogy to Gupta-Bleuler QED quantization. In Gupta-Bleuler quantization physical states are required to have equal numbers of time-like and longitudinal photons thus canceling their physical effects. Similarly, right-handed tachyon quarks are required to be bound to other quarks by quark confinement to avoid continuous emission of

particles.[130] **Since interactions are allowed for right-handed tachyon quarks the Higgs mechanism can be used to change their mass.**

The doublet nature of the left-handed sector implies at least local SU(2) symmetries. The appearance of σ_3 in the right-handed term in expression 12.11 explicitly breaks SU(2) invariance if the left-handed covariant derivative (eq. 12.12 below) were substituted for $D_{q\mu}$ in the right-handed term. Thus the right-handed fields can only have a U(1) local gauge field interaction, and are SU(2) singlets. We thus obtain a left-handed covariant derivative implementing local SU(2)⊗U(1) covariance:

$$D_{qL\mu} = D_{q\mu} + \tfrac{1}{2} i g_2 \boldsymbol{\sigma}\cdot\mathbf{W}_\mu + i g_1 B_\mu/6 \qquad (12.12)$$

where $D_{q\mu}$ is given by eq. 9.27 and a right-handed covariant derivative[131]

$$D_{qR\mu} = D_{q\mu}\sigma_3 + i g_1 B_\mu |Q| \qquad (12.13)$$

where $|Q|$ is the absolute value of the charge operator (with u eigenvalue 2/3 and d eigenvalue 1/3). The absolute value is used in order to compensate for the minus sign in front of the right-handed tachyon (d quark) term. As a result expression 12.11 becomes

$$\Psi_{CL}^\dagger(x)\gamma^0 i\gamma^\mu D_{qL\mu}\Psi_{CL}(x) + \Psi_{CR}^\dagger(x)\gamma^0 i\gamma^\mu D_{qR\mu}\Psi_{CR}(x)$$

Thus the quark sector of the lagrangian[132] is

$$\mathcal{L}_{quark\,1} = \Psi_{CL}^\dagger\gamma^0 i\gamma^\mu D_{qL\mu}\Psi_{CL} + \Psi_{CR}^\dagger\gamma^0 i\gamma^\mu D_{qR\mu}\Psi_{CR} \qquad (12.14)$$

$$= \Psi_{CL}^\dagger\gamma^0 i\gamma^\mu D_{qL\mu}\Psi_{CL} + \bar{\psi}_{CuR} i\gamma^\mu D_{qR\mu}\psi_{CuR} - \bar{\psi}_{CdR} i\gamma^\mu D_{qR\mu}\psi_{CdR}$$

$$= \Psi_{CL}^\dagger\gamma^0 i\gamma^\mu D_{qL\mu}\Psi_{CL} + \bar{\psi}_{CuR} i\gamma^\mu(D_{q\mu} + \tfrac{2}{3} i g_1 B_\mu)\psi_{CuR} - $$
$$- \bar{\psi}_{CdR} i\gamma^\mu(D_{q\mu} + \tfrac{1}{3} i g_1 B_\mu)\psi_{CdR}$$

where we *provisionally* identify the tachyon as a d-type quark and the Dirac-type particle as a u-type quark. Our quark sector lagrangian is now the usual Standard Model quark sector lagrangian modulo

---

[130] Therefore quark confinement is required in order to have a properly formulated quark sector. Another interaction – the strong interaction – is required for quark confinement. Presently there is only one accepted mechanism for quark confinement – through a non-abelian gauge coupling. (Higher derivative theories with quark confinement are in disfavor.) An additional non-abelian symmetry must be introduced for quarks. As discussed in chapter 19, SU(3) appears to be a natural choice.

[131] The quark SU(2) coupling constant is, by gauge invariance, required to have the same value as the leptonic SU(2) coupling constant. The U(1) coupling constants are not required to be the same in both sectors and, in fact, are different. The coupling constants here are defined by $e = g_1\cos\theta_W = g_2\sin\theta_W$.

[132] Note that the gauge fields do not have a tachyon equivalent since they are initially massless prior to spontaneous symmetry breaking.

complex 3-momenta, and modulo the strong interaction, terms, except that d-type quarks are tachyons.

E. The SU(2) gauge field sector has the usual Yang-Mills lagrangian terms, and the B field is a U(1) abelian gauge field.

F. Spontaneous symmetry breaking of gauge symmetry, and of L(C) covariance, via the Higgs mechanism can be implemented in such a way as to give the quarks, and massive vector bosons, their "known" masses. Since spontaneous symmetry breaking breaks L(C) covariance of the dynamical equations of motion to Lorentz covariance it is a moot point whether the Higgs sector (if there is one) is manifestly L(C) covariant or not.

Thus L_C covariance of the quark equations of motion generate most of the "unusual" features of the quark sector of the Standard Model.

12.4 Color SU(3)

Our derivation yields the general form of the ElectroWeak quark sector of the Standard Model with a dynamics modified by the presence of complex spatial 3-momenta. This difference is a positive for the theory. In the discussion of eqs. 5.143-5.154 we showed that complexons have an SU(3) symmetry. The same discussion applies to tachyonic complexons.

Thus, although the discussion in this chapter has hitherto dealt with color scalar quarks,[133] we need only introduce a color index on the quarks to obtain quark fields in the fundamental SU(3) representation $\underline{3}$. Eqs. 5.153, 5.240 and 5.251 contain free color quark field expansions of the $\underline{3}$ representation.

The remaining step is to introduce local SU(3) Yang-Mills gauge fields that interact with the quarks through gauge covariant derivatives. Thus our quark sector has the ElectroWeak and Strong interactions of the Standard Model flowing naturally from L_C and Reality group covariance of the initially free, quark dynamical equations.

It has long been known that the strong interaction must have a non-abelian symmetry group in order to have quark confinement – a feature required by our derivation and in apparent complete agreement with experimental data.[134] If we require that spin ½ bound quark states exist as they do, then strict quark confinement would rule out an SU(2) color strong interaction. Therefore the appearance of SU(3) symmetry in complexon fields, and the requirement of the "most minimal" (Leibniz Minimax Principle) non-abelian symmetry group for the color interaction, provide a logical basis for SU(3) as the color symmetry group.

As shown earlier quarks have a complex spatial 3-momentum – they are "normal" complexons and tachyonic complexons. If we think of quark-gluon interactions, and in particular, perturbative diagrams with quark-gluon loops such as the simplest quark self-energy diagram, then imaginary momentum should "flow" around a loop as well as real momentum.

Thus we postulate that gluons are massless spin 1 complexons (although this is not required by our derivation). Therefore we assume an SU(3) non-abelian, massless, version of vector complexons. The gluon lagrangian term is

---

[133] Dark quarks are SU(3) singlets in our theory.

[134] Evidence is scanty at the present time for five quark states and other exotics that would seem to differ from strict confinement in three quark or quark-antiquark bound states.

$$\mathcal{L}_{CG} = -\tfrac{1}{4} F_C{}^{a\mu\nu}(x) F_C{}^{a}{}_{\mu\nu}(x) \quad (12.15)$$

where x has a complex 3-momentum and

$$F_C{}^{a}{}_{\mu\nu} = D_\nu A_C{}^{a}{}_{\mu} - D_\mu A_C{}^{a}{}_{\nu} + g f^{abc} A_C{}^{b}{}_{\mu} A_C{}^{c}{}_{\nu} \quad (12.16)$$

with D_μ is defined by eq. 5.162, g the coupling constant, and the constants f^{abc} are the SU(3) structure constants.

The theory of the strong color interaction should be developed within a path integral framework that takes account of Faddeev-Popov gauge fixing. Before doing that we must consider the effect of the complex spatial 3-vector on the non-abelian gauge formalism.

12.5 Pure Complexon Gauge Groups

Consider a local Lie group G with the generators of its algebra satisfying the commutation relations:

$$[t^a, t^b] = i f^{abc} t^c \quad (12.17)$$

If a complexon field ψ_{Ca} transforms as some representation of G then a transformed field has the form:

$$\psi_{Ca}'(x) = \Theta_{ab}(x)\psi_{Cb}(x) \quad (12.18)$$

where $x = (x^0, \mathbf{x} = \mathbf{x}_r + i\mathbf{x}_i)$ and $\Theta_{ab}(x)$ is an element of G. Since G is a local group we can represent $\Theta_{ab}(x)$ in the form:

$$\Theta_{ab}(x) = [\exp(-i\varphi_c(x)t^c)]_{ab} \quad (12.19)$$

where $\varphi_c(x)$ is a local parameter dependent on x, and t^c is the algebraic matrix in the representation of G under consideration.

The gauge covariant derivative in the case of complexon fermions is

$$\mathcal{D}_\nu = D_\nu - igA_{C\nu} \quad (12.20)$$

where

$$A_{C\nu} = A_C{}^{a}{}_{\nu} t^a \quad (12.21)$$

Applying a gauge transformation to the gauge covariant derivative of a complexon fermion field $\mathcal{D}_\nu\psi_C(x)$:

$$\Theta\mathcal{D}_\nu\psi_C(x) = \Theta D_\nu\psi_C(x) - ig\Theta A_{C\nu}\Theta^{-1}\Theta\psi_C(x) \quad (12.22)$$

$$= D_\nu\psi_C'(x) - igA_C'{}_\nu\psi_C'(x) = (D_\nu\psi_C(x))'$$

where

$$\psi_C'(x) = \Theta(x)\psi_C(x) \quad (12.23)$$

we find

$$A_C'{}_\nu = (-i/g)(D_\nu\Theta(x))\Theta^{-1}(x) + \Theta(x)A_{C\nu}(x)\Theta^{-1}(x) \quad (12.24)$$

The reader will note that the form of eqs. 12.15 – 12.24 is identical to those of a conventional non-abelian gauge field with the replacement:

$$\partial/\partial x^{\nu} \rightarrow D_{\nu} \tag{12.25}$$

with D_{ν} given by eq. 5.162. Thus we see that $\varphi_c(x)$, the local parameter in eq. 12.19 is dependent in general, on time, and the real and imaginary parts of the complex spatial 3-vector. Thus the formalism differs only in small ways from the conventional non-abelian gauge theory. However, perturbation theory and non-perturbative phenomena such as instantons exhibit significant differences.

The commutator of of covariant derivatives

$$F_{C\mu\nu} = F_{C}{}^{a}{}_{\mu\nu}t^{a} = (i/g)[\mathcal{D}_{\mu}, \mathcal{D}_{\nu}] \tag{12.26}$$

is itself covariant under gauge transformations:

$$F'_{C\mu\nu} = \Theta F_{C\mu\nu}\Theta^{-1} \tag{12.27}$$

The strong interaction part of the action has terms of the form

$$I = \int d^7x[\mathcal{L}_{CG} + \bar{\Psi}_q i\gamma^{\mu}\mathcal{D}_{\mu}\Psi_q + \ldots] \tag{12.28}$$

where the "extra" three integrations are over the imaginary spatial coordinates, and Ψ_q represents a generic quark (complexon) field. The action must be supplemented with a constraint similar to those seen earlier for complexons, which ensures the reality of the lagrangian term $\mathcal{L}_{CG}$ (and thus the corresponding Hamiltonian terms and ultimately the unitarity of the theory.) The constraint simply specifies the imaginary part of $\mathcal{L}_{CG}$ is zero:

$$[\partial A_C{}^{a\mu}/\partial x_{ik}]\mathrm{Re}\, F_C{}^{a}{}_{\mu k} = [\partial A_C{}^{a\mu}/\partial x_{ik}](\partial A_C{}^{a}{}_{\mu}/\partial x_r{}^{k} - \partial A_C{}^{a}{}_{k}/\partial x_r{}^{\mu} + gf^{abc}A_C{}^{b}{}_{\mu}A_C{}^{c}{}_{k}) = 0 \tag{12.29}$$

where $\mathbf{x}_r$ indicates the real part of the spatial coordinates $\mathbf{x} = \mathbf{x}_r + i\mathbf{x}_i$, and $\partial/\partial x_i{}^{k}$ is the derivative with respect to the kth component of the imaginary coordinate 3-vector $\mathbf{x}_i$. It is not a choice of gauge but rather a restriction on the dependence of the $A_C{}^{\nu}$ field on the real and imaginary spatial 3-vectors.

If we assume that we can integrate by parts in the imaginary coordinates then we can re-express the constraint as

$$A_C{}^{a\mu}\ \mathrm{Re}\ \partial F_C{}^{a}{}_{\mu k}/\partial x_i{}^{k} = 0 \tag{12.30}$$

which is explicitly

$$A_C{}^{a\mu}[\partial^2 A_C{}^{a}{}_{\mu}/\partial x_r{}^{k}\partial x_i{}^{k} - \partial^2 A_C{}^{a}{}_{k}/\partial x_r{}^{\mu}\partial x_i{}^{k} + gf^{abc}\partial(A_C{}^{b}{}_{\mu}A_C{}^{c}{}_{k})/\partial x_i{}^{k}] = 0 \tag{12.31}$$

by eq. 12.16 where $x_r{}^0 = x^0$. This restriction can be implemented within the framework of the path integral formalism by using the Faddeev-Popov Method with the introduction of ghosts in a manner similar to that of the Faddeev-Popov Method to implement a gauge condition. The form of the restriction stated in eqs.

12.30 and 12.31 leads to second order derivative Faddeev-Popov ghost terms in the path integral while eq. 12.29 would lead to a third order derivative ghost terms in the path integral with attendant unitarity (and possibly other) issues.

12.6 Pure Gauge Complexon Path Integral Formulation and Faddeev-Popov Method

The path integral formalism for complexon non-abelian, pure, Yang-Mills fields differs significantly from the conventional gauge field path integral formalism. The path integral for a complexon gauge field can be written symbolically as:

$$Z(J^{\mu}) = N\int DA_C \Delta_{FP}(A_C)\delta(F(A_C))\Delta_C(A_C)\delta(F_C(A_C))\exp\{i\int d^7y[\mathscr{L} + J^{\mu}(y)A_{C\mu}(y)]\} \tag{12.32}$$

where $\delta(F(A_C))$ specifies the gauge, $\Delta_{FP}(A_C)$ is its Faddeev-Popov determinant; and $\delta(F_C(A_C))$ specifies the complexon condition (eq. 12.31) with $\Delta_C(A_C)$ the Faddeev-Popov determinant for the complexon condition. In both cases the Faddeev-Popov determinant can be calculated in the standard way.[135]

First we consider the gauge fixing delta function. Note that it can be written as a delta function in the gauge times a determinant:

$$\delta(F(A_C^{\omega})) = \delta(\omega - \omega_0)|\det \delta F(A_{C\mu}^{\ \omega}(x))/\delta\omega(x)|^{-1}|_{F(A_C)=0} \tag{12.33}$$

where ω_0 is a reference gauge, where

$$\begin{aligned} A_{C\ \mu}^{\ a\ \omega}(x) &= A_{C\ \mu}^{\ a}(x) - g^{-1}D_{\mu}\omega^{a} + f^{abc}\,\omega^{b}(x)A_{C\ \mu}^{\ c}(x) \\ &= A_{C\ \mu}^{\ a}(x) + \delta A_{C\ \mu}^{\ a\ \omega}(x) \end{aligned} \tag{12.34}$$

and

$$\begin{aligned} \mathrm{Re}\, F_{C\ \mu k}^{\ a\ \ \omega} &= \mathrm{Re}\, F_{C\ \mu k}^{\ a} + f^{abc}\,\omega^{b}(x)F_{C\ \mu k}^{\ c} \\ &= \mathrm{Re}\, F_{C\ \mu k}^{\ a} + \delta(\mathrm{Re}\, F_{C\ \mu k}^{\ a\ \ \omega}) \end{aligned} \tag{12.35}$$

under an infinitesimal gauge transformation, and where

$$\Delta_{FP}(A_C) = |\det \delta F(A_{C\mu}^{\ \omega}(x))/\delta\omega(x)||_{F(A_C) = 0,\ \omega = 0} \tag{12.36}$$

We will choose the complexon Lorentz gauge to evaluate the Faddeev-Popov determinant:

$$F^{a}(A_C) = D_{\mu}A_C^{a\mu}(x) = 0 \tag{12.37}$$

We find

$$\begin{aligned} F^{a}(A_{C\mu}^{\ \omega}(x)) &= D^{\mu}(A_{C\ \mu}^{\ a}(x) - g^{-1}D_{\mu}\omega^{a}(x) + f^{abc}\,\omega^{b}(x)A_{C\ \mu}^{\ c}(x)) \\ &= -g^{-1}D^{\mu}D_{\mu}\omega^{a}(x) + f^{abc}A_{C\ \mu}^{\ c}(x)D^{\mu}\omega^{b}(x) \end{aligned} \tag{12.38}$$

Thus

---

[135] See for example Huang (1992).

$$\delta F^a(A_{C\mu}{}^{\omega}(x))/\delta\omega^b(x) = -g^{-1}\,\delta^{ab}D^{\mu}D_{\mu} + f^{abc}A_C{}^{c\mu}(x)D_{\mu} \tag{12.39}$$

and

$$\Delta_{FP}(A_C) = |\det\,(g^{-1}\delta^{ab}D^{\mu}D_{\mu} - f^{abc}A_C{}^{c\mu}(x)D_{\mu})| \tag{12.40}$$

where | … | represent absolute value.

We can rewrite the Faddeev-Popov determinant as a path integral over anti-commuting c-number fields with a ghost Lagrangian:

$$\Delta_{FP}(A_C) = \int D\chi^* D\chi \exp[\, i\int d^7x\, \mathscr{L}^{ghost}(x)] \tag{12.41}$$

where

$$\mathscr{L}^{ghost}(x) = \chi^{a*}(x)[\delta^{ab}D^{\mu}D_{\mu} - gf^{abc}A_C{}^{c\mu}(x)D_{\mu}]\chi^b(x) \tag{12.42}$$

Faddeev-Popov Application to Complexon Condition

The complexon condition can also be implemented within the path integral formalism using the Faddeev-Popov Mechanism. Using the identity

$$1 = \int DA_C\Delta_C(A_C)\delta(F_C(A_C)) \tag{12.43}$$

we see that an infinitesimal gauge transformation yields eqs. 12.34 and 12.35. This enables us to relate $\Delta_C(A)$ to the determinant

$$\delta(F_C(A_C{}^{\omega})) = |\det \delta F_C(A_{C\mu}{}^{\omega}(x))/\delta\omega(x)|^{-1}|_{F_C(A_C)=0,\ \omega=0} \tag{12.44}$$

and

$$\Delta_C(A_C) = |\det \delta F_C(A_{C\mu}{}^{\omega}(x))/\delta\omega(x)||_{F_C(A_C)=0,\ \omega=0} \tag{12.45}$$

From eq. 12.31 we see

$$F_C(A_{C\mu}(x)) = A_C{}^{a\mu}[\partial^2 A_C{}^a{}_{\mu}/\partial x_r{}^k\partial x_i{}^k - \partial^2 A_C{}^a{}_k/\partial x_r{}^{\mu}\partial x_i{}^k + gf^{abc}\partial(A_C{}^b{}_{\mu}A_C{}^c{}_k)/\partial x_i{}^k] \tag{12.46}$$

with

$$F_C{}^a(A_{C\mu}) = 0 \tag{12.47}$$

Inserting eq. 12.34 and 12.35 we find

$$\begin{aligned}[\delta F_C(A_{C\mu}{}^{\omega}(x))/\delta\omega^a(x)]|_{F_C(A_C)=0,\ \omega=0} &= \delta[\delta A_C{}^{b\mu\omega}\,\mathrm{Re}\,\partial F_C{}^b{}_{\mu k}/\partial x_{ik} + A_C{}^{b\mu}\,\partial\delta(\mathrm{Re}\,F_C{}^b{}_{\mu k}{}^{\omega})/\partial x_{ik}\,]/\delta\omega^a(x)|_{\omega=0} \\ &= -g^{-1}(\mathrm{Re}\,\partial F_C{}^a{}_{\mu k}/\partial x_{ik})D^{\mu} - f^{abc}A_C{}^{b\mu}(\mathrm{Re}\,F_C{}^c{}_{\mu k})\partial/\partial x_{ik}\end{aligned} \tag{12.48}$$

Thus

$$\Delta_C(A_C) = |\det\,(g^{-1}(\mathrm{Re}\,\partial F_C{}^a{}_{\mu k}/\partial x_{ik})D^{\mu} + f^{abc}A_C{}^{b\mu}(\mathrm{Re}\,F_C{}^c{}_{\mu k})\partial/\partial x_{ik}| \tag{12.49}$$

where | … | represent absolute value.

We can rewrite this Faddeev-Popov determinant as a path integral over anti-commuting c-number fields with a ghost Lagrangian:

$$\Delta_C(A_C) = \lim_{r\to\infty} \int D\chi_C^* D\chi_C \exp[ir^{-2}\int d^7x\, \mathscr{L}_C^{ghost}(x)] \tag{12.50}$$

where r is a constant that is taken to the limit ∞, and where

$$\mathscr{L}_C^{ghost}(x) = \chi_C^*(x)\{D^\mu D_\mu + r^2 t^a[(\mathrm{Re}\ \partial F_{C\,\mu k}^{\ a}/\partial x_{ik})D^\mu + gf^{abc}A_C^{b\mu}(\mathrm{Re}\ F_{C\,\mu k}^{\ c})\partial/\partial x_{ik}]\}\chi_C(x) \tag{12.51}$$

where t^a is a 3 × 3 matrix of the $\underline{3}$ representation of color SU(3) and $\chi_C(x)$ is a three row field in the $\underline{3}$ representation. *The introduction of $D^\mu D_\mu$ is based on consistency with the complexon formalism. It is needed to establish a perturbative expansion of the path integral. Its effect vanishes in the limit $r \to \infty$ reducing ghost loops of this type to point interactions.* The reader will note that second order and third order derivative terms appear in the interaction in $\mathscr{L}_C^{ghost}(x^\mu)$ and raise the issue of non-renormalizable divergences. If one uses the Two-Tier approach to quantum field theory developed by Blaha (2005a) then all potential divergences disappear. *The Two-Tier formulation of the pure, complexon, Yang-Mills theory that we are discussing is finite.*

The complete pure complexon, Yang-Mills path integral is

$$Z(J^\mu) = N \int DA_C D\chi^* D\chi D\chi_C^* D\chi_C \Delta_{FP}(A_C)\delta(F(A_C))\Delta_C(A_C)\delta(F_C(A_C))\exp\{i\int d^7y\, [\mathscr{L} + J^\mu A_{C\mu}]\} \tag{12.52}$$

where

$$\mathscr{L} = \mathscr{L}_{CG} + \mathscr{L}^{ghost} + \mathscr{L}_C^{ghost} \tag{12.53}$$

with the lagrangian terms specified by eqs. 12.15, 12.42, and 12.51.

12.7 Complexon Quark-Gluon Perturbation Theory

In this section we will give the flavor of complexon strong interaction theory by considering a simple diagram. Fig. 12.1 is a self-energy diagram for a quark in which a gluon is emitted and absorbed. If one considers momentum then complexon particles have complex spatial momentum. From the form of the propagators and wave functions previously considered it is clear that the real part of the 3-momentum and the imaginary part of the 3-momentum will be separately conserved at each vertex. The general form of the self-energy integral corresponding to Fig. 12.1 is

$$I_{ab}(p) = \int d^7q\, P_{ab}(q, p)\exp(\mathrm{Gaussian})/Q(q, p) \tag{12.54}$$

where a and b are SU(3) color indices, $P_{ab}(q, p)$ is a polynomial in q and p together with gamma matrix factors and color SU(3) matrix factors, and Q(q, p) is the product of a quark and a gluon propagator denominator factor. A Gaussian exponential factor appears if we use a Two-Tier formulation[136] of the complexon quark and gluon theory. This exponential factor guarantees the convergence of the integral resulting in a finite result.

Thus while perturbation theory is only partly useful in calculating strong interaction phenomena due to the "largeness" of the coupling constant in general, we see complexon perturbation theory is well defined, finite, and sensible if one uses Two-Tier quantum field theory.

---

[136] See Blaha (2005a) and the discussion of the Two-Tier formalism later.

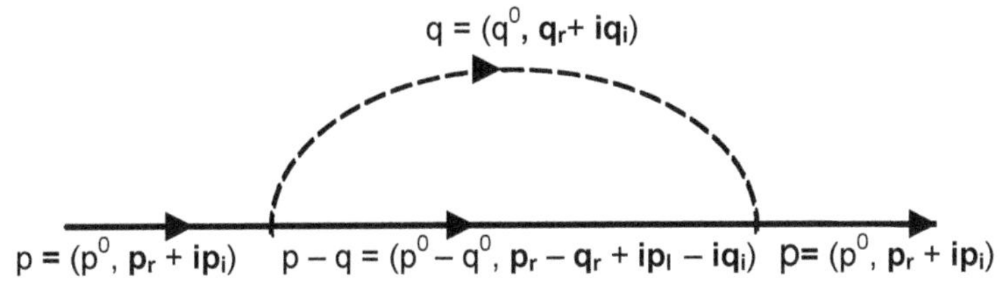

Figure 12.1. Quark self-energy diagram in which a quark emits and subsequently absorbs a gluon. The quark momentum is $p = (p^0, p_r + ip_i)$ and the gluon momentum is $q = (q^0, q_r + iq_i)$ with conservation of energy, and separate conservation of real spatial momentum and imaginary spatial momentum at each vertex.

The integral corresponding to Fig. 12.1 has the form:

$$I_{1ab}(p) = \int dq^0 d^3q_r d^3q_i\, P_{ab}(q, p)\exp(\text{Gaussian})/Q(q, p) \tag{12.55}$$

and includes integrations over both real and imaginary 3-momenta.

12.8 Complexon Quark ElectroWeak Interactions

In our theory quarks are complexons with complex spatial 3-momenta. On the other hand, the ElectroWeak bosons: $W^{\pm}_{\mu}$, Z_{μ} and A_{μ} are not complexons and so have totally real momenta. In this section we address the issue of quark ElectroWeak perurbation theory. *The general perturbation theory rule in this situation is that real and imaginary momenta are separately conserved at each vertex.*

Fig. 12.2 shows a quark self-energy diagram in which a quark emits a photon and subsequently reabsorbs it. Since ElectroWeak bosons have real energies and momenta and since real and imaginary momenta are separately conserved at each vertex by the above stated rule, the photon momentum is real while the quark spatial 3-momentum is complex.

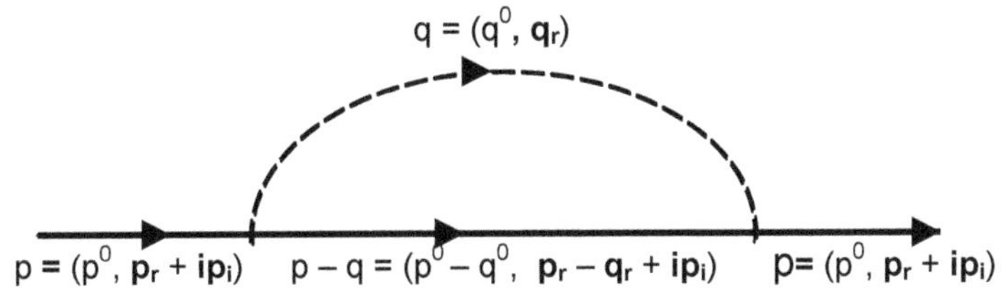

Figure 12.2. Quark self-energy diagram in which a quark emits and subsequently absorbs a photon. The quark momentum is $p = (p^0, p_r + ip_i)$ and the photon momentum is $q = (q^0, q_r)$ with conservation of energy, and separate conservation of real spatial momentum and imaginary spatial momentum at each vertex.

The integral corresponding to Fig. 12.2 has the form:

$$I_{2ab}(p) = \int dq^0 d^3q_r d^3q_i \delta^3(\mathbf{q}_i) P_{ab}(q, p)\exp(\text{Gaussian})/Q(q, p) \tag{12.56}$$

$$= \int dq^0 d^3q_r P_{ab}(q, p)\exp(\text{Gaussian})/Q(q, p)$$

However, in cases where there are complexon quark loops the imaginary quark 3-momenta affects the result. For example, Fig. 12.3 shows a complexon quark loop contribution to the photon self-energy in which the imaginary quark 3-momenta integration contributes.

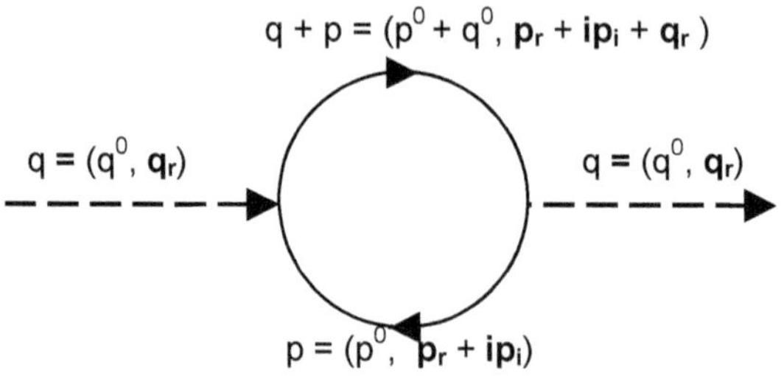

Figure 12.3. Photon self-energy diagram with a quark loop. The photon momentum is $q = (q^0, q_r)$ and quark loop momentum is $p = (p^0, p_r + ip_i)$.

12.9 Complexon Quarks with Color SU(3)

In this chapter we have seen that the complex Lorentz and Reality groups lead to the form of quark ElectroWeak theory and also leads naturally to color SU(3) with complexon gluons.

13. SU(3)⊗SU(2)⊗U(1)⊗SU(2)⊗U(1) Extended Standard Model

BEGINNING OF AMENDED CHAPTER 6 EXTRACT FROM BLAHA (2013b)

We derived a set of Standard Models in Blaha (2011c). Since then experiment has revealed new features of Dark Matter that lead us to extend the models to include a Dark sector. The extended Complexon Standard Model which we will simply call The Extended Standard Model embodies the (broken) symmetry group SU(3)⊗SU(2)⊗U(1)⊗SU(2)⊗U(1).

In section 13.1 of this book we will outline a form of Quantum Field Theory called Two-Tier Quantum Field Theory in section 14.1 that eliminates infinities and anomalies that crop up in conventional Quantum Field Theory. Two-Tier Quantum Field Theory was described in detail in Blaha (2005a) as well as in section 9.5 of Blaha (2011c), and chapter 5 of (2012b). We will describe aspects of it (chapters 18 and 19) later in regard to Dark Energy, The Big Bang, anomalies and Baryon number conservation.

It begins by extending real-valued coordinates by an imaginary-valued quantum field: $X^{\mu}(z) = z^{\mu} + i\ Y^{\mu}(z)/M_c^2$ where M_c is a very large mass. These coordinates are then used to develop Two-Tier quantum field theory. and show that it eliminates infinities in perturbation calculations to any order.

In our Conplexon Standard Model all fields will be functions of $X^{\mu}(z)$ and consequently all perturbation theory calculations will yield finite results including fermion triangle "anomaly" calculations.

13.1 Two-Tier Formulation of One Generation Standard Model

The general ideas of Two-Tier formulations of the Standard Model were presented in Blaha (2003) and Blaha (2005a). In this section we will develop a Two-Tier formulation of the one generation Standard Model described in the previous section. This formulation goes beyond our earlier books by using complex spatial coordinates which necessitate a slightly more complicated Two-Tier formulation.

The basic ansatz of the Two-Tier formalism is to replace every appearance of a real-valued coordinate x with a variable that is, in part, a quantum field Y:[137]

$$x^{\mu} \rightarrow X^{\mu} = (y^0, \mathbf{y} + \mathbf{Y}(y^0, \mathbf{y})) \tag{13.1}$$

where $\mathbf{Y}(y^0, \mathbf{y})$ is the spatial part of a free massless vector field identical in form to the free QED field.

Then one finds that the momentum space free field Feynman propagators G(k) of all particles acquires a Gaussian factor exp(h(k)):

$$G(k) \rightarrow G(k)\exp(h(k)) \tag{13.2}$$

[137] Later we will show that the Y field quanta may be the driving force of the Big Bang giving Y a deeper role in Cosmology.

so that all perturbation theory diagrams are finite. The result is a finite perturbative result in all calculations to any order in perturbation theory. Blaha (2005a) shows that Two-Tier theories are finite, Poincare covariant, and unitary.

Simple Two-Tier Formalism

In this subsection we will describe the basic Two-Tier formalism.[138] Replace a lagrangian $\mathscr{L}_F$ (y) with the lagrangian:

$$\mathscr{L}(y) = \mathscr{L}_F (X_\mu(y))J + \mathscr{L}_C(X^\mu(y), \partial X^\mu(y)/\partial y^\nu, y) \tag{13.3}$$

where

$$X_\mu(y) = y_\mu + i\, Y_\mu(y)/M_c^2 \tag{13.4}$$

with M_c being a large mass scale and $Y_\mu(y)$ a vector quantum field similar to the electromagnetic field, and where J is the absolute value of the Jacobian of the transformation from X to y coordinates:

$$J = |\partial(X)/\partial(y)| \tag{13.5}$$

The lagragian term $\mathscr{L}_C$ is

$$\mathscr{L}_C = +\tfrac{1}{4}\, M_c^4 F^{\mu\nu}F_{\mu\nu} \tag{13.6}$$

with

$$\begin{aligned} F_{\mu\nu} &= \partial X_\mu/\partial y^\nu - \partial X_\nu/\partial y^\mu \\ &\equiv i\,(\partial Y_\mu/\partial y^\nu - \partial Y_\nu/\partial y^\mu)/M_c^2 \end{aligned} \tag{13.7}$$

The sign in eq. 13.3 is not negative – superficially contrary to the conventional electromagnetic Lagrangian. The reason for this difference is that the quantum field part of X^μ is imaginary. Thus $\mathscr{L}_C$ winds up having the correct sign after taking account of the factor of i in the field strength $F_{\mu\nu}$.

Defining

$$F_{Y\mu\nu} = (\partial Y_\mu/\partial y^\nu - \partial Y_\nu/\partial y^\mu) \tag{13.8}$$

we see the Lagrangian assumes the form of the conventional electromagnetic Lagrangian:

$$\mathscr{L}_C = -\tfrac{1}{4}\, F_Y^{\mu\nu} F_{Y\mu\nu} \tag{13.9}$$

The action of this theory has the form

$$I = \int d^4y\, \mathscr{L}(y) \tag{13.10}$$

The further development of the theory is described in chapter 19 and Blaha (2005a).

---

[138] We note that the lagragian formlation presented in Blaha (2005a) relies on a new form of the Calculus of Variations – not one of the three forms previously known and documented in the literature.

Two-Tier L_C-based Standard Model Theory

In the present case we will need two variables X_r^μ and X_i^μ for the Standard Model based on L_C covariance since quarks have complex spatial 3-coordinates. We define them similarly to the previous case:

$$X_{r\mu}(y_r) = y_{r\mu} + i\, Y_{r\mu}(y_r)/M_c^2 \tag{13.11}$$

$$X_{i\mu}(y_i) = y_{i\mu} + i\, Y_{i\mu}(y_i)/M_c^2 \tag{13.12}$$

where we choose the same mass scale for both the "real" and "imaginary" variables. The Two-Tier, single generation, version of the Standard Model given in section 13.2 then has an action of the form

$$I_{SM1tt} = \int dy^0 d^3y_r d^3y_i\, (\mathscr{L}_{SM1}(X_r^\mu(y_r), \mathbf{X}_i^k(y_i))J_2)\big|_{y_i^0 = 0,\, Y_r^0 = Y_i^0 = 0} + \int dy_r^0 d^3y_r\, \mathscr{L}_C(X_r^\mu(y_r), \partial X_r^\mu(y_r)/\partial y_r^\nu, y_r) +$$

$$+ \int dy_i^0 d^3y_i\, \mathscr{L}_C(X_i^\mu(y_i), \partial X_i^\mu(y_i)/\partial y_i^\nu, y_i) \tag{13.13}$$

where the replacements

$$x^\mu \equiv x_r^\mu \rightarrow X_r^\mu(y_r) \tag{13.14}$$

$$x_i^k \rightarrow X_i^k(y_i) \tag{13.15}$$

for $\mu = 0, 1, 2, 3$ and $k = 1, 2, 3$ are made in $\mathscr{L}_{SM1}$ (described in section 13.2 below) followed by defining $y_r^0 = y^0$ and making an L_C transformation to a frame where $y_i^0 = 0$, and where J_2 is the absolute value of the Jacobian of the transformation from (X_r, X_i) to (y_r, y_i) coordinates:

$$J_2 = |\partial(X_r, X_i)/\partial(y_r, y_i)| \tag{13.16}$$

We also choose gauges where $Y_r^0 = Y_i^0 = 0$. These transformations and gauge choices are discussed in detail later. The lagrangian terms $\mathscr{L}_C(X_r^\mu(y_r), \partial X_r^\mu(y_r)/\partial y_r^\nu, y_r)$ and $\mathscr{L}_C(X_i^\mu(y_i), \partial X_i^\mu(y_i)/\partial y_i^\nu, y_i)$ have the form:

$$\mathscr{L}_C = +\tfrac{1}{4} M_c^4 F^{\mu\nu} F_{\mu\nu} \tag{13.17}$$

with

$$F_{\mu\nu} = \partial X_\mu/\partial y^\nu - \partial X_\nu/\partial y^\mu \tag{13.18}$$

$$\equiv i\,(\partial Y_\mu/\partial y^\nu - \partial Y_\nu/\partial y^\mu)/M_c^2$$

or defining

$$F_{Y\mu\nu} = (\partial Y_\mu/\partial y^\nu - \partial Y_\nu/\partial y^\mu) \tag{13.19}$$

we see each lagrangian assumes the form of the conventional electromagnetic Lagrangian:

$$\mathscr{L}_C = -\tfrac{1}{4} F_Y^{\mu\nu} F_{Y\mu\nu} \tag{13.20}$$

The lagrangian is supplemented with the following condition on all complexon fields $\Phi_{\dots}$:

$$(\partial/\partial X_r^k(y_r))\,(\partial/\partial X_i^k(y_i))\Phi\ldots = 0 \tag{13.21}$$

summed over k = 1, 2, 3. Non-complexon fields $\Omega\ldots$ in our left-handed formulation satisfy the subsidiary condition:

$$\{(\partial/\partial X_r^{\,k}(y_r))\,(\partial/\partial X_i^{\,k}(y_i)) - ((\partial/\partial X_r^{\,k}(y_r))^2\,(\partial/\partial X_i^{\,m}(y_i))^2)^{1/2}\}\Omega\ldots = 0 \tag{13.22}$$

summed over k = 1, 2, 3 and over m = 1, 2, 3.

Feynman Propagators

The momentum space free field Feynman propagators G…(k) of all particles and ghosts in our Two-Tier Standard Model acquires a Gaussian factor exp(h(k)):

$$G\ldots(k) \rightarrow G\ldots(k)\,\exp(h(k)) \tag{13.23}$$

so that all perturbation theory diagrams are finite. The result is a finite perturbative result in all calculations to any order in perturbation theory. Blaha (2005a) shows that Two-Tier theories are finite, Poincare covariant, and unitary. (See chapter 19 for a more detailed description and Blaha (2005a) for a complete discussion.)

An example of the Two-Tier effect on propagators is the case of the Two-Tier photon propagator. The Two-Tier photon propagator[139] is:

$$iD_F^{\,TT}(y_1 - y_2)_{\mu\nu} = -i\int \frac{d^4p\, e^{-ip\cdot z}\, g_{\mu\nu} R(\mathbf{p}, z)}{(2\pi)^4\,(p^2 + i\varepsilon)} \tag{13.24}$$

(since the imaginary parts can be taken to be zero: $y_{1i}^{\;\mu} - y_{2i}^{\;\mu} = 0$) where

$$z^\mu = y_{1r}^{\;\mu} - y_{2r}^{\;\mu} \tag{13.25}$$

$$R(\mathbf{p}, \mathbf{z}) = \exp[-p^i p^j \Delta_{Tij}(z)/M_c^{\,4}] \tag{13.26}$$

$$= \exp\{-\mathbf{p}^2[A(v) + B(v)\cos^2\theta] \,/\, [4\pi^2 M_c^{\,4}|\mathbf{z}|^2]\} \tag{13.27}$$

with i, j = 1, 2, 3, and with $\Delta_{Tij}(z)$ the commutator of the positive frequency part $Y^+{}_k(y)$ and the negative frequency part $Y^-{}_k(y)$ of $Y_k(y)$:

$$\Delta_{Tij}(z) = [Y^+{}_j(y_{1r}), Y^-{}_k(y_{2r})] = \int d^3k\, e^{ik\cdot(y_{1r} - y_{2r})}\,(\delta_{jk} - k_j k_k/\mathbf{k}^2)/[(2\pi)^3 2\omega_k] \tag{13.28}$$

and

$$v = |z^0|/|\mathbf{z}| \tag{13.29}$$

$$A(v) = (1 - v^2)^{-1} + .5v\,\ln[(v - 1)/(v + 1)] \tag{13.30}$$

$$B(v) = v^2(1 - v^2)^{-1} - 1.5v\,\ln[(v - 1)/(v + 1)] \tag{13.31}$$

$$\mathbf{p}\cdot\mathbf{z} = |\mathbf{p}|\,|\mathbf{z}|\cos\theta \tag{13.32}$$

---

[139] Blaha (2005a).

with $|\mathbf{p}|$ denoting the length of a spatial vector $\mathbf{p}$, $|\mathbf{z}|$ denoting the length of a spatial vector $\mathbf{z}$, and with $|z^0|$ being the absolute value of z^0.

The gaussian factors $R(\mathbf{p}, z)$ which appear in all Two-Tier propagators damp the large momentum behavior of all perturbation theory integrals producing a completely finite perturbation theory and yet give the usual results of perturbation theory at energies that are small compared to the mass scale M_c.

Complexon Feynman Propagator

In the case of complexons the Two-Tier Feynman propagator differs from the non-complexon case by having an integration over imaginary spatial 3-momenta, a derivative of a delta function embodying the orthogonality of the real and imaginary 3-momenta, and two factors of $R(\mathbf{p}, z)$: one factor being $R(\mathbf{p_r}, z_r)$ and the other factor being $R(\mathbf{p_i}, z_i)$ (where the time components $z_r^0 = z^0$ and $z_i^0 = 0$ since there is only one real time coordinate[140]) thus providing large momentum convergence for both real and imaginary 3-momentum integrations.

For a scalar complexon particle we previously found the Feynman propagator:

$$i\Delta_{CTF}(x - y) = \theta(x^+ - y^+)\langle 0|\phi_{CT}(x)\, \phi_{CT}(y)|0\rangle + \theta(y^+ - x^+)\langle 0|\phi_{CT}(y)\phi_{CT}(x)|0\rangle$$
$$= i\int d^4p_r d^3p_i (2\pi)^{-7}\delta'(\mathbf{p_r}\cdot\mathbf{p_i}/m^2) e^{-ip^+(x^- - y^-) - ip^-(x^+ - y^+) + ip_\perp\cdot(x_\perp - y_\perp) - ip_i\cdot(x_i - y_i)}/(p^2 + m^2 + i\varepsilon)$$

in conventional quantum field theory.

In the case of Two-Tier quantum field a scalar complexon particle has the the Feynman propagator

$$i\Delta_{CTFtt}(x - y) = i\int d^4p_r d^3p_i (2\pi)^{-7}\delta'(\mathbf{p_r}\cdot\mathbf{p_i}/m^2)\, R(\mathbf{p_r}, z_r) R(\mathbf{p_i}, z_i)\cdot$$
$$\cdot\, e^{-ip^+(x^- - y^-) - ip^-(x^+ - y^+) + ip_\perp\cdot(x_\perp - y_\perp) - ip_i\cdot(x_i - y_i)}/(p^2 - m^2 + i\varepsilon) \qquad (13.33)$$

where the time components $z_r^0 = z^0$ and $z_i^0 = 0$ since there is only one time coordinate and where $p^2. = p^{0\,2} - p_r^2 + p_i^2$.

Propagators for other types of particles are similarly modified in the Two-Tier formalism (See Blaha 2005a).

13.2 The New Extended Standard Model Lagrangian

In chapters 10 - 12 we derived the form of a one generation Extended Standard Model that included the known parts of the Standard Model (excepting the Higgs sector) and an SU(2)⊗U(1) part for Dark Matter. Dark Matter was linked to normal matter with a simple scalar gauge field. In the next sections we will display the new Extended Standard Model that results.[141]

---

[140] We can arrange for $z_i^0 = 0$ by making a L_C transformation to an inertial frame where z is real.

[141] It is based on the three principles: 1) The only connecting interaction is a weak interaction, 2) The form of ElectroWeak theory remains unchanged, and 3) Dark Matter parallels normal matter in its general characteristics: three (possibly four) generations, SU(3) singlets, an SU(2)⊗U(1) symmetry analogous to ElectroWeak symmetry, SU(2)⊗U(1) dark lepton and dark quark doublets.

13.2.1 Two-Tier Lepton Sector with Dark Leptons

We begin with the definition of a quadruplet of leptons – a pair of doublets, one normal and one Dark, instead of a single doublet. We define left and right lepton quadruplets with[142]

$$\Psi_{L,R}(X) = \begin{bmatrix} \psi_{DL,R}(X) \\ \psi_{NL,R}(X) \end{bmatrix} \tag{13.34}$$

where $\psi_{NL,R}(X)$ is a "normal" ElectroWeak-like lepton doublet, and where $\psi_{DL,R}(X)$ is a Dark ElectroWeak-like lepton doublet consisting of a Dark electron-like fermion and a Dark neutrino-like fermion.

We define covariant derivative terms which we express in matrix form are

$$D_{L,R}(X) = \begin{bmatrix} \gamma^\mu D_{DL,R\mu} & 0 \\ 0 & \gamma^\mu D_{NL,R\mu} \end{bmatrix} \tag{13.35}$$

where the normal matter left-handed covariant derivative is

$$D_{NL\mu} = \partial/\partial X^\mu - \tfrac{1}{2} ig\boldsymbol{\sigma}\cdot\mathbf{W}_\mu + \tfrac{1}{2} ig'B_\mu \tag{13.36}$$

and where the Dark matter left-handed covariant derivative is

$$D_{DL\mu} = \partial/\partial X^\mu - \tfrac{1}{2} ig_D\boldsymbol{\sigma}\cdot\mathbf{W}'_\mu + \tfrac{1}{2} ig_D'B'_\mu + \tfrac{1}{2} ig_D''B_\mu \tag{13.37}$$

with $\boldsymbol{\sigma}$ a vector composed of the Pauli matrices. The right-handed covariant derivatives have a simpler form. The normal matter right-handed covariant derivative is

$$D_{NR\mu} = \partial/\partial X^\mu + ig'B_\mu \tag{13.38}$$

and the Dark matter right-handed covariant derivative is

$$D_{DR\mu} = \partial/\partial X^\mu + \tfrac{1}{2} ig_D'B'_\mu + \tfrac{1}{2} ig_D''B_\mu \tag{13.39}$$

The normal and Dark electroweak fields above are functions of a Two-Tier X. The Faddeev-Popov mechanism operative for these types of fields is described in appendix '19-A of Blaha (2011c) and in chapter 12.

13.2.2 Quark Sector with Dark Quarks

In the *quark* sector we define left and right quark quadruplets with

142 The X's are Two-Tier coordinates.

$$\Psi_{qL,R}(X_c) = \begin{bmatrix} \psi_{DqL,R}(X_c) \\ \psi_{NqL,R}(X_c) \end{bmatrix} \tag{13.40}$$

where $\psi_{NqL,R}(X_c)$ is a "normal" ElectroWeak-like quark doublet, and where $\psi_{DqL,R}(X_c)$ is a Dark ElectroWeak-like quark doublet consisting of a SU(3) singlet Dark up-quark of unit Dark charge and a SU(3) singlet Dark down-quark of zero Dark charge.

The covariant derivative terms are contained in $D_q(X_c)$ which we express in matrix form as

$$D_{qL,R}(X_c) = \begin{bmatrix} \gamma^\mu D_{qDL,R\mu}(X_c) & 0 \\ 0 & \gamma^\mu D_{qNL,R\mu}(X_c) \end{bmatrix} \tag{13.41}$$

where the normal quark matter left-handed covariant derivative is

$$D_{qNL\mu} = \partial/\partial X_c{}^\mu - \tfrac{1}{2} ig\boldsymbol{\sigma}\cdot\mathbf{W}_\mu - ig'B_\mu/6 + ig_C\tau\cdot A_{C\mu} \tag{13.42}$$

where $A_{C\mu}$ is the color gauge field and where the Dark quark left-handed covariant derivative is

$$D_{qDL\mu} = \partial/\partial X_c{}^\mu - \tfrac{1}{2} ig_D\boldsymbol{\sigma}\cdot\mathbf{W}'_\mu + \tfrac{1}{2} ig_D'B'_\mu + \tfrac{1}{2} ig_D''B_\mu \tag{13.43}$$

since Dark quarks are SU(3) singlets with unit or zero Dark charge. The right-handed quark covariant derivatives have a simpler form. The normal quark right-handed covariant derivative is

$$D_{qNR\mu} = \partial/\partial X_c{}^\mu + ig'B_\mu/3 + ig_C\tau\cdot A_{C\mu} \tag{13.44}$$

and the Dark quark right-handed covariant derivative is

$$D_{qDR\mu} = \partial/\partial X_c{}^\mu + \tfrac{1}{2} ig_D'B'_\mu + \tfrac{1}{2} ig_D''B_\mu \tag{13.45}$$

The normal and Dark gauge boson fields are functions of X_c. = $(X_{r\mu}(y_r), X_{i\mu}(y_i))$ of eqs. 13.11 and 13.12. The Faddeev-Popov mechanism is operative for gauge boson fields and is described in appendix 19-A of Blaha (2011c).[143] The *complexon* quark Standard Model ElectroWeak Sector covariant derivatives in quadruplet matrix form are

$$D_{qL,R}(X_c) = \begin{bmatrix} \gamma^\mu D_{qDL,R\mu} & 0 \\ 0 & \gamma^\mu D_{qNL,R\mu} \end{bmatrix} \tag{13.46}$$

---

[143] Those who might be concerned about the propagator term $<W_i(X), W_j(X_c)>$ and similar propagators where one field is a function of X and the other field is a function of X_c should note that such terms are to very good approximation equal to $<W_i(X), W_j(X)>$ for energies much less than M_c (which could be as large as the Planck energy.)

The remaining parts of the complexon Standard Model are described in chapter 23 of Blaha (2011) and summarized below. The addition of singlet Dark quark Higgs terms is also required.

The lagrangian density and action is[144]

$$\mathcal{L}_{CSM} = \Psi_L^{a\dagger}\gamma^0 i\gamma^\mu D_{L\mu}\Psi_L^a - \Psi_R^{a\dagger}\gamma^0 i\gamma^\mu D_{R\mu}\Psi_{3R}^a + \Psi_{CL}^{a\dagger}\gamma^0 i\gamma^\mu \mathcal{D}_{qL\mu}\Psi_{CL}^a + \Psi_{CR}^{a\dagger}\gamma^0 i\gamma^\mu \mathcal{D}_{qR\mu}\Psi_{CR}^a - \\ - \mathcal{L}_{BareMasses} + \mathcal{L}_{Gauge} + \mathcal{L}_{Mass} \tag{13.47}$$

where a is the generation index. $\mathcal{L}_{BareMasses}$ contains the fermion bare mass terms. Also,

$$\mathcal{L}_{Gauge} = \mathcal{L}_{GaugeEW} + \mathcal{L}_{GaugeC} + \mathcal{L}_{GaugeEWD} \tag{13.48}$$

with

$$\mathcal{L}_{GaugeEW} = -\tfrac{1}{4} F_W{}^{a\mu\nu} F_W{}^a{}_{\mu\nu} - \tfrac{1}{4} F_B{}^{\mu\nu} F_{B\mu\nu} + \mathcal{L}_{EW}{}^{ghost} \tag{13.49}$$

$$\mathcal{L}_{GaugeEWD} = -\tfrac{1}{4} F'_W{}^{a\mu\nu} F'_W{}^a{}_{\mu\nu} - \tfrac{1}{4} F_{B'}{}^{\mu\nu} F_{B'\mu\nu} + \mathcal{L}_{W'}{}^{ghost} \tag{13.50}$$

and

$$\mathcal{L}_{GaugeC} = \mathcal{L}_{CCG} + \mathcal{L}_C{}^{ghost} + \mathcal{L}_{CC}{}^{ghost} \tag{13.51}$$

The ElectroWeak gauge bosons W_μ^a, B_μ and B'_μ field tensors are:

$$F_W{}^a{}_{\mu\nu} = \partial W^a{}_\mu/\partial x^\nu - \partial W^a{}_\nu/\partial x^\mu + g_2 f^{abc} W^b{}_\mu W^c{}_\nu \tag{13.52}$$

$$F_{B\mu\nu} = \partial B_\mu/\partial x^\nu - \partial B_\nu/\partial x^\mu \tag{13.53}$$

and the Dark ElectroWeak gauge bosons W'^a_μ and B'_μ field tensors are:

$$F_{B'\mu\nu} = \partial B'_\mu/\partial x^\nu - \partial B'_\nu/\partial x^\mu$$

$$F'_W{}^a{}_{\mu\nu} = \partial W'^a{}_\mu/\partial x^\nu - \partial W'^a{}_\nu/\partial x^\mu + g_2 f^{abc} W'^b{}_\mu W'^c{}_\nu \tag{13.54}$$

$\mathcal{L}_{EW}{}^{ghost}$ contains the Faddeev-Popov ghost terms for the ElectroWeak W_μ^a gauge bosons. The complexon color gluon lagrangian $\mathcal{L}_{CCG}$ is defined by

$$\mathcal{L}_{CCG} = -\tfrac{1}{4} F_{CC}{}^{a\mu\nu}(x) F_{CC}{}^a{}_{\mu\nu}(x) \tag{13.55}$$

where

$$F_{CC}{}^a{}_{\mu\nu} = \partial/\partial X_c{}^\nu A_C{}^a{}_\mu - \partial/\partial X_c{}^\mu A_C{}^a{}_\nu + g f_{su(3)}{}^{abc} A_C{}^b{}_\mu A_C{}^c{}_\nu \tag{13.56}$$

---

144 The lagrangian below is much the same as that of Blaha (2011c) except for the change necessitated by the additional group SU(2)⊗U(1) in place of a U(1) part.

where $A_{C\,\nu}^{\;a}$ is the color gluon gauge field, g is the color coupling constant, and the $f_{su(3)}^{\;abc}$ are the SU(3) structure constants.

In addition $\mathcal{L}_C^{ghost}$ is the color SU(3) Faddeev-Popov ghost terms defined in appendix 19-A of Blaha (2011c) for the complexon Lorentz gauge and $\mathcal{L}_{CC}^{ghost}$ is the complexon color SU(3) constraint ghost terms defined through the Faddeev-Popov mechanism. The mass sector $\mathcal{L}_{Mass}$ is presumably based on the Higgs Mechanism.which creates the fermion and ElectroWeak vector boson masses, and generation mixing.

The lagrangian is supplemented with the following condition on all complexon fields $\Phi_{\dots}$:[145]

$$\nabla_r \cdot \nabla_i \Phi\ldots = 0 \tag{13.57}$$

Non-complexon fields Ω... in the left-handed formulation under consideration satisfy the subsidiary condition:

$$[\nabla_r \cdot \nabla_i - (\nabla_r^2 \nabla_i^2)^{1/2}]\Omega\ldots = 0 \tag{13.58}$$

which guarantees a complexon's real momentum is parallel to its imaginary momentum.

END OF AMENDED CHAPTER 6 EXTRACT FROM BLAHA (2013b)

We note that W'_μ and B'_μ are the Dark gauge fields with the Dark symmetry group SU(2)⊗U(1). The B_μ term in eqs. 13.43 and 13.45 above give a minimal weak coupling of normal and Dark Matter.

13.3 Renormalization and Divergence Issues

The theory derived from $\mathcal{L}_{SM1}$ when calculated perturbatively in conventional quantum field theory has divergences that are not renormalizable due to its 7-dimensional nature. This issue does not appear to be curable by known renormalization methods. However if the theory is reformulated in the Two-Tier formalism developed by Blaha[146] then a finite theory results. This is also true for the three (four) generation Standard Model to be described later.

The integrals of the terms of the lagrangian can be divided into two sets: integrals of terms that do not contain complexon field factors can be integrated $\int d^4y$ to create one type of contribution to the total action, and integrals of terms that do contain complexon field factors can be integrated $\int dy^0 d^3y_1 \int d^3y_2$ to create the other contributions to the total action. Eq. 13.13 exemplifies the definition of a Two-Tier action in general.

---

[145] These conditions implement the orthogonality of the real and imaginary parts of complexon 3-momentum.

[146] Blaha (2005a) and Blaha (2003).

Appendix 13-A Multi-Quark Particles and Quark Chemistry

13-A.1 New Penta-Quark Particles

The CERN LHC has found evidence for penta-quark particles – particles consisting of five quarks.[147] This discovery is a step beyond the tetraquark (four quark) particle named $Z_C(3900)$ found, and announced,[148] in 2013. The Z_C particle could be viewed as two D-mesons bound together by the strong color force in a color singlet "hadron molecule."

Similarly, the pentaquark could be viewed as a particle consisting of a two quark meson and a three quark baryon formed into a type of molecule by the strong interaction.

13-A.2 Multi-Quark Molecules – Quark Chemistry

These recent discoveries are the initial steps to a spectrum of multi-quark particles consisting of

$$2m + 3b \tag{13-A.1}$$

quarks where m is the number of "meson-like" constituents and b is the number of "baryon-like" constituents. The pentaquark has m = b = 1 constituents.

We are now confronted by a new type of "molecule" where the binding force is not electromagnetic but, instead, the strong color force. The tight binding of these "molecules" by the strong force makes them effectively into particles. But they are perhaps more comparable to the binding of protons and neutrons into atomic nuclei by the nuclear force.

The development of a quark molecule chemistry would, if molecules were produced in quantity, lead to quark materials with extremely important physical characteristics – quark matter – with both new and important "chemical" properties. However, quark molecules have an extraordinarily short lifetime and so the creation of a number of quark molecules, and their fabrication into materials, does not appear to be feasible. Nevertheless, their theoretical and experimental study might serve to drive conventional chemistry and materials science forward.

13-A.3 Large Multi-Quark "Molecules" and Eventually Quark Stars

One can envision the eventual creation of very "large" multi-quark particles/molecules. Taken to the limit of extremely large "molecules" of extraordinary size it appears possible to develop a dynamics of quark stars where the dominant force is not the nuclear force and electromagnetism but rather the strong force. Astronomy has yet to discover an unambiguous quark star.

---

[147] Announced July 15, 2015. See arXiv.org/abs/1507.03414.

[148] June 20, 1013. Discovery by the Belle Collaboration, High Energy Accelerator Research Organization, in Tsukuba, Japan; and BESIII Collaboration at the Beijing Electron Positron Collider in China. Phys. Rev. Lett. **110**, 252001, 2013 and **110**, 252002, 2013 .

14. Dark Matter Features: Particle Spectrum and Basic Chemistry of Dark Matter, Dark Matter Bodies

In chapter 39 of Blaha (2012a) we initially described Dark Matter particles, Dark atoms, the Dark Periodic Table, and Dark basic chemistry – all based on the additional SU(2)⊗U(1) symmetry in the Extended Standard Model that more or less mirrors normal ElectroWeak symmetry. The major differences were 1) that Dark quarks are assumed to be SU(3) singlets as suggested by the known weakness of Dark Matter interactions with normal matter and 2) that *physical* hadron charges – both electric charge and Dark electric charge – are quantized with whole number values. (Dark quarks are the emasculated "hadrons" of this theory since they do not experience the Strong Interaction.) This chapter contains material from chapter 39 of Blaha (2012a) for completeness as well as some additional thoughts.

Another important issue is the effect of gravitation on Dark Matter—Can Dark matter aggregate under the force of gravity to form galactic clusters, galaxies, and smaller objects such as Dark suns and perhaps Dark planets? Sections 14.3 and 14.4 discuss these possibilities.

This chapter is based on the assumption of four generations. Chapter 18 presents a strong theoretical case for four generations of fermions.

14.1 Fundamental Dark Particles

We now consider the Dark particles that are associated with our new SU(2)⊗U(1) symmetry. Since Dark Matter only interacts weakly with known matter, Dark particles must be color singlets. Thus there will be 12 (in the three generation case) or 16 (in the four generation case) Dark particles (plus their antiparticles.) Recent experiments suggest Dark particles have extremely large masses of the order of 8.6 GeV/c or larger. Dark quarks are color singlets with complex 3-momenta in our Extended Standard Model.

"Periodic Table" of Fermions

NORMAL ELECTROWEAK FERMION DOUBLETS

Generation	**Leptons** Real-Valued 3-Momenta			**Color Triplet Quarks** Complex-Valued 3-Momenta	
1	e	ν_e		$u_1\ u_2\ u_3$	$d_1\ d_2\ d_3$
2	μ	ν_μ		$c_1\ c_2\ c_3$	$s_1\ s_2\ s_3$
3	τ	ν_τ		$t_1\ t_2\ t_3$	$b_1\ b_2\ b_3$
4?	ω	ν_ω	Proposed	$v_1\ v_2\ v_3$	$w_1\ w_2\ w_3$

DARK SU(2)⊗U(1) FERMION DOUBLETS

Generation	Dark Leptons			Color Singlet Dark Quarks	
	Real-Valued 3-Momenta			Complex-Valued 3-Momenta	
1	e_D	ν_{eD}		u_D	d_D
2	μ_D	$\nu_{\mu D}$	Proposed	c_D	s_D
3	τ_D	$\nu_{\tau D}$		t_D	b_D
4?	ω_D	$\nu_{\omega D}$		v_D	w_D

Figure 14.1 "Periodic Table" of four generations of normal and Dark fundamental fermions.

In addition to Dark fermions there will be four SU(2)⊗U(1) Dark gauge bosons – also with large masses:

Dark Particle Gauge Bosons

U(1): $W'_0{}^{\mu}$ SU(2): $W'_1{}^{\mu}$ $W'_2{}^{\mu}$ $W'_3{}^{\mu}$

14.2 Dark Particle Chemistry

The chemistry of Dark Matter will be different from the chemistry of known matter due to the absence of the color interaction. Dark particles cannot combine through a strong interaction to form a hadron spectrum, or atomic nuclei, such as we see in normal matter. Thus Dark particle atoms can be like hydrogen atoms, and consist of a Dark quark particle bound to a Dark lepton particle by the Dark electric force assuming the Dark charge is quantized and has equal integer absolute values for Dark quarks and leptons. More complex Dark molecules are also possible as we will see later.

Suppose all Dark charged particles have charge $\pm 1e_D$ (e_D is the Dark unit of charge which may possibly be equal to e, the electromagnetic charge), and each Dark doublet has a Dark particle with unit Dark charge and a neutral Dark particle. Then we would expect that (in the case of four generations) Dark Matter would consist of

1. Lepton-like fundamental particles: Four Dark charged and four Dark charge neutral particles and their anti-particles.

2. Quark-like fundamental particles: Four Dark charged and four Dark charge neutral particles and their anti-particles.

3. There would be a total of eight neutral Dark quarks and leptons.

4. Atoms are composed of oppositely charged Dark particles of different types. There are 16 Dark "atoms" of the form leptoDark particle - quarkDark particle, e_Du_D, e_Dc_D, e_Dt_D, μ_Du_D, μ_Dc_D, μ_Dt_D, τ_Du_D, τ_Dc_D, τ_Dt_D, e_Dv_D, μ_Dv_D, τ_Dv_D, ω_Dv_D, ω_Du_D, ω_Dc_D, and ω_Dt_D plus their anti-matter equivalents. There are 12 "quasi-stable" particle-anti-particle combinations: six leptoDark - antileptoDark particle combinations, and six quarkDark - antiquarkDark particle combinations. (There is no attractive nuclear force.) All of these atoms are bound by the Dark electric force.

5. Simple molecules of the type of Figs. 14.2 – 14.4 below, and so on, based on Dark dipole interactions, Dark van der Waals forces and other Dark electromagnetic interactions are possible.

After a sufficiently long time, collisions would lead perhaps to the dominance of Dark particles and the "disappearance" of antiDark particles if the number of Dark particles is overwhelmingly dominant in a fashion similar to normal matter.[149] (The other possibility is not excluded.) The Dark Periodic Table is:

Periodic Table of Simple Dark Particle Atoms
(Assuming Dark anti-particles are Annihilated)

e_Du_D	μ_Du_D	τ_Du_D	ω_Du_D	d_D
e_Dc_D	μ_Dc_D	τ_Dc_D	ω_Dc_D	s_D
e_Dt_D	μ_Dt_D	τ_Dt_D	ω_Dt_D	b_D
e_Dv_D	μ_Dv_D	τ_Dv_D	ω_Dv_D	w_D

plus their similar antiparticle atoms. Bound states are assumed bound, perhaps into hydrogen-like atoms, through a Dark electromagnetic force. The last column consists of Dark charge neutral quarks. Antiparticle atoms of these states might also exist or be created through the Dark ElectroWeak interactions. One could extend the table with a column of Dark neutrinos. The decays, and mixing between generations, remains to be determined in the unknown Dark Higgs sector.

The periodic table that we constructed is based on an analogy with the features of normal matter: quarks are much heavier than leptons and leptons "revolve" around the Dark quark nuclei. If this view is correct then one can conceive of a chemistry of Dark Matter with molecules bound by Dark electromagnetic forces. Pair bonding of Dark leptons would be possible and so one could conceive of a fairly complex Dark chemistry bounded by the fact that a Dark particle atom has only one Dark lepton. Thus there would only be 64 bound pairs of atoms[150] similar to H_2.

$$
\begin{array}{ccc}
 & e_D & \\
u_D & & u_D \\
 & e_D &
\end{array}
$$

Figure 14.2. Example of two Dark atoms binding in a manner similar to the binding of hydrogen molecules in H_2. Considering the possible combinations of quarks and leptons there are 64 bound varieties of this type. Chains of these molecules are also possible in principle.

Dark particles can be combined into chains and more complex molecules. A simple chain of Dark particles is

---

[149] The study of electron-positron production through Weak Interaction with Dark Matter by M. Aguilar et al, Phys. Rev. Letters **110**, 141102 (2013) does not seem to clarify this issue.

[150] Plus anti-particle equivalents.

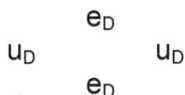

e_D u_D

u_D u_D e_D e_D

e_D u_D

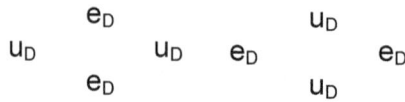

Figure 14.3. Example of a simple Dark Matter chain.

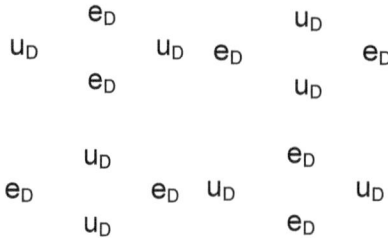

e_D u_D

u_D u_D e_D e_D

e_D u_D

u_D e_D

e_D e_D u_D u_D

u_D e_D

Figure 14.4. A segment of two strands interacting with each other. Other combinations are possible using other Dark charged fermions. This diagram is suggestive of a Dark form of DNA. Dark Life?

More complex chains as well as two and three dimensional bound aggregates are also possible. These considerations suggest something approaching the complexity of simple life forms may be possible.

Dark dipole effects could lead to the (weaker) binding of larger assemblages of Dark atoms. For example, if the masses of the leptonic and quark Dark particles are not too dissimilar, crystalline Dark Matter appears possible.

We thus arrive at a Dark Matter sector with much less variety than normal matter. It may, or may not, preclude the existence of Dark life forms, and possibly of Dark solids composed of Dark Matter. Solid Dark planets may exist. The detection of planet-like and star-like Dark Matter objects within the Dark Matter "cloud" surrounding a galaxy would be of great importance.

14.3 Gravitation and Dark Matter Bodies: Galactic Clusters, Galaxies

Dark Matter is thought to equal 85% of the total matter in the universe. It is known to form halos around galaxies and to form filaments between galaxies in galactic clusters. It appears to influence the orbits of stars in galaxies causing the rotational speed of stars at large radial distances from galactic centers to be roughly constant and independent of the radius.

When two galaxies collide it appears that the Dark Matter drags between the galaxies as they separate. Thus Dark Matter has forces as we suggested earlier.

We see these effect in the Dark Matter halos around galaxies. Galaxies are permeated with Dark Matter. Our galaxy, for example, appears to have ten times as much Dark Matter as normal matter.[151]

---

[151] Therefore one would expect more gravitational phenomena reflecting the presence of Dark Matter locally in our galaxy than its effect on galactic rotation. Since Dark Matter hardly interacts with normal matter it would intersperse with normal matter throughout the galaxy forming a hidden part of the galaxy – present but almost completely not

14.4 Dark Stars, Dark Planets, Dark Globular Clusters—Or Just Darkened?

However we have not seen (as yet?) clumps of Dark Matter within our galaxy similar in size to stars and planets or globular clusters. Nor do we see the earth, or other Solar System planets, having a discrepancy in mass due to Dark Matter clumps within these bodies.

Yet there is no apparent reason why Dark Matter should not clump within planets (and our sun) causing a discrepancy in solar system dynamics and ostensibly the force of gravity. The ten to one dominance of Dark Matter in the galaxy creates a striking discrepancy with the apparent absence of Dark Matter gravitational effects in the Solar System.

14.4.1 Open Questions on Dark Matter in Our Solar System

1. Why does not the earth's mass have a significant contribution from the Dark Matter within the earth since our galaxy has 10 times more Dark Matter than normal matter? Does not Dark Matter clump somewhat around planet size objects?

2. Same question as 1 but applied to the sun. Where is the effect of the Dark Matter within the sun?

3. Why is there no Dark Jupiter in our Solar System with normal moons circling it?

4. Why are not the planetary dynamics of the Solar System affected by the presence of Dark Matter?

14.4.2 Open Questions on Dark(ened) Stars and Dark(ened) Globular Clusters of Stars

There are many stars in the galaxy composed of normal matter apparently. However the ten to one ratio of Dark Matter to normal matter in our galaxy should be reflected in stellar dynamics which is a supposedly well understood field. Stellar dynamics depends significantly on the mass of stars.

1. Why does not stellar dynamics take account of the Dark Matter contribution to the mass of stars?
2. When a star contracts generating gravitational energy why does not the contraction of the Dark Matter clumped within the star impact gravitationally on the contraction of the normal matter?
3. Why is not the evolution and structure of globular clusters of stars affected by their presumably large Dark Matter content?
4. Will we find a Dark star circled by some large normal matter planets?
5. Can we detect a Dark cluster by its gravitational effects?

14.4.3 Comments on these Questions

The only currently apparent resolution of these questions is a lack of significant Dark chemical interactions/reactions between Dark Matter atoms that may prevent the formation of Dark solids and

---

detectable. The hidden part would occupy the same space as normal galactic matter rather like an evanescent ghost in a horror movie.

liquids.[152] Dark nuclear decays similar to the normal nuclear reactions that power stars may also not exist. Thus no Dark stars. But Dark aggregates of the mass of the sun or more should be possible.

Dark Matter in galaxies appears to be rather like a distributed gas within a gravitational well formed by the combined masses of a galaxy. It also forms a filament between galaxies.[153] Within a galaxy Dark Matter appears dispersed in a somewhat uniform way with only very minor gravitational clumping by clusters, stars and planets.

---

[152] Thus Dark atoms may not have Dark dipole forces or other multipole forces that would induce the formation of Dark Matter liquids and solids.

[153] A galaxy composed overwhelmingly of Dark Matter has recently been found.

15. Baryon, Lepton, Dark Baryon, and Dark Lepton Conservation Laws and New U(4) Gauge Symmetry

15.1 Baryon Number Conservation and a Possible Baryonic Force

We have considered baryon number conservation and a possible baryonic force in Blaha (2014a) and (2014b). Much of this section contains selections from these books.

The primary forces involved in the interactions and collisions of baryons are the forces of The Standard Model, the force of gravity, and a fifth force which we take to be the baryonic force, a much discussed force that depends on the baryon numbers of ubjects experiencing it. The force (neglecting Standard Model interactions) between two clumps of baryonic matter containing baryons and other particles: clump1 being of mass m_1 and baryon number n_1, and clump2 being of mass m_2 and baryon number n_2 is

$$F = -Gm_1m_2/r^2 + (\beta^2/4\pi)n_1n_2/r^2 \tag{15.1}$$

where G is the gravitational constant, β is the baryonic constant and r is the distance between widely separated clumps. Experimentally a baryonic force between baryons has not been detected with any degree of certainty. Sakurai (1964) discusses early efforts in detail. Eőtvős experiments on the ratio of the observed gravitational mass to the inertial mass showed that that it is constant to within one part in 100,000,000 as far back as 1922 indicating the baryonic force, if it exists as we believe it does, is extremely weak compared to the gravitational force. Eőtvős et al[154] found

$$(\beta^2/4\pi)/(Gm_p^2) < 10^{-5}$$

where m_p is the proton mass.

Since then, the experiment has been redone with improved accuracy by Dicke and collaborators.[155] They have improved the accuracy to one part in 100 billion. A further analysis showed a very small discrepancy that suggested the ratio, while small, was non-zero, implying the equivalence principle might not be exact and that the discrepancy changed with the material used in the experiment – just what one might expect if a very small baryonic force was present – often called the "fifth force." At present the existence and amount of the discrepancy is unclear. Nevertheless, we will assume a fifth force.

The primary rationale for the fifth force is the apparent conservation of baryon number. The conservation of baryon number has been repeatedly investigated by experimenters and found to be true to

---

[154] Eőtvős, R. V., Pekár, D., Fekete, E., Ann. d. Physik **68**, 11 (1922).

[155] P. G. Roll, R. Krotkov, R. H. Dicke, *Annals of Physics*, 26, 442, 1964.

extremely high accuracy. For decades theorists have suggested that a baryon conservation law[156] follows from the existence of a gauge field in a manner much like electric charge conservation follows from the properties of the electromagnetic gauge field.

15.1.1 Estimate of the Baryonic Coupling Constant

The baryonic force, and coupling constant, is known to be very small in comparison to gravity and the other known forces. Measurements of the gravitational constant G are significantly different.[157,158] The reason(s) for these discrepancies is not known. We will assume that both the 2010 and 2013 measurements of G are experimentally correct but disagree because of the baryonic force term in eq. 15.1 that would create a difference in effective G values if the experiments used different masses and thus baryon numbers. Quinn et al found a value for the gravitational constant of $G_1 = 6.67545\times10^{-11}$ $m^3kg^{-1}s^{-2}$. The combined 2010 CODATA value for the gravitational constant was $G_2 = 6.67384\times10^{-11}$ $m^3kg^{-1}s^{-2}$. Both values are subject to estimated uncertainties.

Suppose these values are correct and due to a difference in the chemical composition (metals) of the test masses used in the experiment. Quinn et all use 1.2 kg test masses composed of Cu-0.7% Te free machining alloy. The CODATA value being a composite of many experiments does not have an effective equivalent test mass value or composition specified.[159] Suppose the test mass value is $N_1^2m_1^2 + N_{1e}^2m_e^2$ for the G_1 result giving

$$-(N_1^2m_1^2 + N_{1e}^2m_e^2)G_1 = [-G(m_1^2N_1^2 + N_{1e}^2m_e^2) + (\beta^2/4\pi)N_1^2] \tag{15.2}$$

where G is the real value of the gravitational constant. The total test mass is $(m_1^2N_1^2 + N_{1e}^2m_e^2)$ with N_1 baryons of average mass m in each test mass and N_{1e} leptons of average mass m_e.

Suppose further the test mass value is $N_2^2m_2^2 + N_{2e}^2m_e^2$ for the G_2 result giving

$$-(N_2^2m_2^2 + N_{2e}^2m_e^2)G_2 = [-G(m_2^2N_2^2 + N_{2e}^2m_e^2) + (\beta^2/4\pi)N_2^2] \tag{15.3}$$

where G is the real value of the gravitational constant. The total test mass is $(m_2^2N_2^2 + N_{2e}^2m_e^2)$ with N_2 baryons of average mass m_2 in each test mass and N_{2e} leptons of average mass m_e. Since the test masses are electrically neutral and there are approximately equal numbers of protons and neutrons in a test mass it follows approximately that

$$N_{1e} = ½N_1 \quad \text{and} \quad N_{2e} = ½N_2 \tag{15.4}$$

Subtracting eq. 15.2 from eq. 15.3 after some algebra[160] we find

---

[156] See Gell-Mann, M. and Levy, M. *Nuovo Cimento* 16, 705 (1960) for a proof and Sakurai (1964) for a discussion of the relation of the baryonic gauge field to gravity experimentally.

[157] T. Quinn et al, Phys. Rev. Lett. **111**, 101102 (2013).

[158] P. J. Mohr, B.N. Taylor, and D. B. Newell, Rev. Mod. Phys. 84, 1527 (2012).

[159] The Eötvös' experiment used a 0.1 gm test mass of $RaBr_2$. R. v. Eötvös, D. Pekár, E. Fekete, Annalen der Physik (Leipzig) 68, 11, 1922.

[160] The reduction of the calculation to algebra reminds the author of Nobelist Hans Bethe's remark that he only felt he understood a physical phenomenon when he could reduce it to algebra. This was quite evident when the author collaborated with Professor Bethe on a study of pion condensation in neutron stars some years ago.

$$\Delta G = -G_2 + G_1 = (\beta^2/4\pi)/(m_2^2 + m_e^2/2) - (\beta^2/4\pi)/(m_1^2 + m_e^2/2)$$
$$\simeq (\beta^2/4\pi)(1/m_2^2 - 1/m_1^2) \quad (15.5)$$

The masses m_1 and m_2 can differ. For example, if m_H is mass of the hydrogen atom, then $m^{-1} = 1.0m_H^{-1}$ for hydrogen, for carbon $m^{-1} = 1.00782m_H^{-1}$, for copper $m^{-1} = 1.00895m_H^{-1}$, and for lead $m^{-1} = 1.00794m_H^{-1}$.[161] Thus using the Quinn et al and CODATA results and assuming copper and lead test masses we find the order of magnitude *estimate*:

$$\alpha_B = \beta^2/4\pi \simeq \Delta G/[(1.00895^2 - 1.00794^2)\, m_H^2]$$
$$\simeq \Delta G/G\, G\, m_H^2/.002037$$
$$\simeq (0.000241/0.002037)Gm_H^2$$
$$\simeq .118\, Gm_H^2 \quad (15.6)$$

indicating a very weak baryonic force consistent with our general view of the Megaverse. The baryon fine structure constant is minute in comparison to the electromagnetic fine structure constant $\alpha \simeq 1/137$.

Due to our assumptions in the calculation of α_B which makes it merely an order of magnitude estimate at best we suggest that an experimental group measure G with differing test masses in the same apparatus to obtain a better value for α_B.

15.1.2 A Baryonic Gauge Field

Based on a reasonable value for α_B we assumed in Blaha (2014) that a baryonic gauge field exists that is similar to the electromagnetic field except for features due to its existence in the 16-dimensional universe that we called the Megaverse.[162] This gauge field couples extremely weakly[163] to individual baryons as well as to aggregates of baryons due to their non-zero baryon number. We called the baryonic gauge field particle a *planckton*. Its electromagnetic analogue is the photon.

Plancktons propagate in the Megaverse, both within universes, and exterior to universes. So the planckton field must be defined in 16-dimensional Megaverse coordinates. They will interact with baryons within a universe with Megaverse coordinates mapped to the curved coordinates of the universe. Since a planckton field in 16-dimensional conventional coordinates would lead to divergences we use 16-dimensional quantum coordinates:[164]

$$Y^i(y) = y^i + i\, Y_u^i(y)/M_u^8 \quad (15.7)$$

with quantum coordinate derivatives defined by

$$\partial_i = \partial/\partial Y^i(y) = \partial/\partial(y^i - Y_u^i(y)/M_u^8) \quad (15.8)$$

to obtain a completely finite theory of planckton interactions with elementary particles and universe particles.

---

161 "One Hundred Years of the Eötvös Experiment", l. Bod, E. Fischbach, G. Marx and Maria Náray-Ziegler, August, 1990.

162 The Megaverse is a 16-dimension complex space within which universes are embedded. We will discuss its features chapter 23.

163 Compared to gravity.

164 See Blaha (2005a) for a discussion of this new method to eliminate infinities in quantum field theory calculations.

Plancktons and the $Y_u^i(y)$ field of quantum coordinates are the only fields in the space between universes in the Megaverse. Since the mass-energy and charge of universes is zero, Standard Model fields are zero in the space between universes.[165] It is reasonable to assume that the vacuum between universes does have fermion and boson seas for Standard Model particles. And we will propose new forces: Dark baryon force, lepton force and Dark lepton force corresponding to three other particle numbers in section 15.2 and in chapter 16. These four forces will have gravitational effects in the Megaverse.

15.1.3 Planckton Second Quantization

The second quantization of the free planckton field $B_u^i(y)$ is similar to the second quantization of the electromagnetic field, and also of the quantum part of the Megaverse quantum coordinates $Y_u^i(y)$. The purpose and role of these fields is quite different: the planckton field generates an interaction between baryons while the $Y_u^i(y)$ field serves as the quantum part of 4-dimensional quantum coordinates giving us a finite quantum field theory of The New Standard Model and gravitation as well as a finite Big Bang for our universe.

We begin by noting that Megaverse quantum coordinates are defined by eqns. 15.7 and 15.8 above. The lagrangian density terms for the free $B_u^i(Y(y))$ fields is

$$\mathscr{L}_{Bu} = -\tfrac{1}{4} F_{Bu}{}^{\mu\nu}(Y(y))F_{Bu\mu\nu}(Y(y)) \tag{15.9}$$

with Y(y) given by eq. 15.7. The lagrangian is

$$L_{Bu} = \int d^{15}y\, \mathscr{L}_{Bu}(Y(y)) \tag{15.10}$$

with

$$F_{Bu\mu\nu} = \partial B_{u\mu}(Y(y))/\partial Y^{\nu}(y) - \partial B_{u\nu}(Y(y))/\partial Y^{\mu}(y) \tag{15.11}$$

where the values of μ and ν range from 1 to 16 in this section.

The equal time commutation relations, derived in the usual way, are:

$$[B_u^{\mu}(Y(\mathbf{y}, y^0)), B_u^{\nu}(Y(\mathbf{y}', y^0))] = [\pi_u^{\mu}(Y(\mathbf{y}, y^0)), \pi_u^{\nu}(Y(\mathbf{y}', y^0))] = 0 \tag{15.12}$$
$$[\pi_{uj}(Y(\mathbf{y}, y^0)), B_{uk}(Y(\mathbf{y}', y^0))] = -i\,\delta^{15tr}{}_{jk}(Y(\mathbf{y},0) - Y(\mathbf{y}',0)) \tag{15.13}$$

where

$$\pi_u{}^k = \partial L_u\,(B_u(Y(y)))/\partial B_{uk}{}'(Y(y)) \tag{15.14}$$
$$\pi_u{}^0 = 0 \tag{15.15}$$

and

$$\delta^{tr}{}_{jk}(\mathbf{y} - \mathbf{y}') = \int d^{15}k\, e^{i\,\mathbf{k}\bullet(Y(\mathbf{y},0) - Y(\mathbf{y}',0))}\,(\delta_{jk} - k_j k_k/\mathbf{k}^2)/(2\pi)^{15} \tag{15.16}$$
$$B_{uk}{}'(Y(y)) = \partial B_{uk}(Y(y))/\partial y^{16} \tag{15.17}$$

for j, k = 1, 2, … , 15.

If we choose the Coulomb gauge for $B_{uk}(Y(y))$:

$$B_u{}^{16}(Y(y)) = 0$$

[165] The vacuum energy of the baryonic field and the $Y_u^i(y)$ fields being uniform throughout the Megaverse do not exert forces or cause gravitational effects except possibly through baryonic Casimir forces between universes.

$$\partial B_u^j(Y(y))/\partial Y^j(y) = 0$$

for j = 1, 2, … , 15 then fourteen degrees of freedom (polarizations) are present in the vector potential.[166] The Fourier expansion of the vector potential $B_u^i(Y(y))$ is:

$$B_u^i(Y(y)) = \int d^{15}k\; N_{0B}(k) \sum_{\lambda=1}^{14} \varepsilon^i(k, \lambda)[a_B(k,\lambda) :e^{-ik\cdot Y(y)}: + a_B^\dagger(k,\lambda) :e^{ik\cdot Y(y)}:] \quad (15.18)$$

for i = 1, … , 15 where

$$N_{0B}(k) = [(2\pi)^{15} 2\omega_k]^{-\frac{1}{2}} \quad (15.19)$$

and (since the field is massless)

$$k^{16} = \omega_k = (\mathbf{k}^2)^{\frac{1}{2}} \quad (15.20)$$

where k^{16} is the energy, and where the $\varepsilon^i(k, \lambda)$ are the polarization unit vectors for $\lambda = 1, \ldots, 14$ and $k^\mu k_\mu = k^{16\,2} - \mathbf{k}^2 = 0$.

The commutation relations of the Fourier coefficient operators are:

$$[a_B(k,\lambda), a_B^\dagger(k',\lambda')] = \delta_{\lambda\lambda'}\delta^{15}(\mathbf{k} - \mathbf{k'}) \quad (15.21)$$
$$[a_B^\dagger(k,\lambda), a_B^\dagger(k',\lambda')] = [a_B(k,\lambda), a_B(k',\lambda')] = 0 \quad (15.22)$$

and the polarization vectors satisfy

$$\sum_{\lambda=1}^{14} \varepsilon_i(k, \lambda)\varepsilon_j(k, \lambda) = (\delta_{ij} - k_i k_j/\mathbf{k}^2) \quad (15.23)$$

The B_u^μ Feynman propagator is

$$iD_F^{trTT}(y_1 - y_2)_{jk} = <0|T(B_{uj}(Y(y_1))B_{uk}(Y(y_2)))|0> \quad (15.24)$$

$$= -ig_{jk} \int \frac{d^{16}k\; e^{-ik.(y_1 - y_2)}\, R(\mathbf{k}, y_1 - y_2)}{(2\pi)^{16}\,(k^2 + i\varepsilon)} \quad (15.25)$$

where g_{jk} is the 16-dimensional Lorentz metric and where $R(\mathbf{k}, y_1 - y_2)$ is given by

$$R(\mathbf{k}, y_1 - y_2) = \exp[-k^i k^j \Delta_{Tij}(y_1 - y_2)/M_u^{16}] \quad (15.26)$$
$$= \exp\{-k^2[A(v) + B(v)\cos^2\theta] / [(2\pi)^{14} M_u^4 z^2]\}$$

where k^2 is the sum of the squares of the 15 spatial components with

$$z^\mu = y_1^\mu - y_2^\mu$$
$$z = |\mathbf{z}| = |\mathbf{y}_1 - \mathbf{y}_2|$$

---

[166] Note we use the Coulomb gauge for Y(y) also.

$$k = |\mathbf{k}|$$
$$v = |z^0|/z$$
$$A(v) = (1 - v^2)^{-1} + .5v \ln[(v - 1)/(v + 1)]$$
$$B(v) = v^2(1 - v^2)^{-1} - 1.5v \ln[(v - 1)/(v + 1)]$$
$$\mathbf{k}\cdot\mathbf{z} = kz \cos\theta$$

and $|\mathbf{k}|$ denoting the length of a spatial 15-vector **k** while $|z^0|$ is the absolute value of $z^0 \equiv z^{16}$.

As eq. 15.26 indicates, the Gaussian damping factor R(k, z) for all large spatial momentum k^j is the same for both the positive and negative frequency parts of the (Two-Tier) B_u Feynman propagator. We are assuming the spatial momentum is real-valued in this discussion. It is also important to note that R(k, z) does not depend on $k^0 = k^{16}$ (in the B_u and Y_u Coulomb gauges) and thus the integration over k^0 proceeds in the usual way to produce time-ordered positive and negative frequency parts.

The Gaussian exponential factor in *all* spatial coordinates causes the Feynman propagator to be finite and, together with the Gaussian factor in universe particle propagators, causes all perturbation theory calculations when interactions are introduced to be finite as we have seen earlier in The Extended Standard Model.

For small momentum much less than M_u then $R(\mathbf{k}, y_1 - y_2) \to 1$ and the Feynman propagator is the "normal" propagator of conventional 16-dimensional quantum field theory. For large momentum the corresponding potential approaches r^{13} in contrast to the electromagnetic Coulomb potential r^{-1}. The B_u potential is highly non-singular at large energies.

15.1.4 Bary-Electric Fields and Bary-Magnetic Fields

As in electromagnetism there is an antisymmetric tensor of the second rank that appears in the free part of the baryonic field $F_{Bu\mu\nu}(y)$ lagrangian:[167]

$$\mathscr{L}_{Bu} = -\tfrac{1}{4} F_{Bu}{}^{ij}(y)F_{Buij}(y) \tag{15.27}$$

where

$$F_{Buij}(y) = \partial B_{ui}(y)/\partial y^j - \partial B_{uj}(y)/\partial y^i \tag{15.28}$$

and i, j = 1, 2, … , 16. The 16th coordinate corresponds to the time coordinate. While the coordinates are complex in general we will treat the 15 spatial coordinates as real and the 16th coordinate as pure imaginary with the resulting invariant interval

$$ds^2 = dy_1{}^2 + dy_2{}^2 + \ldots + dy_{15}{}^2 - c^2dy_{16}{}^2 \tag{15.29}$$

which is invariant under 16 dimensional Lorentz transformations. The coordinates can be transformed into complex-valued coordinates using the Reality group defined in Blaha (2014) and earlier books.

The tensor F_{Buij} is conveniently separated into an baryon electric part and a baryon magnetic part in a manner similar to the separation of the electromagnetic fields into electric and magnetic fields.

---

[167] Parts of the following appear in Blaha (2014a). They are somewhat modified since we are dealing with the classical, low energy, large distance baryonic field where the quantum coordinate fields Y(y) are well approximated by the classical (non-quantum) Megaverse coordinates y.

However the 15 spatial dimensions changes the forms of the baryon fields. Analogously, to electromagnetism the baryon force is given by

$$f_i = F_{Buij}(y)J_B^{\,j}/c \qquad (15.30)$$

where $J_B^{\,j}$ is the jth baryonic current.

The baryon "electric" field is

$$E_{Bui} = -F_{Bui0}(y)/c \qquad (15.31)$$

while the baryon "magnetic" field is

$$B_{Bui} = \varepsilon_{ijk}F_{Bu}^{\;jk}(y) \qquad (15.32)$$

where i, j, k = 1, 2, … , 15 and where ε_{ijk} is a totally anti-symmetric tensor with component values ±1. If i < j < k then ε_{ijk} is +1. Even permutations of these three indices yield a value of +1 for the tensor components. Odd permutations of these three indices yield a value of –1. For example, $\varepsilon_{246} = +1$, $\varepsilon_{426} = -1$, $\varepsilon_{642} = -1$, $\varepsilon_{264} = -1$, $\varepsilon_{462} = +1$, $\varepsilon_{624} = +1$.

With these definitions of the $\mathbf{E_{Bu}}$ and $\mathbf{B_{Bu}}$ fields we can easily derive the 16-dimensional generalization of the *Lorentz force law* for a baryon of charge q and 15-velocity v_j:

$$F_i = qE_{Bui} + q\varepsilon_{ijk}v_jB_{Buk}/c \qquad (15.33)$$

for i = 1, 2, … , 15. One important difference from the 4-dimensional case is the forms of the $\mathbf{E_{Bu}}$ and $\mathbf{B_{Bu}}$ fields

$$E_{Bui} = -F_{Bui0}(y)/c = [-\partial B_{u0}(y)/\partial y^i - \partial B_{ui}(y)/\partial y^0] \qquad (15.34)$$

or, expressed as a 15-vector,

$$\mathbf{E_{Bu}} = [-\nabla_{15}\phi(y) - \dot{\mathbf{B}}_{\mathbf{u}}(y)]/c \qquad (15.35)$$

where ϕ is the baryonic Coulomb potential $B_{u16}(y)$, $\nabla 15$ is the 15-dimensional grad operator, and $B_u(y)$ is the baryonic 15-vector potential with the "dot" above it signifying a time (y_{16}) derivative.

The 15-dimensional baryon magnetic field has the form of eqn. 15.32. A specific illustrative case shows the baryon magnetic field exhibits more complexity than the 3-dimensional magnetic field of electromagnetism:

$$B_{Bu1} = \varepsilon_{1jk}F_{Bu}^{\;jk}(y)/c = [F_{Bu}^{\;23}(y) + F_{Bu}^{\;24}(y) + \ldots + F_{Bu}^{\;215}(y) + F_{Bu}^{\;34}(y) + F_{Bu}^{\;35}(y) + \ldots + F_{Bu}^{\;315}(y) + F_{Bu}^{\;45}(y) + \ldots + F_{Bu}^{\;14,15}(y)]/c \qquad (15.36)$$

Thus each component of the baryon magnetic field impacts on all fifteen spatial directions of the Megaverse. For this reason we use spinning rings, mass configurations and uniships to generate baryon magnetic field interactions to enable uniships to escape from our universe's three spatial dimensions. We consider this possibility in more detail in the following sections.

15.1.5 The Baryonic "Coulombic" Gauge Field

The baryonic gauge field has a "Coulombic" potential part $\phi(y)$, just as the electromagnetic field does. Consequently the total potential between two electromagnetically neutral masses of mass M_1 and M_2, and baryon numbers N_1 and N_2 is

$$V_{tot} = -GM_1M_2/r + (\beta^2/4\pi)\, N_1N_2/r \tag{15.37}$$

where G is the gravitational constant, and β is analogous to the electric charge e in the electromagnetic Coulomb potential. If both masses are composed of the same substance and have the same mass, then we can set $M_1 = M_2 = M = Nm$ where m is the average mass of the baryons in the masses.[168] In addition we can set $N_1 = N_2 = N$. Then eq. 15.37 becomes

$$V = [-Gm^2 + (\beta^2/4\pi)]N^2/r \tag{15.38}$$

Note that the gravitational potential term is attractive, and the baryonic potential term is repulsive between baryons.

In considering eq. 15.1 we have approximated the baryonic potential with only our universe's spatial coordinates. In reality we should be using the spatial separation in all Megaverse coordinates. However since our universe is close to flat, the distance between two objects that are not too far apart is approximately the same in both coordinate systems. The baryonic potential in Megaverse coordinates is actually

$$\phi(y_1, y_2, \ldots, y_{15}) = (\beta^2/4\pi)N_1N_2/(y_1^{\,2} + y_2^{\,2} + \ldots + y_{15}^{\,2})^{1/2} \tag{15.39}$$

15.1.6 The Baryonic Force on Baryonic Objects

The baryonic force on a moving baryon mass is given by the baryon Lorentz force for a baryon of baryon charge q and 15-velocity v_j:

$$F_i = qE_{Bui} + q\varepsilon_{ijk}v_jB_{Buk}/c \tag{15.40}$$

for i = 1, 2, … , 15. The 16-dimensional baryonic Coulombic potential is

$$V = N\phi(y_1, y_2, \ldots, y_{15}) = (\beta^2/4\pi)N/(y_1^{\,2} + y_2^{\,2} + \ldots + y_{15}^{\,2})^{1/2} \tag{15.41}$$

where N is the baryon number of the baryon mass. The baryon Coulombic force is

$$F_i = N\nabla_{15i}\phi(y) \tag{15.42}$$

where ∇_{15i} is the ith component of the 15-dimensional grad operator $\boldsymbol{\nabla}_{15}$.

---

168 We neglect lepton masses since they are negligible relative to the baryon masses.

15.2 Lepton Number, Dark Baryon Number, and Dark Lepton Number Conservation Laws

In the previous section we raised the possibility of an ultra-weak Baryonic force and an associated Baryon number conservation law. This conservation law has been repeatedly tested and found to be satisfied to great accuracy. One possible cause for concern is the Adler-Bell-Jackiw fermion triangle anomaly which follows from a three fermion loop that diverges in conventional quantum field theory. This type of anomaly raises the possibility of baryon number non-conservation. In Two-Tier Quantum Field Theory the triangle graph is convergent – no infinities – and anomalies are not present. Thus our Two-Tier implementation of the Extended Standard Model is anomaly-free and the issue of baryon number non-conservation disappears. (See chapter 19 and Blaha (2005a) for more detailed discussions.)

We will assume Baryon number conservation holds.

15.2.1 Other Conserved Particle Numbers

Another conserved number is Lepton Number, denoted L. Again, repeated attempts to find lepton number violation have failed. On that basis *we will assume lepton number conservation.*

If Baryon Number and Lepton Number are both conserved quantities then any linear combination of them is also conserved. Therefore

$$B' = aB + bL \tag{15.43}$$

is also conserved.

If we consider the Dark Matter sector of the Extended Standard Model it is reasonable to assume that *Dark Baryon Number B_D and Dark Lepton Number L_D are conserved* also although there is no experimental evidence available as yet to confirm (or deny) these assumptions.

Thus we have four conserved particle Numbers. Linear combinations of these numbers are also conserved:

$$\begin{aligned} B' &= aB + bL + cB_D + dL_D \\ L' &= eB + fL + gB_D + hL_D \\ B_D' &= iB + jL + kB_D + lL_D \\ L_D' &= mB + nL + oB_D + pL_D \end{aligned} \tag{15.44}$$

or

$$N' = AN \tag{15.45}$$

Where N and N' are 4-veetors composed of particle numbers and A is a 4×4 matrix. The number of fermion particle types, 4, is determined by the two families of normal fermions, leptons and quarks, and the number of families of Dark fermions, which is ultimately determined by the Reality group study of fermion types in chapter 5.

The constants appearing in linear equations of the form of eq. 15.45 seem arbitrary. However if we want the new 'primed' set of conserved Numbers to be an independent set of numbers then the determinant of the constants must be non-zero. Thus the matrix, A, is invertible.

The set of 4×4 matrices of the type of eq. 15.44 form an U(4) group[169] *if we wish to perform these transformations within lagrangians of the type of the Extended Standard Model. The choice of U(4)*

[169] If we wish to further limit the values of the 'primed' Numbers to integers assuming the unprimed Numbers are integers then the group of the transformation is the set of permutations of four entities – the Symmetric group S_4.

rather than SU(4) is required Since there are four independent particle Numbers and U(4) has four diagonal matrices in its algebra while SU(4) only has three diagonal matrices. U(4) preserves the independence of the four independent particle Numbers.

15.3 U(4) Number Symmetry

At this point we note the observations of Yang and Mills that Numbers can be local and generalize the U(4) Number symmetry to a Yang-Mills symmetry. The transformations then become functions of position A(X) where X represents the space-time coordinates in Two-Tier Quantum Field Theory.

The U(4) rotations of the four Numbers changes the interpretation of the number operators applied to particle fields. Thus we have a symmetry operation induced on particle fields, which in the absence of symmetry brealing terms, becomes a symmetry of the lagrangian.

To implement this symmetry in the particles' lagrangian, all covariant derivatives must acquire another interaction term with 16 U(4) fields corresponding to the 16 generators of U(4). In addition we must add another index to each fermion field specifying its generation. Lastly a set of initially massless gauge field dynamic terms must be added to the Extended Standard Model lagrangian.

The initially massless U(4) gauge transformation symmetry is broken by the Higgs Mechanism with the gauge fields acquiring masses (with two exceptions) and the fermions of the four generations acquiring masses which are generation dependent. The symmetry breaking is described in the next chapter.

However the 'primed' numbers can be integer or not. There is no apparent physical principle requiring integer Numbers. Quarks are usually assigned Baryon Number 1/3. Also Numbers are not necessarily positive valued.

16. Fermion Generations and Broken U(4)

In sections 15.2 and 15.3 we showed that a local U(4) symmetry based on conserved fermion particle numbers existed in Nature and added a new assumption to our construction of the Extended Standard Model. U(4) has 16 generators, which we denote G_i for i = 1, 2, ... , 16. Its fundamental representation has 4×4 matrices. When we introduce this U(4) symmetry directly into the one generation Extended Standard Model each fermion acquires a new index and becomes a four generation set of fermions. Symmetry breaking via a Higgs Mechanism for the U(4) gauge fields gives a different mass to each of the mambers of each set. The Higgs particle lagrangian for U(4) breaking will be described in section 16.3.

16.1 Four Generation Extended Standard Model

In chapter 14 we derived the form of a one generation Extended Standard Model that included the known parts of the Standard Model (excepting the Higgs sector) and an SU(2)⊗U(1) part for Dark Matter. Dark Matter was linked to normal matter with a simple scalar gauge field.

In this section we generalize to the four generation Extended Standard Model that results.[170] Covariant derivatives acquire another interaction term with 16 U(4) fields U_i^{μ}. In addition we add another index to each fermion field specifying its generation. Lastly a set of initially massless gauge field dynamics terms is added to the Extended Standard Model lagrangian to specify U(4) gauge field evolution.

16.1.1 Two-Tier Lepton Sector

We begin with the definition of a quadruplet of leptons – a pair of doublets, one normal and one Dark, instead of a single doublet. We define left and right lepton quadruplets with[171]

$$\Psi_{L,Ra}(X) = \begin{bmatrix} \psi_{DL,Ra}(X) \\ \psi_{NL,Ra}(X) \end{bmatrix} \qquad (16.1)$$

where a is a generation index ranging from 1 to 4, $\psi_{NL,R}(X)$ is a "normal" ElectroWeak-like lepton doublet, and where $\psi_{DL,R}(X)$ is a Dark ElectroWeak-like lepton doublet consisting of a Dark electron-like fermion and a Dark neutrino-like fermion.

We define covariant derivative terms which we express in matrix form are

[170] It is based on the three principles based on Ockham's Razor ("The simplest choice is often the best."): 1) The only connecting interaction is a weak interaction, 2) The form of ElectroWeak theory remains unchanged, and 3) Dark Matter parallels normal matter in its general characteristics: four generations, SU(3) singlets, an SU(2)⊗U(1) symmetry analogous to ElectroWeak symmetry, SU(2)⊗U(1) dark lepton and dark quark doublets.

[171] The X's are Two-Tier coordinates.

$$D_{L,R}(X) = \begin{bmatrix} \gamma^\mu D_{DL,R\mu} & 0 \\ 0 & \gamma^\mu D_{NL,R\mu} \end{bmatrix} \quad (16.2)$$

where the normal matter left-handed covariant derivative is

$$D_{NL\mu} = \partial/\partial X^\mu - ½ig\boldsymbol{\sigma}\cdot\mathbf{W}_\mu + ½ig'B_\mu - ½ig_G\mathbf{G}\cdot\mathbf{U}_\mu \quad (16.3)$$

where g_G is an ultra-weak generational coupling constant, $\mathbf{G}\cdot\mathbf{U}_\mu$ is the sum of the inner product of 16 U(4) generators G_i and gauge fields $U_i(X)$, and where the Dark matter left-handed covariant derivative is

$$D_{DL\mu} = \partial/\partial X^\mu - ½ig_D\boldsymbol{\sigma}\cdot\mathbf{W}'_\mu + ½ig_D'B'_\mu + ½ig_D''B_\mu - ½ig_G\mathbf{G}\cdot\mathbf{U}_\mu \quad (16.4)$$

with $\boldsymbol{\sigma}$ a vector composed of the Pauli matrices. The right-handed covariant derivatives have a simpler form. The normal matter right-handed covariant derivative is

$$D_{NR\mu} = \partial/\partial X^\mu + ½ig'B_\mu - ½ig_G\mathbf{G}\cdot\mathbf{U}_\mu \quad (16.5)$$

and the Dark matter right-handed covariant derivative is

$$D_{DR\mu} = \partial/\partial X^\mu + ½ig_D'B'_\mu + ½ig_D''B_\mu - ½ig_G\mathbf{G}\cdot\mathbf{U}_\mu \quad (16.6)$$

The normal and Dark electroweak fields above are functions of a Two-Tier X. The Faddeev-Popov mechanism operative for these types of fields is described in appendix 19-A of Blaha (2011c) and in chapter 12.

16.1.2 Quark Sector

In the *quark* sector we define left and right quark quadruplets with

$$\Psi_{qL,Ra}(X_c) = \begin{bmatrix} \psi_{DqL,Ra}(X_c) \\ \psi_{NqL,Ra}(X_c) \end{bmatrix} \quad (16..7)$$

where $\psi_{NqL,Ra}(X_c)$ is a "normal" ElectroWeak-like quark doublet, and where $\psi_{DqL,Ra}(X_c)$ is a Dark ElectroWeak-like quark doublet consisting of a SU(3) singlet Dark up-quark of unit Dark charge and a SU(3) singlet Dark down-quark of zero Dark charge in the ath generation.

The covariant derivative terms are contained in $D_q(X_c)$ which we express in matrix form as

$$D_{qL,R}(X_c) = \begin{bmatrix} \gamma^\mu D_{qDL,R\mu}(X_c) & 0 \\ 0 & \gamma^\mu D_{qNL,R\mu}(X_c) \end{bmatrix} \quad (16.8)$$

where the normal quark matter left-handed covariant derivative is

$$D_{qNL\mu} = \partial/\partial X_c^{\mu} - \tfrac{1}{2}ig\boldsymbol{\sigma}\cdot\mathbf{W}_{\mu} - ig'B_{\mu}/6 - \tfrac{1}{2}ig_G\mathbf{G}\cdot\mathbf{U}_{\mu} + ig_C\tau\cdot A_{C\mu} \tag{16.9}$$

and where the Dark quark left-handed covariant derivative is

$$D_{qDL\mu} = \partial/\partial X_c^{\mu} - \tfrac{1}{2}ig_D\boldsymbol{\sigma}\cdot\mathbf{W}'_{\mu} + \tfrac{1}{2}ig_D'B'_{\mu} + \tfrac{1}{2}ig_D''B_{\mu} - \tfrac{1}{2}ig_G\mathbf{G}\cdot\mathbf{U}_{\mu} \tag{16.10}$$

since Dark quarks are SU(3) singlets with unit or zero Dark charge. The right-handed quark covariant derivatives have a simpler form. The normal quark right-handed covariant derivative is

$$D_{qNR\mu} = \partial/\partial X_c^{\mu} + \tfrac{1}{2}ig'B_{\mu}/3 - \tfrac{1}{2}ig_G\mathbf{G}\cdot\mathbf{U}_{\mu} + ig_C\tau\cdot A_{C\mu} \tag{16.11}$$

and the Dark quark right-handed covariant derivative is

$$D_{qDR\mu} = \partial/\partial X_c^{\mu} + \tfrac{1}{2}ig_D'B'_{\mu} + \tfrac{1}{2}ig_D''B_{\mu} - \tfrac{1}{2}ig_G\mathbf{G}\cdot\mathbf{U}_{\mu} \tag{16.12}$$

The normal and Dark gauge boson fields are functions of X_c. = $(X_{r\mu}(y_r), X_{i\mu}(y_i))$ of eqs. 14.11 and 14.12. The Faddeev-Popov mechanism is operative for gauge boson fields and is described in appendix 19-A of Blaha (2011c).[172] The *complexon* quark Extended Standard Model ElectroWeak Sector covariant derivatives in quadruplet matrix form are

$$D_{qL,R}(X_c) = \begin{bmatrix} \gamma^{\mu}D_{qDL,R\mu} & 0 \\ 0 & \gamma^{\mu}D_{qNL,R\mu} \end{bmatrix} \tag{16.13}$$

The remaining parts of the complexon Standard Model are described in chapter 23 of Blaha (2011) and summarized below. The addition of singlet Dark quark Higgs terms is also required.

The lagrangian density and action is

$$\mathcal{L}_{CSM} = \Psi_{L^a}^{\dagger}\gamma^0 i\gamma^{\mu}D_{L\mu}\Psi_{L^a} - \Psi_{R^a}^{\dagger}\gamma^0 i\gamma^{\mu}D_{R\mu}\Psi_{3R^a} + \Psi_{CL^a}^{\dagger}\gamma^0 i\gamma^{\mu}\mathcal{D}_{qL\mu}\Psi_{CL^a} + \Psi_{CR^a}^{\dagger}\gamma^0 i\gamma^{\mu}\mathcal{D}_{qR\mu}\Psi_{CR^a} - $$
$$- \mathcal{L}_{BareMasses} + \mathcal{L}_{Gauge} + \mathcal{L}_{Mass} + \mathcal{L}_{Ufields} \tag{16.14}$$

where a is the generation index. $\mathcal{L}_{BareMasses}$ contains the fermion bare mass terms. Also,

$$\mathcal{L}_{Gauge} = \mathcal{L}_{GaugeEW} + \mathcal{L}_{GaugeC} + \mathcal{L}_{GaugeEWD} \tag{16.15}$$

with

$$\mathcal{L}_{GaugeEW} = -\tfrac{1}{4} F_W^{a\mu\nu}F_{W\,\mu\nu}^{a} - \tfrac{1}{4} F_B^{\mu\nu}F_{B\mu\nu} + \mathcal{L}_{EW}^{ghost} \tag{16.16}$$

[172] Those who might be concerned about the propagator term <$W_i(X)$, $W_j(X_c)$> and similar propagators where one field is a function of X and the other field is a function of X_c should note that such terms are to very good approximation equal to <W $_i(X)$, W $_j(X)$> for energies much less than M_c (which could be as large as the Planck energy.)

$$\mathcal{L}_{GaugeEWD} = -\tfrac{1}{4} F'_W{}^{a\mu\nu} F'_W{}^a{}_{\mu\nu} - \tfrac{1}{4} F_{B'}{}^{\mu\nu} F_{B'\mu\nu} + \mathcal{L}_{W'}{}^{ghost} \tag{16.17}$$

and

$$\mathcal{L}_{GaugeC} = \mathcal{L}_{CCG} + \mathcal{L}_C{}^{ghost} + \mathcal{L}_{CC}{}^{ghost} \tag{16.18}$$

$$\mathcal{L}_{Ufields} = -\tfrac{1}{4} F_U{}^{a\mu\nu} F_{U\mu\nu\nu} + \mathcal{L}_U{}^{ghost} + \mathcal{L}_U{}^{UHiggs} \tag{16.19}$$

where $\mathcal{L}_U{}^{UHiggs}$ is discussed in section 16.4. The ElectroWeak gauge bosons $W_\mu{}^a$, B_μ and B'_μ field tensors are:

$$F_W{}^a{}_{\mu\nu} = \partial W^a{}_\mu/\partial X^\nu - \partial W^a{}_\nu/\partial X^\mu + g_2 f^{abc} W^b{}_\mu W^c{}_\nu \tag{16.20}$$

$$F_{B\mu\nu} = \partial B_\mu/\partial X^\nu - \partial B_\nu/\partial X^\mu \tag{16.21}$$

and the Dark ElectroWeak gauge bosons $W'_\mu{}^a$ and B'_μ field tensors are:

$$F_{B'\mu\nu} = \partial B'_\mu/\partial X^\nu - \partial B'_\nu/\partial X^\mu$$

$$F'_W{}^a{}_{\mu\nu} = \partial W'^a{}_\mu/\partial X^\nu - \partial W'^a{}_\nu/\partial X^\mu + g_2 f^{abc} W'^b{}_\mu W'^c{}_\nu \tag{16.22}$$

The U fields tensor is:

$$F_U{}^a{}_{\mu\nu} = \partial U^a{}_\mu/\partial X^\nu - \partial U^a{}_\nu/\partial X^\mu + g_G f_4{}^{abc} U^b{}_\mu U^c{}_\nu \tag{16.23}$$

where $f_4{}^{abc}$ are the U(4) algebra commutator constants.

$\mathcal{L}_{EW}{}^{ghost}$ contains the Faddeev-Popov ghost terms for the ElectroWeak $W_\mu{}^a$ gauge bosons. The complexon color gluon lagrangian $\mathcal{L}_{CCG}$ is defined by

$$\mathcal{L}_{CCG} = -\tfrac{1}{4} F_{CC}{}^{a\mu\nu}(X) F_{CC}{}^a{}_{\mu\nu}(X) \tag{16.24}$$

where

$$F_{CC}{}^a{}_{\mu\nu} = \partial/\partial X_c{}^\nu A_C{}^a{}_\mu - \partial/\partial X_c{}^\mu A_C{}^a{}_\nu + g f_{su(3)}{}^{abc} A_C{}^b{}_\mu A_C{}^c{}_\nu \tag{16.25}$$

where $A_C{}^a{}_\nu$ is the color gluon gauge field, g is the color coupling constant, and the $f_{su(3)}{}^{abc}$ are the SU(3) structure constants.

In addition $\mathcal{L}_C{}^{ghost}$ is the color SU(3) Faddeev-Popov ghost terms defined in appendix 19-A of Blaha (2011c) for the complexon Lorentz gauge and $\mathcal{L}_{CC}{}^{ghost}$ is the complexon color SU(3) constraint ghost terms defined through the Faddeev-Popov mechanism. The mass sector $\mathcal{L}_{Mass}$ is presumably based on the Higgs Mechanism.which creates the fermion and ElectroWeak vector boson masses, and generation mixing.

The lagrangian is supplemented with the following condition on all complexon fields $\Phi_{\ldots}$:[173]

[173] These conditions implement the orthogonality of the real and imaginary parts of complexon 3-momentum.

$$\nabla_r \cdot \nabla_i \Phi \ldots = 0 \qquad (16.26)$$

Non-complexon fields Ω... in the left-handed formulation under consideration satisfy the subsidiary condition:

$$[\nabla_r \cdot \nabla_i - (\nabla_r^2 \nabla_i^2)^{1/2}]\Omega \ldots = 0 \qquad (16.27)$$

which guarantees a complexon's real momentum is parallel to its imaginary momentum.

16.2 U(4) Gauge Symmetry Breaking and Long Range Forces

In chapter 15 we showed that there was good experimental evidence for a conserved Baryon Number B and we proceeded to develop a simple U(1) gauge theory that would imply Baryon Number conservation in a manner analogous to QED's implying electric charge conservation. In section 16.1 we used a new symmetry group local U(4) to generalize the one generation Extended Standard Model to a four generation Extended Standard Model based on four conserved particle numbers: B, L, B_D, and L_D.[174]

We now assume in our construction that the four generation Extended Standard Model has a local U(4) symmetry that is broken by mass terms gewnerated by the Higgs Mechanism.

Further, we will assume that the Higgs breakdown yields two massless (long range) fields which we associate with Baryon Number B and Dark Baryon Number B_D. The remaining fields acquire masses and generate short range forces.

We use the following U(4) diagonal matrices:

$$\begin{aligned} G_1 &= \mathrm{diag}(1, 1, 1, 1) \\ G_2 &= \mathrm{diag}(0, 1, 0, 0) \\ G_3 &= \mathrm{diag}(0, 0, 1, 0) \\ G_4 &= \mathrm{diag}(0, 0, 0, 1) \end{aligned} \qquad (16.28)$$

The U(4) algebra has 16 hermitean matrices that satisfy

$$G_i^\dagger = G_i \qquad (16.29)$$

The particle numbers can be expressed in terms of the diagonal generators as

$$\begin{aligned} B &= G_1 - G_2 - G_3 - G_4 \\ B_D &= G_2 \\ L &= G_3 \\ L_D &= G_4 \end{aligned} \qquad (16.30)$$

The covariant derivatives have the general form:

$$D_{\ldots\mu} = \partial/\partial X^\mu + \ldots - \tfrac{1}{2} i g_G \mathbf{G} \cdot \mathbf{U}_\mu \qquad (16.31)$$

---

[174] Charge, although a conserved number, is a part of the ElectroWeak sector, account of which has already been taken.

where the ellipses indicate the other details of the particular covariant derivative. We now wish to express the four gauge fields $U_i(X)$ for i = 1, 2, 3, 4 corresponding to the diagonal generators in terms of the fields of the four particle number gauge fields: B_μ, L_μ, $B_{D\mu}$, and $L_{D\mu}$.

$$U_{i\mu} = A_{ik} N_{k\mu} \tag{16.32}$$

where A_{ik} are the elements of a matrix of constants and

$$N_\mu = \begin{bmatrix} B_\mu(X) \\ L_\mu(X) \\ B_{D\mu}(X) \\ L_{D\mu}(X) \end{bmatrix} \tag{16.33}$$

is a column vector consisting of the gauge fields corresponding to each of the conserved particle numbers.

The matrix A must have non-zero determinant so that eq. 16.32 can be inverted to express the particle number fields in terms of the four $U_i(X)$ gauge fields:

$$N_\mu = A^{-1}U_\mu \tag{16.34}$$

Resulting in

$$\begin{aligned} B_\mu(X) &= U_{1\mu} \\ L_\mu(X) &= U_{1\mu} + U_{2\mu} \\ B_{D\mu}(X) &= U_{1\mu} + U_{3\mu} \\ L_{D\mu}(X) &= U_{1\mu} + U_{4\mu} \end{aligned} \tag{16.35}$$

Then

$$D_{\ldots\mu} = \partial/\partial X^\mu + \ldots - \tfrac{1}{2} i g_G\Big[\sum_{i=5}^{16} \mathbf{G}_i \mathbf{U}_{i\mu} + BB_\mu(X) + LL_\mu(X) + B_D B_{D\mu}(X) + L_D L_{D\mu}(X)\Big] \tag{16.36}$$

where the particle numbers, which are analogous to the charges Q and Q' in ElectroWeak theory, are B, L, B_D, and L_D. They are expressed in terms of U(4) generators by eqs. 16.30.

16.3 Higgs Mass Mechanism for U(4) Gauge Fields

We now require that there are two massless fields, one coupled to Baryon number and one coupled to Dark Baryon number. The Dark sector is assumed to be analogous to the normal particle sector in this respect. There are fourteen remaining fields that acquire a mass and longitudinal components. These fields become short range, ultra-weak generational forces. The masses they acquire through the Higgs Mechanism are presumably very large as these gauge particles have not been found experimentally.[175]

We assume that a scalar Higgs field exists which is a U(4) vector with four components corresponding to the fermion generations. It is an SU(2)⊗U(1)⊗SU(3) ElectroWeak scalar. Its lagrangian density is

$$\mathcal{L}_U^{\text{UHiggs}} = (\partial\eta^\dagger/\partial X^\mu)(\partial\eta/\partial X^\mu) - \lambda(\eta^\dagger\eta - \rho^2)^2 + \mathcal{L}_U^{\text{UHiggs}}{}_{\text{FermionMasses}}$$

[175] Section 16.4 discusses this topic in more detail.

where $\mathcal{L}_U{}^{UHiggs}{}_{FermionMasses}$ are the fermion masses produced by the U Higgs Mechanism and where we choose a unitary gauge in which

$$\rho = \begin{bmatrix} 0 \\ \rho_1 \\ 0 \\ \rho_2 \end{bmatrix} \tag{16.37}$$

where ρ_1 and ρ_2 are real fields. Then the covariant derivative of ρ is

$$D_{\ldots\mu}\,\rho = \{\partial/\partial X^\mu + \ldots - \tfrac{1}{2}ig_G[\Sigma\mathbf{G}_i\mathbf{U}_{i\mu} + BB_\mu(X) + LL_\mu(X) + B_DB_{D\mu}(X) + L_DL_{D\mu}(X)]\}\begin{bmatrix} 0 \\ \rho_1 \\ 0 \\ \rho_2 \end{bmatrix} \tag{16.38}$$

The sum over i is from 5 through 16, and $[\mathbf{G}_i]_{jk}$ is the jk element of $\mathbf{G}_i$. Then

$$D_{\ldots\mu}\,\rho = \begin{bmatrix} -\tfrac{1}{2}ig_G\{\rho_1\Sigma[\mathbf{G}_i]_{12}\mathbf{U}_{i\mu} + \rho_2\Sigma[\mathbf{G}_i]_{14}\mathbf{U}_{i\mu}\} \\ \partial\rho_1/\partial X^\mu - \tfrac{1}{2}ig_G\rho_1L_\mu - \tfrac{1}{2}ig_G\{\rho_1\Sigma[\mathbf{G}_i]_{22}\mathbf{U}_{i\mu} + \rho_2\Sigma[\mathbf{G}_i]_{24}\mathbf{U}_{i\mu}\} \\ -\tfrac{1}{2}ig_G\{\rho_1\Sigma[\mathbf{G}_i]_{32}\mathbf{U}_{i\mu} + \rho_2\Sigma[\mathbf{G}_i]_{34}\mathbf{U}_{i\mu}\} \\ \partial\rho_2/\partial X^\mu - \tfrac{1}{2}ig_G\rho_2L_{D\mu} - \tfrac{1}{2}ig_G\{\rho_1\Sigma[\mathbf{G}_i]_{42}\mathbf{U}_{i\mu} + \rho_2\Sigma[\mathbf{G}_i]_{44}\mathbf{U}_{i\mu}\} \end{bmatrix} \tag{16.39}$$

$$= \begin{bmatrix} -\tfrac{1}{2}ig_G\Sigma\{\rho_1[\mathbf{G}_i]_{12} + \rho_2[\mathbf{G}_i]_{14}\}\mathbf{U}_{i\mu} \\ \partial\rho_1/\partial X^\mu - \tfrac{1}{2}ig_G\rho_1L_\mu - \tfrac{1}{2}ig_G\rho_2\Sigma[\mathbf{G}_i]_{24}\mathbf{U}_{i\mu} \\ -\tfrac{1}{2}ig_G\Sigma\{\rho_1[\mathbf{G}_i]_{32} + \rho_2[\mathbf{G}_i]_{34}\}\mathbf{U}_{i\mu} \\ \partial\rho_2/\partial X^\mu - \tfrac{1}{2}ig_G\rho_2L_{D\mu} - \tfrac{1}{2}ig_G\rho_1\Sigma[\mathbf{G}_i]_{42}\mathbf{U}_{i\mu} \end{bmatrix} \tag{16.40}$$

since the generators $\mathbf{G}_i$ have zeroes along their diagonals for i = 5, … , 16.

From eq. 16.39 we find the corresponding Higgs field kinetic terms in the lagrangian are

$$(D_{\ldots\mu}\,\rho)^\dagger D_{\ldots}{}^{\mu}\,\rho = \partial\rho_1/\partial X^\mu\partial\rho_1/\partial X_\mu + \partial\rho_2/\partial X^\mu\,\partial\rho_2/\partial X_\mu + g_G{}^2\rho_1{}^2L_\mu\,L^\mu/4 + g_G{}^2\rho_2{}^2L_{D\mu}\,L_D{}^\mu/4 + \ldots \tag{16.41}$$

Note there are differing mass squared terms for the Lepton ($g_G{}^2\rho_1{}^2/4$) and Dark Lepton ($g_G{}^2\rho_2{}^2/4$) gauge fields making them short range fields with the likelihood of very large masses much beyond ElectroWeak gauge field masses, and with an ultra weak coupling constant g_G as suggested by the "experimental" coupling for the Baryonic force given in eq. 15.6.

The Baryonic and Dark Baryonic gauge fields are massless and thus long range although their coupling constant appears to be ultra weak – much below the gravitational coupling constant G.

We now turn to calculating the remaining terms in eq. 16.41 that determine the masses of the remaining 14 gauge fields. We begin by assigning matrix elements for the remaining hermitean U(4) generators:

$$
\begin{aligned}
[G_5]_{ik} &= \delta_{i1}\delta_{k2} + \delta_{i2}\delta_{k1} \\
[G_6]_{ik} &= -i\delta_{i1}\delta_{k2} + i\delta_{i2}\delta_{k1} \\
[G_7]_{ik} &= \delta_{i1}\delta_{k3} + \delta_{i3}\delta_{k1} \\
[G_8]_{ik} &= -i\delta_{i1}\delta_{k3} + i\delta_{i3}\delta_{k1} \\
[G_9]_{ik} &= \delta_{i1}\delta_{k4} + \delta_{i4}\delta_{k1} \\
[G_{10}]_{ik} &= -i\delta_{i1}\delta_{k4} + i\delta_{i4}\delta_{k1} \\
[G_{11}]_{ik} &= \delta_{i2}\delta_{k3} + \delta_{i3}\delta_{k2} \\
[G_{12}]_{ik} &= -i\delta_{i2}\delta_{k3} + i\delta_{i3}\delta_{k2} \\
[G_{13}]_{ik} &= \delta_{i2}\delta_{k4} + \delta_{i4}\delta_{k2} \\
[G_{14}]_{ik} &= -i\delta_{i2}\delta_{k4} + i\delta_{i4}\delta_{k2} \\
[G_{15}]_{ik} &= \delta_{i3}\delta_{k4} + \delta_{i4}\delta_{k3} \\
[G_{16}]_{ik} &= -i\delta_{i3}\delta_{k4} + i\delta_{i4}\delta_{k3}
\end{aligned}
\qquad (16.42)
$$

Then completing eq. 16.41 using eq. 16.40 we find

$$
\begin{aligned}
(D_{\ldots\mu}\,\rho)^\dagger D_{\ldots}^{\ \mu}\,\rho = {} & \partial\rho_1/\partial X^\mu \partial\rho_1/\partial X_\mu + \partial\rho_2/\partial X^\mu\, \partial\rho_2/\partial X_\mu + g_G^2\rho_1^2 L_\mu\, L^\mu/4 + g_G^2\rho_2^2 L_{D\mu}\, L_D^{\ \mu}/4 + \\
& + (g_G/2)^2\rho_1^2(U_5^2 + U_6^2) + (g_G/2)^2\rho_2^2(U_9^2 + U_{10}^2) + (g_G/2)^2\rho_1^2(U_{11}^2 + U_{12}^2) + + \\
& (g_G/2)^2(\rho_1^2 + \rho_2^2)(U_{13}^2 + U_{14}^2) + \ + (g_G/2)^2\rho_2^2(U_{15}^2 + U_{16}^2)
\end{aligned}
\qquad (16.43)
$$

up to total divergences which generate surface terms which we discard and assuming that all fields satisfy the gauge condition

$$\partial U_i^{\ \mu}/\partial X^\mu = 0 \qquad (16.44)$$

Note that there are no mass terms for $U_7(X)$ and $U_8(X)$ as well as $B_\mu(X)$ and $B_{D\mu}(X)$ due to our choice of unitary gauge eq. 16.37. Consequently there are four massless long range fields and 12 gauge fields that acquire masses of three different values: $(g_G/2)\rho_{10}$, $(g_G/2)\rho_{20}$, and $(g_G/2)(\rho_{10}^{\ 2} + \rho_{20}^{\ 2})^{½}$ where ρ_{10} and ρ_{20} ar the vacuum expectation values of ρ_1 and ρ_2 respectively. The fields $U_7(X)$ and $U_8(X)$ are not "diagonal" and thus appear in the fermion sector as terms connecting fermions in different generations within the four species of normal fermions and within the four species of Dark fermions.[176] Therefore they do not change the values of any of the four types of particle numbers.

Based on the estimate of eq. 15.6 the ultra weak value of the coupling constant is

$$g_G = (4\pi\alpha_B)^{½} \approx 1.218\ (Gm_H^{\ 2})^{½} \qquad (16.45)$$

The ultra weak value of the coupling constant implies that the baryonic force with gauge field $B_\mu(X)$ which is now part of a quadruplet of fields. It is a massless, long range field that corresponds to th at of chapter 15 with the exception that chapter 15 looks ahead to later chapters where we discuss a 16-dimensional space that we call the *Megaverse* in which our universe resides where the baryonic force and force associated with the Dark Baryon force exist beyond our universe and act with other possible "island" universes. (The leptonic and Dark leptonic forces are short range and thus do not extend beyond our universe.)

The two non-diagonal long range forces, being between different generations of a species and having an ultra weak coupling are not of great consequence because of the short lifetime of the higher

---

[176] Neutral lepton, charged lepton, up-type quark and down-type quark plus the four corresponding Dark species..

generations of a species. Therefore,,despite their long range, they have only the "shortest" time to exert an inter-generation force before a higher generation particle decays.

Since we expect the other massive fields to have very large masses (and thus very large Higgs field vacuum expectation values) and ultra weak coupling they are not likely to be experimentally found for the foreseeable future.

16.4 Impact of this U(4) Higgs Mechanism on Fermion Generation Masses

The fermion masses of the charged lepton, and the up-type quark, and down-type quark species' generations all show a rapid increase of mass with the generation. For example the u quark mass is a few MeV while the t quark (third generation) has a mass of about 170 GeV/c. The ratio of these masses is about 170,000. While one can account for this great difference by the judicious choices of Higgs' parameter values, when one considers the generational group and its associated numerical quantities: ultra weak coupling, very large U particle masses – perhaps of the order of hundreds or thousands of GeV/c, and the corresponding very large Higgs particle vacuum expectation values in this U gauge field sector[177] then the differences in fermion masses within a species become more understandable and natural from a Leibniz Principle perspective.

Thus the popular view that the ElectroWeak gauge field symmetry breaking is solely via ElectroWeak Higgs fields is not part of our Extended Standard Model unless the U(4) sector is removed. In our model there are two sets of contributions to fermion symmetry breaking: ElectroWeak Higgs particles symmetry breaking, and Generation group U(4) Higgs particles symmetry breaking. The Generation group causes each species to break into four generations.

In the conventional Standard Model the breakup of species into generations is inserted "by hand." It is not a consequence of the existence of SU(2)⊗U(1) symmetry or symmetry breaking. In our approach the U(4) Generation group causes the appearance of generations. We base the existence of the Generation group[178] on the four conserved particle numbers. Leibniz' Principle and Ockham's Razor then lead to the above construction/derivation.

16.5 Generation Group Higgs Mechanism for Fermion Masses

We now consider the Generation group Higgs Mechanism for the eight species of fermions (four species of "normal" matter and four species of Dark Matter). We shall consider the mass terms for the four normal species which is the same as that of the four Dark species except for the values in the various species mass matrices. Therefore we define the initial 4-vector for the generations of the normal species by

---

[177] They are not the Higgs particles of the SU(2)⊗U(1) ElectroWeak sector.

[178] In earlier books we suggested the fermion generations might be the result of a wormhole to another 4-dimensional universe. The new approach is simpler and more consistent with known facts – thus more consistent with the Leibniz Minimax Principle.

$$\Psi_s = \begin{bmatrix} \psi_{11} \\ \psi_{12} \\ \psi_{13} \\ \psi_{14} \\ \dots \\ \psi_{41} \\ \psi_{42} \\ \psi_{43} \\ \psi_{44} \end{bmatrix} \qquad (16.46)$$

where ψ_{ki} is the generation index for the ith generation of the kth species. ψ_{k1} is the wave function for the 1st generation, ψ_{k4} is the 4th generation member of the kth species, and we omit other indices in the interests of clarity. The normal fermion species are ordered charged lepton (k = 1), up-type quark, neutral lepton, and down-type quark (k = 4). Other indices of these wave functions are surpressed in the interests of clarity. A 4th generation fermion of any species is yet to be found experimentally. The lagrangian density mass term for the four normal fermion species is

$$\mathcal{L}_U{}^{UHiggs}{}_{FermionMasses} = \Sigma_{\alpha,\beta}\, \bar{\psi}_{kL\alpha}\, \eta m_{k\alpha\beta}\, \psi_{kR\beta} + \text{c.c.} \qquad (16.47)$$

where $m_{k\alpha\beta}$ is complex constant matrix, and where α, β = 1, … , 4. The total fermion lagrangian mass terms are

$$\mathcal{L}^{Higgs}{}_{FermionMasses} = \mathcal{L}_U{}^{UHiggs}{}_{FermionMasses} + \mathcal{L}_{EW}{}^{Higgs}{}_{FermionMasses} \qquad (16.48)$$

where $\mathcal{L}_{EW}{}^{Higgs}$ is the contribution of ElectroWeak Higgs Mechanism to the fermion masses (discussed in the following chapter). Using the vacuum expectation value of η in eq. 16.37 we find

$$\mathcal{L}_U{}^{UHiggs}{}_{FermionMasses} = \Sigma_{\alpha,\beta}\, \{\bar{\psi}_{2L\alpha}\, \rho_1 m_{2\alpha\beta}\psi_{2R\beta} + \bar{\psi}_{4L\alpha}\, \rho_2 m_{4\alpha\beta}\psi_{4R\beta}\} + \text{c.c.} \qquad (16.49)$$

giving mass terms for the up-type and down-type quark species but not for lepton species. There is an implicit color summation over the color quarks in each generation and quark species. *Qualitatively eq. 16.49 could be viewed as corresponding to the experimentally determined largeness of quark masses relative to lepton masses in each generation of normal matter.*

The mass matrices $m_2 = [m_{2\alpha\beta}]$ and $m_4 = [m_{4\alpha\beta}]$ are both complex, constant mass matrices. They can be brought to diagonal form with non-negative values by U(4) matrices A_k and B_k:

$$\begin{aligned} A_2 m_2 B_2^{-1} &= D_2 \\ A_4 m_4 B_4^{-1} &= D_4 \end{aligned} \qquad (16.50)$$

or

$$m_2 = A_2^{-1} D_2 B_2 \qquad (16.51)$$
$$m_4 = A_4^{-1} D_4 B_4$$

We now note, that although, both D_2 and D_4 have non-negative real values, down-type quarks are all tachyonic and up-type quarks are all non-tachyonic due to their lagrangian kinetic terms as seen in chapter 5.

We further note that $m_2^\dagger m_2$ and $m_4^\dagger m_4$ are hermitean, and A_k and B_k are members of U(4) as is D_k for k = 2,4, with the result that m_2 and m_4 are also both members of the U(4) group. Thus

$$m_2^{-1} = m_2^\dagger \qquad (16.52)$$
$$m_4^{-1} = m_4^\dagger$$

We can express the mass matrices in terms of U(4) generators

$$m_2 = \Sigma G_i m_{2i} \qquad (16.53)$$
$$m_4 = \Sigma G_i m_{4i}$$

$$m_2^{-1} = m_2^\dagger = \Sigma G_i m_{2i}^* \qquad (16.54)$$
$$m_4^{-1} = m_4^\dagger = \Sigma G_i m_{4i}^*$$

since the matrices G_i are all hermitean, where $\{m_{2i}\}$ and $\{m_{4i}\}$ are each a set of sixteen complex constants.

While we do not as yet know the 4th generation fermions or their masses, the third generation quarks have masses that are far greater than the 1st and 2nd generation quarks or their sum suggesting that the trace of m_2 and m_4.is dominated by the 4th generation mass of the two quark species with a similar situation holding, perhaps, for the two Dark quark species. Therefore if we take the trace of m_2 and m_4 then it seems probable based on the trend of the generations that the 4th generation mass dominates the trace:

$$D_{24} \approx \text{tr } D_2 \qquad (16.55)$$
$$D_{44} \approx \text{tr } D_4$$

We can use these A_k and B_k U(4) transformations to define the eight "physical" (up to further ElectroWeak Higgs Mehanism effects) up-type and down-type quark generations fields:

$$\psi_{2L\alpha}\, \rho_1 m_{2\alpha\beta} \psi_{2R\beta} + \psi_{4L\alpha}\, \rho_2 m_{4\alpha\beta} \psi_{4R\beta} = (\psi_{2L} A_2^{-1})_\alpha\, \rho_1 D_{2\alpha\beta} (B_2 \psi_{2R})_\beta + (\psi_{4L} A_4^{-1})_\alpha\, \rho_2 D_{4\alpha\beta} (B_4 \psi_{4R})_\beta$$
$$= \psi_{2Lphys\alpha}\, \rho_1 D_{2\alpha\beta} \psi_{2Rphys\beta} + \psi_{4Lphs\alpha}\, \rho_2 D_{4\alpha\beta} \psi_{4Rphys\beta} \qquad (16.56)$$

Species: up-type quarks down-type quarks

The preceding discussion with changes in the values of constants and constant matrices holds for Dark Matter also where the Dark quarks acquire mass terms but the Dark leptons do not. The Dark Matter species mass terms, with the subscript D signifying Dark Matter, are

$$= \psi_{D2Lphys\alpha}\, \rho_{D1} D_{D2\alpha\beta} \psi_{D2Rphys\beta} + \psi_{D4Lphs\alpha}\, \rho_{D2} D_{D4\alpha\beta} \psi_{D4Rphys\beta} \qquad (16.57)$$

Dark Species: up-type quarks down-type quarks

17. Higgs Mechanism in the ElectroWeak Sector

The discussion of the ElectroWeak Higgs Mechanism for the generation of ElectroWeak gauge field masses and fermion masses is analogous to the discussions of mass generation in sections 16.3 – 16.5 for the U(4) Generation group.[179]

17.1 ElectroWeak Higgs Mechanism Generation of ElectroWeak Gauge Field Masses

We now require that there is one massless field, the electromagnetic field coupled to electric charge and three massive fields that receive masses via the Higgs Mechanism which breaks SU(2)⊗U(1) symmetry. *The Dark sector is assumed to be analogous to the normal particle sector in this respect since it has a SU(2)⊗U(1) symmetry in the Extended Standard Model.* The massive fields become the short range Weak interactions. These assumptions in the construction of the Extended Standard Model conform to the Leibniz Minimax Principle.

We assume that a doublet Higgs field exists with two components:

$$\eta = \begin{bmatrix} \varphi_+ \\ \varphi_0 \end{bmatrix} \tag{17.1}$$

with conjugate Higgs doublet

$$\eta' = \begin{bmatrix} \varphi_0 \\ -\varphi_- \end{bmatrix} \tag{17.2}$$

The Higgs sector lagrangian has the form:

$$\mathcal{L}_{EW}{}^{Higgs} = (\partial\eta^\dagger/\partial X^\mu)(\partial\eta/\partial X^\mu) - \lambda(\eta^\dagger\eta - \rho^2)^2 + \mathcal{L}_{EW}{}^{Higgs}{}_{EWMasses} \tag{17.3}$$

where the symmetry breaking follows from the choice of unitary gauge

$$\rho = \begin{bmatrix} 0 \\ \rho \end{bmatrix} \tag{17.4}$$

where ρ is a real field. Then the covariant derivative of η is

[179] These discussions are similar to those in Huang (1992), and other books and papers.

$$D_{\ldots\mu}\,\eta = \{\partial/\partial X^{\mu} + \ldots + ig\mathbf{t}\cdot\mathbf{W}_{\mu} + + ig't_0 W_{0\mu}\}\begin{bmatrix} 0 \\ \rho \end{bmatrix} \tag{17.5}$$

where g and g' are coupling constants, t_0.is a ½ the identity matrix, and the **t** matrices are ½ the vector of Pauli matrices.The ellipses indicate additional indices and additional terms respectively.

Then

$$D_{\ldots\mu}\,\eta = \begin{bmatrix} \tfrac{1}{2}ig\rho(W_{1\mu} - iW_{2\mu}) \\ \partial\rho/\partial X^{\mu} - \tfrac{1}{2}ig\rho(\cos\theta_W)^{-1}Z_{\mu} \end{bmatrix} \tag{17.6}$$

where θ_W is the Weinberg angle and

$$\begin{aligned} W_3{}^{\mu} &= Z^{\mu}\cos\theta_W + A^{\mu}\sin\theta_W \\ W_0{}^{\mu} &= -Z^{\mu}\sin\theta_W + A^{\mu}\cos\theta_W \end{aligned} \tag{17.7}$$

From eq. 17.6 we find the corresponding Higgs field kinetic terms in the lagrangian are

$$(D_{\ldots\mu}\,\eta)^{\dagger} D_{\ldots}{}^{\mu}\,\eta = \partial\rho/\partial X^{\mu}\partial\rho/\partial X_{\mu} + g^2\rho^2[W_1{}^{\mu}W_{1\mu} + W_2{}^{\mu}W_{2\mu}]/4 + g^2\rho^2 Z^{\mu}Z_{\mu}/(2\cos\theta_W)^2 \tag{17.8}$$

with the $W_1{}^{\mu}$ and $W_2{}^{\mu}$ and Z^{μ} gauge bosons acquiring masses and the electromagnetic field A^{μ} massless.

17.2 ElectroWeak Higgs Mechanism Generation of Fermion Masses

We now consider the ElectroWeak Higgs Mechanism for the eight species of fermions (four species of "normal" matter and four species of Dark Matter). We shall consider the mass terms for the four normal species which is the same as that of the four Dark species except for the values in the various species mass matrices. Therefore we define the initial 4-vector for the generations of the normal species by

$$\Psi_s = \begin{bmatrix} \psi_{11} \\ \psi_{12} \\ \psi_{13} \\ \psi_{14} \\ \ldots \\ \psi_{41} \\ \psi_{42} \\ \psi_{43} \\ \psi_{44} \end{bmatrix} \tag{17.9}$$

$\Psi_s =$

where ψ_{ki} is the generation index for the ith generation of the kth species. ψ_{k1} is the wave function for the 1st generation, ψ_{k4} is the 4th generation member of the kth species, and we omit other indices in the interests of clarity. The normal fermion species are ordered: charged lepton (k = 1), up-type quark, neutral lepton, and down-type quark (k = 4). Other indices of these wave functions are surpressed in the interests of clarity. A 4th generation fermion of any species is yet to be found experimentally.

We assume that a "double" doublet Higgs field exists with four components:

$$\eta = \begin{bmatrix} \varphi_{1+} \\ \varphi_{10} \\ \varphi_{2+} \\ \rho_{20} \end{bmatrix} \tag{17.10}$$

with conjugate Higgs doublet

$$\eta' = \begin{bmatrix} \varphi_{10} \\ -\varphi_{1-} \\ \varphi_{20} \\ -\rho_{2-} \end{bmatrix} \tag{17.11}$$

The Higgs sector lagrangian has the form:

$$\mathcal{L}_{EW}{}^{Higgs} = (\partial\eta^{\dagger}/\partial X^{\mu})(\partial\eta/\partial X^{\mu}) - \lambda(\eta^{\dagger}\eta - \rho^{2})^{2} + \mathcal{L}_{EW}{}^{Higgs}{}_{FermionMasses} \tag{17.12}$$

where the symmetry breaking follows from the choice of unitary gauge (similar in form to eq. 16.37)

$$\rho = \begin{bmatrix} 0 \\ \rho_{1} \\ 0 \\ \rho_{2} \end{bmatrix} \qquad \rho' = \begin{bmatrix} \rho_{1} \\ 0 \\ \rho_{2} \\ 0 \end{bmatrix} \tag{17.13}$$

where ρ is a real field quadruplet.

The lagrangian density mass term for the four normal fermion species is

$$\mathcal{L}_{EW}{}^{Higgs}{}_{FermionMasses} = \Sigma_{\alpha,\beta}\ \{\bar{\psi}_{kL\alpha}\eta m_{k\alpha\beta}\ \psi_{kR\beta} + \bar{\psi}_{kL\alpha}\ \eta' m'_{k\alpha\beta}\psi_{kR\beta}\} + \text{c.c.} \tag{17.14}$$

where $m_{k\alpha\beta}$ and $m'_{k\alpha\beta}$ are complex constant matrices, where α, β = 1, … , 4, and where the second term is the double conjugation doublet used to produce a total mass term invariant under weak hypercharge. The total fermion lagrangian mass terms are

$$\mathcal{L}^{Higgs}{}_{FermionMasses} = \mathcal{L}_{U}{}^{UHiggs}{}_{FermionMasses} + \mathcal{L}_{EW}{}^{Higgs}{}_{FermionMasses} \tag{16.48}$$

plus the iota mass which is negligible except perhaps for neutrinos. $\mathcal{L}_{EW}{}^{Higgs}{}_{FermionMasses}$ is the contribution of ElectroWeak Higgs Mechanism to the fermion masses. Using the vacuum expectation value of η in eq. 17.13 we find

$$\mathcal{L}_{EW}{}^{Higgs}{}_{FermionMasses} = \Sigma_{\alpha,\beta}\, \{\bar{\psi}_{2L\alpha}\, \rho_1 m_{2\alpha\beta}\psi_{2R\beta} + \bar{\psi}_{4L\alpha}\, \rho_2 m_{4\alpha\beta}\psi_{4R\beta} + + \psi_{1L\alpha}\, \rho_1 m'_{1\alpha\beta}\psi_{1R\beta} + \psi_{3L\alpha}\, \rho_2 m'_{3\alpha\beta}\psi_{3R\beta}\} + \text{c.c.} \tag{17.15}$$

giving mass terms for all four species. **There is an implicit color summation over the color quarks in each generation and quark species.** The values of ρ_1 and ρ_2 are unlikely to be the same as those appearing in the Generation group mass terms of chapter 16.

The four mass matrices m_1 , … , m_4 are all complex, constant mass matrices. They can be brought to diagonal form D_k with non-negative values by U(4) matrices A_k and B_k:

$$A_k m_k B_k^{-1} = D_k \tag{17.16}$$

or

$$m_k = A_k^{-1} D_k B_k \tag{17.17}$$

for k= 1, …, 4.

We now note, that although, D_k has non-negative real values, down-type quarks are all tachyonic and up-type quarks are all non-tachyonic, and neutral leptons are all tachyonic and charged leptons are all Dirac non-tachyonic leptons, due to their lagrangian kinetic terms as seen in chapter 5.

We further note that $m_k^\dagger m_k$ is hermitean, and A_k and B_k are members of U(4) as is D_k for k = 1, 2, 3, 4, with the result that all matrices m_k are members of the U(4) group.

We can use these U(4) transformations A_k and B_k to define the sixteen "physical" fermion fields:

$$\begin{aligned}
&\psi_{2L\alpha}\, \rho_1 m_{2\alpha\beta}\psi_{2R\beta} + \psi_{4L\alpha}\, \rho_2 m_{4\alpha\beta}\psi_{4R\beta} + \psi_{1L\alpha}\, \rho_1 m_{1\alpha\beta}\psi_{1R\beta} + \psi_{3L\alpha}\, \rho_2 m_{3\alpha\beta}\psi_{3R\beta} \\
&= (\psi_{2L}A_2^{-1})_\alpha \rho_1 D_{2\alpha\beta}(B_2\psi_{2R})_\beta + (\psi_{4L}A_4^{-1})_\alpha\, \rho_2 D_{4\alpha\beta}(B_4\psi_{4R})_\beta + \\
&\quad + (\psi_{1L}A_1^{-1})_\alpha\, \rho_1 D_{1\alpha\beta}(B_1\psi_{1R})_\beta + (\psi_{3L}A_3^{-1})_\alpha \rho_2 D_{3\alpha\beta}(B_3\psi_{3R})_\beta \\
&= \psi_{2Lphys\alpha}\rho_1 D_{2\alpha\beta}\psi_{2Rphys\beta} + \psi_{4Lphs\alpha}\rho_2 D_{4\alpha\beta}\psi_{4Rphys\beta} + \psi_{1Lphys\alpha}\rho_1 D_{1\alpha\beta}\psi_{1Rphys\beta} + \psi_{3Lphs\alpha}\rho_2 D_{3\alpha\beta}\psi_{3Rphys\beta}
\end{aligned}$$

Species: Up-type quarks down-type quarks charged leptons neutral leptons

(17.18)

The preceding discussion with changes in the values of constants and constant matrices holds for Dark Matter also where the Dark quarks and leptons acquire ElectroWeak mass terms. The Dark Matter species ElectroWeak mass terms, with the subscript D signifying Dark Matter, are[180]

$$\psi_{D2Lphys\alpha}\rho_{D1}D_{D2\alpha\beta}\psi_{D2Rphys\beta} + \psi_{D4Lphs\alpha}\rho_{D2}D_{D4\alpha\beta}\psi_{D4Rphys\beta} + \psi_{D1Lphys\alpha}\rho_{D1}D_{D1\alpha\beta}\psi_{D1Rphys\beta} + \psi_{D3Lphs\alpha}\rho_{D2}D_{D3\alpha\beta}\psi_{D3Rphys\beta}$$

Dark (17.19)

Species: Up-type quarks down-type quarks charged leptons neutral leptons

---

[180] Dark quarks are Color SU(3) singlets.

17.3 Combined Effect of Generation Group and ElectroWeak Higgs Mechanisms on Fermion Masses

The Generation group Higgs Mechanism and the ElectroWeak Higgs Mechanism combine to give masses to the fermion species and the Dark fermion species. The combined mass terms for normal fermions is

$$\psi_{2Lphys\alpha}(\rho_{1U}D_{U2\alpha\beta} + \rho_1 D_{2\alpha\beta})\psi_{2Rphys\beta} + \psi_{4Lphs\alpha}(\rho_{2U}D_{U4\alpha\beta} + \rho_2 D_{4\alpha\beta})\psi_{4Rphys\beta} + \psi_{1Lphys\alpha}\rho_1 D_{1\alpha\beta}\psi_{1Rphys\beta} + $$
$$+ \psi_{3Lphs\alpha}\rho_2 D_{3\alpha\beta}\psi_{3Rphys\beta} \quad (17.20)$$

showing that the quark species acquire Higgs contributions from both the Generation group and ElectroWeak group symmetry breaking. A similar situation occurs for Dark Matter. Its combined mass terms have the same form as eq. 17.20 with a subscript "D" inserted to indicate it is for Dark Matter.

There is one possible source of ambiguity. The order of the rotations on the generation 4-vectors of the two quark species and the two Dark quark species to produce the 4-vectors of physical quark species due to Generation and ElectroWeak symmetry breaking appears to be an issue. For example in eq. 16.52 we see Generation symmetry breaking necessitates the rotation $\psi_{U2Rphys} = B_{U2}\psi_{U2R}$ to diagonalize the up-type quark 4-vector while in eq. 17.18 we see that ElectroWeak symmetry breaking necessitates the rotation $\psi_{EW2Rphys} = B_{EW2}\psi_{EW2R}$ to diagonalize the up-type quark 4-vector.[181]

Consequently we must choose between the order of rotations since they do not commute. Based on Ockham's Razor we choose to do Generation group rotations first followed by ElectroWeak symmetry breaking rotations because the Generation group is more primary due to its role as the origin of the generations of the various species. Thus

$$\psi_{2Rphys} = B_{EW2}B_{U2}\psi_{U2R} \quad (17.21)$$

with a corresponding expression for the Left-handed species Generation 4-vector.

This issue does not arise for lepton and Dark lepton species as their 4-vectors only undergo ElectroWeak symmetry breaking rotations.

17.4 Generalization of the ElectroWeak Higgs Mechanism for Gauge Field Masses to Include Dark Gauge Fields

In section 17.2 we introduced a double doublet (a quadruplet) to derive the symmetry breaking fermion mass spectrum of the four normal fermion families and saw that a similar derivation should be operative for Dark fermions by Ockham's Razor – it is the simplest possible approach to handling the broken SU(2)⊗U(1) symmetry that the Dark ElectroWeak sector has in the Extended Standard Model.

Now we generalize the SU(2)⊗U(1) ElectroWeak symmetry breaking via the Higgs Mechanism that gives mass to three SU(2)⊗U(1) gauge bosons to include Dark SU(2)⊗U(1) symmetry breaking via the Higgs Mechanism that will give mass to three Dark SU(2)⊗U(1) gauge bosons and also yield a massless gauge boson analogous to the electromagnetic field. This assumption is consistent both with the Leibniz Minimax Principle and Ockham's Razor having both simplicity within the context of the

[181] We introduce the subscripts "U" and "EW" to distinguish the Generation group rotation from the ElectroWeak rotation.

Extended Standard Model and achieving a maximal effect with a minimal extension of the formalism. We will therefore be constructing an SU(2)⊗U(1)⊗SU(2)⊗U(1) Higgs Mechanism symmetry breaking derivation using the double doublets of section 17.2.

We assume that a "double" doublet Higgs field exists with four components:

$$\eta = \begin{bmatrix} \varphi_{1+} \\ \varphi_{10} \\ \varphi_{2+} \\ \rho_{20} \end{bmatrix} \tag{17.22}$$

with conjugate Higgs doublet

$$\eta' = \begin{bmatrix} \varphi_{10} \\ -\varphi_{1-} \\ \varphi_{20} \\ -\rho_{2-} \end{bmatrix} \tag{17.23}$$

The Higgs sector lagrangian has the form:

$$\mathcal{L}_{EW}^{Higgs} = (\partial\eta^{\dagger}/\partial X^{\mu})(\partial\eta/\partial X^{\mu}) - \lambda(\eta^{\dagger}\eta - \rho^2)^2 + \mathcal{L}_{EW}^{Higgs}{}_{EWMasses} \tag{17.24}$$

where the symmetry breaking follows from the choice of unitary gauge (similar in form to eq. 17.13)

$$\rho = \begin{bmatrix} 0 \\ \rho_1 \\ 0 \\ \rho_2 \end{bmatrix} \qquad \rho' = \begin{bmatrix} \rho_1 \\ 0 \\ \rho_2 \\ 0 \end{bmatrix} \tag{17.25}$$

where ρ is a real field quadruplet.

The covariant derivative of η in the unitary gauge is

$$D_{\ldots\mu}\,\eta = \begin{bmatrix} \{\partial/\partial X^{\mu} + \ldots + ig\mathbf{t}\cdot\mathbf{W}_{\mu} + + ig't_0W_{0\mu}\} & 0 \\ 0 & \{\partial/\partial X^{\mu} + \ldots + ig_D\mathbf{t}\cdot\mathbf{W}_{D\mu} + + ig_D't_0W_{D0\mu}\} \end{bmatrix} \begin{bmatrix} 0 \\ \rho_1 \\ 0 \\ \rho_2 \end{bmatrix} \tag{17.26}$$

where g and g' are coupling constants with g_D and g_D' their Dark equivalents, t_0.is a ½ the identity matrix, the **t** matrices are ½ the vector of Pauli matrices, and the zeros and derivative terms are all 2×2 submatrices.The ellipses indicate additional indices and additional terms respectively.

Then

$$D_{\dots\mu}\eta = \begin{bmatrix} \frac{1}{2}ig\rho_1(W_{1\mu} - iW_{2\mu}) \\ \partial\rho_1/\partial X^\mu - \frac{1}{2}ig\rho_1(\cos\theta_W)^{-1}Z_\mu \\ \frac{1}{2}ig_D\rho_2(W_{D1\mu} - iW_{D2\mu}) \\ \partial\rho_2/\partial X^\mu - \frac{1}{2}ig_D\rho_2(\cos\theta_{WD})^{-1}Z_{D\mu} \end{bmatrix} \tag{17.27}$$

where θ_{WD} is the Dark Weinberg angle and

$$\begin{aligned} W_3{}^\mu &= Z^\mu\cos\theta_W + A^\mu\sin\theta_W \\ W_0{}^\mu &= -Z^\mu \sin\theta_W + A^\mu\cos\theta_W \\ W_{D3}{}^\mu &= Z_D{}^\mu\cos\theta_{WD} + A_D{}^\mu\sin\theta_{WD} \\ W_{D0}{}^\mu &= -Z_D{}^\mu \sin\theta_{WD} + A_D{}^\mu\cos\theta_{WD} \end{aligned} \tag{17.28}$$

From eq. 17.27 we find the corresponding Higgs field kinetic terms in the lagrangian are

$$\begin{aligned} (D_{\dots\mu}\eta)^\dagger D_{\dots}{}^\mu\eta &= \partial\rho_1/\partial X^\mu\partial\rho_1/\partial X_\mu + g^2\rho_1{}^2[W_1{}^\mu W_{1\mu} + W_2{}^\mu W_{2\mu}]/4 + g^2\rho_1{}^2\, Z^\mu Z_\mu/(2\cos\theta_W)^2 + \\ &+ \partial\rho_2/\partial X^\mu\partial\rho_2/\partial X_\mu + g_D{}^2\rho_2{}^2[W_{D1}{}^\mu\, W_{D1\mu} + W_{D2}{}^\mu W_{D2\mu}]/4 + g_D{}^2\rho_2{}^2\, Z_D{}^\mu Z_{D\mu}/(2\cos\theta_{WD})^2 \end{aligned} \tag{17.29}$$

with the $W_1{}^\mu$ and $W_2{}^\mu$ and Z^μ gauge bosons acquiring masses and the electromagnetic field A^μ massless, and with the Dark $W_{D1}{}^\mu$ and $W_{D2}{}^\mu$ and $Z_D{}^\mu$ gauge bosons acquiring masses and the Dark electromagnetic field $A_D{}^\mu$ massless.

Thus the SU(2)⊗U(1)⊗SU(2)⊗U(1) Higgs Mechanism symmetry breaking yields the desired results. Chapter 14 discusses Dark Matter Chemistry. Its basis is consistent with a massless Dark electromagnetic field.

17.5 The Reality Group Compared to the Generations Group U(4)

It appears to be a somewhat remarkable coincidence that the Reality group, which maps complex coordinates to real-valued coordinates, consists of the tensor product of U(4) subgroups, and the generations group consists of (broken) U(4). However the "fourness" of the groups can be traced back to the dimension of space-time, and the consequent four boosts that generate the four families of fermions: charged leptons, neutral leptons, up-type quarks and down-type quarks and the corresponding four Dark fermion families, which in turn yield four conserved particle numbers B, L, B_D, and L_D that can undergo U(4) transformations and yield conserved linear combinations.

Thus the dimensionality of complex space-time ultimately is the origin of the "fourness" of the Reality group and the Generations group.

17.6 Completion of Construction/Derivation of the Form of The Extended Standard Model and Quantum Gravity - Unification

This concludes our construction/derivation of the form of the Extended Standard Model. Its lies primarily in the assumption of a complex-valued space-time that is mapped to the physical reality of real space-time coordinates by the Reality group, to multi-generation fermions by the Generation group, and to a Two-Tier quantum field thery formulation that eliminates infinities in calculations. The basis of the form of The Extended Standard Model in complex space-time is clear. The basis of General Relativity also is space-time geometry.

The geometrical origin of both, and the finiteness of the theory of The Extended Standard Model and Quantum Gravity (sections 18-6 – 18.8 below), shows that we have achieved a fundamental unified theory of Nature.

Our construction/derivation does not lead to the determination of constants that appear in The Extended Standard Model: the many coupling constants and masses; and does not determine the value of the gravitational constsnt G.. These issues remain to be resoved.

Appendix 17-A. Higgs Mechanism for Tachyons

The Higgs mechanism is currently the favored mechanism for spontaneous symmetry breaking and to give masses to fermions and bosons. The nature of the free tachyon lagrangian terms,

$$\mathcal{L}_{\text{free}} = \psi_T^{\dagger} i\gamma^0\gamma^5(\gamma^\mu\partial/\partial x^\mu + m_0)\psi_T(x) \tag{17-A.1}$$

where m_0 is a possible bare mass, requires a Higgs sector that contributes to a tachyon mass through spontaneous symmetry breaking having the general form:

$$\mathcal{L}_{\text{Higgs}} = ½\partial\phi/\partial x^\mu\partial\phi/\partial x_\mu - \psi_T^{\dagger} i\gamma^0\gamma^5\psi_T\phi - V(\phi) \tag{17-A.2}$$

where

$$V(\phi) = g^2(\phi^2 - (\delta m)^2)^2 \tag{17-A.3}$$

Note the quadratic term in ϕ – the "mass" term has the negative sign of a tachyon – again showing that tachyons are a feature of modern physics. In the present case the quartic term "stabilizes" the tachyon field which then can be shifted to the minimum of the potential.

The spontaneous symmetry breaking resulting from the potential $V(\phi)$ causes the mass of the tachyon to change to

$$m = m_0 \pm \delta m \tag{17-A.4}$$

A choice of vacuum state corresponding to the positive sign in eq. 17-A.4 causes the tachyon mass to increase. A choice of the vacuum state corresponding to the negative sign causes the tachyon mass to decrease, and could cause m to become negative. However this event would not make the tachyon into a normal particle. Rather it would essentially transform the tachyon into its antiparticle.

END OF AMENDED AND EXPANDED CHAPTERS 7 – 11 EXTRACT FROM BLAHA (2007b)

18. Two-Tier Quantum Field Theory and Renormalization

In chapter 13 and in section 16.1 we outlined our approach to eliminating the ultra-violet divergences that would appear in the Extended Standard Model. Previously the divergences in ElectroWeak theory were handled by methods developed by t'Hooft and collaborators with contributions by other theoretical physicists. These methods are not sufficient for the Extended Standard Model and for Quantum Gravity.

The Two-Tier Quantum Field Theory that we presented initially in Blaha (2002) eliminates all divergences in perturbation theory calculations in the Extended Standard Model and Quantum Gravity, and in the extension of our theories to 16-dimensional space that will be presented in the subsequent chapters. In this chapter we will consider major features of Two-Tier Quantum Field Theories using material abstracted from Blaha (2005a) leaving the interested reader to pursue additional details in that book.

Two-Tier Quantum Field Theory is based on our new paradigm in the Calculus of Variations that we present in Appendix 18-A.

18.1 Perturbation Theory Divergences

In chapter 4 we suggested that complex coordinates and the complex Loreatz group are the true space-time of our universe (plus some small curvature due to General Relativistic effects.) Subsequently we showed that the form of The Standard Model, embedded in a larger Extended Standard Model, follows directly from the complexity of space-time upon the introduction of the Reality group.

We now address an issue in the Extended Standard Model – divergences that appear when perturbation theory Feynman diagram calculations are undertaken. The elimination of these divergences in certain sectors of The Standard Model was a major achievement. But in the Extended Standard Model divergencs remain in the Standard Model secor as well as the extended sector. In addition the unification of the (Extended) Standard Model with quantized General Relativity is prevented by the divergences that appear in Quantum General Relativity despite attempts to get around them through novel approaches.

In the spirit of Ockham's Razor we proposed a new approach to eliminating all divergences[182] in all quantum field theories including Quantum Gravity. We called this approach Two-Tier Quantum Field Theory. Chapter 13 and section 16.1 discussed applications of this approach. In this chapter we present a detailed discussion which is a slightly amended version of parts of Blaha (2005) and Blaha (2002).

The essence of Two-Tier Quantum Field Theory is to take the real coordinate systems that result after the application of Reality group transformations to the underlying complex coordinate system and to introduce an addition to the real coordinates of a q-number gauge field similar to Quantum Electrodynamics as an imaginary part of the now "slightly" complex q-number coordinates. With this

---

[182] The notorious divergence in the fermion triangle diagram is eliminated by our approach eliminating the Adler-Bell-Jackiw anomaly.

addition, and the use of our new paradigm for the Calculus of Variations, we can develop finite quantum field theories for the Extended Standard Model and Quantum Gravity (using the Einstein or other lagrangian).

Turning to the larger question of a unified theory of the Extended Standard Model and Quantum Gravity we note that 1) our extended Standard Model is firmly grounded in space-time geometry just like Quantum Gravity, and 2) we eliminate divergences in both these through the Two-Tier formalism. Thus we fulfill the criteria of Ockham's Razor by providing the "simplest" solution to the divergence problems that plagued both theories; and Leibniz's Minimax Principle by providing a minimal change in quantum field theory that gives a maximal benefit – finite, divergence-free theories when in perturbative calculations with the low energy behavior of theories well-approximated by "normal" perturbation theory.[183]

18.2 Quantization of Coordinate Systems

18.2.1 Non-commuting Coordinates

Field theories with non-commuting coordinates are an active field of study.[184] Investigators are studying gauge theories, and in particular Quantum Electrodynamics, with non-commuting coordinates. Non-commuting coordinates are usually implemented quantum mechanically by positing non-zero commutators for coordinates:

$$[x^i, x^j] = i\theta^{ij} \qquad (18.2.1)$$

18.2.2 New Approach to Non-Commuting Coordinates

In this book we will consider an alternative approach that postulates a q-number coordinate system X^μ with which all particle fields are defined. This coordinate system is realized as a mapping from a more fundamental c-number coordinate system y^ν, which we will call the subspace for want of a better term. We will treat X^μ as a vector of quantum fields, thus realizing a new type of non-commutative coordinates at unequal subspace times.

This approach is radically different from the non-commutative coordinate realizations hitherto discussed in the literature. It has a number of beneficial results to recommend it – the main result is the finiteness of quantum field theories that are defined within its framework. We will explore some of these results in the following chapters.

The X^μ coordinate system, as we define it, has a c-number real part and a q-number imaginary part. Thus particle fields which are normally defined on four-dimensional real space-time will now be defined on a complex four-dimensional space-time where four imaginary dimensions will appear as *Quantum Dimensions*™ embodied in a vector quantum field $Y^\mu(y)$.

$$X_\mu(y) = y_\mu + i\, Y_\mu(y)/M_c^2$$

---

[183] Thus the magnificent higher order calculations of T. Kinoshita and others remain correct in Two-Tier QED.

[184] M. R. Douglas and N. A. Nekrasov, Rev. Mod. Phys. **73**, 977 (2002) and references therein; J. Harvey, hep-th/0102076; M. Hamanaka and K. Toda, hep-th/0211148; N. Seiberg and E. Witten, hep-th/9908142; R. J. Szabo, hep-th/0109162; G. Berrino, S. L. Cacciatori, A. Celi, L. Martucci, and A. Vicini, hep-th/0210171; S. Godfrey and A. Doncheski, DESY eprint 02-195; M. Caravati, A. Devoto, and W. W. Repko, hep-th/0211463; and references within these papers.

The $Y^\mu(y)$ field is a function of the subspace y coordinates. The real part of the space-time dimensions will be taken to be the subspace y coordinates.[185]

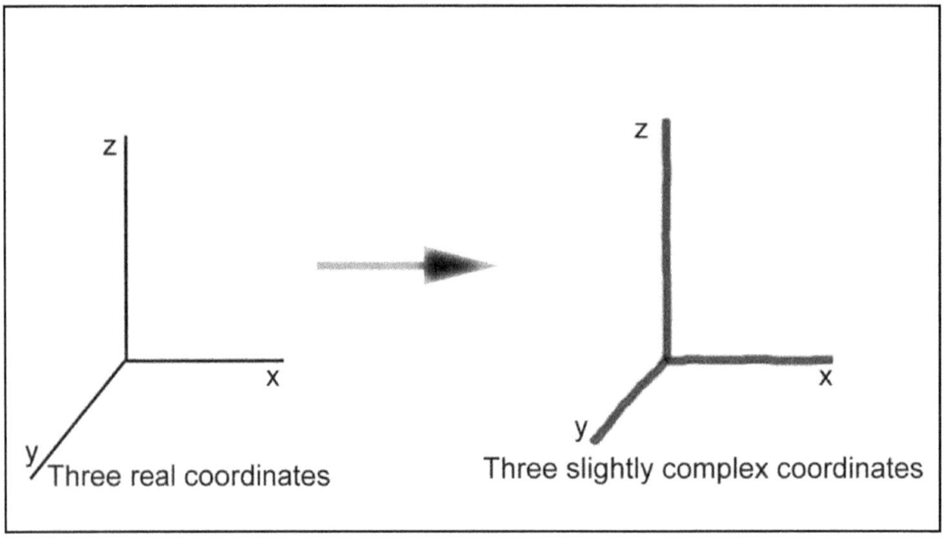

Figure 18.2.1. The change from purely real space to a slightly complex space with imaginary quantum fluctuations for each spatial axis in the Coulomb gauge of the Y field.

The imaginary part of space-time will simply be the quantum fluctuations of a massless vector quantum field that are suppressed further by a large mass scale – perhaps of the order of the Planck mass – that reduces the imaginary Quantum Dimensions™ to the infinitesimal. The effects of Quantum Dimensions only become appreciable in quantum field theory at energies of the order of M_c. At these energies the exponential Gaussian factor in each particle (and ghost) propagator that is generated by the Quantum Dimensions serves to make perturbation theory calculations ultra-violet finite – including calculations in Quantum Gravity.

The formalism that we will describe introduces a new form of interaction that does not have the form of the simple polynomial interactions that have hitherto dominated quantum field theories. This form of interaction takes place via the composition of quantum fields and can be called a *Dimensional Interaction* or an *Interdimensional Interaction* since it affects particle behavior through Quantum Dimensions.

18.2.3 Quantization Using a C-Number X^μ

We will begin by considering the case of a scalar quantum field theory. We assume a real underlying y subspace. Since X^μ is a set of coordinates, we choose to define a scalar field ϕ as a function

[185] In a deeper theory the real part might also be a quantum field that undergoes a condensation to generate c-number coordinates. We will not consider this possibility in this book.

of X^μ, which in turn is a function of the y^ν coordinates. We will provisionally second quantize ϕ treating X^μ as c-number coordinates using a conventional approach.[186]

We assume a Lagrangian, with the momentum conjugate to ϕ:

$$\pi_\phi = \partial L_F / \partial\phi' \equiv \partial L_F / \partial(\partial\phi/\partial X^0) \tag{18.2.2}$$

Following the canonical quantization procedure, π and ϕ become hermitian operators with equal time ($X^0 = X^{0\prime}$) commutation rules:

$$[\phi(X), \phi(X')] = [\pi_\phi(X), \pi_\phi(X')] = 0 \tag{18.2.3}$$

$$[\pi_\phi(X), \phi(X')] = -i\,\delta^3(\mathbf{X} - \mathbf{X'}) \tag{18.2.4}$$

The hamiltonian is defined by eq. 18-A.112. (Appendix 18-A contains a detailed development of the formalism for the scalar particle case. It was placed there because there are many formal similarities to conventional quantum field and this approach allows us to proceed more quickly to the main points of difference between conventional quantum field theory and Two-Tier quantum field theory in the present chapter. Appendix 18-A also describes a new type of method – the composition of extrema – for the Calculus of Variations. *Equations numbered 18-A.xxx are in Appendix 18-A.*) We assume a metric $\eta_{\mu\nu}$ where $\eta_{00} = +1$, $\eta_{0i} = 0$, and $\eta_{ij} = -1$ for i, j = 1,2,3.

The standard Fourier expansion of the solution to the Klein-Gordon equation (eq. 18-A.34) is:

$$\phi(X) = \int d^3p\, N_m(p)\, [a(p)\, e^{-ip\cdot X} + a^\dagger(p)\, e^{ip\cdot X}] \tag{18.2.5}$$

where

$$N_m(p) = [(2\pi)^3 2\omega_p]^{-1/2} \tag{18.2.6}$$

and

$$\omega_p = (\mathbf{p}^2 + m^2)^{1/2} \tag{18.2.7}$$

The commutation relations of the Fourier coefficient operators are:

$$[a(p), a^\dagger(p')] = \delta^3(\mathbf{p} - \mathbf{p'}) \tag{18.2.8}$$

$$[a^\dagger(p), a^\dagger(p')] = [a(p), a(p')] = 0 \tag{18.2.9}$$

The reader will recognize the quantization procedure is formally identical to the standard canonical quantization procedure of a free scalar quantum field.

In the case of spin ½, spin 1 and spin 2 fields the standard quantization procedure *in terms of the X coordinate system* can also be followed in a way similar to the procedure in standard texts. We will see

---

[186] Some texts are: Bogoliubov, N. N., Shirkov, D. V., *Introduction to the Theory of Quantized Fields* (Wiley-Interscience Publishers Inc., New York, 1959); Bjorken, J. D., Drell, S. D., *Relativistic Quantum Fields* (McGraw-Hill, New York, 1965); Huang, K., *Quarks, Leptons & Gauge Fields Second Edition* (World Scientific, River Edge, NJ, 1992); Kaku, M., *Quantum Field Theory* (Oxford University Press, New York, 1993); Weinberg, S., *The Quantum Theory of Fields* (Cambridge University Press, New York, 1995).

these quantization procedures in the following chapters. In the next section we will quantize the transformation from the y coordinate system to the X coordinate system.

The procedures developed in this section and the following sections may disturb some readers since we are placing operators with Dirac delta functions and using other unusual operator expressions. These concerns should be put at rest when we show that a path integral formulation presented later gives precisely the same results as the present development.

18.2.4 Coordinate Quantization

In this section we will quantize the coordinates X^μ as a vector field defined on a fundamental c-number coordinate system y^ν of the same dimensionality. We will assume the y^ν space is a "normal" flat Minkowski space with three spatial and one time dimensions. Generalizations to spaces with more dimensions are straightforward but will not be considered here.

Thus we will assume X^μ has three spatial dimensions and one time dimension. For reasons primarily of simplicity (primarily to avoid multiple time coordinates) we will assume the X^μ fields are similar to the free electromagnetic vector potential A^μ with the Lagrangian:

$$\mathscr{L}_C = +\tfrac{1}{4}\, M_c^{\,4} F^{\mu\nu} F_{\mu\nu} \qquad (18.2.10)$$

$$F_{\mu\nu} = \partial X_\mu/\partial y^\nu - \partial X_\nu/\partial y^\mu \qquad (18.2.11)$$

where $M_c^{\,4}$ is a mass scale to the fourth power that is required on dimensional grounds and serves to set the scale for new Physics as we will see later. *Note the sign in eq. 18.2.10 is not negative – superficially contrary to the conventional electromagnetic Lagrangian. The reason for this difference is that the field part of X^μ is imaginary.* Thus L_C winds up having the correct sign after taking account of the factors of i in the field strength $F_{\mu\nu}$.

We assume X^μ is complex[187] with the form:

$$X_\mu(y) = y_\mu + i\, Y_\mu(y)/M_c^{\,2} \qquad (18.2.12)$$

where $Y_\mu(y)$ is a quantum field, M_c is a mass scale, and the real part is the c-number 4-vector y_μ. If X^μ has this form, then

$$F_{\mu\nu} = i\,(\partial Y_\mu/\partial y^\nu - \partial Y_\nu/\partial y^\mu)/M_c^{\,2} \qquad (18.2.13)$$

Defining

$$F_{Y\mu\nu} = (\partial Y_\mu/\partial y^\nu - \partial Y_\nu/\partial y^\mu) \qquad (18.2.14)$$

---

[187] Theories of quantum mechanics, and quantum fields, in complex and quaternion spaces have been considered by numerous authors. For example see C. M. Bender, D. C. Brody and H. F. Jones, "Complex Extension of Quantum Mechanics" Phys. Rev. Letters **89**, 270401-1 (2002) and references therein; S. L. Adler and A. C. Millard, "Generalized Quantum Dynamics as Pre-Quantum Mechanics", Princeton Univ. preprint arXiv:hep-th/9508076 (1995) and references therein. These theories are all very different from the theories presented herein.

we see the Lagrangian assumes the form of the conventional electromagnetic Lagrangian:

$$L_C = -\tfrac{1}{4} F_Y{}^{\mu\nu} F_{Y\mu\nu} \tag{18.2.15}$$

This Lagrangian can be used to develop field equations and a canonical quantization that is completely analogous to Quantum Electrodynamics.

18.2.5 Gauge Invariance

The gauge invariance of the Lagrangian allows us to choose a convenient gauge. The gauge invariance of the full Lagrangian

$$\mathcal{L}_s = L_F(\phi(X), \partial\phi/\partial X^\mu)\, J + \mathcal{L}_C(X^\mu(y), \partial X^\mu(y)/\partial y^\nu) \tag{18-A.96}$$

is based on the standard gauge invariance of $\mathcal{L}_C$, and the gauge invariance of $J\mathcal{L}_F$ in the form of translational invariance

$$X^\mu(y) \rightarrow X^\mu(y) + \delta X^\mu(y) \tag{18-A.97}$$

for the special case of a translation of X with the form of a gauge transformation:

$$\delta X^\mu(y) = \partial\Lambda(y)/\partial y_\mu$$

In this case eq. 18-A.106 implies

$$\int d^4y\, \Lambda(y)\, \partial\, [\, J\, \partial/\partial X^\mu\, \mathcal{T}_{F\mu\nu}\,]/\partial y_\nu \;= 0$$

after a partial integration and so gives the differential conservation law:

$$\partial\, [\, J\, \partial\mathcal{T}_{F\mu\nu}/\partial X^\mu]/\partial y_\nu \;= 0 \tag{18.2.16}$$

since $\Lambda(y)$ is arbitrary. This conservation law is trivially obeyed since, by eq. 18-A.108:

$$\partial/\partial X^\mu\, \mathcal{T}_{F\mu\nu} \;= 0 \tag{18-A.108}$$

Thus translational invariance in the $\mathcal{L}_F$ sector together with standard gauge invariance in the $\mathcal{L}_C$ sector automatically guarantees Y field gauge invariance of the total Lagrangian. Basically we use the separate invariance of each term of

$$L = \int d^4y\, [\mathcal{L}_F J + \mathcal{L}_C] = \int d^4X\, \mathcal{L}_F + \int d^4y\, \mathcal{L}_C = L_F + L_C$$

under a constant translation $X^\mu \rightarrow X^\mu + \delta X^\mu$ where δX^μ is constant to establish eq. 18-A.108. Then we consider a position dependent translation/gauge transformation to derive eq. 18.2.16, which taken together with eq. 18-A.108, establishes the invariance under the position dependent translation/gauge transformation eq. 18-A.97.

An alternate approach that leads to the same result is to start with the particle part of the Lagrangian $\mathscr{L}_F$ rewritten to be invariant under general coordinate transformations as it must when we generalize to include General Relativity. Since position dependent translations are a form of general coordinate transformation the full theory must be invariant under position dependent translations due to invariance under general coordinate transformations.

Having established invariance under gauge transformations we now choose to use the most convenient gauge – the Coulomb gauge[188]:

$$\partial Y^i/\partial y^i = 0 \tag{18.2.17a}$$

which, in the absence of external sources, allows us to set

$$Y^0 = 0 \tag{18.2.17b}$$

since Y^0 does not have a canonically conjugate momentum. A conventional treatment leads to the equal time commutation relations:

$$[Y^\mu(\mathbf{y}, y^0), Y^\nu(\mathbf{y}', y^0)] = [\pi^\mu(\mathbf{y}, y^0), \pi^\nu(\mathbf{y}', y^0)] = 0 \tag{18.2.18}$$

$$[\pi^j(\mathbf{y}, y^0), Y_k(\mathbf{y}', y^0)] = -i\,\delta^{tr}{}_{jk}(\mathbf{y} - \mathbf{y}') \tag{18.2.19}$$

(Note the locations of the j indexes in eq. 18.2.19 introduce a minus sign.) where

$$\pi^k = \partial \mathscr{L}_C \,/\, \partial Y_k' \tag{18.2.20}$$

$$\pi^0 = 0 \tag{18.2.21}$$

$$\delta^{tr}{}_{jk}(\mathbf{y} - \mathbf{y}') = \int d^3k\, e^{i\,\mathbf{k}\cdot(\mathbf{y} - \mathbf{y}')}(\delta_{jk} - k_j k_k/\mathbf{k}^2)/(2\pi)^3 \tag{18.2.22}$$

$$Y_k' = \partial Y_k/\partial y^0 \tag{18.2.23}$$

The Coulomb gauge reveals the two degrees of freedom that are present in the vector potential. The Fourier expansion of the vector potential is:

---

[188] It is also possible to quantize using an indefinite metric that preserves manifest Lorentz covariance as was done by Gupta and Bleuler for the electromagnetic field. We will use the Gupta-Bleuler approach later to establish covariance under special relativity later. Now we opt for manifest positivity and use the Coulomb gauge.

$$Y^i(y) = \int d^3k\, N_0(k) \sum_{\lambda=1}^{2} \varepsilon^i(k, \lambda)[a(k,\lambda)\, e^{-ik\cdot y} + a^\dagger(k,\lambda)\, e^{ik\cdot y}] \tag{18.2.24}$$

where

$$N_0(k) = [(2\pi)^3 2\omega_k]^{-\frac{1}{2}} \tag{18.2.25}$$

and (since m = 0)

$$\omega_k = (\mathbf{k}^2)^{\frac{1}{2}} = k^0 \tag{18.2.26}$$

with $\vec{\epsilon}(k, \lambda)$ being the polarization unit vectors for $\lambda = 1,2$ and $k^\mu k_\mu = 0$.

The commutation relations of the Fourier coefficient operators are:

$$[a(k,\lambda), a^\dagger(k',\lambda')] = \delta_{\lambda\lambda'}\delta^3(\mathbf{k} - \mathbf{k'}) \tag{18.2.27}$$
$$[a^\dagger(k,\lambda), a^\dagger(k',\lambda')] = [a(k,\lambda), a(k',\lambda')] = 0 \tag{18.2.28}$$

and the polarization vectors satisfy

$$\sum_{\lambda=1}^{2} \varepsilon_i(k, \lambda)\varepsilon_j(k, \lambda) = (\delta_{ij} - k_i k_j/\mathbf{k}^2) \tag{18.2.29}$$

It will be convenient to divide the Y field into positive and negative frequency parts:

$$Y^+_i(y) = \int d^3k\, N_0(k) \sum_{\lambda=1}^{2} \varepsilon_i(k, \lambda)\, a(k,\lambda)\, e^{-ik\cdot y} \tag{18.2.30}$$

and

$$Y^-_i(y) = \int d^3k\, N_0(k) \sum_{\lambda=1}^{2} \varepsilon_i(k, \lambda)\, a^\dagger(k,\lambda)\, e^{ik\cdot y} \tag{18.2.31}$$

For later use we note the commutator between the positive and negative frequency parts is:

$$[\, Y^-_j(y_1), Y^+_k(y_2)] = -\int d^3k\, e^{ik\cdot(y_1 - y_2)}\, (\delta_{jk} - k_j k_k/\mathbf{k}^2)/[(2\pi)^3 2\omega_k] \tag{18.2.32}$$

18.2.6 Bare ϕ Particle States

We now turn to the ϕ particle states. The creation and annihilation operators can be used to define "bare" free particle states. Bare free particle states are states that are not dressed with coherent states of Y quanta. For example a bare one-particle state of momentum p is

$$|p\rangle = a^\dagger(p)|0_\phi\rangle \tag{18.2.33}$$

with corresponding bare bra state

$$\langle p| = \langle 0_\phi|a(p) \tag{18.2.34}$$

where the vacuum is defined as usual:

$$a(p)|0_\phi> = 0 \quad (18.2.35)$$

$$<0_\phi|a^\dagger(p) = 0 \quad (18.2.36)$$

Multi-particle bare states can also be defined in the conventional way with products of creation and annihilation operators applied to the vacuum.

18.2.7 Y Fock Space Imaginary Coordinate States

States can also be defines for the quantized Y field. These states will be similar in form to electromagnetic photon states but play a different role in our approach since they are in fact coordinate excitation states for the imaginary part of X^μ. Thus the scalar field (and other particle fields) will exist in a real four-dimensional space with quantum excitations into imaginary Quantum Dimensions. These excitations become significant at high energies. At the low energies with which we are familiar, space-time appears real; at very high energies space-time becomes slightly complex.

There are two types of imaginary coordinate excitations: 1.) Quantum excitations into Fock states consisting of superpositions of states with a definite finite number of Y "particles" and 2.) Imaginary coordinate excitations into coherent Y states with an "infinite" number of particles. Coherent states can be viewed as representing "classical" fields.

In this section we will consider Y field states with a definite number of excitations ("particles"). The creation and annihilation operators of the Y field can be used to define free particle states. For example a one particle state can be defined by

$$|k, \lambda> = a^\dagger(k, \lambda)|0_Y> \quad (18.2.37)$$

with corresponding bra state

$$<k, \lambda| = <0_Y|a(k, \lambda) \quad (18.2.38)$$

where the "coordinate vacuum" is defined as usual:

$$a(k, \lambda)|0_Y> = 0 \quad (18.2.39)$$

$$<0_Y|a^\dagger(k, \lambda) = 0 \quad (18.2.40)$$

Multi-particle states can also be defined in the conventional way with products of the creation and annihilation operators applied to the vacuum. The set of all states containing a finite number of "particles" constitutes a Fock space.

A state with a finite number of Y "particles" represents a quantum fluctuation into imaginary Quantum Dimensions. Such states do not appear in Two-Tier quantum field theory since the Y field is a free field and has no source. Thus they appear only as part of normal particles. A normal particle, such as a ϕ particle, has a coherent state of Y quanta associated with it, which play a role in interactions. The

Y coherent state part of a normal particle can be viewed as boring an infinitesimal "hole" into an extra pair of imaginary dimensions in a neighborhood of the particle of a radial extent set by the length M_c^{-1}.

18.2.8 Y Coherent Imaginary Coordinate States

Coherent Y states bring us closer what we might consider to be "classical" imaginary dimensions – dimensions that we can, in principle, experience as we do normal dimensions. Let us define the coherent state[189]

$$| y, p> = e^{-\mathbf{p}\cdot\mathbf{Y}^-(y)/M_c^2}|0_Y> \tag{18.2.41}$$

This state is an eigenstate of the coordinate operator $Y^+(y')$:

$$Y^+_j(y_1)\ |y_2, p> = -[Y^+_j(y_1), \mathbf{p}\cdot\mathbf{Y}^-(y_2)]/M_c^2\ |y, p> \tag{18.2.42}$$

$$= -\int d^3k\ [N_0(k)]^2\ e^{ik\cdot(y_2 - y_1)}\ (p_j - k_j\mathbf{p}\cdot\mathbf{k}/\mathbf{k}^2)/M_c^2\ |y, p> \tag{18.2.43}$$

$$= p^i\Delta_{Tij}(y_1 - y_2)/M_c^2\ |y, p> \tag{18.2.44}$$

where $p^i\Delta_{Tij}(y_1 - y_2)/M_c^2$ is the eigenvalue of $Y^+_j(y_1)$. As we will see in the next chapter, the eigenvalue of Y^+ becomes large as $(y_1 - y_2)^2 \to 0$. Thus the imaginary Quantum Dimensions become significant at very short distances, and significantly modify the high-energy behavior of quantum field theories. In particular, Quantum Dimensions have a significant effect when

$$(y_1 - y_2)^2 \lesssim (4\pi^2 M_c^2)^{-1} \tag{18.2.45}$$

according to eq. 18.3.13 in the next chapter. We are assuming the mass scale M_c is very large – perhaps of the order of the Planck mass (1.221×10^{19} GeV/c^2). Thus imaginary Quantum Dimensions™ are far from detectable in today's "low" energy experiments. Their effect are significant in the analysis of the first instants after the Big Bang.[190]

18.2.9 The Dynamical Generation of New Dimensions

Effectively, the imaginary dimensions that we have constructed raise the total number of real and Quantum Dimensions™ to 8 with 6 space dimensions and two time dimensions. As we will see later the requirement of gauge invariance for the quantized Y field reduces the number of time dimensions to one and constrains the six space dimensions to five degrees of freedom giving a 5+1 dimensional space. Since X is a function of y we can also view the four dimensional world that we live in as a four-dimensional surface in a 6-dimensional space-time.

---

[189] Coherent states are well known in the physics literature. See for example T. W. B. Kibble, J. Math. Phys. **9**, 315 (1968) and references therein; V. Chung, Phys. Rev. **140**, B1110 (1965); J. R. Klauder, J. McKenna, and E. J. Woods, J. Math. Phys. **7**, 822 (1966) and references therein.
[190] Blaha (2004).

18.2.10 Generation of Quantum Dimensions by the ϕ (X) field

The ϕ(X) field generates Quantum Dimensions via coherent states from the vacuum. From eq. 18.2.5 and 18.2.12 we see

$$\phi(X) = \int d^3p\, N_m(p)\, [a(p)\, e^{-ip\cdot(y + iY/M_c^2)} + a^\dagger(p)\, e^{ip\cdot(y + iY/M_c^2)}] \qquad (18.2.46)$$

with the result

$$\phi(X)|0\rangle = \int d^3p\, N_m(p)\, a^\dagger(p)\, e^{ip\cdot(y + iY/M_c^2)}|0\rangle \qquad (18.2.47)$$

is a superposition of coherent Y states plus one scalar particle. The vacuum state |0> is the product of the ϕ and Y vacuum states $|0\rangle = |0_Y\rangle|0_\phi\rangle$. We will use |0> in most of the following discussions.

We can also define coherent Y states with total momentum q using the expression:

$$|q\ Y\rangle = \int d^4y\, e^{iq\cdot X(y)}|0\rangle = \int d^4y\, e^{iq\cdot(y + iY/M_c^2)}|0\rangle \qquad (18.2.48)$$

Expanding the Y part of the exponential in eq. 18.2.48 gives

$$|q\ Y\rangle = \sum_{n=0}^{\infty}(-1)^n(n!)^{-1}\prod_{j=1}^{n}\left(\int d^3k_j N_0(k_j)\right)\delta^4(q - \sum_{s=1}^{n} k_s)\prod_{r=1}^{n}\sum_{\lambda_r=1}^{2} \mathbf{q}\cdot \boldsymbol{\varepsilon}(k_r, \lambda_r)\, a^\dagger(k_r,\lambda_r)|0\rangle$$

(18.2.49)

which indicates that the sum of the Y particle momenta for each term in the expansion is q.

18.2.11 Hamiltonian for Particle and Coordinate States

The hamiltonian for the separable (field hamiltonian term separate from the Y hamiltonian term – see Appendix 18-A), coordinate quantized, scalar quantum field theory is:

$$\mathcal{H}_s = \int d^3y\, \mathcal{H}_s \qquad (18\text{-A}.79)$$

with

$$\mathcal{H}_s = J\mathcal{H}_F + \mathcal{H}_C \qquad (18\text{-A}.82)$$

$$\mathcal{H}_F(\phi(X), \pi_\phi, \partial\phi/\partial X^i) = \pi_\phi\, \phi' - \mathcal{L}_F \qquad (18\text{-A}.83)$$

$$\mathcal{H}_C(X^\mu(y), \pi_X{}^\mu, \partial X^\mu(y)/\partial y^j, y^\nu) = \pi_X{}^\mu X_\mu{}' - \mathcal{L}_C \qquad (18\text{-A}.84)$$

$$\mathcal{L}_F = \tfrac{1}{2}\, [\, (\partial\phi/\partial X^\nu)^2 - m^2\phi^2\,] \qquad (18\text{-A}.33)$$

$$\mathcal{L}_C = -\tfrac{1}{4}\, M_c^4 F_Y{}^{\mu\nu} F_{Y\mu\nu} \qquad (18.2.15)$$

We note

$$\mathscr{H}_F = \tfrac{1}{2}\,[\pi_\phi^{\,2} + (\partial\phi/\partial X^i)^2 + m^2\phi^2] \qquad (18.2.50)$$

is the conventional scalar particle hamiltonian when viewed as a function of the X coordinates. $\mathscr{H}_C$ has the same form as the conventional electromagnetic hamiltonian when eq. 18.2.12 is used to specify X in terms of the Y fields.

$$\mathscr{H}_C = \tfrac{1}{2}\,(E_Y^{\,2} + B_Y^{\,2}) \qquad (18.2.51)$$

where

$$E_Y^{\,i} = -\partial Y^i/\partial y^0 \qquad (18.2.52)$$

$$B_Y^{\,i} = \varepsilon^{ijk}\,\partial Y_j/\partial y^k \qquad (18.2.53)$$

Using the fourier expansions of ϕ and X^μ (eqs. 18.2.5 and 18.2.24) we obtain the following expression for the normal-ordered hamiltonian $\mathscr{H}_S$:

$$P_s^{\,0} \equiv \mathscr{H}_s = \int :\mathscr{H}_s:\, d^3y \qquad (18.2.54)$$

$$\mathscr{H}_s = \int d^3p\,(\mathbf{p}^2 + m^2)^{1/2} a^\dagger(p)a(p) + \int d^3k \sum_{\lambda=1}^{2} (\mathbf{k}^2)^{1/2} a^\dagger(k, \lambda)a(k, \lambda) \qquad (18.2.55)$$

where : : indicates normal ordering and where we perform a functional integration over X (Note the Jacobian is present within H_s.) for the particle part of the hamiltonian H_F. The hamiltonian is manifestly positive definite.

The spatial momentum is specified by

$$P_s^{\,j} = -\int d^3X\, :\pi_\phi(X)\partial\phi(X)/\partial X_j: + \int d^3y\, :E_Y^{\,i}\partial Y^i/\partial y_j: \qquad (18.2.56)$$

$$= \int d^3p\, p^j\, a^\dagger(p)a(p) + \int d^3k \sum_{\lambda=1}^{2} k^j\, a^\dagger(k, \lambda)a(k, \lambda) \qquad (18.2.57)$$

where the first term in eq. 18.2.57 follows because of $\int d^3X$ in eq. 18.2.56. The momentum operator generates displacements in ϕ

$$[P_s^{\,\mu}, \phi(X)] = -\,i\partial\phi/\partial X_\mu \qquad (18.2.58)$$

18.2.12 Second Quantized Coordinates

At this point we have developed a formalism for a scalar particle quantum field theory based on our non-commutative coordinates. In the following sections we will proceed to use this formalism to develop a unified quantum field theory of the known forces of nature.

18.3 Scalar Two-Tier Quantum Field Theory

18.3.1 Introduction

In this section we will examine a new formulation of quantum field theory that we call *Two-Tier quantum field theory* in more detail for the case of a free scalar particle. This type of quantum field theory incorporates a structure similar to a string-like substructure within a quantum field theoretic framework.

18.3.2 "Two-Tier" Space

In the preceding section we developed quantized coordinates X^μ defined on an underlying c-number coordinates y^ν with the equations:

$$X_\mu(y) = y_\mu + iY_\mu(y)/M_c^2 \tag{18.2.12}$$

$$Y^i(y) = \int d^3k\, N_0(k) \sum_{\lambda=1}^{2} \varepsilon^i(k, \lambda)[a(k,\lambda)\, e^{-ik\cdot y} + a^\dagger(k,\lambda)\, e^{ik\cdot y}] \tag{18.2.24}$$

We also developed a free scalar quantum field theory with the Fourier expansion:

$$\phi(X) = \int d^3p\, N_m(p)\, [a(p)\, e^{-ip\cdot X} + a^\dagger(p)\, e^{ip\cdot X}] \tag{18.2.5}$$

We will now consider the implications of the separable Lagrangian:

$$\mathscr{L}_s = \mathscr{L}_F(\phi(X), \partial\phi/\partial X^\mu)\, J + \mathscr{L}_C(X^\mu(y), \partial X^\mu(y)/\partial y^\nu) \tag{18-A.96}$$

where

$$\mathscr{L}_F = \tfrac{1}{2}\, [\, (\partial\phi/\partial X^\nu)^2 - m^2\phi^2\,] \tag{18-A.33}$$

and

$$\mathscr{L}_C = -\tfrac{1}{4}\, M_c^4 F_Y^{\mu\nu} F_{Y\mu\nu} \tag{18.2.10}$$

with

$$F_{Y\mu\nu} = \partial Y_\mu/\partial y^\nu - \partial Y_\nu/\partial y^\mu \tag{18.2.14}$$

M_c is the mass that sets the scale at which the imaginary part of X^μ becomes significant.

This quantum field theory behaves as a conventional quantum field theory until energies reach the magnitude of M_c. At energies of the order of M_c, and above, the imaginary part of X^μ becomes significant and alters the high-energy behavior of the theory in a major way. This modification leads to the elimination of divergences that normally appear in perturbation theory when interactions are introduced. Yet the low energy behavior of the theory remains the same remains the same as conventional scalar quantum field theory. Thus the precise calculations of QED that have been verified to an amazing degree of accuracy remain valid when a Two-Tier formulation of QED is created. And the "low energy" results found in other conventional quantum field theories such as Electroweak Theory and the Standard Model also are closely approximated by their corresponding Two-Tier versions.

The straightforward use of the above equations[191] (and the canonical quantization described in the preceding chapters) leads to a scalar quantum field with the Fourier expansion:

$$\phi(X) = \int d^3p\, N_m(p)\, [a(p)e^{-ip\cdot(y + iY/M_c^2)} + a^\dagger(p)e^{ip\cdot(y + iY/M_c^2)}] \qquad (18.3.1)$$

using eq. 18.2.5 above. We note the equal time commutation relations of ϕ and π_ϕ are the same as the conventional equal time commutation relations of a scalar field despite the fact that X^μ and Y^μ are themselves quantum fields since $[Y^\mu(\mathbf{y}, y^0), Y^\nu(\mathbf{y}', y^0)] = 0$ for $\mathbf{y} \neq \mathbf{y}'$. In addition, we note the ϕ and π_ϕ fields are not hermitean.

The Fourier expansion of ϕ does require one refinement – the exponential terms in X^μ must be *normal ordered* to avoid infinities in the unequal time commutation relations:

$$\phi(X) = \int d^3p\, N_m(p)\, [a(p) :e^{-ip\cdot(y + iY/M_c^2)}: + a^\dagger(p) :e^{ip\cdot(y + iY/M_c^2)}:] \qquad (18.3.2)$$

Since the hamiltonian as well as other quantities are normal ordered in quantum field theory the additional requirement of normal ordering in the field operator is merely an extension of a standard procedure to a more complex situation and is not disturbing. The unequal time commutation relation of the normal ordered ϕ field is:

$$[\phi(X^\mu(y_1)), \phi(X^\mu(y_2))] = i\Delta(y_1 - y_2) + O(1/M_c^2) \qquad (18.3.3)$$

where

$$\Delta(y_1 - y_2) = -i \int d^3k\, (e^{-ik\cdot(y_1 - y_2)} - e^{ik\cdot(y_1 - y_2)})/[(2\pi)^3 2\omega_k] \qquad (18.3.4)$$

is a familiar c-number invariant singular function. The additional terms in eq. 18.3.3 are q-number terms that become significant at very short distances of the order M_c^{-1}. Thus precise measurements of field strengths at larger distances are limited by standard quantum effects as indicated by the commutation relation.

The principle of *microscopic causality* is violated at extremely short distances of the order M_c^{-1} since the commutator (eq. 18.3.3) is non-zero, in general, for space-like distances of the order of M_c^{-1} due to the q-number terms. This violation is not experimentally measurable now – and for the foreseeable future – and reflects a type of non-locality at extremely short distances.

We will see that the short distance behavior of Two-Tier quantum field theory leads to the elimination of divergences resulting in finite interacting quantum field theories.

18.3.3 Vacuum Fluctuations

While the expectation value of a *conventional* free scalar field $\phi_{conv}(X)$ is zero in a conventional quantum field theory:

---

[191] The use of functionals in quantum field theory is, of course, far from new as one can see in texts such as Bogoliubov (1959) (see for example pp. 198-226).

$$<0|\phi_{conv}(X)|0> = 0 \tag{18.3.5}$$

the vacuum fluctuations of *conventional* scalar quantum field theory are quadratically divergent:

$$<0|\phi_{conv}(X)\phi_{conv}(X)|0> = \int d^3p/[(2\pi)^3 2\omega_p] \tag{18.3.6}$$

In "Two-Tier" quantum field theory we find the vacuum expectation value of a free field is zero (like eq. 18.3.5) *and the expectation value of the square of the field is also zero:*

$$<0|\phi(X)\phi(X)|0> = \int d^3p\, e^{-p^i p^j \Delta_{Tij}(0)/Mc^4}/[(2\pi)^3 2\omega_p] = 0 \tag{18.3.7}$$

since the exponential factor in the integral is $-\infty$. The exponent contains

$$\Delta_{Tij}(z) = \int d^3k\, e^{-ik\cdot z}\,(\delta_{ij} - k_i k_j/\mathbf{k}^2)/[(2\pi)^3 2\omega_k] \tag{18.3.8}$$

where "T"is for "Two-Tier". Thus *vacuum fluctuations are zero in Two-Tier quantum field theory.* Correspondingly, we will see that renormalization constants are finite in the Two-Tier versions of QED, Electroweak Theory, the Standard Model and Quantum Gravity.

18.3.4 The Feynman Propagator

The Feynman propagator for a Two-Tier free scalar quantum field is:

$$i\Delta_F^{TT}(y_1 - y_2) = <0|T(\phi(X(y_1)),\phi(X(y_2)))|0> \tag{18.3.9}$$

$$\equiv <0|\phi(X(y_1))\phi(X(y_2))|0>\, \theta(y_1^0 - y_2^0) + \phi(X(y_2))\phi(X(y_1))|0>\, \theta(y_2^0 - y_1^0) \tag{18.3.10}$$

Since $X^0 = y^0$ in the Coulomb gauge of the X^μ field there is no ambiguity in the choice of the relevant time variable. A straightforward calculation shows:

$$i\Delta_F^{TT}(y_1 - y_2) = i\int d^4p\, e^{-ip\cdot(y_1 - y_2)}\, R(\mathbf{p}, y_1 - y_2)/[(2\pi)^4(p^2 - m^2 + i\varepsilon)] \tag{18.3.11}$$

where

$$R(\mathbf{p}, y_1 - y_2) = \exp[-p^i p^j \Delta_{Tij}(y_1 - y_2)/M_c^4] \tag{18.3.12}$$

$$= \exp\{-p^2[A(v) + B(v)\cos^2\theta] / [4\pi^2 M_c^4 z^2]\} \tag{18.3.13}$$

with

$$z^\mu = y_1^\mu - y_2^\mu \tag{18.3.14}$$

$$z = |\mathbf{z}| = |\mathbf{y}_1 - \mathbf{y}_2| \tag{18.3.15}$$

$$p = |\mathbf{p}| \tag{18.3.16}$$

$$v = |z^0|/z \tag{18.3.17}$$

$$A(v) = (1 - v^2)^{-1} + .5v \ln[(v - 1)/(v + 1)] \tag{18.3.18}$$
$$B(v) = v^2(1 - v^2)^{-1} - 1.5v \ln[(v - 1)/(v + 1)] \tag{18.3.19}$$

$$\mathbf{p{\cdot}z} = pz \cos\theta \tag{18.3.20}$$

and $|\mathbf{p}|$ denoting the length of a spatial vector **p** while $|z^0|$ is the absolute value of z^0.

As eq. 18.3.11 indicates, the Gaussian damping factor R(p, z) for large momentum p is the same for both the positive and negative frequency parts of the Two-Tier Feynman propagator. It is also important to note that R(p, z) does not depend on p^0 (in the Y Coulomb gauge) and thus the integration over p^0 proceeds in the usual way to produce time-ordered positive and negative frequency parts.

18.3.5 Large Distance Behavior of Two-Tier Theories

The large distance behavior of the Two-Tier Feynman propagator approaches the behavior of the conventional Feynman propagator since

$$R(\mathbf{p}, y_1 - y_2) \to 1 \tag{18.3.21}$$

when $(y_1 - y_2)^2$ becomes much larger than M_c^{-2} as eq. 18.3.13 shows. Thus the behavior of a conventional quantum field theory naturally emerges at large distance. We will see that the conventional Standard Model is the large distance limit of the Two-Tier Standard Model thus *realizing a form of Correspondence Principle for Quantum Field Theory.* Some features of the conventional Standard Model that depend specifically on the existence of divergences, such as the axial anomaly, will be different in the Two-Tier Standard Model since it is a divergence-free theory.

18.3.6 Short Distance Behavior of Two-Tier Theories

At short distances the Gaussian factor dominates and radically changes the behavior of the Feynman propagator eliminating its short distance singular behavior, and thus paving the way to finite quantum field theories. Near the light cone, $M_c^{-2} \gg -(y_1 - y_2)^2 \to 0$, we can approximate eq. 18.3.11 with

$$i\Delta_F^{TT}(y_1 - y_2) \approx \int d^3p\, [N(p)]^2 R(\mathbf{p}, y_1 - y_2) \tag{18.3.22}$$

since $e^{-ip\cdot(y_1 - y_2)}$ is approximately unity for small $(y_1 - y_2)$. We assume the mass of the ϕ particle is zero or is negligible at high energies so we set m = 0 to study the high energy behavior of eq. 18.3.22. Upon performing the integrations in eq. 18.3.22 for space-like $(y_1 - y_2)^2$ (and analytically continuing to the time-like regions[192,193]) we find

$$i\Delta_F^{TT}(y_1 - y_2) \approx [z^2M_c^4/(4i\sqrt{A}\sqrt{B})] \ln[(\sqrt{A} + i\sqrt{B})/(\sqrt{A} - i\sqrt{B})] \tag{18.3.23}$$

---

[192] See S. Blaha, "Relativistic Bound State Models with Quasi-Free Constituent Motion", Phys. Rev. **D12**, 3921 (1975) and references therein.

[193] It should be noted that A and B in eq. 4.23 have the same sign for $0 \le v < 1.1243$ thus making for easy analytic continuation across the light cone (which corresponds to v = 1 in eqs. 4.18 and 4.19).

with A and B defined in eqs. 18.3.18 and 18.3.19. As $(y_1 - y_2)^2 \rightarrow 0$ from the space-like or time-like side of the light cone we find eq. 18.3.23 becomes:

$$i\Delta_F^{TT}(y_1 - y_2) \rightarrow \pi M_c^4 |(y_1 - y_2)^\mu (y_1 - y_2)_\mu|/8 \qquad (18.3.24)$$

Eq. 18.3.24 has several noteworthy points:

1. The propagator is well behaved on the light cone and approaches zero smoothly from both space-like and time-like directions. In contrast, the conventional scalar Feynman propagator diverges as $[(y_1 - y_2)^\mu (y_1 - y_2)_\mu]^{-2}$. This good behavior near the light cone will be seen later for other particle propagators with the net result that the usual infinities found in conventional quantum field theory are absent in Two-Tier quantum field theories.

2. The quadratic form of the propagator in eq. 18.3.24 is suggestive of attempts to formulate a relativistic harmonic oscillator model of elementary particles[194] and more recent attempts to achieve quark confinement. The fact that the absolute value of the quadratic term appears in eq. 18.3.24 neatly avoids the common pitfall seen in fully relativistic harmonic oscillator attempts.

3. The quadratic behavior *in coordinate space* of the propagator at short distances is equivalent to a high-energy behavior of

$$p^{-6} \qquad (18.3.25)$$

in momentum space. Thus we get the equivalent *of a higher derivative theory* in Two-Tier quantum field theory at high energies while retaining a positive definite energy spectrum. The problems of negative metric states that have plagued conventional higher derivative quantum field theories are avoided.[195]

18.3.7 String-like Substructure of the Theory

Imaginary Quantum Dimensions endow a particle with an extended structure that resembles to some extent the extended structure seen in bosonic string and Superstring theories. For example, Bailin (1994) use the operator[196]

$$V_\Lambda(k) = \int d^2\sigma \sqrt{-h}\, W_\Lambda(\tau, \sigma)\, e^{-ik\cdot X} \qquad (18.3.26)$$

where X^μ is a quantized fourier expansion of the string fields (see eq. 7.22 of Bailin (1994)).

We note our X^μ coordinate-field has two transverse degrees of freedom due to gauge invariance, which also invites comparison to the bosonic string. A point of difference is that we will create a well-

---

[194] H. Yukawa, H., Phys. Rev. **91**, 416 (1953); Y. S. Kim and M. E. Noz, Phys. Rev. **D8**, 3521 (1973) and references therein.

[195] S. Blaha, Phys.Rev. **D10**, 4268 (1974); S. Blaha, Phys.Rev. **D11**, 2921 (1975); S. Blaha, Nuovo Cim. **A49**, :113 (1979); S. Blaha, "Generalization of Weyl's Unified Theory to Encompass a Non-Abelian Internal Symmetry Group" SLAC-PUB-1799, Aug 1976; S. Blaha, "Quantum Gravity and Quark Confinement" Lett. Nuovo Cim. **18**, 60 (1977); Nakanishi, N., Suppl. Prog. Theo. Phys. **51**, 1 (1972); and references therein.

[196] D. Bailin and A. Love, *Supersymmetric Gauge Field Theory and String Theory* (Institute of Physics Publishing, Philadelphia, PA, 1994) page 272.

defined quantum field theoretic formulation in conventional space-time that has the Standard Model as its "large distance" behavior thus introducing a note of reality that is not (yet?) very apparent in Superstring theories. We see that the interacting quantum field theories based on this approach also have good, finite, short distance behavior just as string theories.

The scalar, and other particles', Feynman propagators can be viewed as describing the propagation of a particle cloaked (accompanied) by a cloud of Y particles (which generates the $R(\mathbf{p}, y_1 - y_2)$ factor in the propagator of eq. 18.3.11). If we examine the fourier transform of R(p, z) we see:

$$(2\pi)^4 R(\mathbf{p}, q) = \int d^4z\, e^{iq\cdot z}\, R(\mathbf{p}, z) = \int d^4z\, e^{iq\cdot z} \exp[-p^i p^j \Delta_{Tij}(z)/M_c^4] \qquad (18.3.27)$$

and we find

$$R(\mathbf{p},q) = \sum_{n=0}^{\infty} [i(2\pi M_c)^4]^{-n} (n!)^{-1} \prod_{j=1}^{n} [\int d^4k_j\, \theta(k_j^0)(\mathbf{p}^2 - (\mathbf{p}\cdot\mathbf{k}_j)^2/\mathbf{k}_j^2)/(k_j^2 + i\varepsilon)]\, \delta^4(q - \Sigma\, k_r) \qquad (18.3.28)$$

which can be interpreted as a "cloud" of Y particles dressing the "bare" particle propagator. (The manifest divergences in eq. 18.3.28 for R(p, q) are an artifact of the expansion and the subsequent fourier transformation. They are not present in the $R(\mathbf{p}, y_1 - y_2)$ factor in the propagator of eq. 18.3.11.) See Fig. 18.3.1 for the Feynman diagram of the Two-Tier cloaked propagator as compared to the normal scalar particle Feynman propagator. The Two-Tier Feynman propagator is basically a conventional scalar propagator that is modified by coherent Y particle emission.[197]

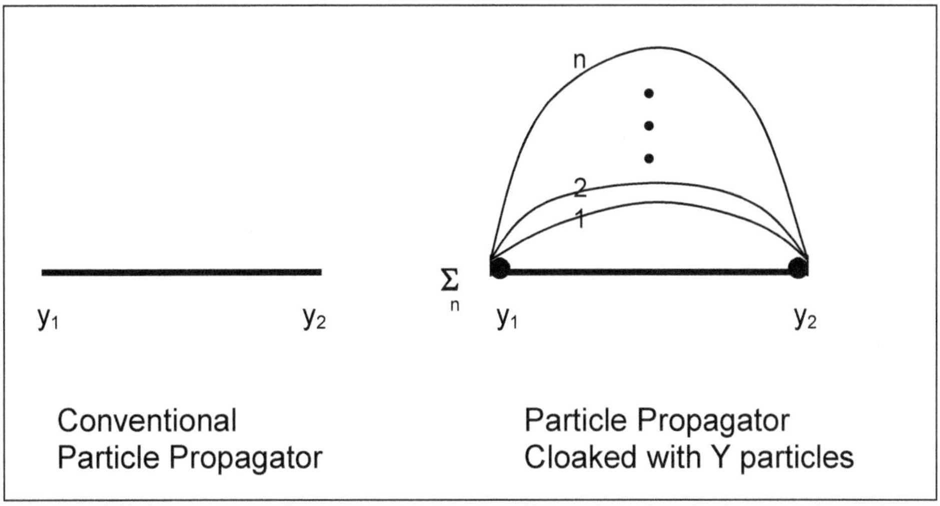

Figure 18.3.1. Feynman diagram for conventional and cloaked Two-Tier propagators.

We note that R(p, q) satisfies the convolution theorem:

[197] T. W. B. Kibble, Phys. Rev. **173**, 1527 (1968) and references therein. In particular see p. 1532 of Kibble's paper.

$$\int d^4k\ R(\mathbf{p}, k)\ R(\mathbf{p}, q - k) = [R(\mathbf{p}, q)]^2 \quad (18.3.29a)$$

or

$$(2\pi)^4 \int d^4z\ e^{iq\cdot z}\ R(\mathbf{p}, z)\ R(\mathbf{p}, z) = [\ \int d^4z\ e^{iq\cdot z}\ R(\mathbf{p}, z)\ ]^2 \quad (18.3.29b)$$

The proof follows from eq. 18.3.28 and the Binomial theorem.

18.3.8 Parity

Parity can appear in two guises within the framework of Two-Tier quantum field theory. One can consider a parity operation where the space parts of X^μ are reversed while y^μ is unchanged. Or one can consider a second type of parity where the space parts of y^μ are reversed.

18.3.9 X Parity

Under this form of parity operation y^μ is unchanged while the arguments of ϕ *appear* to change by

$$X^i(y) \rightarrow -X^i(y) \quad (18.3.30)$$

$$X^0(y) \rightarrow X^0(y) \quad (18.3.31)$$

We will denote the parity operator of this type P_X. Under P_X the arguments of the scalar quantum field operator ϕ change according to eqs. 18.3.30-1 so that ϕ transforms as

$$\mathscr{P}_X\phi(\mathbf{X}(y), X^0(y))\mathscr{P}_X^{-1} = \phi(-\mathbf{X}(y), X^0(y)) \quad (18.3.32)$$

From the form of ϕ in eq. 4.2 we see we can implement eq. 18.3.32 by requiring:

$$\mathscr{P}_X a(\mathbf{p}, p^0)\mathscr{P}_X^{-1} = a(-\mathbf{p}, p^0) \quad (18.3.33)$$

$$\mathscr{P}_X a^\dagger(\mathbf{p}, p^0)\mathscr{P}_X^{-1} = a^\dagger(-\mathbf{p}, p^0) \quad (18.3.34)$$

$$\mathscr{P}_X X^0(y)\mathscr{P}_X^{-1} = X^0(y) \quad (18.3.35)$$

$$\mathscr{P}_X X^i(y)\mathscr{P}_X^{-1} = X^i(y) \quad (18.3.36)$$

$$\mathscr{P}_X Y^i(y)\mathscr{P}_X^{-1} = Y^i(y) \quad (18.3.37)$$

where i = 1,2,3.

This parity transformation is analogous to the standard form of parity transformation in conventional quantum field theory. The separable Lagrangian in eq. 18-A.96 (and listed at the beginning of this chapter) is invariant under this parity transformation.

18.3.10 y Parity

This form of parity transformation in which $y^i \rightarrow -y^i$ has significant differences from the normal parity transformation. We specify this parity transformation for a scalar quantum field by:

$$\mathscr{P}_y \phi(\mathbf{X}(\mathbf{y}, y^0), X^0(\mathbf{y}, y^0))\mathscr{P}_y^{-1} = \phi(\mathbf{X}(-\mathbf{y}, y^0), X^0(-\mathbf{y}, y^0)) \tag{18.3.38}$$

This transformation can be implemented through the following set of transformations:

$$\mathscr{P}_y a(\mathbf{p}, p^0)\mathscr{P}_y^{-1} = a(-\mathbf{p}, p^0) \tag{18.3.39}$$

$$\mathscr{P}_y a^\dagger(\mathbf{p}, p^0)\mathscr{P}_y^{-1} = a^\dagger(-\mathbf{p}, p^0) \tag{18.3.40}$$

$$\mathscr{P}_y X^0(\mathbf{y}, y^0)\mathscr{P}_y^{-1} = X^0(-\mathbf{y}, y^0) \tag{18.3.41}$$

$$\mathscr{P}_y Y^i(\mathbf{y}, y^0)\mathscr{P}_y^{-1} = -Y^i(-\mathbf{y}, y^0) \tag{18.3.42a}$$

$$\mathscr{P}_y a(\mathbf{k}, k^0,1)\mathscr{P}_y^{-1} = a(-\mathbf{k}, k^0,1) \tag{18.3.42b}$$

$$\mathscr{P}_y a(\mathbf{k}, k^0,2)\mathscr{P}_y^{-1} = -a(-\mathbf{k}, k^0,2) \tag{18.3.42c}$$

where i = 1,2,3 and assuming: $\boldsymbol{\varepsilon}(\mathbf{k},k^0,1)=-\boldsymbol{\varepsilon}(-\mathbf{k},k^0,1)$ and $\boldsymbol{\varepsilon}(\mathbf{k},k^0,2)=+\boldsymbol{\varepsilon}(-\mathbf{k},k^0,2)$.

18.3.11 Forms of the Parity Transformations

The parity transformations for a scalar particle are

$$\mathscr{P}_X = \exp\{-i\pi\int d^3p\, [a^\dagger(\mathbf{p}, p^0)a(\mathbf{p}, p^0) - a^\dagger(\mathbf{p}, p^0)a(-\mathbf{p}, p^0)]/2\} \tag{18.3.43a}$$

$$\mathscr{P}_y = \mathscr{P}_X \exp\{-i\pi \int d^3k\, [\sum_{\lambda=1}^{2} a^\dagger(\mathbf{k},k^0,\lambda)a(\mathbf{k},k^0,\lambda) - a^\dagger(\mathbf{k}, k^0, 1)a(-\mathbf{k},k^0,1) + $$
$$+ a^\dagger(\mathbf{k}, k^0, 2)a(-\mathbf{k},k^0,2)]/2\} \tag{18.3.43b}$$

The separable Lagrangian of eq. 18-A.96 is invariant under these parity transformations.

18.3.12 Charge Conjugation

Charge conjugation is implemented in a way similar to that of conventional quantum field theory. In particular

$$\mathscr{C}\, X^{\mu}(\mathbf{y}, y^0)\mathscr{C}^{-1} = X^{\mu}(\mathbf{y}, y^0) \tag{18.3.44}$$

18.3.13 Time Reversal

Since $X^0 = y^0$ in the Y Coulomb gauge in Two-Tier quantum theory the only non-trivial form of time reversal transformation $\mathscr{T}$ is based on $y^0 = -y^0$. This time reversal transformation is similar in part to to the conventional time reversal transformation in conventional quantum field theory. Therefore we will define $\mathscr{T}$ as the product of the operation of taking the complex conjugate of all c-numbers times a unitary operator $\mathscr{U}_y$. Under $\mathscr{T}$ a scalar quantum field operator ϕ transforms as

$$\mathscr{T}\phi(\mathbf{X}(\mathbf{y}, y^0), X^0(\mathbf{y}, y^0))\mathscr{T}^{-1} = \phi(\mathbf{X}(\mathbf{y}, -y^0), X^0(\mathbf{y}, -y^0)) \tag{18.3.45}$$

From the form of in ϕ eq. 4.2 we see that

$$\mathscr{T}a(\mathbf{p}, p^0)\mathscr{T}^{-1} = a(-\mathbf{p}, p^0) \tag{18.3.46}$$

$$\mathscr{T}a^{\dagger}(\mathbf{p}, p^0)\mathscr{T}^{-1} = a^{\dagger}(-\mathbf{p}, p^0) \tag{18.3.47}$$

$$\mathscr{T}X^{i}(\mathbf{y}, y^0)\mathscr{T}^{-1} = X^{i}(\mathbf{y}, -y^0) \tag{18.3.48}$$

$$\mathscr{T}Y^{i}(\mathbf{y}, y^0)\mathscr{T}^{-1} = -Y^{i}(\mathbf{y}, -y^0) \tag{18.3.49a}$$

$$\mathscr{T}a(\mathbf{k}, k^0,1)\mathscr{T}^{-1} = a(-\mathbf{k}, k^0,1) \tag{18.3.49b}$$

$$\mathscr{T}a(\mathbf{k}, k^0,2)\mathscr{T}^{-1} = -a(-\mathbf{k}, k^0,2) \tag{18.3.49c}$$

where i = 1,2,3 and assuming: $\boldsymbol{\varepsilon}(\mathbf{k},k^0,1)=-\boldsymbol{\varepsilon}(-\mathbf{k},k^0,1)$ and $\boldsymbol{\varepsilon}(\mathbf{k},k^0,2)=+\boldsymbol{\varepsilon}(-\mathbf{k},k^0,2)$.

The unitary operator U_y is given by

$$\mathscr{U}_X = \exp\{-i\pi\int d^3p\, [a^{\dagger}(\mathbf{p}, p^0)a(\mathbf{p}, p^0) - a^{\dagger}(\mathbf{p}, p^0)a(-\mathbf{p}, p^0)]/2\} \tag{18.3.50a}$$

and

$$\mathscr{U}_y = \mathscr{U}_X \exp\{-i\pi \int d^3k\, [\sum_{\lambda=1}^{2} a^\dagger(\mathbf{k},k^0,\lambda)a(\mathbf{k},k^0,\lambda) - a^\dagger(\mathbf{k}, k^0, 1)a(-\mathbf{k},k^0,1) + $$
$$+ a^\dagger(\mathbf{k}, k^0, 2)a(-\mathbf{k},k^0,2)]/2\} \qquad (18.3.50b)$$

The separable Klein-Gordon Lagrangian (eq. 18-A.96) is invariant under our definition of time reversal.

We note

$$\mathscr{U}_y = \mathscr{P}_y \qquad (18.3.50c)$$

Although the present theory is somewhat more complicated than conventional quantum field theory the overall nature of the $\mathscr{P}$, $\mathscr{C}$, and $\mathscr{T}$ transformations is the same.

18.4 Interacting Quantum Field Theory – Perturbation Theory

18.4.1 Introduction

The form of quantum field theory that we have developed in sections 18.2 and 18.3 can be used as the basis for new formulations of QED, Electroweak Theory, the Standard Model and a divergence-free, unified theory of all the known interactions. The development of these theories requires a number of topics be addressed. This section covers perturbation theory. As much as possible, we attempt to retain the features of the standard approach so that the reader will more readily follow the discussion and more readily accept this new formalism. In physics originality is secondary to reality. The perturbation theory that we will develop will be shown to be identical to the perturbation theory that we develop later using a path integral formalism.

18.4.2 An Auxiliary Asymptotic Field

The definition of the asymptotic "free" in and out states is an issue in Two-Tier quantum field theory because the "free particle field" of the theory $\phi(X(y))$ is a "dressed" particle, ab initio, since it is cloaked in a cloud of Y particles as discussed in the passage following eq. 18.3.27.

While one could use $\phi(X(y))$ directly to define in and out asymptotic states it is more convenient initially to introduce a "fictitious" auxiliary asymptotic quantum field $\Phi(y)$ that will represent the equally fictitious "bare ϕ particle" in and out states.

We will consider the case of a scalar field. We define a free, scalar Klein-Gordon particle field with the physical mass m of the physical $\phi(X(y))$ particle.

$$\Phi(y) = \int d^3p\, N_m(p)\, [a(p)\, e^{-ip\cdot y} + a^\dagger(p)\, e^{ip\cdot y}] \qquad (18.4.1)$$

using the creation and annihilation operators of $\phi(X(y))$ (in eq. 18.3.2). The set of particle states of $\Phi(y)$ has the familiar Fock space form

$$|\, p_1, p_2, \ldots\, p_n\rangle = a^\dagger(p_1,)a^\dagger(p_2) \ldots a^\dagger(p_n)|0\rangle \qquad (18.4.2)$$

with powers of creation operators allowed since Φ particles are bosons. The set of particle states constitutes a complete orthonormal set of states. The corresponding bra states are defined by hermitean conjugation:

$$\langle p_1, p_2, \ldots p_n| = (| p_1, p_2, \ldots p_n\rangle)^\dagger \tag{18.4.3}$$

We note that the energy spectrum of these states is positive definite with the hamiltonian

$$H_\Phi = P_\Phi{}^0 = \int d^3y \, \tfrac{1}{2}[\pi_\Phi{}^2 + (\partial\Phi/\partial X^i)^2 + m^2\Phi^2] \tag{18.4.4a}$$

$$= \int d^3p \, (\mathbf{p}^2 + m^2)^{1/2} a^\dagger(p)a(p) \tag{18.4.4b}$$

and momentum vector:

$$\mathbf{P}_\Phi = \int d^3p \, \mathbf{p} \, a^\dagger(p)a(p) \tag{18.4.5}$$

We will use this set of energy-momentum eigenstates to define asymptotic "in" and "out" states in perturbation theory.

18.4.3 Transformation Between Φ(y) and ϕ(X(y))

For later use in the definition of the perturbation theory expansion, we will determine the transformations between the in and out Φ(y) fields, and the in and out ϕ(X(y)) fields. Let us define a transformation $W_a(y)$ that transforms in and out Φ(y) fields to in and out ϕ(X(y)) fields respectively:

$$\phi_a(X(y)) = :W_a(y)\Phi_a(y)W_a{}^{-1}(y): \tag{18.4.6}$$

where the label a = "in" or a = "out", where : … : signifies normal ordering, and where

$$\Phi_{in}(y) = \int d^3p \, N_m(p) \, [a_{in}(p) \, e^{-ip\cdot y} + a_{in}{}^\dagger(p) \, e^{ip\cdot y}] \tag{18.4.7}$$

$$\Phi_{out}(y) = \int d^3p \, N_m(p) \, [a_{out}(p) \, e^{-ip\cdot y} + a_{out}{}^\dagger(p) \, e^{ip\cdot y}] \tag{18.4.8}$$

$$\phi_{in}(X) = \int d^3p \, N_m(p) \, [a_{in}(p) :e^{-ip\cdot(y + iY/M_c^2)}: + a_{in}{}^\dagger(p) :e^{ip\cdot(y + iY/M_c^2)}:] \tag{18.4.9}$$

$$\phi_{out}(X) = \int d^3p \, N_m(p) \, [a_{out}(p) :e^{-ip\cdot(y + iY/M_c^2)}: + a_{out}{}^\dagger(p) :e^{ip\cdot(y + iY/M_c^2)}:] \tag{18.4.10}$$

Note that the transformation eq. 18.4.6 includes normal ordering. While this transformation may seem strange it is no stranger than the time reversal operator, in which the complex conjugate of all c-number terms is taken in addition to applying a unitary transformation.

In the Coulomb gauge of Y it is easy to show that

$$W_a(y) = \exp(-\mathbf{Y}(y)\cdot\mathbf{P}_{\Phi a}/M_c{}^2) \tag{18.4.11}$$

and

$$W_a^{-1}(y) = \exp(\mathbf{Y}(y)\cdot\mathbf{P}_{\Phi a}/M_c^2) \quad (18.4.12)$$

where the label a = "in" or a = "out", where the inner products in the exponentials are the usual spatial vector inner product, and where

$$\mathbf{P}_{\Phi a} = -\int d^3y\ \partial\Phi_a(y)/\partial y^0\ \nabla\Phi_a(y) = \int d^3p\ \mathbf{p}\ a_a^\dagger(p)a_a(p) \quad (18.4.12a)$$

is a spatial vector (the Φ spatial momentum operator) that is written solely in terms of $\Phi_a(y)$'s creation and annihilation operators.

In addition to performing the transformation in eq. (18.4.6 $W_a(y)$ also performs a "translation" in Y^μ:

$$W_a(y)Y^i(y')W_a^{-1}(y) = Y^i(y') + i\Delta^{trij}(y' - y)P_{\Phi a}^j/M_c^2 \quad (18.4.13a)$$

where

$$i\Delta^{trij}(y' - y) = \int d^3k\ (e^{-ik\cdot(y' - y)} - e^{ik\cdot(y' - y)})(\delta_{jk} - k_jk_k/\mathbf{k}^2)/[(2\pi)^3 2\omega_k] \quad (18.4.13b)$$

We note that $W_a(y)$ is not a unitary operator but it is pseudo-unitary:

$$W_a^{-1}(y) = V\ W_a^\dagger(y)\ V^{-1} = V\ W_a(y)\ V^{-1} \quad (18.4.14)$$

where

$$V = \exp(-i\pi \sum_{\lambda=1}^{2}\int d^3k\ a^\dagger(k, \lambda)a(k, \lambda)) \quad (18.4.15)$$

is a unitary operator with the property

$$V\ Y^j(y)\ V^{-1} = -Y^j(y) \quad (18.4.16)$$

for j = 1,2,3. We note

$$V^\dagger = V^{-1} = V \quad (18.4.17)$$

and thus

$$V^2 = I \quad (18.4.18)$$

V will be shown to be a metric operator in the following discussion.[198] We note that the Y "particle" (hermitean) number operator appears in eq. 18.4.9 in the expression for V:

$$N_Y = \sum_{\lambda=1}^{2}\int d^3k\ a^\dagger(k, \lambda)a(k, \lambda) \quad (18.4.19)$$

Thus states with an even number of Y "particles" have a V eigenvalue of one, and states with an odd number of Y "particles" have a V eigenvalue of minus one.

---

[198] P. A. M. Dirac, Proc. R. Soc. London A **180**, 1 (1942); T. D. Lee and G. C. Wick, Nucl. Phys. **B9**, 209 (1969); C. M. Bender, D. C. Brody and H. F. Jones, "Complex Extension of Quantum Mechanics" Phys. Rev. Letters **89**, 270401-1 (2002) and references therein.

18.4.4 Model Lagrangian with ϕ^4 Interaction

We will develop our perturbation theory using a scalar Lagrangian model with a ϕ^4 interaction term:

$$\mathcal{L}_s = \text{JL}_F + \text{L}_C \tag{18.4.20}$$

with

$$\mathcal{L}_F = \tfrac{1}{2}\,[\,(\partial\phi/\partial X^\nu)^2 - m^2\phi^2] + \mathcal{L}_{Fint} \tag{18.4.21}$$

and

$$\mathcal{L}_C = -\tfrac{1}{4}\, F_Y{}^{\mu\nu} F_{Y\mu\nu} \tag{18.4.22}$$

with

$$F_{Y\mu\nu} = \partial Y_\mu/\partial y^\nu - \partial Y_\nu/\partial y^\mu \tag{18.4.23}$$

and

$$\mathcal{L}_{Fint} = \tfrac{1}{4!}\, \chi_0\, \phi(X(y))^4 + \tfrac{1}{2}\,(m^2 - m_0{}^2)\phi^2 \tag{18.4.24}$$

where J is the Jacobian (as in Appendix 18-A), χ_0 is the bare coupling constant, and m_0 is the bare mass.

The conserved momentum operator is:

$$P_{F\beta} = \int d^3X\; T_{F0\beta} \tag{18.4.25}$$

where

$$\mathcal{T}_{F\mu\nu} = -\, g_{\mu\nu} L_F + \partial L_F / \partial(\partial\phi/\partial X_\mu)\; \partial\phi/\partial X^\nu \tag{18.4.26}$$

is the ϕ field energy-momentum tensor with conservation law (eq. 18-A.110):

$$\partial P_{F\beta}/\partial X^0 = 0 \tag{18.4.27}$$

due to eq. 18-A.108.

The hamiltonian density (eq. 18-A.83) is

$$\mathcal{H}_F = \mathcal{T}_{F0\beta} = \mathcal{H}_{F0} + \mathcal{H}_{Fint} \tag{18.4.28}$$

with

$$\mathcal{H}_{F0} = \tfrac{1}{2}\,[\pi_\phi{}^2 + (\partial\phi/\partial X^i)^2 + m^2\phi^2] \tag{18.4.29}$$

$$\mathcal{H}_{Fint} = -\,\tfrac{1}{4!}\,\chi_0\, \phi(X(y))^4 + \tfrac{1}{2}\,(m^2 - m_0{}^2)\phi(X(y))^2 \tag{18.4.30}$$

18.4.5 In-states and Out-States

In this section we will develop properties of in-fields and out-fields. We will use a somewhat more complicated procedure to set up the perturbation theory for the S matrix due to the introduction of imaginary coordinates. The procedure can be schematized as:

$$\Phi_{in}(y) \Rightarrow \phi_{in}(X(y)) \Rightarrow \phi(X(y)) \Rightarrow \phi_{out}(X(y)) \Rightarrow \Phi_{out}(y) \tag{18.4.31}$$

In-states are constructed using the auxiliary field Φ_{in} which are then effectively transformed into $\phi_{in}(X(y))$ expressions in order to make contact with our Lagrangian formalism. Then $\phi_{in}(X(y))$ is related to the interacting field $\phi(X(y))$ as a limit ($y^0 \rightarrow -\infty$). Similarly out-states are constructed using the auxiliary field Φ_{out} which are then expressed in terms of $\phi_{out}(X(y))$. Then $\phi_{out}(X(y))$ is related to the interacting field $\phi(X(y))$ using the LSZ limiting process ($y^0 \rightarrow +\infty$).

Since much of the development differs only trivially from the standard treatment in textbooks we will simply "list" relevant equations and let the reader pursue them further in quantum field theory textbooks.

18.4.6 ϕ In-Field

In order to define a perturbation theory for particle scattering we will next specify features of the in-field $\phi_{in}(X(y))$ and in-field states – the field and states representing physical particles as $X^0 = y^0 \rightarrow -\infty$.

A. The in-field $\phi_{in}(X(y))$ satisfies the Klein-Gordon equation in the X variable:

$$(\Box_X + m^2)\, \phi_{in}(X) = 0 \tag{18.4.32}$$

where

$$\Box_X = (\partial/\partial X^\nu)(\partial/\partial X_\nu)$$

B. Under coordinate displacements and Lorentz transformations $\Phi_{in}(y)$, $\phi_{in}(X(y))$, and $\phi(X(y))$ transform in the same way:

$$[P^\mu, \Phi_{in}(y)] = -\, i\partial\Phi_{in}/\partial y_\mu \tag{18.4.33a}$$
$$[P^\mu, \phi_{in}(X)] = -\, i\partial\phi_{in}/\partial y_\mu \tag{18.4.33b}$$

$$[P^\mu, \phi(X)] = -\, i\partial\phi/\partial y_\mu \tag{18.4.34}$$

with the energy-momentum vector P^μ specified by eq. 18-A.57.

C. We can relate the asymptotic in-field $\phi_{in}(X(y))$ to the interacting field $\phi(X(y))$ using the equation of motion of $\phi(X(y))$

$$(\Box_X + m^2)\, \phi(X) = j(X) \tag{18.4.35}$$

where j(X) embodies the interaction. Using the physical mass m we find

$$(\Box_X + m^2)\, \phi(X) = j(X) + (m^2 - m_0{}^2)\phi(X) = j_{tot}(X) \tag{18.4.36}$$

If the current is taken to be the source of the scattered waves we may write

$$\sqrt{Z}\ \phi_{in}(X(y)) = \phi(X(y)) - \int d^4X(y')\ \Delta_{ret}(y - y')\ j_{tot}(X(y')) \quad (18.4.37)$$

$$= \phi(X(y)) - \int d^4y'\ J\ \Delta_{ret}(y - y')\ j_{tot}(X(y')) \quad (18.4.38)$$

where Z is a wave function renormalization constant, J is the Jacobian, and Δ_{ret} is a retarded Green's function.

D. We can define Φ_{in} in-field states with expressions like

$$|\ p_1, p_2, \ldots\ p_n\ \text{in}\rangle = a_{in}^{\dagger}(p_1)a_{in}^{\dagger}(p_2)\ \ldots\ a_{in}^{\dagger}(p_n)|0\rangle \quad (18.4.39)$$

with powers of creation operators allowed since Φ_{in} is a boson field. The set of all particle states constitutes a complete orthonormal set of states. The corresponding bra states are defined by hermitean conjugation:

$$\langle p_1, p_2, \ldots\ p_n\ \text{in}| = (|\ p_1, p_2, \ldots\ p_n\ \text{in}\rangle)^{\dagger} \quad (18.4.40)$$

18.4.7 ϕ Out-Field

In order to define a perturbation theory for particle scattering we begin by listing aspects of the out-field $\phi_{out}(X(y))$ and out-field states – the field and states representing physical particles as $X^0 = y^0 \to -\infty$.

A. The out-field $\phi_{out}(X(y))$ satisfies the Klein-Gordon equation in the X variable:

$$(\Box_X + m^2)\ \phi_{out}(X) = 0 \quad (18.4.41)$$

where

$$\Box_X = (\partial/\partial X^{\nu})(\partial/\partial X_{\nu})$$

B. Under coordinate displacements and Lorentz transformations $\Phi_{out}(y)$, $\phi_{out}(X(y))$, and $\phi(X(y))$ transform in the same way:

$$[P^{\mu}, \Phi_{out}(y)] = -\ i\partial\Phi_{out}/\partial y_{\mu} \quad (18.4.42a)$$

$$[P^{\mu}, \phi_{out}(X)] = -\ i\partial\phi_{out}/\partial y_{\mu} \quad (18.4.42b)$$

$$[P^{\mu}, \phi(X)] = -\ i\partial\phi/\partial y_{\mu} \quad (18.4.43)$$

with the energy-momentum vector P^{μ} specified by eq. 18-A.57.

C. We can relate the asymptotic out-field $\phi_{out}(X(y))$ to the interacting field $\phi(X(y))$ using the equation of motion of $\phi(X(y))$ specified by eq. (18.4.36:

$$\sqrt{Z}\,\phi_{out}(X(y)) = \phi(X(y)) - \int d^4X(y')\,\Delta_{adv}(y - y')\,j_{tot}(X(y')) \tag{18.4.44}$$

$$= \phi(X(y)) - \int d^4y'\,J\,\Delta_{adv}(y - y')\,j_{tot}(X(y')) \tag{18.4.45}$$

where Z is a wave function renormalization constant, J is the Jacobian, and Δ_{adv} is an advanced Green's function.

D. We can define Φ_{out} out-field states with expressions like

$$|\,p_1, p_2, \ldots\, p_n\ out\rangle = a_{out}^{\dagger}(p_1,)a\Phi_{out}^{\dagger}(p_2)\ \ldots\ a\Phi_{out}^{\dagger}(p_n)|0\rangle \tag{18.4.46}$$

with powers of creation operators allowed since Φ_{out} is a boson field. The set of all particle states constitutes a complete orthonormal set of states. The corresponding bra states are defined by hermitean conjugation:

$$\langle p_1, p_2, \ldots\, p_n\ out| = (|\,p_1, p_2, \ldots\, p_n\ out\rangle)^{\dagger} \tag{18.4.47}$$

18.4.8 The Y Field

The Y field in the present model Lagrangian (eq. 18.4.20) is a free field and thus:

$$Y_{in}(y) = Y_{out}(y) = Y(y) \tag{18.4.48}$$

The states of the Y field have two general forms: 1) States in a Fock space consisting of particle states that are eigenstates of the Y particle number operator (eq. 18.4.19); and 2) Coherent states in a non-Fock space of generalized coherent states in an infinite tensor product space.[199]

The coherent ket states that arise in Two-Tier quantum field theory have the general form (eq. 18.2.41):

$$|y,\ p\rangle = e^{-\mathbf{p}\cdot\mathbf{Y}^{-}(y)/M_c^{\,2}}|0\rangle \tag{18.2.41}$$

as can be seen from an examination of $\phi_{in}(X(y))$. The corresponding bra state is:

$$\langle y, p| = (V|\,y, p\rangle)^{\dagger} = \langle 0|e^{+\mathbf{p}\cdot\mathbf{Y}^{+}(y)/M_c^{\,2}} \tag{18.4.49}$$

with V, the metric operator, reversing the sign of Y in the exponential. The inner product of coherent states is:

$$\langle y_1\,,\, p_1|\, y_2, p_2\rangle = \exp[\,-p_1{}^i p_2{}^j \Delta_{Tij}(y_1 - y_2)/M_c{}^4] \tag{18.4.50}$$

showing the set of coherent states is not orthonormal and, in fact, is overcomplete. Comparing eq. 18.4.50 to eq. 18.3.12 gives

---

199 See Kibble and other references on coherent states.

$$<y_1\,, p|\, y_2, p> = R(p, y_1 - y_2) \tag{18.4.50a}$$

The completeness of the set of states for each time y^0 can be verified by examining the projection operator:

$$\mathscr{R}_Y(y^0) = \because \exp[-i \int d^3y\, Y^-_i(y)|0><0|\pi^{+j}(y)] \because \tag{18.4.51}$$

where

$$\pi^{+j}(y) = -\partial Y^{+j}(y)/\partial y^0 \tag{18.4.52}$$

and where $\because$ represents an extended normal ordering operator:

$$\because \ldots \because$$

which is defined as placing creation operators to the left, projection operators in the center, and annihilation operators to the right. Thus eq 18.4.51 can be written

$$\mathscr{R}_Y = \sum_n (-i/n!)^n \int d^3y_1 \ldots \int d^3y_n Y^{-j_1}(y_1)Y^{-j_2}(y_2)\ldots Y^{-j_n}(y_n)|0><0|\pi^+_{j_1}(y_1)\pi^+_{j_2}(y_2)\ldots\pi^+_{j_n}(y_n) \tag{18.4.53}$$

where we have used the fact that $|0><0|$ is a projection operator, and reduced $|0><0|\ |0><0| \ldots |0><0|$ to $|0><0|$ in eq. 18.4.53. The vacuum state is the product of the Y and ϕ vacuum states:

$$|0> = |0_Y>|0_\phi> \tag{18.4.53a}$$

We note

$$\mathscr{R}_Y(y^0)|\mathbf{y},y^0\ p> = |\mathbf{y},y^0\ p> \tag{18.4.54}$$

using eq. 18.2.22 and $\int d^3y_2\, p^i\, \Delta^{tr}_{ij}(y_1 - y_2)Y^{+j}(y_2) = \mathbf{p}\cdot\mathbf{Y}^+(y_1)$. Also

$$\mathscr{R}_Y(y^0)|n> = |n> \tag{18.4.55}$$

where $|n>$ is any Y particle Fock state of finite particle number. In view of eqs. 18.4.54 and 18.4.55, we see that $\mathscr{R}_Y$ is the identity operator on the Fock space and the space of generalized coherent Y field states. Thus the set of Y coherent states forms an overcomplete set of states. We will define the S matrix for any combination of Φ Fock space states and coherent Y states. The R_Y operator can be generalized to include Φ Fock space states:

$$\mathscr{R}_{\Phi Y}(y^0) = \because \exp[-i \int d^3y\, Y^-_j(y)\mathscr{R}_\Phi \pi^{+j}(y)] \because \tag{18.4.56}$$

with

$$\mathscr{R}_\Phi = \sum_n |n><n| \tag{18.4.57}$$

is a sum over all Φ Fock space states with vacuum state given by eq. 18.4.53a. Since R_Φ is a projection:

$$[\mathscr{R}_\Phi]^N = \mathscr{R}_\Phi$$

for any power N, we find:

$$\mathscr{R}_{\Phi Y}(y^0) = \sum_n (-i)^n \int d^3y_1 \ldots \int d^3y_n Y^{-j_1}(y_1) Y^{-j_2}(y_2) \ldots Y^{-j_n}(y_n) R_\Phi \pi^+_{j_1}(y_1) \pi^+_{j_2}(y_2) \ldots \pi^+_{j_n}(y_n) \quad (18.4.58)$$

As a result we have

$$\mathscr{R}_{\Phi Y}(y^0)|y, p; n_\Phi> = |y, p; n_\Phi> \quad (18.4.59)$$

for any combination of Y coherent states and Φ Fock space states n_Φ. Also

$$\mathscr{R}_{\Phi Y}(y^0)|n_\Phi> = |n_\Phi> \quad (18.4.60)$$

Thus $\mathscr{R}_{\Phi Y}$ is the identity operator on this space – the (over) complete space of in and out states which we will use to define the S matrix of the scalar field theory specified by the Lagrangian eq. 18.4.20.

18.4.9 S Matrix

Following the standard definition of the S matrix we have:

$$S_{\alpha\beta} = <\alpha\ \text{out}|\beta\ \text{in}> \quad (18.4.61)$$

$$= <\alpha\ \text{in}|S|\beta\ \text{in}> \quad (18.4.62)$$

$$|0> = |0\ \text{in}> = |0\ \text{out}> = S|0\ \text{in}> \quad (18.4.63)$$

$$\Phi_{in}(y) = S\Phi_{out}(y)S^{-1} \quad (18.4.64)$$

and the other standard properties of the S matrix with the sole exception being the form of the unitarity relation (discussed later).

18.4.10 LSZ Reduction for Scalar Fields

In this section we will determine the reduction formula for the S matrix for scalar ϕ fields. Consider the S matrix element corresponding to an in state of particles β plus one ϕ particle of momentum p, and an out state α:

$$S_{\alpha\beta p} = <\alpha\ \text{out}|\beta p\ \text{in}> \quad (18.4.65)$$

After standard manipulations we have:

$$S_{\alpha\beta p} = \langle \alpha - p\ \text{out}|\beta\ \text{in}\rangle - i\langle \alpha\ \text{out}|\int d^3y\ f_p(y)\ \overleftrightarrow{\partial_0}\ [\Phi_{in}(y) - \Phi_{out}(y)]\ |\beta\ \text{in}\rangle \tag{18.4.66}$$

where $\langle \alpha - p\ \text{out}|$ is an out state with a particle of momentum p removed (if present) and where

$$f(y^0)\ \overleftrightarrow{\partial_0}\ g(y^0) = f(y^0)\ \partial g(y^0)/\partial y^0 - \partial f(y^0)/\partial y^0\ g(y^0) \tag{18.4.67}$$

and

$$f_p(y) = N_m(p)e^{-ip\cdot y} \tag{18.4.68}$$

with $N_m(p)$ specified by eq. 18.2.6.

We now express

$$S_{\alpha\beta p} = S_{\alpha-p\beta} - i\langle \alpha\ \text{out}|\ \int d^3y\ f_p(y)\ \overleftrightarrow{\partial_0}\ W^{-1}[\phi_{in}(X(y)) - \phi_{out}(X(y))]W|\beta\ \text{in}\rangle \tag{18.4.69}$$

using $W(y) = W_{in}(y)$ with

$$\Phi_a(y) = W_a^{-1}(y)\phi_a(X(y))W_a(y) \tag{18.4.70}$$

where the label a = "in" or a = "out", and where

$$W_a(y) = \exp(-\mathbf{Y}(y)\cdot\mathbf{P}_{\Phi a}/M_c^2) \tag{18.4.71}$$

and

$$W_a^{-1}(y) = \exp(\mathbf{Y}(y)\cdot\mathbf{P}_{\Phi a}/M_c^2) \tag{18.4.72}$$

in the Coulomb gauge of Y with $\mathbf{P}_{\Phi a}$ the momentum spatial vector defined by eq. 18.4.12a.

We note that the interacting $\phi(X(y))$ approaches the in and out fields $\phi_{in}(X(y))$ and $\phi_{out}(X(y))$ in the limit that $y^0 \to -\infty$ and $y^0 \to +\infty$ respectively in the sense of Lehmann, Symanzik and Zimmermann[200] which we *symbolize* as:

$$\phi(X(y)) \to \sqrt{Z}\ \phi_{in}(X(y)) \quad \text{as} \quad y^0 \to -\infty \tag{18.4.73}$$

$$\phi(X(y)) \to \sqrt{Z}\ \phi_{out}(X(y)) \quad \text{as} \quad y^0 \to +\infty \tag{18.4.74}$$

with $\sqrt{Z}$ defined in eqs. 18.4.37 and 18.4.44. Thus we can rewrite eq. 18.4.69 as

$$S_{\alpha\beta p} = S_{\alpha-p\beta} + iZ^{-1/2}\ (\lim_{y^0\to+\infty} - \lim_{y^0\to-\infty})\langle \alpha\ \text{out}|\ \int d^3y\ f_p(y)\ \overleftrightarrow{\partial_0}\ W^{-1}\phi(X(y))W|\beta\ \text{in}\rangle \tag{18.4.75}$$

which after standard manipulations becomes

$$S_{\alpha\beta p} = S_{\alpha-p\beta} + iZ^{-1/2}\ \int d^4y\ f_p(y)(\Box_y + m^2)\langle \alpha\ \text{out}|\ W(y)^{-1}\phi(X(y))W(y)|\beta\ \text{in}\rangle \tag{18.4.76}$$

---

[200] H. Lehmann, K. Symanzik and W. Zimmermann, Nuov. Cim., **1**, 1425 (1955); W. Zimmermann, Nuov. Cim., **10**, 567 (1958); O. W. Greenberg, Doctoral Dissertation, Princeton University 1956.

Eq. 18.4.76 is similar to the usual LSZ reduction formula except for the appearance of the W(y) operator and its inverse. We note that $W(y) = W_{in}(y)$ still because $\mathbf{P}_{\Phi in}$ is independent of y^0.

Similarly an out ϕ particle can be reduced from an S matrix part. For example,

$$<\alpha\ out|W^{-1}(y)\phi(X(y))W(y)|\beta\ in>=<\alpha-p'\ out|W^{-1}(y)\phi(X(y))W(y)|\beta-p'\ in> -$$
$$- i<\alpha-p'\ out|\int d^3y'\ [W^{-1}(y')\phi_{in}(X(y'))W(y')W^{-1}(y)\phi(X(y))W(y) -$$
$$- W^{-1}(y)\phi(X(y))W(y)W^{-1}(y')\phi_{out}(X(y'))W(y')]|\beta\ in>\overleftrightarrow{\partial_0}\ f_{p'}^{\ *}(y') \quad (18.4.77)$$

which becomes

$$<\alpha\ out|W^{-1}(y)\phi(X(y))W(y)|\beta\ in> = <\alpha-p'\ out|\varphi(y)|\beta-p'\ in> +$$
$$+ iZ^{-1/2}\int d^4y'\ <\alpha-p'\ out|T(\varphi(y')\varphi(y))|\beta\ in>\ (\overleftarrow{\Box}_{y'} + m^2)\ f_{p'}^{\ *}(y') \quad (18.4.78)$$

where the time ordered product is defined with respect to ordering with respect to y^0 and where

$$\varphi(y) = W^{-1}(y)\phi(X(y))W(y) \quad (18.4.79)$$

These results directly generalize to multi-particle in and out states:

$$<p_1, p_2, \ldots p_n\ out|\ q_1, q_2, \ldots q_m\ in> = \ldots <0|T(\varphi(y'_1) \ldots \varphi(y'_n)\varphi(y_1) \ldots \varphi(y_m))|0> \ldots \quad (18.4.80)$$

thus reducing the development of the perturbation theory of the S matrix to the evaluation of time ordered products such as

$$<0|T(\varphi(y_1) \ldots \varphi(y_n))|0> \quad (18.4.81)$$

18.5 The U Matrix in Perturbation Theory

The U matrix for a Two-Tier theory is developed in a way similar to conventional field theory starting from the defining relations:

$$\phi(X(y)) = U^{-1}\phi_{in}(X(y))U \quad (18.5.1)$$

$$\pi_\phi(X(y)) = U^{-1}\pi_{\phi in}(X(y))U \quad (18.5.2)$$

From eq. 18.4.29 we define the free field hamiltonian

$$H_{F0in}(\ \phi_{in}, \pi_{\phi in}) = \int d^3X\ \mathscr{H}_{F0}(\ \phi_{in}, \pi_{\phi in}) \quad (18.5.3)$$

Noting $X^0 = y^0$ in the Y Coulomb gauge we find

$$\partial\phi_{in}/\partial y^0 = i[H_{F0in}, \phi_{in}(X)] \quad (18.5.4)$$

$$\partial\pi_{\phi in}/\partial y^0 = i[H_{F0in}, \pi_{\phi in}(X)] \quad (18.5.5)$$

For the entire hamiltonian (eq. 18.4.28) we have

$$\partial\phi/\partial y^0 = i[H_F, \phi(X)] \tag{18.5.6}$$

$$\partial\pi_\phi/\partial y^0 = i[H_F, \pi_\phi(X)] \tag{18.5.7}$$

with

$$H_F(\phi, \pi_\phi) = :\int d^3X\, \mathscr{H}_F(\phi, \pi_\phi): \tag{18.5.8}$$

(Note the *entire* interaction term is normal ordered since d^3X is a q-number. Combining the above equations in the standard way yields a familiar differential equation for the U matrix:

$$i\partial U(y^0)/\partial y^0 = (H_{Fint} + E_0(t))U(y^0) \tag{18.5.9}$$

where $E_0(t)$ is a c-number function of y^0 that we can set equal to 0 (as it would be cancelled later in any case), and where

$$H_{Fint}(\phi_{in}, \pi_{\phi in}) = :\int d^3X\, \mathscr{H}_{Fint}(\phi_{in}, \pi_{\phi in}): \tag{18.5.10}$$

with $\mathscr{H}_{Fint}$ given by eq. 18.4.30. Solving for U gives the familiar time ordered exponential:

$$U(y^0) = T(\exp[-i\int_{-\infty}^{t} dy^0\, H_{Fint}]) \tag{18.5.11a}$$

which is a symbolic notation for:

$$U(y^0) = 1 + \sum_{n=1}^{\infty} (-i)^{-n}(n!)^{-1}\int_{-\infty}^{y^0} dy_1^0 \ldots \int_{-\infty}^{y^0} dy_n^0\, T(H_{Fint}(y_1^0) \ldots H_{Fint}(y_n^0)) \tag{18.5.11b}$$

We note for later use that the hermiticity of H_{Fint} is not used in the derivation of eq. 18.5.11. Thus eq. 18.5.11 would still hold if H_{Fint} were not hermitean.

18.5.1 Reduction of Time Ordered φ Products

In the previous chapter we reduced the calculation of the S matrix to the evaluation of time ordered products of the form

$$\tau(y_1, \ldots, y_n) = \langle 0|T(\varphi(y_1) \ldots \varphi(y_n))|0\rangle \tag{18.5.12}$$

where $\varphi(y)$ is specified by eq. 18.4.79. Expanding the terms within eq. 18.5.12 using eq. 18.4.79 we find

$$\varphi(y_1) \ldots \varphi(y_n) = W^{-1}(y_1)\phi(X(y_1))W(y_1)W^{-1}(y_2)\phi(X(y_2))W(y_2) \ldots W^{-1}(y_n)\phi(X(y_n))W(y_n) \tag{18.5.13}$$

which can be re-expressed as

$$W^{-1}(y_1)U^{-1}(y_1{}^0)\phi_{in}(X(y_1))U(y_1{}^0)W(y_1)W^{-1}(y_2)U^{-1}(y_2{}^0)\phi_{in}(X(y_2))U(y_2{}^0)W(y_2)\ldots \quad (18.5.14)$$

using eq. 18.5.1 and denoting $W_{in}(y)$ as $W(y)$. Defining

$$\mathscr{U}(y_1, y_2) = U(y_1{}^0)W(y_1)W^{-1}(y_2)U^{-1}(y_2{}^0) \quad (18.5.15)$$

we see eq. 18.5.14 can be rewritten as

$$W^{-1}(y_1)U^{-1}(y_1{}^0)\phi_{in}(X(y_1))U(y_1, y_2)\phi_{in}(X(y_2))\ U\ (y_2, y_3)\phi_{in}(X(y_3))\ \ldots\phi_{in}(X(y_n))U(y_n{}^0)W(y_n) \quad (18.5.16)$$

From eqs. 18.4.71 and 18.4.72

$$\mathscr{U}(y_1, y_2) = U(y_1{}^0)\exp((\mathbf{Y}(y_2) - \mathbf{Y}(y_1))\cdot\mathbf{P}_{\Phi a}/M_c{}^2)U^{-1}(y_2{}^0) \quad (18.5.17)$$

Defining

$$W(\ y_1, y_2) = \exp((\mathbf{Y}(y_2) - \mathbf{Y}(y_1))\cdot\mathbf{P}_{\Phi a}/M_c{}^2) \quad (18.5.18)$$

and looking ahead to the Wick expansion of the time ordered product of eq. 18.5.12 we note that the only time ordered products involving $W(y_1, y_2)$ that would appear in the expansion are

$$<0|T(\phi_{in}(X(y))W(y_1, y_2))|0> = 0 \quad (18.5.19a)$$

$$<0|T(Y(y)W(y_1, y_2))|0> = 0 \quad (18.5.19b)$$

$$<0|T(\partial Y(y)/\partial y^\mu\ W(y_1, y_2))|0> = 0 \quad (18.5.19c)$$

$$<0|T(\partial Y(y)/\partial y^\mu\ \phi_{in}(X(y)))|0> = 0 \quad (18.5.19d)$$

$$<0|T(W(y_1, y_2)W(y_3, y_4))|0> = 1 \quad (18.5.19e)$$

due to the factor of $\mathbf{P}_{\Phi a}$ that appears in $W(y_1, y_2)$. Also

$$<0|T(\phi_{in}(X(y))Y(y_1))|0> = 0 \quad (18.5.20)$$

due to the $a_{in}(p)$ and $a_{in}{}^\dagger(p)$ factors appearing in $\phi_{in}(X(y))$.

Thus the $W(y_1, y_2)$ factor in eq. 18.5.17 may be set to the value one with the result

$$\mathscr{U}(y_1, y_2) \equiv U(y_1{}^0)U^{-1}(y_2{}^0) = U(y_1{}^0, y_2{}^0) \quad (18.5.21)$$

where $U(y_1{}^0, y_2{}^0)$ is the conventionally defined U matrix satisfying

$$i\partial\ U(y_1{}^0, y_2{}^0)/\partial y_1{}^0 = iH_{Fint}\ U(y_1{}^0, y_2{}^0) \quad (18.5.22)$$

with the boundary condition

$$U(y^0, y^0) = 1 \qquad (18.5.23)$$

This result would still be true if the $W(y_1, y_2)$ exponentials were expanded in their "power series" form.

Then, paralleling the standard approach we find an expression for the U matrix:

$$U(y_1^0, y_2^0) = T(\exp[-i \int_{y_2^0}^{y_1^0} dy'^0{:}d^3X(y')\ \mathscr{H}_{Fint}(\phi_{in}(X(y')), \pi_{\phi in}(X(y'))){:}]) \qquad (18.5.24)$$

The $U(y_1^0, y_2^0)$ matrix satisfies the conventional multiplication rule:

$$U(y_1^0, y_3^0) = U(y_1^0, y_2^0)U(y_2^0, y_3^0) \qquad (18.5.25)$$

The inverse of $U(y_1, y_2)$ is

$$U^{-1}(y_1^0, y_2^0) = U(y_2^0, y_1^0) \qquad (18.5.26)$$

We now return to eq. 18.5.16, which can now be written in the form:

$$U^{-1}(y^0)U(y^0, y_1^0)\phi_{in}(X(y_1))U(y_1^0, y_2^0)\phi_{in}(X(y_2))U(y_2^0, y_3^0) \ldots \phi_{in}(X(y_n))U(y_n^0, -y^0)U(-y^0) \qquad (18.5.27)$$

where y^0 is a reference time that is later than all other times, and $-y^0$ is earlier than all the other times, in the time-ordered product. As a result the time-ordered product in eq. 18.4.80 can be expressed in a symbolic notation as:

$$<0|U^{-1}(y^0)T(\phi_{in}(X(y_1))\phi_{in}(X(y_2)) \ldots \phi_{in}(X(y_n))U(y^0, -y^0))U(-y^0)|0> \qquad (18.5.28)$$

The analysis of eq. 18.5.28 as $y^0 \to \infty$ follows the standard path, which begins by noting

$$U(-y)|0> = \lambda_-|0> \qquad \text{when } y^0 \to \infty \qquad (18.5.29a)$$
$$U(y)|0> = \lambda_+|0> \qquad \text{when } y^0 \to \infty \qquad (18.5.29b)$$

following a standard textbook proof, which, in turn, leads to:

$$\lambda_-\lambda_+{}^* = <0|T(\exp[\ +i \int_{-\infty}^{\infty} dy'^0{:}d^3X(y')H_{\ Fint}(\phi_{in}(X(y')), \pi_{\phi in}(X(y'))){:}])|0> \qquad (18.5.30)$$

$$= [<0|T(\exp[\ -i \int_{-\infty}^{\infty} dy'^0 d^3X(y')\ \mathscr{H}_{\ Fint}(\phi_{in}(X(y')), \pi_{\phi in}(X(y')))])|0>]^{-1} \qquad (18.5.31)$$

Thus the time ordered product of eq. 18.5.12, which appears in the evaluation of the S matrix element in eq. 18.4.80, can be symbolically written as:

$$\tau(y_1, \ldots, y_n) = \frac{<0|T(\phi_{in}(X(y_1)) \ldots \phi_{in}(X(y_n))U(\infty, -\infty))|0>}{<0|T(\exp[-i\int dy'^0 : d^3X(y')\mathscr{H}_{Fint}(\phi_{in}(X(y')), \pi_{\phi in}(X(y'))):])|0>} \quad (18.5.32)$$

in the limit $y^0 \to \infty$.

18.5.2 The $\int d^3X$ Integration

The integration over the X space coordinates presents the difficulty of a functional integration of a q-number that needs to be properly defined. Since

$$X^\mu(y) = y^\mu + i\, Y^\mu(y)/M_c^2 \quad (18.2.12)$$

by definition and since, in the Y Coulomb gauge we have $X^0(y) = y^0$ due to $Y^0 = 0$, the classical Jacobian for the transformation from y to X coordinates is the absolute value:

$$J = \left|\varepsilon^{ijk}\left(\delta^{1i} + \frac{i}{M_C^2}\frac{\partial Y^1}{\partial y^i}\right)\left(\delta^{2j} + \frac{i}{M_C^2}\frac{\partial Y^2}{\partial y^j}\right)\left(\delta^{3k} + \frac{i}{M_C^2}\frac{\partial Y^3}{\partial y^k}\right)\right| \quad (18.5.33)$$

The Jacobian appears in a change of integration variables:

$$\int d^3X = \int d^3y\, J \quad (18.5.34)$$

$$\int d^4X = \int d^4y\, J \quad (18.5.35)$$

in the Y Coulomb gauge.

A change of variables for c-number coordinate transformations is well known. The situation changes when one set of coordinates are in fact q-numbers. The second quantization of the Y field requires the definition of J to be clarified since the product of fields at the same position is normally undefined. The normal ordering of the interaction hamiltonian term in eqs. 18.5.34 and 18.5.32 resolves the issue. Therefore eq. 18.5.33 must be considered as inserted within a normal ordered expression.

While normal ordering eliminates the infinities that would otherwise be present, J still presents a problem because it is still effectively part of the interaction term. This situation appears to be unsatisfactory in the present, scalar quantum field theory in which Y is not intended to play a direct dynamical role but rather a passive role as a coordinate. The normal ϕ field is supposed to be the only in, out, and interacting field.

The problem of J is resolved by eqs. 18.5.19c and 18.5.19d, which reduces the effect of the derivative terms in eq. 18.5.33 to zero in the Wick expansion of the time ordered product in eq. 18.5.32 if no Y quanta appear in or out S matrix states. Thus

$$J \equiv 1 \tag{18.5.36}$$

As a result the time ordered product (eq. 18.5.32) becomes:

$$\tau(y_1,\ldots, y_n) = \frac{<0|T(\phi_{in}(X(y_1))\ldots\phi_{in}(X(y_n))\exp[-i\int d^4y' \mathscr{H}_{Fint}(\phi_{in}(X(y')))])\,|0>}{<0|T(\exp\,[-i\int d^4y' \mathscr{H}_{Fint}(\phi_{in}(X(y')))])|0>} \tag{18.5.37}$$

18.5.3 Y In and out states

The Y fields have no interactions and are thus free fields in the model Lagrangian under consideration and in the Two-Tier quantum field theories that we will construct later. Therefore "in" Y quanta are the same as "out" Y quanta.

Since the Lagrangians that we consider do not have interaction terms explicitly containing Y field factors, the S matrix is "block diagonal" in the sense that if an in-state does not contain Y quanta, (or Y coherent states) then out-states will not contain Y quanta (or coherent Y states). The proof is based on the expansion of S matrix elements using Wick's theorem in products of time ordered products of pairs of in field operators. Eqs. 18.5.19, 18.5.20 and 18.5.36, and in particular,

$$<0|\ T(\phi_{in}(X(y_1))Y^j(y_2))|0> = 0 \tag{18.5.39}$$

$$<0|T(\phi_{in}(X(y_1))e^{-\mathbf{q}\cdot\mathbf{Y}^-(y)/M_c^2})|0> = <0|T(\phi_{in}(X(y_1))e^{+\mathbf{q}\cdot\mathbf{Y}^+(y)/M_c^2})|0> = 0 \tag{18.5.40}$$

prove S matrix elements with no incoming Y quanta or coherent states will have zero matrix elements to produce outgoing Y quanta or coherent states. In addition any non-zero S matrix element with n incoming Y quanta must have n outgoing Y quanta. For example an incoming state with 5 Y quanta and 2 ϕ particles can only become an outgoing state with 5 Y quanta and two or more ϕ particles. Therefore we have proved the general result:

Theorem 18.5.I: ***Any non-zero S matrix element has the same number of incoming Y quanta and outgoing Y quanta.***

This theorem is true in any Two-Tier quantum field theory. In order to have a tractable theory we will require all in-states and out-states not to contain Y quanta or coherent states. All normal in-state and out-state particles will contain factors of $:e^{\pm p\cdot Y/M_c^2}:$ *in the fourier expansions of their corresponding fields.*

18.5.4 Unitarity

For many years it has been evident that modified field theories[11, 17, 201] might offer some hope of avoiding the divergences of conventional quantum field theory. Usually these theories suffer from

---

[201] S. Blaha, Phys.Rev. **D10**, 4268 (1974); S. Blaha, Phys.Rev. **D11**, 2921 (1975); S. Blaha, Nuovo Cim. **A49**, :113 (1979); S. Blaha, "Generalization of Weyl's Unified Theory to Encompass a Non-Abelian Internal Symmetry Group"

unitarity problems: negative norms and negative probabilities. In the absence of a physically acceptable interpretation of negative probabilities, these theories have been thought to be unsatisfactory.

The Two-Tier type of quantum field theory *superficially* also appears to have a unitarity problem due to the non-hermitean nature of Two-Tier hamiltonians. The lack of hermiticity is due entirely to the appearance of iY^μ in the X^μ field coordinates. *In fact Two-Tier quantum field theories satisfy unitarity for physical states. Physical states are defined to consist of any number of normal Two-Tier particles and NO Y quanta.*

Two-tier interaction hamiltonians, such as the one in eq. 18.5.37, are not hermitean. For example,

$$H_{Fint} = \int d^3y' \mathscr{H}_{Fint}(\phi_{in}(y' + iY(y')/M_c^2)) \qquad (18.5.41)$$

$$H_{Fint}^\dagger = \int d^3y' \mathscr{H}_{Fint}(\phi_{in}(y' - iY(y')/M_c^2)) \neq H_{Fint} \qquad (18.5.42)$$

The relation between H_{Fint} and its hermitean conjugate is

$$H_{Fint} = V\, H_{Fint}^\dagger\, V \qquad (18.5.43)$$

where $V^2 = I$ is the metric operator defined in eqs. 18.4.15 – 18.4.18. Thus the S matrix is not unitary; the S matrix is *pseudo-unitary*:

$$S^{-1} = V\, S^\dagger\, V \qquad (18.5.44)$$

$$VS^\dagger\, VS = I \qquad (18.5.45)$$

We will now show that the S matrix is *unitary between physical states.* To prove this point, consider eq. 18.5.45 between physical states $|i>$ and $<f|$ – each consisting of a number of ϕ particles and no Y quanta.

$$\delta_{fi} = <f\,|I|\,i> = <f\,|VS^\dagger VS|\,i>$$

$$= \sum_{n,\,m,\,p} <f\,|V|p><p|S^\dagger|n><n|V|m><m|S|\,i>$$

$$= \sum_{n,\,m,\,p} <f\,|S^\dagger|m><m|S|\,i> \qquad (18.5.46)$$

since V has the eigenvalue 1 between states consisting of no Y quanta. Due to eqs. 18.5.19a – 18.5.19e and 18.5.20 since there are no incoming Y quanta there are no outgoing Y quanta.

The block diagonality of S (and the diagonality of V) limits the intermediate states $|n>$ and $|m>$ to states containing ϕ particles and no Y quanta – although normalization factors R(**p**, z) will appear (described later) due to the presence of $:e^{\pm p\cdot Y/M_c^2}:$ factors within quantum field fourier expansions that embody Y coherent state effects. Thus

$$S_{phys}^\dagger\, S_{phys} = I \qquad (18.5.47)$$

$$S_{phys}^\dagger = S_{phys}^{-1} \qquad (18.5.48)$$

proving unitarity between physical states – states consisting of ϕ particles and no Y quanta that are properly normalized. A detailed example is presented starting on page 67.

SLAC-PUB-1799, Aug 1976; S. Blaha, "Quantum Gravity and Quark Confinement" Lett. Nuovo Cim. **18**, 60 (1977); S. Blaha, "The Local Definition of Asymptotic Particle States" Nuovo Cim. **A49**, 35 (1979) and references therein.

18.5.5 Finite Renormalization of External Legs

In the previous section we showed the theory satisfies unitarity for states that are properly normalized. However the use of the non-unitary operator W(y) (eq. 18.4.6) to transform $\Phi_{in}(y)$ fields into $\phi_{in}(X(y))$ fields in the LSZ procedure in eq. 18.4.69, and related equations, does not preserve the norm of input and output ϕ particle legs. Thus a finite renormalization is needed for each external particle leg in order to have a unitary S-matrix.

We define this renormalization of external legs within the framework of a perturbation theory example in section 18.5.9.

18.5.6 Perturbation Expansion

Perturbation theory in Two-Tier quantum field theory is very similar to conventional perturbation theory. The difference is in the form of the propagators, which have a high energy damping factor R(**p**, z) that eliminates infinities that normally appear at high energy in conventional quantum field theories.

In order to develop a feeling for Two-Tier perturbation theory we will calculate a few low order diagrams in the perturbation theory of the model scalar ϕ^4 theory that we have been using as an example in this chapter.

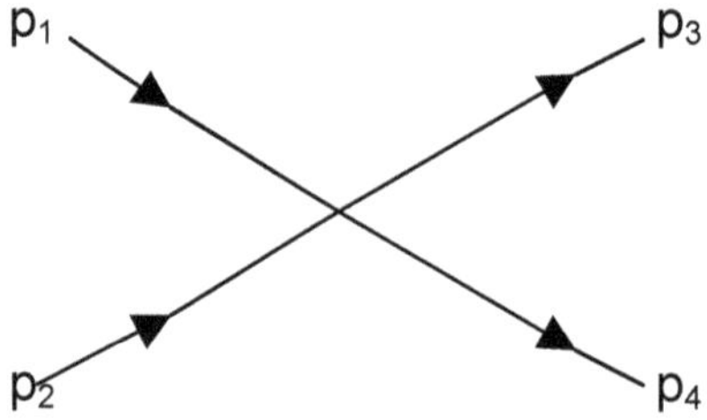

Figure 18.5.1. Lowest order quartic interaction diagram.

Fig. 18.5.1 contains the lowest order diagram for the scattering of two ϕ particles into a two ϕ particle out-state. The S matrix element for this diagram is

$$S_1 = i^4(\tfrac{1}{4!}\, i\chi_0) \prod_{j=1}^{4} \int d^4y_j\, d^4y\, f_{Zp_1}(y_1) f_{Zp_2}(y_2) f_{Zp_3}{}^*(y_3) f_{Zp_4}{}^*(y_4)(\Box_{y_1} + m^2)\cdot$$
$$\cdot(\Box_{y_2}+m^2)(\Box_{y_3}+m^2)(\Box_{y_4}+m^2)<0|T(\phi_{in}(X(y_1))\ldots\phi_{in}(X(y_4)):(\phi_{in}(X(y))^4:)|0> \qquad (18.5.49)$$

with $f_{Zp}(y)$ specified by

$$f_{Zp}(y) = [(2\pi)^3 2p^0 Z_p]^{-\frac{1}{2}} e^{-ip\cdot y} \qquad (18.5.49a)$$

where Z_p is a normalization factor that will be specified later.

Expanding the time ordered product and realizing there are 4! ways of combining the four field factors in the interaction hamiltonian leads to:

$$S_1 = i^4(i\varkappa_0)\prod_{j=1}^{4}\int d^4y_j\, d^4y\, f_{Zp_1}(y_1)f_{Zp_2}(y_2)f_{Zp_3}{}^*(y_3)f_{Zp_4}{}^*(y_4)(\Box_{y_1} + m^2)\cdot$$
$$\cdot(\Box_{y_2}+m^2)(\Box_{y_3}+m^2)(\Box_{y_4}+m^2)i\Delta_F{}^{TT}(y_1-y)i\Delta_F{}^{TT}(y_2-y)i\Delta_F{}^{TT}(y_3-y)i\Delta_F{}^{TT}(y_4-y) \quad (18.5.50)$$

where

$$i\Delta_F{}^{TT}(y_1 - y_2) = \langle 0|T(\phi(X(y_1)),\phi(X(y_2)))|0\rangle \quad (18.5.51)$$

$$= i\int \frac{d^4p\, e^{-ip\cdot(y_1 - y_2)}\, R(\mathbf{p}, y_1 - y_2)}{(2\pi)^4\,(p^2 - m^2 + i\varepsilon)} \quad (18.5.52)$$

with

$$R(\mathbf{p}, y_1 - y_2) = \exp[-p^ip^j\Delta_{Tij}(y_1 - y_2)/M_c{}^4] \quad (18.5.53)$$

(summations are over space indices only in the Y Coulomb gauge) and

$$\Delta_{Tij}(z) = \int d^3k\, e^{-ik\cdot z}(\delta_{ij} - k_ik_j/\mathbf{k}^2)/[(2\pi)^3 2\omega_k] \quad (18.5.54)$$

From section 18.4 we have:

$$R(\mathbf{p}, y_1 - y_2) = \exp\{-p^2[A(v) + B(v)\cos^2\theta] / [4\pi^2M_c{}^4z^2]\} \quad (18.5.55)$$

with

$$z^\mu = y_1{}^\mu - y_2{}^\mu \quad (18.5.56)$$

$$z = |\mathbf{z}| = |\mathbf{y_1} - \mathbf{y_2}| \quad (18.5.57)$$
$$p = |\mathbf{p}| \quad (18.5.58)$$
$$v = |z^0|/ z \quad (18.5.59)$$
$$A(v) = (1 - v^2)^{-1} + .5v \ln[(v - 1)/(v + 1)] \quad (18.5.60)$$
$$B(v) = v^2(1 - v^2)^{-1} - 1.5v \ln[(v - 1)/(v + 1)] \quad (18.5.61)$$
$$\mathbf{p}\cdot\mathbf{z} = pz \cos\theta \quad (18.5.62)$$

and with $|\mathbf{p}|$ denoting the length of the spatial vector **p**, while $|z^0|$ is the absolute value of z^0.

We note

$$R(\mathbf{p}, y) = R(\mathbf{p}, -y) \quad (18.5.62a)$$

for later use.

Letting $y_i = w_i + y$ yields

$$S_1 = i^4(i\varkappa_0)(2\pi)^4\, \delta^4(p_3 + p_4 - p_1 - p_2)N^+(p_4)N^+(p_3)N(p_2)N(p_1) \quad (18.5.63)$$

where

$$N(p) = iZ_p{}^{-1/2}\int d^4w\, f_p(w)(\Box + m^2)\Delta_F{}^{TT}(w) \quad (18.5.64)$$

$$N^+(p) = iZ_p^{-½}\int d^4w\ f_p^*(w)(\Box + m^2)\Delta_F^{TT}(w) \qquad (18.5.65)$$

are "normalizations" of the "external legs" – the in and out states due to the Y field cloud around each particle with $Z^{-½}$ a renormalization factor to be determined later. In the limit of low momentum ($p \ll M_C$):

$$N(p) = N^+(p) \to -iZ_p^{-½}[(2\pi)^3 2p^0]^{-½} \qquad (18.5.66)$$

which the reader will note is the standard normalization factor for external scalar field legs in conventional quantum field theory modulo the $Z_p^{-½}$ factor. The factor $Z_p^{-½}$ performs the finite renormalization of external legs discussed in the preceding unitarity discussion.

18.5.7 Higher Order Diagram With a Loop

We will now consider the simplest one loop scattering diagrams in the scalar ϕ^4 theory.

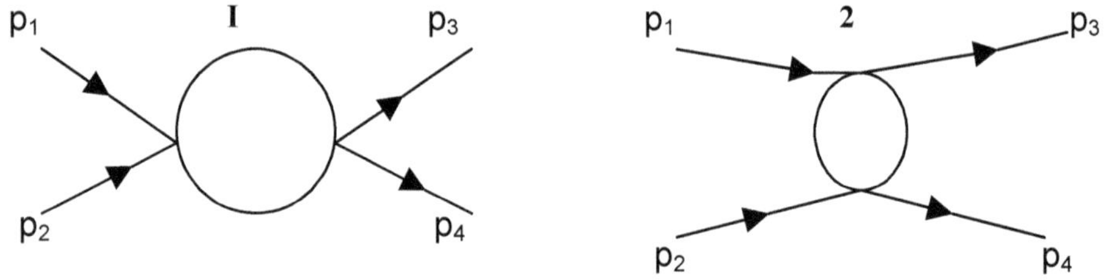

Figure 18.5.2. Lowest order loop scattering diagrams.

The S matrix element for these diagrams (and some other disconnected diagrams) is contained in

$$S_2 = i^4(¼!\, i\chi_0)^2 \prod_{j=1}^{4} \int d^4y_j\, d^4y'_1\, d^4y'_2\, f_{Zp_1}(y_1)f_{Zp_2}(y_2)f_{Zp_3}^*(y_3)f_{Zp_4}^*(y_4)(\Box_{y_1} + m^2)\cdot$$
$$\cdot(\Box_{y_2}+m^2)(\Box_{y_3}+m^2)(\Box_{y_4}+m^2)<0|T(\phi_{in}(X(y_1))\ldots\phi_{in}(X(y_4)):(\phi_{in}(X(y'_1))^4::(\phi_{in}(X(y'_2))^4:)|0>/2! \qquad (18.5.67)$$

together with some other disconnected diagrams.

Expanding the time ordered product and keeping only the terms corresponding to Fig. 18.5.2 gives:

$$S_2 = i^4(i\chi_0)^2/2 \prod_{j=1}^{4} \int d^4y_j\, d^4y'_1\, d^4y'_2\, f_{Zp_1}(y_1)f_{Zp_2}(y_2)f_{Zp_3}^*(y_3)f_{Zp_4}^*(y_4)\cdot$$
$$\cdot(\Box_{y_1} + m^2)(\Box_{y_2}+m^2)(\Box_{y_3}+m^2)(\Box_{y_4}+m^2)\cdot\{i\Delta_F^{TT}(y_1-y'_1)i\Delta_F^{TT}(y_2-y'_1)i\Delta_F^{TT}(y_3-y'_2)i\Delta_F^{TT}(y_4-y'_2) +$$

$$+ i\Delta_F{}^{TT}(y_1-y_1')i\Delta_F{}^{TT}(y_2-y_2')i\Delta_F{}^{TT}(y_3-y_1')i\Delta_F{}^{TT}(y_4-y_2')\}i\Delta_F{}^{TT}(y_1'-y_2')i\Delta_F{}^{TT}(y_1'-y_2') \quad (18.5.68)$$

Following a similar procedure to the previous calculation yields

$$S_2 = i^4[(i\varkappa_0)^2/2](2\pi)^4\delta^4(p_3 + p_4 - p_1 - p_2)N^+(p_4)N^+(p_3)N(p_2)N(p_1) \int d^4z\, [e^{-i(p_1+p_2)\cdot z} + e^{-i(p_1-p_3)\cdot z}]\, [i\Delta_F{}^{TT}(z)]^2 \quad (18.5.69)$$

omentum conserving delta function as in eq. 18.5.63. The loop integrals have the form:

$$I(q) = \int d^4z\, e^{-iq\cdot z}\, [i\Delta_F{}^{TT}(z)]^2 \quad (18.5.70)$$

The behavior of the Two-Tier Feynman propagator $\Delta_F{}^{TT}(z)$ was studied at long and short distance in eqs. 18.3.21-18.3.24. The large distance behavior of the Two-Tier Feynman propagator $\Delta_F{}^{TT}(z)$ approaches the behavior of the conventional Feynman propagator since

$$R(\mathbf{p}, z) \to 1 \quad (18.5.71)$$

as $z^2 = z^\mu z_\mu$ becomes much larger than $M_c{}^{-2}$ ($z^2 \gg M_c{}^{-2}$) (eq. 18.5.55). Thus I(q) approaches the standard one loop expression of conventional field theory at large distance (or small momentum). Again we seee that Two-Tier *quantum field theory realizes a form of Correspondence Principle approaching conventional quantum field theory at large distance.*

At short distances the Gaussian factor $R(\mathbf{p}, z)$ dominates. The Two-Tier Feynman propagator $\Delta_F{}^{TT}(z)$ is radically different from the conventional Feynman propagator at very short distances (or very high momentum). The singular behavior of the conventional Feynman propagator is replaced with a well-behaved, high-energy (short distance) behavior. Near the light cone $M_c{}^{-2} \gg z^2 \to 0$ (or $p^2 \gg M_c{}^2$) we can approximate eq. 18.5.52 with

$$i\Delta_F{}^{TT}(z) \approx \int d^3p\, [N(p)]^2\, R(\mathbf{p}, z) \quad (18.5.72)$$

since $e^{-ip\cdot z}$ is approximately unity for small z. We assume the mass of the ϕ particle is negligible on this scale. Upon performing the integrations (see eq. 18.3.23 for the exact result) we find eq. 18.5.72 approaches:

$$i\Delta_F{}^{TT}(z) \to \pi\, M_c{}^4|z^2|/8$$

as $z^2 = z^\mu z_\mu \to 0$ from the space-like or time-like side of the light cone where | | represents the absolute value.

Therefore I(q) is finite and well-behaved. At high energy ($q^2 \gg M_c{}^2$)

$$I(q) \sim q^{-8}$$

since the fourier transform of $\Delta_F{}^{TT}(z)$ (momentum space) is

$$\Delta_F{}^{TT}(p) = \int d^4z\, e^{-ip\cdot z}\, \Delta_F{}^{TT}(z) \sim p^{-6}$$

for large p ($p^2 \gg M_c{}^2$). (Compare the preceding high energy behavior of I(q) with the conventional logarithmically divergent one loop result $I(q) \sim \ln(q^2/\Lambda^2)$ with Λ a cutoff.)

Thus Two-Tier quantum provides the benefits of a higher derivative theory without its drawbacks.

18.5.8 Finite Renormalization of External Particle Legs & Unitarity Example

The renormalization factor $Z_p^{-1/2}$ appearing in eqs. 18.5.64 and 18.5.65 that is due to the use of the non-unitary operator W(y) (eq. 18.4.6) to transform $\Phi_{in}(y)$ fields into $\phi_{in}(X(y))$ fields in the LSZ procedure in eq. 18.4.69, and related equations, does not preserve the norm of input and output ϕ particle legs. $Z_p^{-1/2}$ performs a finite renormalization for each external particle leg to compensate for the effects of W(y).

The required renormalization is nicely illustrated by considering the unitarity sum in the imaginary part of the preceding example.

The transition matrix T_{fi} is defined in terms of the S matrix by

$$S_{fi} = \delta_{fi} - i\,(2\pi)^4\,\delta^4(P_f - P_i)\,T^{(+)}{}_{fi}$$

The unitarity condition is

$$T^{(+)}{}_{fi} - T^{(-)}{}_{fi} = -i \sum_n (2\pi)^4\,\delta^4(P_n - P_i)\,T^{(-)}{}_{fn}\,T^{(+)}{}_{ni} \tag{18.5.73}$$

Therefore we see that the first term on the right side of eq. 18.5.69 gives a transition matrix term:

$$T^{(+)}{}_{2a} = -\,i[\chi_0{}^2/2]N^+(p_4)N^+(p_3)N(p_2)N(p_1)\int d^4z\, e^{-iP\cdot z}[i\Delta_F{}^{TT}(z)]^2 \tag{18.5.69a}$$

where $P = p_1 + p_2$. Substituting for $i\Delta_F{}^{TT}$ (using eq. 18.3.11) we find that the imaginary part of $T^{(+)}{}_{2a}$ is given by (Note R(**p**, z) is real.)

$$T^{(+)}{}_{2a} - T^{(-)}{}_{2a} = -i[\chi_0{}^2/2]N^+(p_4)N^+(p_3)N(p_2)N(p_1)\int d^4z\, e^{-iP\cdot z}\,[i\int d^4p\,\theta(p_0)\,\delta(p^2 - m^2)e^{-ip\cdot z}\,R(\mathbf{p}, z)/(2\pi)^3]^2$$

If we express the R(**p**, z) factors in terms of their fourier transforms (see eq. 18.3.27):

$$R(\mathbf{p}, z) = \int d^4q\, e^{-iq\cdot z}\,R(\mathbf{p}, q)$$

Then we find

$$T^{(+)}{}_{2a} - T^{(-)}{}_{2a} = -i[\chi_0{}^2/2]N^+(p_4)N^+(p_3)N(p_2)N(p_1)\int d^4z\, e^{-iP\cdot z}\cdot$$
$$\cdot\Big[i\int d^4k_1\,d^4q_1\theta(k_1{}^0)\,\delta(k_1{}^2 - m^2)e^{-ik1\cdot z}\,e^{-iq1\cdot z}\,R(\mathbf{k_1}, q_1)/(2\pi)^3\Big]\cdot$$
$$\cdot\Big[i\int d^4k_2\,d^4q_2\theta(k_2{}^0)\,\delta(k_2{}^2 - m^2)e^{-ik2\cdot z}\,e^{-iq2\cdot z}\,R(\mathbf{k_2}, q_2)/(2\pi)^3\Big]$$

Performing the integral over z gives

$$T^{(+)}{}_{2a} - T^{(-)}{}_{2a} = +i[\chi_0{}^2/2]N^+(p_4)N^+(p_3)N(p_2)N(p_1)(2\pi)^4\cdot$$
$$\cdot\int d^4k_1d^4q_1d^4k_2d^4q_2\theta(k_1{}^0)\,\delta(k_1{}^2 - m^2)\,\theta(k_2{}^0)\,\delta(k_2{}^2 - m^2)\cdot$$

$$\cdot R(\mathbf{k_1}, q_1)R(\mathbf{k_2}, q_2)\,\delta^4(P + k_1 + q_1 + k_2 + q_2)/(2\pi)^6$$

Introducing delta functions enables us to re-express this equation as

$$T^{(+)}{}_{2a} - T^{(-)}{}_{2a} = +i[\chi_0{}^2/2]N^+(p_4)N^+(p_3)N(p_2)N(p_1)\int d^4r_1d^4r_2(2\pi)^4\delta^4(P - r_1 - r_2)\cdot$$

$$\cdot \int d^4k_1d^4q_1\,\delta^4(r_1 + k_1 + q_1)\theta(k_1{}^0)\,\delta(k_1{}^2 - m^2)\,R(\mathbf{k_1}, q_1)\cdot$$

$$\cdot\int d^4k_2d^4q_2\theta(k_2{}^0)\,\delta(k_2{}^2 - m^2)\,\delta^4(r_2 + k_2 + q_2)R(\mathbf{k_2}, q_2)/(2\pi)^6$$

which becomes

$$T^{(+)}{}_{2a} - T^{(-)}{}_{2a} = +i[\chi_0{}^2/2]N^+(p_4)N^+(p_3)N(p_2)N(p_1)\int d^4r_1d^4r_2(2\pi)^4\delta^4(P - r_1 - r_2)\cdot$$

$$\cdot \int d^4k_1\theta(k_1{}^0)\,\delta(k_1{}^2 - m^2)\,R(\mathbf{k_1}, -k_1 - r_1)\cdot$$

$$\cdot\int d^4k_2\theta(k_2{}^0)\,\delta(k_2{}^2 - m^2)R(\mathbf{k_2}, -k_2 - r_2)/(2\pi)^6$$

$R(\mathbf{k_2}, -k_2 - r_2)$ can be expressed in terms of its fourier transform $R(\mathbf{k_2}, z)$ using eq. 18.3.27.
We can now rewrite the above expression in terms of intermediate states:

$$T^{(+)}{}_{2a} - T^{(-)}{}_{2a} = -i\int d^4r_1d^4r_2\,(2\pi)^4\delta^4(P - r_1 - r_2)\cdot$$

$$\cdot\, i\chi_0N^+(p_4)N^+(p_3)\int d^4k_1\theta(k_1{}^0)\,\delta(k_1{}^2 - m^2)\,[R(\mathbf{k_1}, -k_1 - r_1)/(2\pi)^3]\cdot$$

$$\cdot\int d^4k_2\theta(k_2{}^0)\,\delta(k_2{}^2 - m^2)[R(\mathbf{k_2}, -k_2 - r_2)/(2\pi)^3]i\,\chi_0N(p_2)N(p_1)/2$$

which has the form:

$$T^{(+)}{}_{2a} - T^{(-)}{}_{2a} = -i\int d^4r_1d^4r_2(2\pi)^4\delta^4(P - r_1 - r_2)\Big[\int d^4k_1\theta(k_1{}^0)\,\delta(k_1{}^2 - m^2)\,R(\mathbf{k_1}, -k_1 - r_1)/(2\pi)^3\Big]\cdot$$

$$\cdot\Big[\int d^4k_2\theta(k_2{}^0)\,\delta(k_2{}^2 - m^2)R(\mathbf{k_2}, -k_2 - r_2)/(2\pi)^3\Big]T^{(-)}{}_{fn}T^{(+)}{}_{ni}/2!$$

where

$$T^{(-)}{}_{fn} = \chi_0N^+(p_4)N^+(p_3)N(r_2)N(r_1)$$

$$T^{(+)}{}_{ni} = \chi_0N^+(r_2)N^+(r_1)N(p_2)N(p_1)$$

if

$$N^+(p) = N(p) = 1 \qquad (18.5.74)$$

Eq. 18.5.74 implies the (finite) external leg renormalization must be

$$Z_p = -\left[\int d^4w\, f_p(w)(\Box + m^2)\Delta_F^{TT}(w)\right]^2 \qquad (18.5.74a)$$

by eqs. 18.5.64 and 18.5.65. Thus all external legs must be "lopped off."

The result is a theory that satisfies the unitarity condition (eq. 18.5.73) as shown in the above detailed discussion.

If we define

$$N(r) = \int d^4k\, \theta(k^0)\delta(k^2 - m^2)R(\mathbf{k}, -k - r) \qquad (18.5.75a)$$

$$= (2\pi)^{-4}\int d^4k\, d^4z\, \theta(k^0)\delta(k^2 - m^2)\, e^{-i(k+r)\cdot z}\, R(\mathbf{k}, z) \qquad (18.5.75b)$$

then the Two-Tier completeness expression becomes:

$$S_{fi} = \sum_n (2\pi)^{-3n}(n!)^{-1}\int\left(\prod_{j=1}^{n} d^4r_j N(r_j)\right)S_{fn}S_{ni}^{\dagger}\,\delta^4\left(P_n - \sum_{k=1}^{n} r_k\right) \qquad (18.5.75c)$$

This expression reflects the fact that ϕ particles are surrounded by a "cloud" of Y quanta. Thus we have achieved unitarity! For small momenta $r_j \ll M_c$, we find $N(r_j) \simeq \theta(r_j^0)\delta(r_j^2 - m^2)$ (eq. 18.5.75b with R(k, q) $\simeq$ 1.) $\theta(r_j^0)\delta(r_j^2 - m^2)$ is the form seen in conventional quantum field theory. For large momenta $N(r_j)$ is very different.

18.5.9 General Form of Propagators

In this section we have considered a scalar Two-Tier quantum field theory. We have seen that the Two-Tier Feynman propagator is well behaved near the light cone resulting in a finite ϕ^4 theory. This finite ϕ^4 theory approximates the results of conventional ϕ^4 theory at low energy thus implementing a correspondence principle: *At low energy results in Two-Tier quantum field theory approach the corresponding results of the corresponding conventional quantum field theory.*

The observations on Two-Tier field theory made in this chapter generally apply to Two-Tier versions of Quantum Electrodynamics, ElectroWeak Theory and the Standard Model as well as Two-Tier Quantum Gravity:

1. At low energy ($p^2 \ll M_c^2$ or large distances $z^2 \gg M_c^{-2}$) the Two-Tier quantum field theory is the same as the corresponding conventional quantum field theory to good approximation. (Correspondence Principle)

2. At high energy ($p^2 \gg M_c^2$ or short distances: $z^2 \ll M_c^{-2}$) Two-Tier quantum field theories (of physical interest) are well-behaved and finite.

3. Two-Tier quantum field theories (of physical interest) satisfy unitarity and Lorentz invariance (and in the case of quantum gravity their dynamical equations satisfy the requirements of general relativity).

The generality of these results is based on:

1. The expansion of the S matrix in time ordered products of field operators.
2. Wick's Theorem
3. The general form of all particle propagators in Two-Tier quantum field theories. All particle Feynman propagators have the form:

$$iG_F{}^{TT}{}_{\ldots}(y_1 - y_2) = \langle 0|T(\chi_{\ldots}(X(y_1)),\chi_{\ldots}(X(y_2)))|0\rangle \tag{18.5.76}$$

$$= \int d^4p\; iG_{F\ldots}(p)e^{-ip\cdot(y_1 - y_2)}\, R(\mathbf{p}, y_1 - y_2) \tag{18.5.77}$$

where $iG_{F\ldots}(p)$ is the conventional momentum space $\chi_{\ldots}$ particle propagator, and where … represents the relevant tensor and matrix indices. $R(\mathbf{p}, y_1 - y_2)$ introduces a damping factor in each particle propagator that eliminates divergences.

18.5.10 Scalar Particle Propagator

The Two-Tier propagator for the case of a free scalar particle is:

$$i\Delta_F{}^{TT}(y_1 - y_2) = \langle 0|T(\phi(X(y_1)),\phi(X(y_2)))|0\rangle \tag{18.5.51}$$

$$= i\int \frac{d^4p\; e^{-ip\cdot(y_1 - y_2)}\, R(\mathbf{p}, y_1 - y_2)}{(2\pi)^4\,(p^2 - m^2 + i\varepsilon)} \tag{18.5.52}$$

Since the mass m is not relevant at high energy we set m = 0. This enables us to obtain a more tractable expression for the propagator. After some manipulation the massless scalar propagator can be represented as:

$$i\Delta_F{}^{TT}(z) = -\beta[16\pi^3(AB)^{1/2}]^{-1}\int_{-\infty}^{\infty} dy_1 \iint_{-\infty}^{\infty} dy_2\cdot$$
$$\cdot\{\theta(z_0)\exp[-\beta((y_1 - z_0)^2B + (y_2 + z)^2A)/(4AB)] +$$
$$+ \theta(-z_0)\exp[-\beta((y_1 + z_0)^2B + (y_2 - z)^2A)/(4AB)]\}/(y_1{}^2 - y_2{}^2) \tag{18.5.78}$$

with $\beta = 4\pi^2 M_c{}^4\mathbf{z}^2$. Using

$$(y_1{}^2 - y_2{}^2)^{-1} = -0.5\int_0^{\infty} dq_1 \int_{-\infty}^{\infty} dq_2\; \theta(q_1{}^2 - q_2{}^2)\exp[iq_1y_1 - iq_2y_2] \tag{18.5.79}$$

we obtain the representation

$$i\Delta_F{}^{TT}(z^\mu) = (8\pi^2)^{-1}\int_0^{\infty} dq_1 \int_{-\infty}^{\infty} dq_2\; \theta(q_1{}^2 - q_2{}^2)\exp\{iq_1|z_0| + iq_2z - [A'q_1{}^2 + B'q_2{}^2]/[\beta'(\mathbf{z}^2 - z_0{}^2)]\} \tag{18.5.80}$$

where $|z_0|$ is the absolute value of z_0, $z^2 - z_0{}^2 = -z^\mu z_\mu$ and

$$A = A'/(1 - v^2) \tag{18.5.81}$$

$$B = B'/(1 - v^2) \tag{18.5.82}$$
$$\beta = 4\pi^2 M_c^4 \mathbf{z}^2 = \beta' z^2 \tag{18.5.83}$$

with $\mathbf{z} = |\vec{\mathbf{z}}|$ – the magnitude of the spatial vector $\vec{\mathbf{z}}$, and A and B given by eqs. 18.5.60 – 18.5.61.

The representation of $i\Delta_F^{TT}$ in eq. 18.5.80 is particularly useful in determining its low energy ($\ll M_c$), and its high energy ($\gg M_c$) behavior. The low energy behavior is governed by the linear terms in the exponential in eq. 18.5.80 since $\beta'(\mathbf{z}^2 - z_0^2)$ is very large in this limit:

$$i\Delta_F^{TT}(z'')_{low} \simeq (8\pi^2)^{-1} \int_0^\infty dq_1 \int_{-\infty}^\infty dq_2\, \theta(q_1^2 - q_2^2)\exp\{ iq_1|z_0| + iq_2 z\} \tag{18.5.84}$$

$$= [4\pi^2(\mathbf{z}^2 - z_0^2)]^{-1} \tag{18.5.85}$$
$$= i\Delta_F(z'')$$

equaling the exact massless, free, spin 0 Feynman propagator of conventional quantum field theory.

In the high energy limit when $\beta'(\mathbf{z}^2 - z_0^2)$ is small since $\mathbf{z}^2 \approx z_0^2$ (i.e. near the light cone), the quadratic terms in the exponential in eq. 18.5.80 dominate and $A' \simeq B'$. We then find

$$i\Delta_F^{TT}(z'')_{high} \simeq (8\pi^2)^{-1} \int_0^\infty dq_1 \int_{-\infty}^\infty dq_2 \theta(q_1^2 - q_2^2)\exp\{A'(q_1^2 + q_2^2)/[\beta'(\mathbf{z}^2 - z_0^2)]\} \tag{18.5.86}$$
$$= \pi M_c^4 |(\mathbf{z}^2 - z_0^2)|/8 \tag{18.5.87}$$

as in eq. 18.3.24. As pointed out earlier, eq. 18.5.87 corresponds to k^{-6} behavior in momentum space:

$$i\Delta_F^{TT}(k)_{high} \backsim k^{-6} \tag{18.5.87a}$$

18.5.11 Spin ½ Particle Propagator

For the case of a free, spin ½ particle the propagator is:

$$iS_F^{TT}(y_1 - y_2) = \langle 0|T(\bar{\psi}(X(y_1))\psi(X(y_2)))|0\rangle \tag{18.5.88}$$

$$= i \int \frac{d^4p\, e^{-ip\cdot(y_1 - y_2)}\, (\not{p} + m)\, R(\mathbf{p}, y_1 - y_2)}{(2\pi)^4\, (p^2 - m^2 + i\varepsilon)}$$

Again setting m = 0 we find a convenient representation in the form:

$$S_F^{TT}(z'') = i(8\pi^2)^{-1} \int_0^\infty dq_1 \int_{-\infty}^\infty dq_2\, \theta(q_1^2 - q_2^2)(\in(z_0)q_1\gamma_0 - q_2\vec{\mathbf{z}}\cdot\vec{\gamma}/z)\cdot$$
$$\cdot \exp\{iq_1|z_0| + iq_2 z - [A'q_1^2 + B'q_2^2]/[\beta'(\mathbf{z}^2 - z_0^2)]\} \tag{18.5.89}$$

using the same symbols and notation as eq. 18.5.80, and with $\in(z_0) = +1$ if $z_0 \geq 0$ and -1 otherwise.

The representation of S_F^{TT} in eq. 18.5.89 is useful in determining its low energy ($\ll M_c$), and high energy ($\gg M_c$) behavior. The low energy behavior is governed by the linear terms in the exponential in eq. 18.5.89 since $\beta'(\mathbf{z}^2 - z_0^2)$ is large in this limit:

$$S_F^{TT}(z^\mu)_{low} \simeq (8\pi^2)^{-1} \int_0^\infty dq_1 \int_{-\infty}^\infty dq_2\, \theta(q_1^2 - q_2^2)(\in(z_0)q_1\gamma_0 - q_2\vec{\mathbf{z}}\cdot\vec{\gamma}/z)\exp\{iq_1|z_0| + iq_2 z\} \tag{18.5.90}$$

$$= \not{z}[2\pi^2(\mathbf{z}^2 - z_0^2)^2]^{-1} \tag{18.5.91}$$

$$= S_F(z^\mu)$$

equaling the exact massless, spin ½ Feynman propagator of conventional quantum field theory. If we had not set m = 0 initially, we would have obtained the usual massive, spin ½ Feynman propagator.

In the high energy limit when $\beta'(\mathbf{z}^2 - z_0^2)$ is small since $\mathbf{z}^2 \approx z_0^2$ (i.e. near the light cone), the quadratic terms in the exponential in eq. 18.5.89 dominate and $A' \simeq B'$. We then find

$$S_F^{TT}(z^\mu)_{high} \simeq (8\pi^2)^{-1} \int_0^\infty dq_1 \int_{-\infty}^\infty dq_2 \theta(q_1^2 - q_2^2)(\in(z_0)q_1\gamma_0 - q_2\vec{\mathbf{z}}\cdot\vec{\gamma}/z)\exp\{A'(q_1^2 + q_2^2)/[\beta'(\mathbf{z}^2 - z_0^2)]\} \tag{18.5.92}$$

$$= i(8\pi^2)^{-1}\{\mathbf{z}^{-1}(\mathbf{z}^2 - z_0^2)^{3/2}2^{3/2}\pi^{7/2}M_c^6 z_0\gamma_0 - 4i(\mathbf{z}^2 - z_0^2)^2\pi^5 M_c^8 \vec{\mathbf{z}}\cdot\vec{\gamma})\} \tag{18.5.93}$$

The leading momentum dependence of the fourier transform of $S_F^{TT}(z^\mu)_{high}$ is

$$S_F^{TT}(p)_{high} \backsim M_c^6 p^{-7}\gamma_0 \tag{18.5.94}$$

18.5.12 Massless Spin 1 Particle Propagator

The Two-Tier Feynman propagator for the case of a free, massless, spin 1, gauge field particle (coupled to a conserved current) such as a photon is:

$$iD_F^{TT}(z)_{\mu\nu} = -i \int \frac{d^4p\, e^{-ip\cdot z}\, g_{\mu\nu} R(\mathbf{p}, y_1 - y_2)}{(2\pi)^4 (p^2 + i\varepsilon)} \tag{18.5.95}$$

The form of eq. 18.5.95 is the same as the scalar particle propagator multiplied by $-g_{\mu\nu}$. As a result we have the representation:

$$iD_F^{TT}(z)_{\mu\nu} = -(8\pi^2)^{-1} \int_0^\infty dq_1 \int_{-\infty}^\infty dq_2\, \theta(q_1^2 - q_2^2)\, g_{\mu\nu} \exp\{iq_1|z_0| + iq_2 z - [A'q_1^2 + B'q_2^2]/[\beta'(\mathbf{z}^2 - z_0^2)]\} \tag{18.5.96}$$

As before in the scalar particle case, the low energy behavior is governed by the linear terms in the exponential in eq. 18.5.96 since $\beta'(\mathbf{z}^2 - z_0^2)$ is very large in this limit:

$$iD_F^{TT}(z)_{\mu\nu low} \simeq -g_{\mu\nu}(8\pi^2)^{-1} \int_0^\infty dq_1 \int_{-\infty}^\infty dq_2\ \theta(q_1^2 - q_2^2)\exp\{iq_1|z_0| + iq_2 z\} \tag{18.5.97}$$

$$= -g_{\mu\nu}[4\pi^2(\mathbf{z}^2 - z_0^2)]^{-1} \tag{18.5.98}$$

$$= -ig_{\mu\nu}\Delta_F(z)$$

equaling the exact free, massless, spin 1 Feynman gauge field propagator of conventional quantum field theory.

In the high energy limit when $\beta'(\mathbf{z}^2 - z_0^2)$ is small since $\mathbf{z}^2 \approx z_0^2$ (i.e. near the light cone), the quadratic terms in the exponential in eq. 18.5.96 dominate, and $A' \simeq B'$. We then find

$$iD_F^{TT}(z)_{\mu\nu high} \simeq -(8\pi^2)^{-1} \int_0^\infty dq_1 \int_{-\infty}^\infty dq_2 \theta(q_1^2 - q_2^2)g_{\mu\nu}\exp\{ A'(q_1^2 + q_2^2)/[\beta'(\mathbf{z}^2 - z_0^2)]\} \tag{18.5.99}$$

$$= -g_{\mu\nu}\pi\, M_c^4\, |(\mathbf{z}^2 - z_0^2)|/8 \tag{18.5.100}$$

Eq. 18.5.100 corresponds to k^{-6} behavior in momentum space:

$$iD_F^{TT}(k)_{\mu\nu high} \backsim g_{\mu\nu} M_c^4 k^{-6} \tag{18.5.101}$$

18.5.13 Spin 2 Particle Propagator

The Two-Tier propagator for the case of a free, massless, spin 2 particle such as a graviton is:

$$i\Delta_{F2}^{TT}(z)_{\mu\nu\rho\sigma} = i \int \frac{d^4p\, e^{-ip\cdot z}\, b_{\mu\nu\rho\sigma}(p)R(\mathbf{p}, y_1 - y_2)}{(2\pi)^4\,(p^2 + i\varepsilon)} \tag{18.5.102}$$

in an appropriate gauge where $b_{\mu\nu\rho\sigma}(p)$ is a tensor that is independent of the coordinates. We can express eq. 18.5.102 in the form:

$$i\Delta_{F2}^{TT}(z)_{\mu\nu\rho\sigma} = (8\pi^2)^{-1} \int_0^\infty dq_1 \int_{-\infty}^\infty dq_2\ \theta(q_1^2 - q_2^2)\, \tilde{b}(z_0, z, q_1, q_2)_{\mu\nu\rho\sigma} \cdot$$

$$\cdot\exp\{ iq_1|z_0| + iq_2 z - [A'q_1^2 + B'q_2^2]/[\beta'(\mathbf{z}^2 - z_0^2)]\} \tag{18.5.103}$$

where $\tilde{b}(z_0, z, q_1, q_2)_{\mu\nu\rho\sigma}$ is a tensor generated from the $b_{\mu\nu\rho\sigma}(p)$ tensor.

As before in the scalar particle case, the low energy behavior is governed by the linear terms in the exponential in eq. 18.5.103 since $\beta'(\mathbf{z}^2 - z_0^2)$ is very large in this limit and we find that the covariant piece[202] behaves like:

---

[202] S. Weinberg, Phys. Rev. **135**, B1049 (1964); Phys. Rev. **138**, B988 (1965).

$$i\Delta_{F2}{}^{TT}(z)_{\mu\nu\rho\sigma lowCov} \simeq \tilde{\tilde{b}}_{\mu\nu\rho\sigma}(8\pi^2)^{-1}\int_0^\infty dq_1 \int_{-\infty}^\infty dq_2\, \theta(q_1{}^2 - q_2{}^2)\exp\{\, iq_1|z_0| + iq_2 z\} \tag{18.5.104}$$

$$= \tilde{\tilde{b}}_{\mu\nu\rho\sigma}[4\pi^2(\mathbf{z}^2 - z_0{}^2)]^{-1} \tag{18.5.105}$$

$$= i\Delta_F(z^\mu)\, \tilde{\tilde{b}}_{\mu\nu\rho\sigma}$$

where

$$\tilde{\tilde{b}}_{\mu\nu\rho\sigma} = ½\, [\eta_{\mu\rho}\eta_{\nu\sigma} + \eta_{\mu\sigma}\eta_{\nu\rho} - \eta_{\mu\nu}\eta_{\rho\sigma}] \tag{18.5.106}$$

so that the expression in eq. 18.5.105 equals the corresponding covariant piece of the exact free, massless, spin 2 Feynman propagator of conventional quantum field theory.

In the high energy limit when $\beta'(\mathbf{z}^2 - z_0{}^2)$ is small since $\mathbf{z}^2 \approx z_0{}^2$ (i.e. near the light cone), the quadratic terms in the exponential in eq. 18.5.103 dominate, and $A' \simeq B'$. We then find

$$i\Delta_{F2}{}^{TT}(z)_{\mu\nu\rho\sigma high} \simeq (8\pi^2)^{-1}\int_0^\infty dq_1 \int_{-\infty}^\infty dq_2 \theta(q_1{}^2 - q_2{}^2)\ \tilde{b}(z_0, z, q_1, q_2)_{\mu\nu\rho\sigma}\,\cdot$$

$$\cdot\, \exp\{A'(q_1{}^2 + q_2{}^2)/[\,\beta'(\mathbf{z}^2 - z_0{}^2)]\} \tag{18.5.107}$$

and the covariant piece behaves like

$$i\Delta_{F2}{}^{TT}(z)_{\mu\nu\rho\sigma highCov} \simeq \tilde{\tilde{b}}_{\mu\nu\rho\sigma}\pi M_c{}^4|(\mathbf{z}^2 - z_0{}^2)|/8 \tag{18.5.108}$$

The coordinate space behavior of eq. 18.5.108 corresponds to k^{-6} behavior in momentum space:

$$i\Delta_{F2}{}^{TT}(k)_{\mu\nu\rho\sigma highCov} \sim \tilde{\tilde{b}}_{\mu\nu\rho\sigma}k^{-6} \tag{18.5.109}$$

The high-energy behavior of the spin 2 propagator in momentum space results in a Two-Tier theory of quantum gravity that has no high-energy divergences and is thus finite. See chapter 9 for a detailed discussion.

18.6 Finite Quantum Gravity

18.6.1 Introduction - Two-Tier Quantum Gravity: Finite!

There are numerous excellent books and monographs on classical gravity and a large literature on quantum gravity.[203] Therefore our discussion will assume the reader is familiar with classical General Relativity and aware of attempts to create quantum theories of gravity.

---

[203] H. Weyl, *Space, Time, Matter* (Dover, New York, 1950); L. D. Landau and E. M. Lifshitz, *The Classical Theory of Fields*, (Addison-Wesley, New York, 1962); S. Weinberg, *Gravitation and Cosmology*, (John Wiley & Sons, New York, 1972); C. W. Misner, K. S. Thorne and J. A. Wheeler, *Gravitation*, (W. H. Freeman, San Francisco, 1973); B. S. DeWitt, Phys. Rev. **162**, 1239 (1967), **162**, B1195 (1967); R. P. Feynman, Acta Physica Polonica **24**, 697 (1963); S. Deser and P. van Nieuwenhuizen, Phys. Rev. Letters **32**, 245 (1974); S. Deser, H.-S. Tsao and P. van

We will begin by establishing the general form of Two-Tier classical General Relativity and then proceed to define a quantization procedure. We will work in Minkowski space with three space and one time dimension. The flat-space metric $\eta_{\alpha\beta}$ is defined as diagonal with $\eta_{00} = 1$ and $\eta_{ij} = -\delta_{ij}$ for i, j = 1, 2, 3.

18.6.2 Two-Tier General Relativity

In developing Quantum Gravity we will make the same ansatz that we have made throughout our development of Two-Tier quantum field theories: all field expressions are functions of the X coordinate field system, which in turn are functions of the "ordinary" y space-time coordinate system. Two-Tier Theory of Quantum Gravity is invariant under special relativistic transformations. The dynamical field equations, which are strictly functions of the X coordinates, are covariant under general relativistic transformations.

We define the proper time differential $d\tau$ as

$$d\tau^2 = g_{\mu\nu}(X(y))dX^\mu dX^\nu \tag{18.6.1}$$

where, as usual,

$$X^\mu(y) = y^\mu + i\, Y^\mu(y)/M_c^2$$

Thus eq. 18.6.1 could be written:

$$d\tau^2 = g_{\mu\nu}(X(y))(\eta^\mu{}_\alpha + iM_c^{-2}\partial Y^\mu/\partial y^\alpha)(\eta^\nu{}_\beta + iM_c^{-2}\partial Y^\nu/\partial y^\beta)dy^\alpha dy^\beta \tag{18.6.2}$$

The inverse of $g_{\mu\nu}$, denoted $g^{\nu\lambda}$, satisfies

$$g_{\mu\nu}(X(y))g^{\nu\lambda}(X(y)) = \delta_\mu{}^\lambda \tag{18.6.3}$$

Since the algebraic manipulation of the tensor indices is the same as that of the conventional theory of gravitation the Two-Tier affine connection is:

$${}_X\Gamma^\sigma{}_{\lambda\mu} = \tfrac{1}{2}\, g^{\nu\sigma}\{\partial g_{\mu\nu}/\partial X^\lambda + \partial g_{\lambda\nu}/\partial X^\mu - \partial g_{\lambda\mu}/\partial X^\nu\} \tag{18.6.4}$$

The Two-Tier Riemann-Christoffel curvature tensor is:

$${}_XR^\lambda{}_{\mu\nu\kappa} \equiv \partial_X\Gamma^\lambda{}_{\mu\nu}/\partial X^\kappa - \partial_X\Gamma^\lambda{}_{\mu\kappa}/\partial X^\nu + {}_X\Gamma^\alpha{}_{\mu\nu}\, {}_X\Gamma^\lambda{}_{\kappa\alpha} - {}_X\Gamma^\alpha{}_{\mu\kappa}\, {}_X\Gamma^\lambda{}_{\nu\alpha} \tag{18.6.5}$$

Nieuwenhuizen, "One Loop Divergences of the Einstein-Yang-Mills System", Brandeis Univ. preprint (1974); S. Weinberg, Phys. Rev. **138**, B988 (1965); L. Smolin, *Three Roads to Quantum Gravity*, (Basic Books, New York, 2001); L. Smolin, "How Far are We From the Quantum Theory of Gravity", (Univ. Waterloo preprint (2003) and references therein; T. Thiemann, "Lectures on Loop Quantum Gravity", Preprint AEI-2002-087, Albert Einstein Insitute, Golm, Germany (2002) and references therein; A. pais and G. E. Uhlenbeck, Phys. Rev. **79**, 145 (1950); G. E. Uhlenbeck, "Lecture Notes on General Relativity", The Rockefeller University (1967), unpublished; S. Blaha, "Generalization of Weyl's Unified Theory to Encompass a Non-Abelian Internal Symmetry Group" SLAC-PUB-1799, Aug 1976; S. Blaha, "Quantum Gravity and Quark Confinement" Lett. Nuovo Cim. **18**, 60 (1977); R. Utiyama, Phys. Rev. **101**, 1597 (1956); T. W. B. Kibble, J. Math. Phys. **2**, 212 (1961); R. Arnowitt, S. Deser, and C. W. Misner, Phys. Rev. **117**, 1595 (1960); and references therein.

and the Two-Tier Ricci tensor is

$$_X R_{\mu\nu} = {_X R}^{a}{}_{\mu a\nu} \qquad (18.6.6)$$

The Two-Tier curvature scalar is

$$_X R = g^{\mu\nu}{_X R_{\mu\nu}} \qquad (18.6.7)$$

We also define

$$_X R_{\lambda\mu\nu\kappa} = g_{\lambda a}\, {_X R}^{a}{}_{\mu\nu\kappa} \qquad (18.6.8)$$

with the result

$$_X R_{\lambda\mu\nu\kappa} = ½[\partial^2 g_{\lambda\nu}/\partial X^\kappa \partial X^\mu - \partial^2 g_{\mu\nu}/\partial X^\kappa \partial X^\lambda - \partial^2 g_{\lambda\kappa}/\partial X^\nu \partial X^\mu + \partial^2 g_{\mu\kappa}/\partial X^\nu \partial X^\lambda] + g_{a\beta}[\,{_X\Gamma}^{a}{}_{\nu\lambda}\, {_X\Gamma}^{\beta}{}_{\mu\kappa} - {_X\Gamma}^{a}{}_{\kappa\lambda}\, {_X\Gamma}^{\beta}{}_{\mu\nu}] \qquad (18.6.9)$$

We denote the fact that all quantities in eqs. 18.6.4 – 18.6.8 are only functions of X by placing a left subscript X on each quantity.

The algebraic properties, and the Bianchi identities, satisfied by $_X R_{\lambda\mu\nu\kappa}$ in the Two-Tier theory of gravitation are identical to those of the conventional theory with all derivatives being with respect to X.

The Two-Tier version of Einstein's field equations is:

$$_X R_{\mu\nu} - ½\, g_{\mu\nu}\, {_X R} = -8\pi G\, T_{\mu\nu} \qquad (18.6.10)$$

where G is Newton's gravitational constant (6.674×10^{-11} $m^3 kg^{-1} s^{-2}$) and $T_{\mu\nu}$ is the energy-momentum tensor – also strictly a function of X. It is convenient to define the coupling constant

$$\kappa = \sqrt{4\pi G} \qquad (18.6.11)$$

18.6.3 Lagrangian Formulation

We will now formulate a Two-Tier Quantum Gravity theory following the same ansatz that we have used throughout this book.

18.6.4 Unified Standard Model and Quantum Gravity Lagrangian

We define the Lagrangian, and action, for the unified quantum field theory of gravitation and the Standard Model as

$$L_{\text{Unified}} = \int d^4y\, \mathscr{L}_{\text{Unified}} \qquad (18.6.12)$$

$$\mathscr{L}_{\text{Unified}} = J\sqrt{g(X)}\, (\mathscr{L}_F^{\text{Grav}}(X^\mu) + \mathscr{L}_F^{\text{SM}}(X^\mu)) + L_C \qquad (18.6.13)$$

with

$$\mathscr{L}_F^{\text{Grav}}(X^\mu) = (2\kappa^2)^{-1}\, {_X R} \qquad (18.6.14)$$

where $\mathscr{L}_F^{\text{SM}}$ is the complete "normal" Quantum Field theory Lagrangian for the Standard Model version under consideration written in a general covariant form, g(X) is the absolute value of the determinant of

$g_{\mu\nu}$, and J is the Jacobian of eq. A.21. *All particle fields in* $\mathscr{L}_F^{SM}$ *are assumed to be functions of the* X^μ *coordinate only. The dependence of the particle fields on the "underlying" coordinates* y^μ *is assumed to be solely through* X^μ. The Lagrangian $\mathscr{L}_{\text{Unified}}$ is a separable Lagrangian of the type of eq. 18-A.26 embodying the composition of extrema described in Appendix 18-A.

As in all of cases that we have considered, we have specified the coordinate part of the Lagrangian $\mathscr{L}_C$ as

$$\mathscr{L}_C = -\tfrac{1}{4}\, F_Y^{\mu\nu} F_{Y\mu\nu} \qquad (18.2.15)$$

with

$$F_{Y\mu\nu} = \partial Y_\mu/\partial y^\nu - \partial Y_\nu/\partial y^\mu \qquad (18.2.14)$$

and

$$F_Y^{\mu\nu} = \eta^{\mu\alpha}\eta^{\nu\beta} F_{Y\alpha\beta} \qquad (18.6.15)$$

18.6.5 Why Are the Y Field Dynamics Independent of the Gravitational Field?

It is evident from eqs. 18.6.12-3 and 18.6.15 that the Y field is truly free *in a flat universe* and, in particular, does not depend on the gravitational field as represented by $\sqrt{g}$ and $g_{\mu\nu}$. Our rationale for this formulation is described in chapter 10. For the moment it suffices to make the following remarks. The Y field is a quantum field at each point in space-time including regions with ultra-strong gravitational fields such as the neighborhoods of black holes. If Y were to depend on the gravitational field then the Y field could be appreciable in such regions and might even be a "classical" field. In this case we would have new dimensions, allbeit imaginary, for which no evidence exists as yet.

Lastly, the non-invariance of the Y part of the action under general coordinate transformations effectively creates an "absolute" coordinate system – actually a class of "absolute coordinate systems" – namely the class of inertial reference frames that are related to each other by special relativistic transformations. This feature does not conflict with our knowledge of the universe. The universe appears to be almost flat. The large-scale distribution of masses is responsible for this flatness. The flatness, or flattened space if it is slightly curved, together with Mach's Principle (inertial forces are absent in the reference frame determined by the distribution of masses in the universe) selects a preferred class of local reference frames – local inertial reference frames. Since space is almost flat, or flat, these local reference frames occupy a large volume (if we exclude regions with intense gravitational fields.) We can define the Y field within this class of local inertial reference frames in each locale and establish a satisfactory quantum field theory. Thus we have a dynamics defined in the variable X, which we require to be covariant under general coordinate transformations, and a local "ground state" that "breaks" general coordinate invariance down to special relativistic invariance. See chapter 10 for a more complete discussion.

18.6.6 No "Space-time Foam"

The fact that our unified theory of the known forces of Nature *self-consistently* has a weak gravitational field at high energies (the graviton sector is finite to all orders in perturbation theory) supports the formulation of eq. 18.6.12-3. Gravity becomes weaker at ultra-short distances. Therefore space-time is not quantum foam at ultra-short distances but rather smooth and flat a là special relativity – consistent with our formulation.

18.6.7 Quantum Gravity – Scalar Particle Model Lagrangian

While the application of the Two-Tier approach to the unified theory is a straightforward extension of the concepts and approaches described in the preceding chapters, it is useful to consider a simplified model that minimizes the tensorial verbiage so that the concepts and features might better stand out. The procedure differs only in detail from the case of gauge fields.

The introduction of spinor fields requires the use of a Two-Tier vierbein formalism, which is straightforward to develop. A Two-Tier vierbein field $e^{\mu}{}_{a}$ is a function of X, $e^{\mu}{}_{a}(X)$, with $g_{\mu\nu} = e_{\mu a}(X)e_{\nu}{}^{a}(X)$ where the index a is an index of a flat tangent space defined at each space-time point. The Two-Tier formulation of a vierbein theory is similar to the other Two-Tier formulations that we have considered and will not be developed here.

Thus we will consider the Lagrangian model for a scalar particle field interacting with the $g_{\mu\nu}$ gravitational field:

$$L_{GS} = \int d^4y \sqrt{g(X)}\ (\mathscr{L}_{GS} + \mathscr{L}_{C}) \qquad (18.6.16)$$

$$\mathscr{L}_{GS} = J\ \mathscr{L}_{F}^{Grav}(X^{\mu}) + J\ \mathscr{L}_{F\phi}(X^{\mu}) \qquad (18.6.17)$$

with covariant versions of eqs. 5.21 and 5.24:

$$\mathscr{L}_{F\phi} = \tfrac{1}{2}\,[\, g^{\mu\nu}\partial\phi/\partial X^{\mu}\ \partial\phi/\partial X^{\nu} - m^2\phi^2] + \mathscr{L}_{F\phi int} \qquad (18.6.18)$$

$$\mathscr{L}_{F\phi int} = \tfrac{1}{4!}\,\chi_0\ \phi(X(y))^4 + \tfrac{1}{2}\,(m^2 - m_0{}^2)\phi^2 \qquad (18.6.19)$$

18.6.8 A Justifiable Weak Field Approximation for Quantum Gravity

Many discussions of quantizing conventional gravity make a weak field approximation for the gravity sector which, in view of divergences in the resulting quantum field theory, are impossible to justify:

$$g_{\mu\nu} = \eta_{\mu\nu} + \kappa h_{\mu\nu} \qquad (18.6.20)$$

where $\eta_{\mu\nu}$ is the flat space-time metric and $h_{\mu\nu}$ is a "small" deviation ($<h_{\mu\nu}> \ll 1$) from the flat space-time metric.

The Two-Tier formulation of quantum gravity is finite and the effective field becomes increasingly weaker at short distances. Thus the weak field approximation becomes *more accurate* at short distances:

$$g_{\mu\nu}(X(y)) \simeq \eta_{\mu\nu} + \kappa h_{\mu\nu}(X(y)) \qquad (18.6.21a)$$

At short distances space-time can be considered approximately flat (except possibly in the neighborhood of singularities) with quantum fluctuations embodied in $h_{\mu\nu}$. Thus eq. 18.6.21a is reasonable within the context of Two-Tier Quantum Gravity.

To first order in $h_{\mu\nu}$ the square root of the absolute value of the determinant of the metric tensor is:

$$\sqrt{g(X)} \simeq 1 + \tfrac{1}{2}\,\kappa h^{\sigma}{}_{\sigma}(X(y)) \qquad (18.6.21b)$$

18.6.9 Quantization of Quantum Gravity – Scalar Particle Model

We now proceed to quantize gravity based on the linearization of the gravitational field equations in the weak field approximation. Assuming eq. 18.6.21a and keeping terms to first order in $h_{\mu\nu}$ gives the affine connection:

$$_X\Gamma^\sigma{}_{\mu\nu} = \tfrac{1}{2}\,\kappa\eta^{\sigma\alpha}[\partial h_{\alpha\nu}/\partial X^\mu + \partial h_{\alpha\mu}/\partial X^\nu - \partial h_{\mu\nu}/\partial X^\alpha] + O(h^2) \qquad (18.6.22)$$

and the Ricci tensor:

$$_XR_{\mu\nu} = \partial_X\Gamma^\lambda{}_{\lambda\mu}/\partial X^\nu - \partial_X\Gamma^\lambda{}_{\mu\nu}/\partial X^\lambda + O(h^2) \qquad (18.6.23)$$

Thus the linearized gravitation lagarangian terms are

$$L^{Grav} = \int d^4y\,\sqrt{g(X)}\,J\mathcal{L}_F^{Grav}(X^\mu) \rightarrow L^{Grav}{}_{linear} = \int d^4y\,J\mathcal{L}^{Grav}{}_{linear}(X^\mu) \qquad (18.6.24)$$

The scalar particle Lagrangian terms become

$$\begin{aligned} L^\phi = \int d^4y\sqrt{g(X)}J\mathcal{L}_{F\phi} \rightarrow \int d^4yJ\{[\tfrac{1}{2}(\eta^{\mu\nu}\partial_\mu\phi\partial_\nu\phi - m^2\phi^2) + \mathcal{L}_{F\phi int}] + \\ + \tfrac{1}{2}\kappa h^{\mu\nu}\partial_\mu\phi\partial_\nu\phi + \tfrac{1}{4}\kappa h(\eta^{\mu\nu}\partial_\mu\phi\partial_\nu\phi - m^2\phi^2) + \\ + \tfrac{1}{2}\kappa h\mathcal{L}_{F\phi int}\} \end{aligned} \qquad (18.6.25)$$

with the notation $h = h^\sigma{}_\sigma$ and using

$$\partial_\mu \equiv \partial/\partial X^\mu \qquad (18.6.26)$$

$\eta^{\mu\nu}$ and $\eta_{\mu\nu}$ are used to raise and lower indices in keeping with the linearized, weak field approximation.

The Y terms in the Lagrangian are (as previously):

$$L^Y = \int d^4y\mathcal{L}_C = -\tfrac{1}{4}\int d^4y\eta^{\mu\nu}\eta^{\alpha\beta}F_{Y\mu\alpha}F_{Y\nu\beta} \qquad (18.6.27)$$

We will lump the higher order terms (in h) in the gravity part of the Lagrangian, and the scalar particle part of the Lagrangian, into

$$L_{Higher} = \int d^4y\,J\mathcal{L}_{Higher}(h, \phi) \qquad (18.6.28)$$

Thus the complete lagragian for a scalar particle interacting with gravitons is

$$L_{GS} = L^{Grav}{}_{linear} + L^\phi{}_{linear} + L^Y + L_{Higher} \qquad (18.6.29)$$

The linearized gravitational Lagrangian term $L^{Grav}{}_{linear}$ generates the field equations:

$$\Box_X h_{\mu\nu} + \partial_\nu\partial_\mu h - \partial_\alpha\partial_\nu h^\alpha{}_\mu - \partial_\alpha\partial_\mu h^\alpha{}_\nu = \kappa S_{\mu\nu} \qquad (18.6.30)$$

where

$$\partial_\mu S^\mu{}_\nu = \tfrac{1}{2}\, \partial_\nu S^\sigma{}_\sigma \tag{18.6.31}$$

to 0th order in h and where

$$\Box_X = (\partial/\partial X^\nu)(\partial/\partial X_\nu) \tag{18.6.32}$$

The most general coordinate transformation that maintains the weakness of the gravitational field has the form:

$$y^a \rightarrow\ y'^a = y^a + \chi^a(X(y)) \tag{18.6.33}$$

This transformation induces a gauge transformation in $h_{\mu\nu}$ to:

$$h'_{\mu\nu} = h_{\mu\nu} - \partial_\mu \chi_\nu - \partial_\nu \chi_\mu \tag{18.6.34}$$

It is easy to verify that eq. 18.6.30 is satisfied by $h'_{\mu\nu}$ if it is satisfied by $h_{\mu\nu}$.

Let us assume that we perform a gauge transformation making $h_{\mu\nu}$ traceless:

$$h^\sigma{}_\sigma = 0 \tag{18.6.35}$$

and choose the gauge

$$\partial^\mu h_{\mu\nu} = 0 \tag{18.6.36}$$

then eq. 18.6.30 becomes the wave equation:

$$\Box_X\, h_{\mu\nu} = \kappa S_{\mu\nu} \tag{18.6.37}$$

Another gauge transformation of the free field $h_{\mu\nu}$ (if $S_{\mu\nu} = 0$) makes

$$h_{\mu 0} = h_{0\mu} = 0 \tag{18.6.38}$$

while retaining

$$h_{\mu\nu} = h_{\nu\mu} \tag{18.6.39}$$

The general solution[204] for the free field $h_{\mu\nu}$ (with $S_{\mu\nu} = 0$ in eq. 18.6.37) can be expressed as a fourier expansion:

$$h_{\mu\nu}(X(y)) = \int d^3k\, N_0(k) \sum_{\lambda=1}^{2} \varepsilon_{\mu\nu}(k, \lambda)[a(k,\lambda)\, e^{-ik\cdot X} + a^\dagger(k,\lambda)\, e^{ik\cdot X}] \tag{18.6.40}$$

where $\lambda = 1,2$ labels the ± 2 helicity states, and where $N_0(k)$ is specified by eq. 18.2.25. The equal time ($y'^0 = y^0$) commutation relations are:

$$[h_{\mu\nu}(X(y)), h_{\alpha\beta}(X(y'))] = [\pi_{\mu\nu}(X(y)), \pi_{\alpha\beta}(X(y'))] = 0 \tag{18.6.41}$$

$$[h_{\alpha\beta}(X(y')), \pi_{\mu\nu}(X(y))] = i\, D_{\alpha\beta,\mu\nu}(\mathbf{X}(y) - \mathbf{X}(y')) \tag{18.6.42}$$

for μ, ν = 1, 2, 3 and where

[204] S. Weinberg, Phys. Rev. **135**, B1049 (1964); Phys. Rev. **138**, B988 (1965)

$$\pi_{\mu\nu}(X(y)) = \partial h_{\mu\nu}(X(y))/\partial y^0 \qquad (18.6.43)$$

in the Y Coulomb gauge where $X^0 = y^0$. $\mathscr{D}_{\alpha\beta,\mu\nu}$ is specified by:

$$\mathscr{D}_{\alpha\beta,\mu\nu}(X(y) - X(y')) = \int d^3k\, e^{i\,\mathbf{k}\cdot(\mathbf{X}(y) - \mathbf{X}(y'))}\, \Pi_{\alpha\beta\mu\nu}(\mathbf{k})/(2\pi)^3 \qquad (18.6.44)$$

$$\Pi_{\alpha\beta\mu\nu}(\mathbf{k}) = \tfrac{1}{2}\,[(\delta_{\alpha\mu} - k_\alpha k_\mu/\mathbf{k}^2)(\delta_{\beta\nu} - k_\beta k_\nu/\mathbf{k}^2) + (\delta_{\alpha\nu} - k_\alpha k_\nu/\mathbf{k}^2)(\delta_{\beta\mu} - k_\beta k_\mu/\mathbf{k}^2) -$$

$$- (\delta_{\alpha\beta} - k_\alpha k_\beta/\mathbf{k}^2)(\delta_{\mu\nu} - k_\mu k_\nu/\mathbf{k}^2)] \qquad (18.6.45)$$

where $\alpha, \beta, \mu, \nu = 1, 2, 3$.

The "transverse" graviton propagator can be represented as a time-ordered product of field operators:

$$i\Delta_{F2}{}^{TT}(y_1 - y_2)_{\lambda\tau\rho\sigma} = \langle 0|T(h_{\lambda\tau}(X(y_1)), h_{\rho\sigma}(X(y_2)))|0\rangle \qquad (18.6.46)$$

$$= -i \int \frac{d^4k\, e^{-ik\cdot(y_1 - y_2)}\, b_{\lambda\tau\rho\sigma}(k) R(\mathbf{k}, y_1 - y_2)}{(2\pi)^4\, (k^2 + i\varepsilon)}$$

where $R(\mathbf{k}, y_1 - y_2)$ is the gaussian factor appearing in propagators throughout Two-Tier theories, $\mathbf{k}$ is a spatial 3-vector, and where $b_{\mu\nu\rho\sigma}(k)$ is a function of k only:

$$b_{\alpha\beta\mu\nu}(k) = \tfrac{1}{2}[(\eta_{\alpha\mu} - k_\alpha k_\mu/\mathbf{k}^2)(\eta_{\beta\nu} - k_\beta k_\nu/\mathbf{k}^2) + (\eta_{\alpha\nu} - k_\alpha k_\nu/\mathbf{k}^2)(\eta_{\beta\mu} - k_\beta k_\mu/\mathbf{k}^2) - (\eta_{\alpha\beta} - k_\alpha k_\beta/\mathbf{k}^2)(\eta_{\mu\nu} - k_\mu k_\nu/\mathbf{k}^2)] \qquad (18.6.47)$$

where $\alpha, \beta, \mu, \nu = 0, 1, 2, 3$.

The quantum gravitational interaction also has an "instantaneous" part (similar to the instantaneous Coulomb interaction of QED) in addition to the transverse interaction embodied in eq. 18.6.46. This "instantaneous" interaction contains the Newtonian potential (described later) as its large distance limit. The sum of the instantaneous interaction and the transverse interaction gives the total gravitational interaction.

The above graviton propagator has the form given in eq. 6.102. The caculation of the leading behavior is the same as that of the Two-Tier scalar boson propagator except for the presence of factors such as $\eta_{\rho\sigma}$. The leading momentum dependence of the graviton propagator in momentum space is

$$i\Delta_{F2}{}^{TT}(p)_{\lambda\tau\rho\sigma} \backsim p^{-6} \qquad (18.6.48)$$

The graviton vertices in Two-Tier Quantum Gravity will be described within the framework of the path integral formulation.

18.6.10 Quantum Gravity–Scalar Particle Model Path Integral

A path integral formalism can be developed for Two-Tier Quantum Gravity interacting with matter fields. In this section we will consider the case of a matter field consisting of massive scalar bosons with a quartic interaction. The path integral formalism that we develop is similar to that of Yang-Mills theories in the previous chapter.

The Two-Tier path integral for a Quantum Gravity–Scalar Particle Theory can be written as:

$$Z(J, J^{\mu\nu}) = N \int D\phi Dh DY \Delta_{FPG}(h)\delta(F(h)) \exp\{i\mathscr{J}\int d^4y[\mathscr{A}(\mathscr{L}^{Grav}{}_{linear}(X^\mu) +$$

$$+ \mathscr{L}^{\phi}{}_{linear}(X^\mu) + \mathscr{L}_{Higher}(h, \phi)) + \mathscr{L}_C(X, y) +$$

$$+ j_\mu(y)Y^\mu(y) + J(y)\phi(X) + \mathscr{J}J^{\mu\nu}(y)h_{\mu\nu}(X)]\}|_{j_\mu = 0} \tag{18.6.49}$$

where $\delta(F(h))$ specifies the gauge as a functional delta function, and $\Delta_{FPG}(h)$ is the corresponding Fadeev-Popov determinant. $\mathscr{J}$ is the Jacobian for the transformation from y coordinates to X coordinates. The Fadeev-Popov determinant $\Delta_{FPG}(h)$ can be calculated in the standard way. First we note

$$\delta(F(h^\chi)) = \delta(\chi - \chi_0)\ |\det \delta F(h_{\mu\nu}{}^\chi(X))/\delta\chi(X)|^{-1}|_{F(h)=0} \tag{18.6.50}$$

where

$$h_{\mu\nu}{}^\chi = h_{\mu\nu} - \partial_\mu \chi_\nu - \partial_\nu \chi_\mu \tag{18.6.34}$$

Then

$$\Delta_{FPG}(h) = \ |\det \delta F(h^\chi(X))/\delta\chi(X)||_{F(h)=0} \tag{18.6.51}$$

We will choose the gauge of eq. 18.6.36 to evaluate the Fadeev-Popov determinant. Under an infinitesimal gauge transformation of the form:

$$h_{\mu\nu}{}^\chi(X) = h_{\mu\nu}(X) - \partial_\mu \chi_\nu - \partial_\nu \chi_\mu \tag{18.6.52}$$

which preserves the weak field nature of $h_{\mu\nu}$, we find

$$F_\nu(h^\chi) = \partial^\mu (h_{\mu\nu}(X) - \partial_\mu \chi_\nu - \partial_\nu \chi_\mu)$$
$$= - \Box_X \chi_\nu(X) - \partial_\nu\partial^\mu \chi_\mu \tag{18.6.53}$$

Thus

$$\delta F_\mu(h^\chi(X))/\delta\chi^\nu(X) = - \eta_{\mu\nu}\Box_X - \partial_\mu\partial_\nu \tag{18.6.54}$$

and

$$\Delta_{FP}(A) = |\det (- \eta_{\mu\nu}\Box_X - \partial_\mu\partial_\nu)||_{F(h)=0} \tag{18.6.55}$$

We note the Two-Tier Fadeev-Popov determinant is solely a function of the X coordinates. The determinant only introduces an overall multiplicative constant that can be absorbed into the normalization constant N. This fact becomes evident if we follow the standard procedure and rewrite the determinant as a path integral over anti-commuting c-number fields with a ghost Lagrangian. Then we see that the ghost does not interact with the other fields and thus only generates an overall multiplicative constant that can be absorbed in N:

$$\Delta_{FPG}(h) = \int Dc^* Dc \exp[\, i\int d^4X\, \mathscr{L}^{ghost}(X^\mu)] \tag{18.6.56}$$

where

$$\mathscr{L}^{ghost}(X^\mu) = c^{\mu *}(X)[\eta_{\mu\nu}\Box_X + \partial_\mu\partial_\nu]c^\nu(X) \tag{18.6.57}$$

We now go through the same analysis as we did in the ϕ^4 theory path integral example and the Yang-Mills path integral example (with some superficial differences). First we integrate the linear part of the Y field Lagrangian as we did previously. Then we integrate the linear part of the ϕ field Lagrangian as done previously. Lastly we integrate the linear part of the gravitation Lagrangian to obtain the path integral for the perturbative expansion with the result:

$$\begin{aligned} Z(J, J^{\mu\nu}) = N\, \{&\exp[i\int d^4y\, \mathscr{L}_{Higher}(\partial/\partial y^\nu, -i\delta/\delta J^{\mu\nu}(y), -i\delta/\delta J(y))]\cdot \\ &\cdot \exp[-\tfrac{1}{2}\, i\int d^4y_1 d^4y_2\, J^{\mu\nu}(y_1)\Delta_{F2}{}^{TT}(y_1 - y_2, z)_{\mu\nu\rho\sigma} J^{\rho\sigma}(y_2)]\cdot \\ &\cdot \exp[-\tfrac{1}{2}\, i\int d^4y_1 d^4y_2\, J(y_1)\Delta_F{}^{TT}(y_1 - y_2, z)J(y_2)]\}\big|_{z=y_1-y_2} \end{aligned} \tag{18.6.58}$$

There are two issues that arise in the development of eq. 18.6.58:

1.) The integral over y in $\int d^4y\,\mathscr{L}_{Higher}$ which began as the integral $\int d^4y\,\mathscr{J}\,\mathscr{L}_{Higher} = \int d^4X\,\mathscr{L}_{Higher}$ in eq. 18.6.49; and

2.) The handling of derivatives with respect to X in $\mathscr{L}_{Higher}$.

These are resolved by the following respective observations:

1.) See the discussion following eqs. 18.6.34 that apply here as well without change.

2.) We note that the derivative with respect to X of the graviton propagator (eq. 18.6.46-7) is specified by the following:

$$\partial i\Delta_{F2}{}^{TT}(y_1 - y_2)_{\lambda\tau\rho\sigma}/\partial X^\mu(y_1) = \partial[i\Delta_{F2}{}^{TT}(y_1 - y_2, z)_{\lambda\tau\rho\sigma}]/\partial y_1{}^\mu\,\big|_{z = y_1 - y_2} \tag{18.6.59}$$

where

$$i\Delta_{F2}{}^{TT}(y_1 - y_2, z)_{\lambda\tau\rho\sigma} = -i\int \frac{d^4k\, e^{-ik\cdot(y_1 - y_2)}\, b_{\lambda\tau\rho\sigma}(k)R(\mathbf{k}, z)}{\qquad}$$

$$(2\pi)^4\,(k^2 + i\varepsilon) \qquad (18.6.60)$$

Thus

$$\frac{\partial\, i\Delta_{F2}{}^{TT}(y_1 - y_2)_{\lambda\tau\rho\sigma}}{\partial X^\mu(y_1)} = -i\int \frac{d^4k\, e^{-ik\cdot(y_1 - y_2)}\,(-ik_\mu)b_{\lambda\tau\rho\sigma}(k)R(\mathbf{k}, y_1 - y_2)}{(2\pi)^4\,(k^2 + i\varepsilon)} \qquad (18.6.61)$$

Therefore derivatives with respect to X in the interaction Lagrangian terms can be replaced by derivatives with respect to y if the graviton propagator is generalized to eq. 18.6.60. After taking all derivatives with respect to y, we set z equal to the respective $y_1 - y_2$ (actually the difference of the appropriate variables) in each propagator with results similar to eq. 18.6.61.

$$Z(J, J^{\mu\nu}) = N\,\{\exp[i\int d^4y\,\mathscr{L}_{Higher}(\partial/\partial y^\nu, -i\delta/\delta J^{\mu\nu}(y), -i\delta/\delta J(y))]\cdot$$
$$\cdot\exp[-\tfrac{1}{2}\, i\int d^4y_1 d^4y_2\, J^{\mu\nu}(y_1)\Delta_{F2}{}^{TT}(y_1 - y_2, z)_{\mu\nu\rho\sigma}\, J^{\rho\sigma}(y_2)]\cdot$$
$$\cdot\exp[-\tfrac{1}{2}\, i\int d^4y_1 d^4y_2\, J(y_1)\Delta_F{}^{TT}(y_1 - y_2, z)J(y_2)]\}\Big|_{z=y_1-y_2} \qquad (18.6.58a)$$

To be precise eq. 18.6.58a is interpreted as executing the following steps:

1. For a given process take appropriate functional derivatives of Z(J) with respect to J and $J^{\mu\nu}$.

2. Then expand the exponential factors in a perturbation series applying any derivatives with respect to y in L_{Higher}. Do not perform any of the $\int d^4y_1 d^4y_2$ integrals.

3. Then set $z = y_1 - y_2$ in each $\Delta_{Fk}{}^{TT}(y_1 - y_2, z)$ and $\Delta_{F2}{}^{TT}(y_1 - y_2, z)_{\mu\nu\rho\sigma}$ propagator.

4. Lastly perform all $\int d^4y_1 d^4y_2$ integrals.

Thus we achieve a path integral formulation that is very similar to the corresponding expression in conventional field theory – the only difference is in the form of the free field propagators, which each now contain a Gaussian factor. The net consequence is that graviton vertices result in exactly the same polynomials in momenta as the conventional theory.

Thus Two-Tier gravity generates a perturbative expansion identical to conventional quantum gravity except that each graviton propagator has a gaussian damping factor $R(\mathbf{k}, y_1 - y_2)$. At low energies the tree diagrams of conventional gravity theory emerge to good approximation in Two-Tier gravity. All diagrams with loops converge. Thus Two-Tier gravity is finite.

18.6.11 Finiteness of Quantum Gravity–Scalar Particle Model

Two-tier Quantum Gravity perturbation theory is finite. Calculations are highly convergent at large momentum ($\gtrsim M_c$). At low momentum the Two-Tier theory is similar to conventional gravity – particularly for tree diagrams and other convergent diagrams in conventional quantum gravity.

For pure *conventional* Quantum Gravity DeWitt[205] finds the superficial degree of divergence of a diagram to be:

$$D = -2L_i + 2\sum_n V_n + 4K \qquad (18.6.62)$$

where L_i is the number of internal lines, V_n is the number of n-pronged vertices, and K is the number of independent momentum integrations. DeWitt further points out

$$K = L_i - \sum_n V_n + 1 \qquad (18.6.63)$$

Thus the superficial degree of divergence of a conventional Quantum Gravity diagram is:

$$D = 2(K + 1) \qquad (18.6.64)$$

for $K \geq 1$, displaying an ever increasing degree of divergence as the order of the diagram increases.

In the case of *Two-Tier Quantum Gravity* the superficial degree of divergence of a diagram is:

$$D_{TT} = -6L_i + 2\sum_n V_n + 4K \qquad (18.6.65)$$

(from eq. 18.6.48) with the result (taking account of eq. 18.6.63):

$$D_{TT} = -2L_i - 2\sum_n V_n + 2 \qquad (18.6.66)$$

Since any diagram with a loop has $L_i \geq 1$ and $\sum_n V_n \geq 1$ we see that $D \leq -2$. Thus *all* diagrams are convergent and *the Two-Tier formulation of Quantum Gravity theory is finite. The addition of arbitrary species of other Two-Tier fields – matter and gauge fields – does not introduce divergences in the combined Two-Tier theory.*

18.6.12 Unitarity of Quantum Gravity–Scalar Particle Model

The Two-Tier Quantum Gravity – Scalar Particle Model *superficially* appears to have a unitarity problem due to the non-hermitean nature of its hamiltonian. The lack of hermiticity is due entirely to the appearance of iY^μ in the X^μ field coordinates.

Thus interaction Lagrangian is not hermitean:

---

[205] B. S. DeWitt, Phys. Rev. **162**, 1239 (1967).

$$L_{Higher} = \int d^3y' \mathscr{L}_{Higher}(y' + iY(y')/M_c^2) \tag{18.6.67}$$

and

$$L_{Higher}^{\dagger} = \int d^3y' \mathscr{L}_{Higher}(y' - iY(y')/M_c^2) \neq L_{Higher} \tag{18.6.68}$$

The relation between L_{Higher} and its hermitean conjugate is

$$L_{Higher} = V\, L_{Higher}^{\dagger}\, V \tag{18.6.69}$$

where $V^2 = I$ is the metric operator defined in eqs. 5.16 – 5.18. By eq. 6.37 we see as a result that the Two-Tier S matrix is not unitary – it is pseudo-unitary:

$$S^{-1} = V\, S^{\dagger}\, V \tag{18.6.70}$$

Therefore

$$S^{\dagger}\, VS = V \tag{18.6.71}$$

The S matrix satisfies the unitarity condition between physical asymptotic states – states consisting of only scalar ϕ particles and gravitons. The proof is identical in form to eqs. 6.46 – 6.48. The S matrix of the unified theory of the Standard Model and Quantum Gravity can be similarly shown to satisfy the unitarity condition.

18.6.13 The Mass Scale M_c

The mass scale of Two-Tier theories is set by M_c. This mass scale cannot be ascertained with any degree of certainty at current, experimentally accessible, accelerator energies. Cosmic ray data also does not seem to give any clues as to the value of M_c. It appears that M_c is probably above 10^3 GeV/c^2 and may be of the order of (or equal to) the Planck mass:

$$M_{planck} = \sqrt{\hbar c/G} = 1.22\times10^{19}\ \text{GeV/c}^2 \tag{18.6.75}$$

If M_c is of the 1,000 GeV/c^2 or larger the diff erences between its predictions at current accelerator energies and the predictions of conventional renormalized perturbation theory will be negligible. Actually a much lower value of M_c would still be consistent with the current stringent QED theoretical predictions as well as other predictions of conventional renormalized perturbation theory.

18.6.14 Planck Scale Physics

A finite theory of Quantum Gravity can provide information on the issues that have been of concern for many years – including the short distance behavior of the gravitational metric and ultra-small black holes.

18.6.15 Quantum Foam

Some theorists have conjectured that the classical view of smooth, almost flat space-time does not hold in the quantum regime at energies of the order of the Planck mass. Suggestions that space-time dissolves into quantum foam have appeared.

The finite Two-Tier formulation of Quantum Gravity is well-behaved at short distances and suggests that the quantum behavior of gravity and space-time in the short distance limit does not have limitless quantum fluctuations that result in a foam-like space-time picture.

18.6.16 Measurement of the Quantum Gravity Field

A number of conceptual problems have been raised about the effects of quantized General Relativity. Two-tier Quantum Gravity seems to resolve these issues.

18.6.17 Measurement of Time Intervals

Wigner[206] has studied the measurement of time intervals in General Relativity and sees a problem in the measurement of extremely short intervals. According to Wigner: the measurement of a time inteval in a region of space requires the measurement of the length of time required for an event to happen. The measurement requires an accurate clock. But the accuracy of the clock is limited by the energy-time uncertainty relation:

$$\Delta E \Delta t \geq \hbar \qquad (18.6.76)$$

Thus the uncertainty in the clock's time measurement is related to the uncertainty in the clock's energy which is, in turn, related to the uncertainty in the clock's mass:

$$\Delta E = (\Delta m)c^2 \qquad (18.6.77)$$

To obtain "infinite" accuracy the uncertainty (fluctuations) in the clock's mass must be infinite and thus the clock's mass must be infinite. Infinite fluctuations in the clock's mass will produce corresponding infinite fluctuations in the gravitational field.

$$\Delta h \propto \Delta E \qquad \text{(in conventional General Relativity)} \qquad (18.6.78)$$

As a result the notion of space-time and time intervals (which depend on the geometry through General Relativity) become uncertain. Thus, according to Wigner and others, the concept of time intervals and space-time points becomes questionable.

The Two-Tier version of Quantum Gravity offers a potential way out of this dilemma. The gravitational force becomes stronger as one goes to shorter distances (higher energies) down to a distance (up to an energy) whose scale is set by M_c. At shorter distances (higher energies) the gravitational force becomes weaker and declines to zero at zero distance. Thus at very high energy the gravitational field fluctuations (Δh) are at worst inversely proportional to the energy (and probably decline by a higher

---

[206] E. P. Wigner, Rev. Mod. Phys. **29**, 255 (1957); J. Math. Phys. **2**, 207 (1961).

power of inverse energy.) (The same considerations would apply if one chooses to consider fluctuations in the Riemann-Christoffel symbols.)

$$\Delta h < c_1/E < c_1/(\Delta E) \qquad \text{(in Two-Tier Quantum Gravity)} \qquad (18.6.79)$$

Thus Wigner's conclusion does not hold in the Two-Tier version of Quantum Gravity as gravitational fluctuations actually become smaller at energies above a critical energy whose scale is set by M_c.

In fact, combining eqs. 18.6.79 and 18.6.76 we see

$$c_1\Delta t/\Delta h \geq \hbar \qquad (18.6.80)$$

at sufficiently high energy. Therefore the time uncertainty Δt, and the gravitational field fluctuations Δh, can both decrease while maintaining the energy-time uncertainty relation. *Thus the notion of a space-time point "is saved" in Two-Tier quantum gravity.*

18.6.18 Vacuum Fluctuations in the Gravitation Fields

While the expectation value of the free graviton field $h_{\mu\nu\text{conv}}(X)$ is zero in a conventional quantum field theoric approach:

$$<0|h_{\mu\nu\text{conv}}(X)|0> = 0 \qquad (18.6.81)$$

the vacuum fluctuations of the *conventional* quantum graviton field is quadratically divergent since

$$<0|h_{\mu\nu\text{conv}}(X)h_{\alpha\beta\text{conv}}(X)|0> = \int d^3p\, b'_{\mu\nu\alpha\beta}(p)/[(2\pi)^3\, 2\omega_p] = \infty \qquad (18.6.82)$$

where $b'_{\mu\nu\alpha\beta}(p)$ is a rational function of the momentum p.

In "Two-Tier" quantum field theory we find

$$<0|h_{\mu\nu}(X)h_{\alpha\beta}(X)|0> = \int d^3p\, b'_{\mu\nu\alpha\beta}(p)\, e^{-p^ip^j\Delta_{Tij}(0)}/[(2\pi)^3 2\omega_p] = 0 \qquad (18.6.83)$$

since the exponential factor in the integrand is $-\infty$. The exponent contains

$$\Delta_{Tij}(z) = \int d^3k\, e^{-ik\cdot z}(\delta_{ij} - k_ik_j/\mathbf{k}^2)/[(2\pi)^3 2\omega_k] \qquad (16.3.8)$$

Thus the vacuum fluctuations of $h_{\mu\nu}$ are zero in "Two-Tier" quantum field theory.

18.6.19 The Two-Tier Gravitational Potential vs. Newton's Gravitational Potential

The familiar gravitational potential of Newton is:

$$V_{\text{Newton}} = -G/|\mathbf{r}| \qquad (18.6.84)$$

The Two-Tier gravitational potential is:

$$V_{\text{Two-Tier}} = -G\Phi(M_c^2\pi|\mathbf{r}|^2)/|\mathbf{r}| \qquad (18.6.85)$$

where Φ(y) is the error function.[207] It can be calculated in Two-Tier Quantum Gravity from Two-Tier Quantum Gravity propagator terms similar to corresponding terms in the Two-Tier photon propagator that led to the Two-Tier Coulomb potential. At small distances ($\pi r^2 \ll M_c^{-2}$)

$$V_{\text{Two-Tier}} \rightarrow -\,G2\sqrt{\pi}\, M_c^{2}|\mathbf{r}| \qquad (18.6.86)$$

a linear potential, and at large distances ($\pi r^2 \gg M_c^{-2}$)

$$V_{\text{Two-Tier}} \rightarrow V_{\text{Newton}} = -\,G/|\mathbf{r}| \qquad (18.6.87)$$

the Newtonian potential.

The Two-Tier gravitational potential has a minimum at

$$M_c^{2}\pi r_{\text{MIN}}^{2} = 1 \qquad (18.6.88)$$

At the minimum $V_{\text{Two-Tier}}$ has the value:

$$V_{\text{Two-TierMIN}} = -\,.8427G\sqrt{\pi}\, M_c \qquad (18.6.89)$$

Figs.18.6.1 – 18.6.2 display plots of $V_{\text{Two-Tier}}$ for $M_c = 1$ TeV/c^2, and $M_c = 1.22\ 10^{19}$ GeV/$c^2 = G^{-1/2}$ – the Planck mass.

207 W. Magnus and F. Oberhettinger, *Formulas and Theorems for the Special Functions of Mathematical Physics* (Chelsea Publishing Co., New York, 1949) page 96.

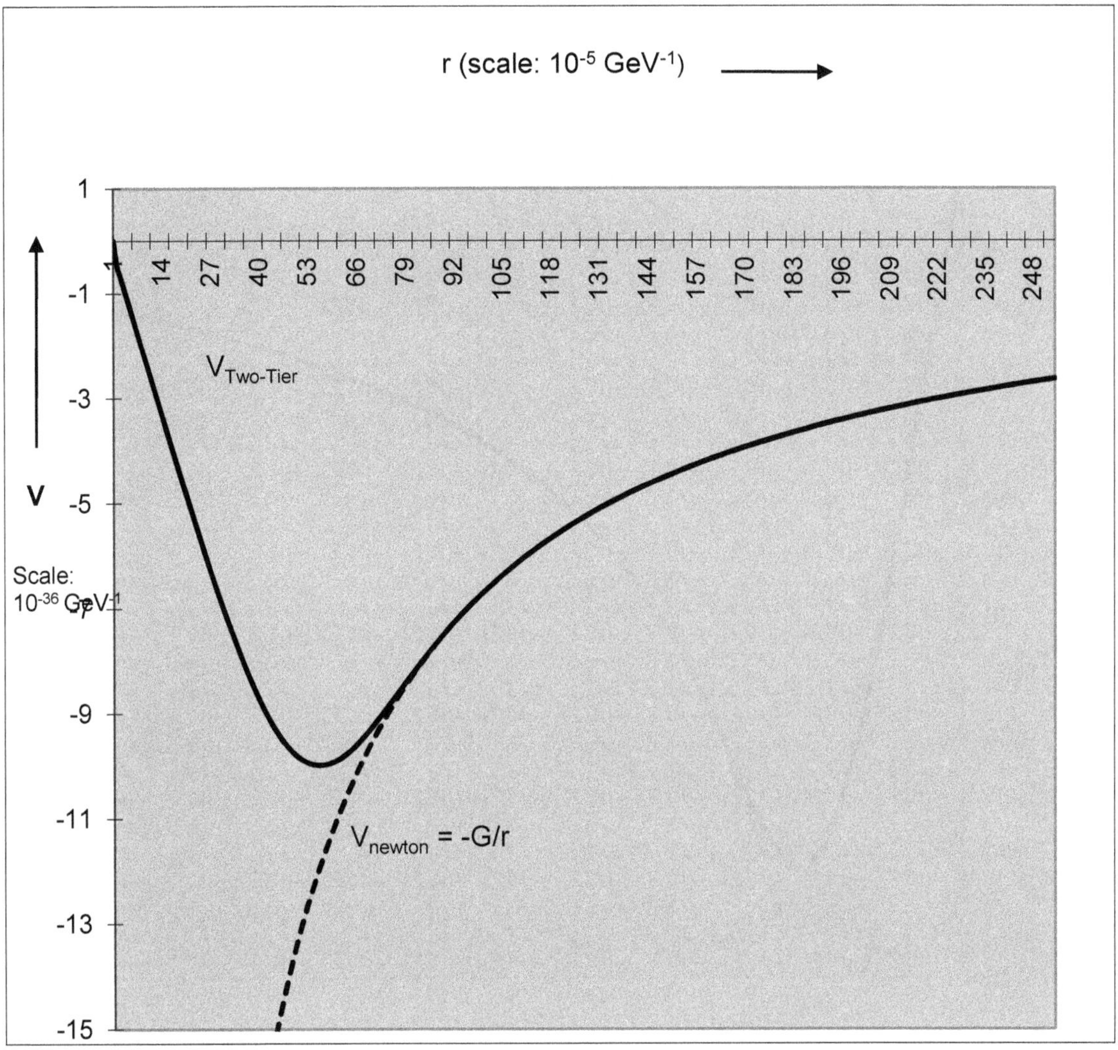

Figure 18.6.1. Plot of Two-Tier gravitational potential for M_c = 1 TeV/c^2 and Newton's gravitational potential. The potentials are measured in units of 10^{-36} GeV^{-1}. The radial distance is measured in units of 10^{-5} GeV^{-1}. The plot of the Two-Tier potential shows the force of gravity is repulsive for small r < 5.7×10^{-4} GeV^{-1}.

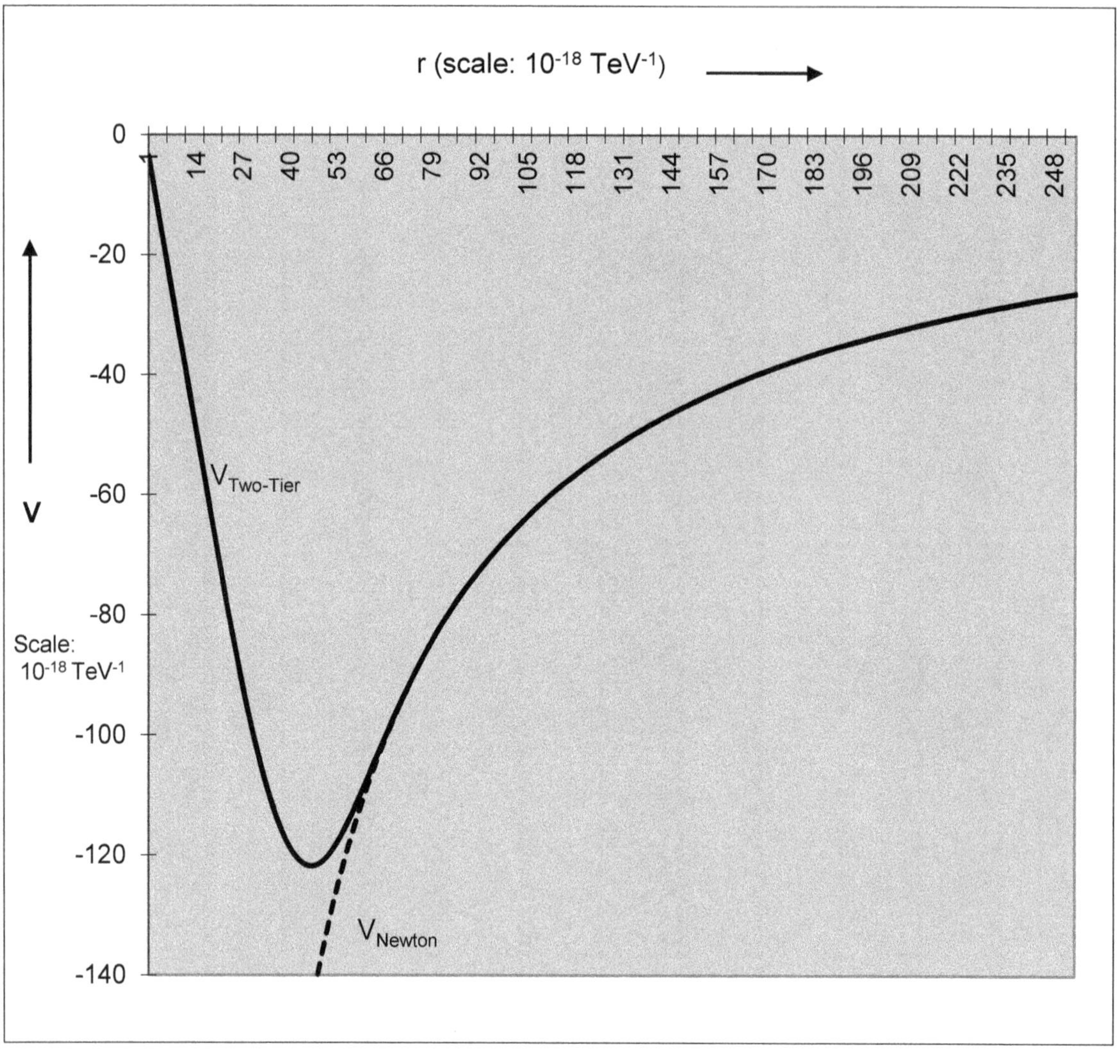

Figure 18.6.2. Plot of Two-Tier gravitational potential for M_c = 1.22 × 10^{19} GeV/c^2 (the Planck mass) and Newton's gravitational potential. Potentials are measured in units of 10^{-18} TeV^{-1}. The radial distance is measured in units of 10^{-18} TeV^{-1}.

18.6.20 Black Holes

The existence of microscopic black holes has been the subject of much speculation. It appears that arbitrarily small black holes can exist in classical General Relativity. The divergences associated with the short distance behavior of its conventional quantization raise the possibility of additional singular behavior at short distances as well.

On the other hand, in Two-Tier Quantum Gravity, at short distances, when the distance scale becomes less than M_c^{-1} (and thus the energy scale becomes greater than M_c), the Two-Tier gravitational force grows smaller and become zero in the limit of zero distance or infinite energy. The preceding figures (Figs. 18.6.1 – 18.6.2) show the Two-Tier gravitational potential linearly approaches zero at short distances unlike the Newtonian gravitational potential which approaches $-\infty$ as r approches zero. (The transverse gravitational propagator also approaches zero at short distances.) Thus the short distance behavior of Two-Tier gravity suggests that black holes of ultra-small size may not exist in Two-Tier Quanum Gravity.

If we examine the Two-Tier gravitational potential we note that it is similar to the Newtonian potential until the separation distance approaches the minimum of the potential. Thus we might expect that conventional classical General Relativity would be approximately valid down to distances of the order of the location of the minimum of the Two-Tier potential. Based on this assumption and on the assumption that M_c equals the Planck mass:

Assumption: $$M_c = M_{Planck} = G^{-1/2} \tag{18.6.90}$$

we can calculate the mass of a black hole whose radius equals the minimum of the Two-Tier potential. From eq. 18.6.88 we obtain

$$r_{MIN} = (G/\pi)^{1/2} = r_{BlackHole} = 2GM_{BlackHoleMIN} \tag{18.6.91}$$

with the result

$$M_{BlackHoleMIN} = (4\pi G)^{-1/2} = \kappa^{-1} \tag{18.6.92}$$

by eq. 18.6.11 and

$$M_{BlackHoleMIN} \cong .282\ M_{Planck} \tag{18.6.93}$$

or 6.15×10^{-6} grams. This lower limit on black hole mass is substantially greater than the collision energy than can be achieved in any current particle accelerator. Thus the production of ultra-small black holes in particle accelerators is unlikely.

Since corrections to conventional quantum gravity are at most of the order of M_c^{-2} it appears that the value of $M_{BlackHoleMIN}$ is consistent with the approximate validity of classical expression for a black hole radius. We note

$$(M_{BlackHoleMIN}/M_c)^2 \cong .0795 \tag{18.6.94}$$

and so corrections to eq. 18.6.93 would be very small.

18.7 Curved Space-time Generalization of Two-tier Quantum Gravity

Thus the preceding chapters developed a divergence-free theory of scalar particles and quantum gravity in a flat space-time. In this section we show that a curved space-time version of Two-Tier quantum field theories including quantum gravity can be developed along the lines pioneered by DeWitt and collaborators. Two-tier curved space-time quantum field theory is based on a mapping from a flat space-time parametrized by y coordinates to a curved space-time parametrized by X coordinates.

The physical picture of the mapping can be visualised using the simple example of a sphere of radius one in three-dimensional space with a coordinate system on the sphere and two planes – one above the sphere and one below it – each with its own flat space coordinate system. Both planes are assumed to be parallel to the disk defined by the crossection of the sphere bounded by the equator of the sphere. A minimum of two coordinate patches are needed to cover a sphere in three dimensions since it necessarily has coordinate singularities.

Let us place a rectangular coordinate system on the top plane. Points on this plane can be mapped onto its northern hemisphere of the sphere in a simple one-to-one fashion. Similarly a rectangular coordinate system can be placed on the bottom plane which can be mapped in a one to one fashion onto the southern hemisphere of the sphere. The top and bottom planes each have a two-dimensional coordinate system that we can choose to be a Cartesian coordinate system in both cases. We will label the coordinates on the top plane x_t^1 and x_t^2, and the points on the bottom plane as x_b^1 and x_b^2. Each plane has a flat space metric $g_{tij} = g_{bij} = \delta_{ij}$ for i, j = 1,2 with δ_{ij} the Kronecker delta.

In addition, just for concreteness, we will place the origin of the top plane coordinate system vertically above the north pole of the sphere, and the origin of the bottom plane coordinate system vertically below the south pole of the sphere.

If we place the sphere at the center of a three dimensional, coordinate system then the points on the sphere (x,y,z) all satisfy:

$$x^2 + y^2 + z^2 = 1 \tag{18.7.3}$$

We can defined coordinates u^1 and u^2 for each hemisphere on the surface of the sphere with equations of the form:

$$x_n = f_{1n}(u_n^1, u_n^2) \tag{18.7.4}$$
$$y_n = f_{2n}(u_n^1, u_n^2) \tag{18.7.5}$$
$$z_n = f_{3n}(u_n^1, u_n^2) \tag{18.7.6}$$

for the northern hemisphere, and

$$x_s = f_{1s}(u_s^1, u_s^2) \tag{18.7.7}$$
$$y_s = f_{2s}(u_s^1, u_s^2) \tag{18.7.8}$$
$$z_s = f_{3s}(u_s^1, u_s^2) \tag{18.7.9}$$

for the southern hemisphere.

In addition, we choose $u_n^1 = u_n^2 = 0$ at the north pole and $u_s^1 = u_s^2 = 0$ at the south pole. The surface of the sphere is curved and each (u^1, u^2) coordinate system has a metric, g_{nij} and g_{sij} for i, j = 1,2 respectively, and a non-zero curvature tensor R_{nijkl} and R_{sijkl}.

Now we are allowed to define a simple map of points on the northern hemisphere of the sphere to points on the top plane such as:

$$x_t^1 = u_n^1 \tag{18.7.10}$$
$$x_t^2 = u_n^2 \tag{18.7.11}$$

and of points on the southern hemisphere of the sphere to points on the bottom plane:

$$x_b^1 = u_s^1 \tag{18.7.12}$$
$$x_b^2 = u_s^2 \tag{18.7.13}$$

Thus we can specify the location of events on the sphere on our planes. Note that eqs. 18.7.4 – 18.7.9 are *not* a coordinate transformation of the (u^1, u^2) coordinate systems on the sphere and thus the plane can have a different (flat) metric from the sphere.

The preceding example can be simplified by using a cylinder enclosing the sphere instead of two planes. The cylinder, which is a flat surface technically, is aligned so that its axis is parallel to, and cenetered on, the north-south axis of the sphere. Then a map can be made from points on the sphere to points on the cylinder that is similar to a Mercator projection, or from points on the sphere to the cylinder that maps the poles to the ends of the cylinder at + and – infinity.

The preceding discussion shows a clear analogy to our map from the y Minkowski space-time to the curved X space-time using

$$X^\mu = y^\mu + i\, Y^\mu(y)/M_c^2 \tag{18.7.14}$$

modulo the imaginary term. The y Minkowski space-time has a flat space-time in which we are allowed to choose the Minkowski metric $\eta_{\mu\nu}$. The curved X space-time has an appropriate metric $g_{\mu\nu}(X)$ that can only be transformed to locally inertial coordinates with perhaps a Minkowski metric in the neighborhood of a point. The additional imaginary term does not alter this picture except that the curved X space-time is now a slightly complex manifold in complex space-time.

Therefore we conclude that our Two-Tier quantum field theoretic formalism that is erected on eq. 18.7.14, where the real part of the X space-time was flat, can be extended to curved space-time while maintaining eq. 18.7.14 if the y space-time consists of coordinate patches analogous to the two planes (or the cylinder) in the example of the sphere. The difference is that we now use a curved space-time background metric $g_{\mu\nu}(X)$ instead of $\eta_{\mu\nu}$ throughout the lagrangian with the exception of L^Y (eq. 18.6.27).

In L^Y we continue to use $\eta_{\mu\nu}$ as the metric. As a result L^Y breaks the invariance of the complete lagrangian under general coordinate transformations. Thus an implicit absolute space-time is implied – as it is implicitly in classical General Relativity and in cosmological experiments. This consequence is not disturbing and is physically acceptable for the following reasons:

1. As Bergmann and Synge point out classical general relativity implicitly embodies an absolute space-time.
2. Experiment shows that space in the large (of the order of the Hubble length) is nearly flat although space does appear to be closed. CBR, and other, experimental data suggests that an absolute reference frame exists.

Thus our universe does appear to be in a state of broken general coordinate transformation invariance. Two-tier quantum field theory in curved space-time is not in contradiction with our previous classical general relativistic theories or with our experimental knowledge of the universe. *The full lagrangian theory L is invariant under special relativity. $L - L^Y$ is formally invariant under general coordinate transformations in the X coordinates.*

18.8 Why Are the Y Field Dynamics Independent of the Gravitational Field?

It is evident that the Y^a field is a truly free field in our formulation. In particular, it does not depend on, or interact directly with, the gravitational field as represented by $\sqrt{g}$ and $g_{\mu\nu}$ factors. On the other hand, these quantities depend on the Y^a field through their dependence on the variable X^μ.

Thus the role of Y^a is strictly that of a coordinate, and of a field that is parametrized by a set of inertial frame coordinates y^μ. The arguments of Mach supplemented by the arguments of Bergmann and Synge show that a de facto absolute reference frame exists (actually it is the set of inertial reference frames). Therefore we can chose to formulate our theory in an inertial reference frame and require that the theory only be invariant under Lorentz transformations to other inertial reference frames.

In this context it is allowed to have one or more fields like Y^a whose dynamics are not invariant under general coordinate transformations. It it is reasonable to require the particle and gravitational dynamical equations be covariant under general coordinate transformations in X. *Thus a part of the dynamics is invariant under Lorentz transformations – the Y^a sector – but this part of the dynamics is not directly observable; and a part of the dynamics – the observable part – is invariant under general coordinate transformations.*

Some reasons for having a free Y^a field are:

1. It is required to avoid divergences that would appear in perturbation theory if the Y^a were allowed to interact with gravitons. For example an hhYY interaction term causes a divergence to appear by generating a Y particle loop in graviton-graviton scattering.
2. If the Y^a particle interacted with gravity then measurable, classical Y^a fields could be generated in regions with ultra-strong gravitational fields such as the neighborhoods of black holes. In this case we would have new dimensions, allbeit imaginary, for which no experimental evidence currently exists.
3. The Principle of Equivalence has only been shown to apply on the classsical level for real coordinates. Any quantization that uses Minkowskian coordinates, or quasi-Minkowskian coordinates, causes general coordinate transformation invariance to be abandoned ab initio in the quantum regime.

18.9 Renormalization of the Standard Model

The Standard Model that we have developed has the same renormalizability as the conventional Standard Model except that we have assigned complex 3-momenta to quarks. This was an arbitrary choice. However it made a point of difference between quarks and leptons, and we see that they differ very much in experiments. In addition it made the introduction of the SU(3) color interaction seem natural.

However, the extra three imaginary spatial dimensions would make our theory non-renormalizable in the quark sector if we followed the standard renormalization approach.

Therefore, we suggest that a different form of quantum field theory[208] be used in which our Standard Model, and a unified theory of the Standard Model and Quantum Gravity, are both fully

[208] Blaha (2005a). This book is the second book in this volume and the second edition of Blaha (2003). Blaha (2005a) discusses this type of quantum field theory in detail including issues such as anomalies and unitarity.

renormalizable. In fact all diagrams calculated in perturbation theory are finite in this new approach. The reader is directed to Blaha (2005) for a detailed account of this new form of quantum field theory.

Thus our form of the Standard Model is fully renormalizable in the new Two-Tier form of quantum field theory.

18.10 Adler-Bell-Jackiw Anomalies

The axial anomaly (Adler-Bell-Jackiw anomaly) follows from the linear divergence of a fermion triangle graph (Fig. 18.10.1) in the conventional Standard Model. All higher order terms are divergence-free. These terms do not contribute to the axial anomaly. Thus the axial anomaly can properly be regarded as an artifact of the regularization of the divergence of the fermion triangle diagram.

In Two-Tier theory the axial anomaly does not appear to be present. Fermion triangle diagrams in Two-Tier quantum field theories are finite. Thus the source of the anomaly in conventional theories is absent in Two-Tier theories.

A massless Dirac field theory is formally invariant under a chiral transformation implying a conserved axial-vector current. The Two-Tier axial-vector current is

$$j_5^\mu(X(y)) = \bar{\psi}(X(y))\gamma^\mu\gamma_5\psi(X(y)) \qquad (18.10.1)$$

with formal conservation law:

$$\partial j_5^\mu(X(y))/\partial X^\mu = 2m\, j_5(X(y)) = 2m\, \bar{\psi}(X(y))\gamma_5\psi(X(y)) \qquad (18.10.2)$$

Eq. 18.10.2 implies

$$\partial j_5^\mu(X(y))/\partial X^\mu = 0 \qquad (18.10.3)$$

in the limit $m \to 0$. The question we now address is whether eq. 18.10.3 holds in Two-Tier perturbation theory – perhaps in the same form as the conventional axial anomaly:

$$\partial j_5^\mu(X(y))/\partial X^\mu = 2m\, j_5(X(y)) + \alpha_0(4\pi)^{-1}\varepsilon^{\mu\nu\alpha\beta}F_{\alpha\beta}F_{\mu\nu} \quad ? \qquad (18.10.4)$$

where α_0 is the unrenormalized fine structure constant.

The simplest manifestation of the axial anomaly in conventional field theory is the fermion triangle diagram, which we will now examine in Two-Tier quantum field theory. As stated earlier, the Two-Tier triangle diagram is finite and zero unlike the conventional quantum field theory result. *Thus the axial anomaly does not appear to exist in Two-Tier quantum field theory. The axial anomaly is a result of the divergence of the triangle diagram in conventional quantum field theory.*

The absence of the anomaly reflects the absence of divergences in Two-Tier quantum field theory, which preserves chiral invariance. Unlike Pauli-Villars regularization, for example, the finiteness of Two-Tier theory follows from the Gaussian factors. Unlike the dimensional regularization approach (where there is no equivalent to γ_5), Two-Tier theory can use the normal γ_5 matrix.

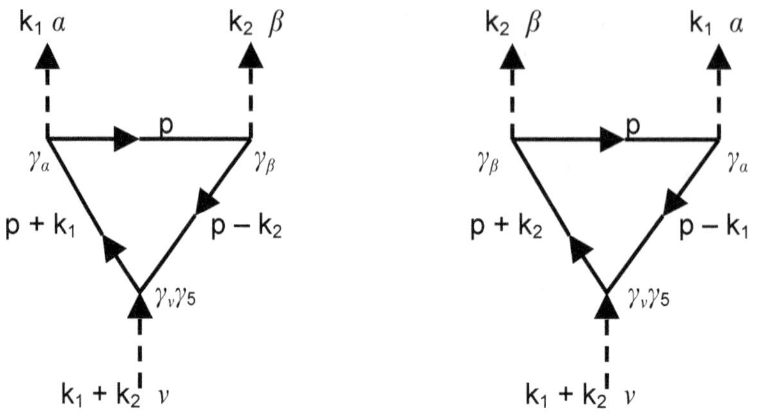

Figure 18.10.1. The V-V-A triangle diagrams.

The expression for the Two-Tier triangle diagrams is:

$$T_{\alpha\beta\nu}(k_1, k_2) = S_{\alpha\beta\nu}(k_1, k_2) + S_{\beta\alpha\nu}(k_2, k_1) \qquad (18.10.5)$$

where

$$S_{\alpha\beta\nu}(k_1, k_2)\delta^4(k_1 + k_2 - q) = -iN\int d^4y_1 d^4y_2 d^4y_3\, e^{ik_1\cdot y_1 + ik_2\cdot y_2 - iq\cdot y_3}\cdot$$
$$\cdot\, Tr\{S_F^{TT}(y_1 - y_3)\gamma_\alpha S_F^{TT}(y_2 - y_1)\gamma_\beta S_F^{TT}(y_3 - y_2)\gamma_\nu\gamma_5\}/(2\pi)^4 \qquad (18.10.6)$$

where N is a constant, and where $S_F^{TT}(z)$ is specified previously. We now define the fourier transform:

$$S_F^{TT}(z) = -i\int d^4p\, e^{-ip\cdot z}\, \mathscr{S}^{TT}(p)/(2\pi)^4 \qquad (18.10.7)$$

where $S^{TT}(p)$ defined previously. We then substitute the fourier transform in eq. 18.10.6 and perform the coordinate integrations to obtain:

$$S_{\alpha\beta\nu}(k_1, k_2) = N\int d^4p\, Tr\{\mathscr{S}^{TT}(p + k_1)\gamma_\alpha \mathscr{S}^{TT}(p)\gamma_\beta \mathscr{S}^{TT}(p - k_2)\gamma_\nu\gamma_5\}/(2\pi)^4$$

We note that

$$k_1^{\alpha}T_{\alpha\beta\nu}(k_1, k_2) \neq 0 \qquad (18.10.8)$$

$$k_2^{\beta}T_{\alpha\beta\nu}(k_1, k_2) \neq 0 \qquad (18.10.9)$$
$$(k_1 + k_2)^{\nu}\, T_{\alpha\beta\nu}(k_1, k_2) \neq 0 \qquad (18.10.10)$$

in Two-Tier quantum field theory because the conservation laws are expressed with respect to the X coordinates – not the y coordinates. Thus since k_1^{α} corresponds $\partial/\partial y_{\alpha}$, and not $\partial/\partial X_{\alpha}$ there is no reason for eqs. 18.10.8-18.10.10 to be zero. However at "large distances" relative to M_c^{-1} we see

$$k_1^{\alpha}T_{\alpha\beta\nu}(k_1, k_2) \cong 0 \tag{18.10.11}$$

$$k_2^{\beta}T_{\alpha\beta\nu}(k_1, k_2) \cong 0 \tag{18.10.12}$$

$$(k_1 + k_2)^{\nu}T_{\alpha\beta\nu}(k_1, k_2) \cong 0 \tag{18.10.13}$$

to very good approximation since the gaussian damping factor in the fermion propagators is approximately unity and thus the Two-Tier expression becomes essentially the same as the conventional field theory expression.

On the other hand at very short distances the anomaly appears to be absent since Two-Tier theory is very well behaved at high energy with

$$\mathscr{S}^{TT}(p) \backsim \gamma^0 M_c^{\ 6}\, p^{-7} + O\,(p^{-9}) \tag{18.10.14}$$

As a result we see

$$k_1^{\alpha}T_{\alpha\beta\nu}(k_1, k_2) \sim p^{4-21} \sim p^{-17} \tag{18.10.15}$$

as $p \to \infty$ is highly convergent. Thus there is no high energy divergence unlike conventional field theory where the integral is linearly divergent. And so no anomaly is generated.

18.11 Gravity and Y^{μ} (y)

Our discussion until this point has been based on the assumption of a flat space-time. The introduction of gravity and the requirement of covariance under coordinate transformations requires the introduction of a $\sqrt{g}$ factor in the Y^{μ} (y) lagrangian kinetic terms:

$$L_C = \int d^4y\, \sqrt{g(X^{\mu}(y)}\; \mathscr{L}_C(X^{\mu}(y), \partial X^{\mu}(y)/\partial y^{\nu}) \tag{18.11.1}$$

Then Y^{μ} (y) interacts with the gravitational field and it is no longer a free field. However the weakness of gravitation and the almost flat nature of space-time makes the effect of this interaction negligible with respect to the Extended Standard Model.

Appendix 18-A. New Paradigm: Composition of Extrema in the Calculus of Variations

This Appendix appeared originally in Blaha (2005a) and earlier books.

18-A.1. A New Paradigm in the Calculus of Variations

The Calculus of Variations has a long and venerable history in Physics and Mathematics. Many problems in Physics and Mathematics have been treated with approaches based on techniques in the Calculus of Variations.[209] In this book we have developed a unified quantum field theory of the known forces of nature based on a new type of problem, or paradigm, in the Calculus of Variations. One way of viewing the spectrum of problems in the calculus of variations is the following progression.

18-A.1.1. A Classification of Variational Problems

1. Variational problems in a Euclidean, or Minkowski, flat space such as the minimal distance between two points or the extrema of a field theory Lagrangian.

2. Variational problems seeking extrema on a curved surface such as the shortest distance between points on the surface of a sphere.

The development in this book suggests a third and fourth, possibility, that to the author's knowledge, has not been addressed in the literature:

3. Variational problems where the extrema are determined on a surface that is itself defined as an extremum. The discussions in this book exemplify this pardigm.

4. Variational problems where the extrema are determined on a surface that is itself defined as an extremum that depends on the extrema on the surface. More simply put the extrema, and the surface upon which they are defined, are jointly determined and are interrelated. Fortunately, our unified theory does not use this paradigm. A future theory might.

In the unified theory that we will develop all particle fields including the graviton field are defined as a mapping of a Minkowski space-time y to a "particle" space-time X with the mapping determined as an extremum of a variation of a fundamental field (a type 3 variational problem in the above classification). Our theory could be generalized to include a back-reaction of the particle fields on

---

[209] See Akhiezer (1962), Blaha (2003), Gelfand (2000), Giaquinta (1996) and (1998), Jost (1998), and Sagan (1993),

the fundamental field (a type 4 variational problem in the above classification). We will not discuss this possibility in this book.

18-A.2. Simple Physical Example – Strings On Springs

In this section we will describe a simple physical example that illustrates a variational problem of type 3 in the Calculus of Variations. We view it as a composition of extrema. (This problem can be addressed using other calculus of Variations techniques.) The approach used in the solution of this problem is similar to the approach used in Two-Tier quantum field theory.

18-A.2.1 A Strings on Springs Mechanics Problem

Consider a long string or bar that can oscillate (undulate) in a direction perpendicular to its length. Further assume that one end of this bar or string is attached to a spring that cause the entire bar or string to oscillate back and forth in a direction parallel to its long side. This configuration is illustrated in Fig. 18-A.1.

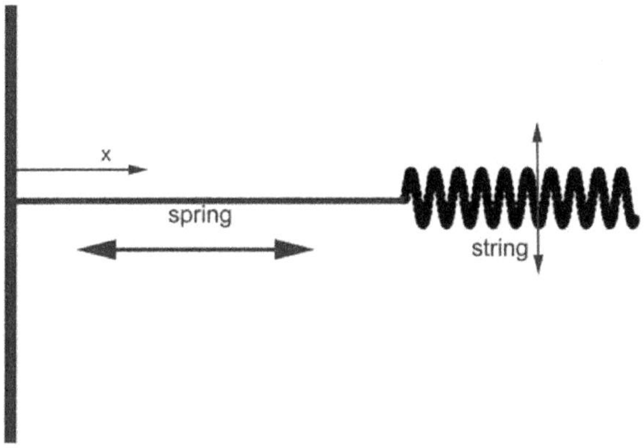

Figure 18-A.1. An oscillating string attached to a spring.

Let x denote the distance to a point on the string when the spring is at equilibrium. If 2π times the frequency of the spring is ω_1, then the location of this point when the spring is oscillating is

$$X(t) = x + A \sin(\omega_1 t + \phi_1) \qquad (18\text{-A.1})$$

where ϕ_1 is a phase, and A is the amplitude of the spring oscillation. Then the vertical displacement of a traveling wave on the *string* can take the form

$$\psi(t) = B \sin(\omega_2 t - k_2(x + A \sin(\omega_1 t + \phi_1)) + \phi_2) \qquad (18\text{-A.2})$$

where B is the amplitude of the string wave, and k_2, ω_2 and ϕ_2 are the parameters of the string wave. These simple mechanical formulae are well known. But they lead to an interesting new application of the ideas of the Calculus of Variations.

Suppose we treat X as an independent variable with X given by eq. (18-A.1), and with eq. (18-A.2) written as:

$$\psi(\mathrm{t}) = \mathrm{B}\ \sin(\omega_2 \mathrm{t} - \mathrm{k}_2 \mathrm{X} + \phi_2) \tag{18-A.3}$$

Defining

$$\psi = \psi(\mathrm{X}(\mathrm{t}), \mathrm{t}) \tag{18-A.4}$$

we can specify the dynamics of the above motion by finding the extrema of

$$\mathrm{I} = \int \mathscr{L}_\psi\, \mathrm{dX(t)} + \int \mathscr{L}_\mathrm{X}\, \mathrm{dt} \tag{18-A.5}$$

where the Lagrangian terms are

$$\mathscr{L}_\psi = \tfrac{1}{2} \{ \mu\ (\partial\psi/\partial \mathrm{t})^2 - \mathrm{Y}\ (\partial\psi/\partial \mathrm{X})^2 \} \tag{18-A.6}$$

with μ and Y being constants, and

$$\mathscr{L}_\mathrm{X} = \tfrac{1}{2} \{ \mathrm{m}(\partial \mathrm{X}/\partial \mathrm{t})^2 - \mathrm{k}(\mathrm{X} - \mathrm{x})^2 \} \tag{18-A.7}$$

where m and k are constants, and where x is a parameter. Applying Hamilton's Principle, and performing independent variations of X and ψ yields the Lagrangian equations:

$$\frac{\partial \mathscr{L}_\psi}{\partial \psi} - \frac{\partial}{\partial \mathrm{X}} \frac{\partial \mathscr{L}_\psi}{\partial(\partial\psi/\partial \mathrm{X})} - \frac{\partial}{\partial \mathrm{t}} \frac{\partial \mathscr{L}_\psi}{\partial(\partial\psi/\partial \mathrm{t})} = 0 \tag{18-A.8}$$

and

$$\frac{\partial \mathscr{L}_\mathrm{X}}{\partial \mathrm{X}} - \frac{\partial}{\partial \mathrm{t}} \frac{\partial \mathscr{L}_\mathrm{X}}{\partial(\partial \mathrm{X}/\partial \mathrm{t})} = 0 \tag{18-A.9}$$

The resulting equations of motion are:

$$\mu\ \partial^2\psi/\partial \mathrm{t}^2 - \mathrm{Y}\ \partial^2\psi/\partial \mathrm{X}^2 = 0 \tag{18-A.10}$$

and

$$\mathrm{m}\ \partial^2 \mathrm{X}/\partial \mathrm{t}^2 + \mathrm{k}(\mathrm{X} - \mathrm{x}) = 0 \tag{18-A.11}$$

with the solutions given in eqs. 18-A.1 and 18-A.2.

The procedure that we use to obtain these results may look a bit strange but they illustrate a type 3 problem in the Calculus of Variations involving the composition of extrema—the composition of an extremum that specifies a manifold in a space (possibly including all of space in a $R^n \longrightarrow R^n$ mapping) with an extremum determining a function on that manifold. The procedure is described in detail in the next section.

18-A.3. The Composition of Extrema – A Lagrangian Formulation

In this section we will explore the general case of the composition of extrema for fields. We will discuss the case of a scalar field ϕ that is a function of a vector field X^μ in a D-dimensional space with coordinate variables that we will denote as y^μ. (The discussion for other types of fields is a straightforward extension of this discussion.) Thus

$$\phi = \phi(X) \tag{18-A.12}$$

and

$$X^\mu = X^\mu(y) \tag{18-A.13}$$

We assume that the dynamics can be described by a Lagrangian formulation using an extension of Hamilton's principle:

$$I = \int \mathscr{L} d^4y \tag{18-A.14}$$

with

$$\mathscr{L} = \mathscr{L}(\phi(X), \partial\phi/\partial X^\nu, X^\mu(y), \partial X^\mu(y)/\partial y^\nu, y) \tag{18-A.15}$$

If we perform a standard variation[210] in ϕ for fixed y (and thus fixed X) we find

$$\delta I = \int [\delta\phi\, \partial\mathscr{L}/\partial\phi + \delta(\partial\phi/\partial X^\nu)\, \partial\mathscr{L}/\partial(\partial\phi/\partial X^\nu)]\, d^4y \tag{18-A.16}$$

We can rewrite the variation in the derivative of ϕ as

$$\delta(\partial\phi/\partial X^\nu) = \partial(\delta\phi)/\partial X^\nu \tag{18-A.17}$$

$$= \partial y^\mu/\partial X^\nu\, \partial(\delta\phi)/\partial y^\mu \tag{18-A.18}$$

with an implied summation over repeated indices. After substituting eq. 18-A.18 in eq. 18-A.16, and performing an integration by parts (and discarding the surface term which is assumed to yield zero in the standard fashion) we obtain:

[210] Bogoliubov, N. N., & Shirkov, D. V., Volkoff, G. M. (tr), *Introduction to the Theory of Quantized Fields* (Wiley-Interscience, New York, 1959); Goldstein H., *Classical Mechanics* (Addison-Wesley, Reading, MA 1965).

$$\delta I = \int \delta\phi \, \{\partial\mathscr{L}/\partial\phi - \partial/\partial y^\mu \, [\partial\mathscr{L}/\partial(\partial\phi/\partial X^\nu) \, \partial y^\mu/\partial X^\nu) \,]\} \, d^4y$$

Since the variation of $\delta\phi$ is arbitrary we conclude

$$\partial\mathscr{L}/\partial\phi - \partial/\partial y^\mu \, [\partial\mathscr{L}/\partial(\partial\phi/\partial X^\nu) \, \partial y^\mu/\partial X^\nu)] = 0 \quad \text{(18-A.19a)}$$

The second term in eq. 18-A.19a shows the effect of the dependence of ϕ on the field X, $\phi = \phi(X)$, rather than directly on the coordinate system y.

Similarly we can perform a variation in X^μ and obtain

$$\partial\mathscr{L}/\partial X^\mu - \partial/\partial y^\nu \, [\partial\mathscr{L}/\partial(\partial X^\mu/\partial y^\nu)] = 0 \qquad \text{(18-A.19b)}$$

The X field defines a "manifold" or, more properly, specifies a transformation from $R^n \rightarrow R^n$. If we make standard assumptions about the mapping: that it is continuous and piece-wise invertible, then we can establish the following lemmas:

Lemma 1: *If the transformation $X^\mu = X^\mu(y)$ is a transformation from $R^n \rightarrow R^n$ that is of class C' and piece-wise invertible, then*

$$\frac{\partial}{\partial y^\nu} \frac{\partial y^\nu}{\partial X^\mu} = -\frac{\partial \ln J}{\partial X^\mu} \qquad \text{(18-A.20)}$$

where

$$J = |\partial(X)/\partial(y)| \qquad \text{(18-A.21)}$$

is the absolute value of the Jacobian of the transformation.

Proof:

Consider two equivalent forms of an integral:

$$I = \int \mathscr{L} J \, d^4y = \int \mathscr{L} \, d^4X$$

where $\mathscr{L}$ is specified as in eq. 18-A.15. Then the first expression for I leads to eq. 18-A.19a which can be written in the form

$$\partial\mathscr{L}/\partial\phi - \partial/\partial X^\mu \, [\partial\mathscr{L}/\partial(\partial\phi/\partial X^\mu)] - \partial\mathscr{L}/\partial(\partial\phi/\partial X^\mu)\{\partial[J\partial y^\nu/\partial X^\mu]/\partial y^\nu\} = 0$$

Using the second expression for I above we obtain the following equation by variation in ϕ:

$$\partial\mathscr{L}/\partial\phi - \partial/\partial X^{\mu}\, [\partial\mathscr{L}/\partial(\partial\phi/\partial X^{\mu})] = 0$$

Comparing these two expressions and realizing that $\partial[J\partial y^{\nu}/\partial X^{\mu}]/\partial y^{\nu}$ is totally independent of ϕ and its derivatives leads us to conclude

$$\partial\, [\, J\partial y^{\nu}/\partial X^{\mu}]/\partial y^{\nu} = 0 \qquad (18\text{-A.22})$$

It is a general relationship for a transformation between X and y based on continuity and piece-wise invertibility. After a few elementary manipulations eq. 18-A.22 can be rewritten in the form of eq. 18-A.20. ■

Lemma 2: *If the transformation $X^{\mu} = X^{\mu}(y)$ is a transformation from $R^n \rightarrow R^n$ that is of class C' and piece-wise invertible and* $\mathscr{L} = \mathscr{L}(\phi(X), \partial\phi/\partial X^{\nu}, X^{\mu}(y), \partial X^{\mu}(y)/\partial y^{\nu}, y)$, *then*

$$\partial\mathscr{L}/\partial(\partial\phi/\partial X^{\nu})\, \partial y^{\mu}/\partial X^{\nu} = \partial\mathscr{L}/\partial(\partial\phi/\partial y^{\mu}) \qquad (18\text{-A.23})$$

Proof:

Let us express $\mathscr{L}$ as a power series in derivatives of ϕ:

$$\mathscr{L} = \sum_{n=0} a_{n\mu_1\mu_2\ldots\mu_n}(\phi(X), X^{\mu}(y), \partial X^{\mu}(y)/\partial y^{\nu}, y) \prod_{j=1}^{n} \partial\phi/\partial X^{\mu_j}$$

which can rewritten using piece-wise invertibility as

$$\mathscr{L} = \sum_{n=0} a_{n\mu_1\mu_2\ldots\mu_n}(\phi(X), X^{\mu}(y), \partial X^{\mu}(y)/\partial y^{\nu}, y) \prod_{j=1}^{n} \partial\phi/\partial y^{\nu_j}\, \partial y^{\nu_j}/\partial X^{\mu_j}$$

Taking the derivative of this equation with respect to $\partial\phi/\partial y^{\mu}$ immediately yields the result. ■

Eq. 18-A.23 enables us to rewrite eq. 18-A.19a as:

$$\partial\mathscr{L}/\partial\phi - \partial/\partial y^{\mu}\, [\partial\mathscr{L}/\partial(\partial\phi/\partial y^{\mu})] = 0 \qquad (18\text{-A.24})$$

which is as one would expect.

In order to get a feeling for the effect of eq. 18-A.19a we will look at a simple example where we specify the relation of the X and y variables directly. Then we will look at the composition of extrema where the transformation between X and y is itself determined as an extremum solution.

18-A.3.1. Example: a hyperplane

We assume eq. 18-A.19b yields the transformation:

$$X^i = ay^i \qquad \text{for } i = 1,2,3$$

$$X^0 = 0$$

Then eq. 18-A.19a becomes

$$\partial\mathscr{L}/\partial\phi - \partial/\partial y^i\,[\partial\mathscr{L}/\partial(\partial\phi/\partial y^i)] = 0 \qquad (18\text{-A}.25)$$

with the time derivative disappearing. Effectively the variation of ϕ on the hyperplane $X^0 = 0$ is determined by the differential equation generated by 18-A.25. On this hyperplane the transformation between the X and y variables is invertible.

18-A.3.2 Coordinate Transformation Determined as an Extremum Solution

We now develop a formalism that determines a mapping from space onto itself as the solution of an extremum problem and also determines the dynamics of one or more fields as a function of this mapping. To this author's knowledge this area in the Calculus of Variations – the determination of an extremum on a manifold where the manifold itself is determined by an extremum – has not been previously explored. We will also develop a hamiltonian formulation. Then we will proceed to quantize the theory.

18-A.3.3. Separable Lagrangian Case

Although there are many forms that the composition of extrema could take, one fairly general form that is directly useful in quantum field theory applications is based on a Lagrangian that can be split into two parts which we will call a *separable Lagrangian*:

$$\mathscr{L} = \mathscr{L}_F J + \mathscr{L}_C(X^\mu(y), \partial X^\mu(y)/\partial y^\nu, y) \qquad (18\text{-A}.26)$$

where J is defined in eq. 18-A.21, where $\mathscr{L}_F$ contains all the dynamics of the fields and their interactions, and where $\mathscr{L}_C$ defines the coordinate mapping as an extremum solution. The procedure to determine the differential equations that specify the mapping, and the field equations that specify field interactions and evolution, is to vary in the coordinates X^μ and in the fields independently, using Hamilton's Principle. The extrema are to be determined for

$$I = \int \mathscr{L}\, d^4y \qquad (18\text{-A}.27)$$

We will begin by considering the case of one scalar field:

$$\mathscr{L}_F = \mathscr{L}_F(\phi(X), \partial\phi/\partial X^\nu) \tag{18-A.28}$$

and

$$\mathscr{L}_C = \mathscr{L}_C(X^\mu(y), \partial X^\mu(y)/\partial y^\nu, y) \tag{18-A.29}$$

Eq. 18-A.27 can be written in the form:

$$I = \int \mathscr{L}_F(\phi(X), \partial\phi/\partial X^\nu)\, dX + \int \mathscr{L}_C(X^\mu(y), \partial X^\mu(y)/\partial y^\nu, y)\, d^4y \tag{18-A.30}$$

using the Jacobian to transform to an integral over dX in the first term. A standard variation of ϕ and the application of Hamilton's Principle yields

$$\partial\mathscr{L}_F/\partial\phi - \partial/\partial X^\mu\, [\partial\mathscr{L}_F/\partial(\partial\phi/\partial X^\mu)] = 0 \tag{18-A.31}$$

reflecting the fact that ϕ is a function of X^μ only, with X^μ a function of the y coordinates.

Next we perform a variation of X^μ determining the mapping from $y \rightarrow X$ as an extremum of the integral in eq. 18-A.27. We note the piece-wise invertibility of the coordinate mapping $X^\mu(y)$ allows us to write the Jacobian J as a function of y^μ only. A standard variation of X^μ and the application of Hamilton's Principle yields

$$\partial\mathscr{L}_C/\partial X^\mu - \partial/\partial y^\nu\, [\partial\mathscr{L}_C/\partial(\partial X^\mu/\partial y^\nu)] = 0 \tag{18-A.32}$$

18-A.3.4 Klein-Gordon Example

The Klein-Gordon scalar field theory furnishes us with a simple example of the application of the preceding development. The Lagrangian is

$$\mathscr{L}_F = ½\, [\, (\partial\phi/\partial X^\nu)^2 - m^2\phi^2\,] \tag{18-A.33}$$

From eq. 18-A.31 we obtain the field equation:

$$(\Box + m^2)\, \phi(X) = 0 \tag{18-A.34}$$

where

$$\Box = \partial/\partial X^\nu\, \partial/\partial X_\nu \tag{18-A.34a}$$

A fourier representation of the solution of eq. 18-A.34 is:

$$\phi(X) = \int dp\, \delta(p^2 - m^2)\theta(p^0)\, [A(p)\, e^{-ip\cdot X} + A(p)^*\, e^{ip\cdot X}] \tag{18-A.35}$$

where A(k) is a function of k and * indicates complex conjugation.

The determination of $X^\mu(y)$ depends on the Lagrangian $\mathscr{L}_C$ and the solutions of eq. 18-A.3A. If we chose

$$\mathscr{L}_C = -\tfrac{1}{2}\,(\partial X^\mu/\partial y^\nu)^2 \tag{18-A.36}$$

Then we obtain the equation

$$\Box\, X^\mu = 0 \tag{18-A.37}$$

with the solution

$$X^\mu = \int dk\, \delta(k^2)\theta(k^0)\, [a^\mu(k)\, e^{-ik\cdot y} + a^\mu(k)^*\, e^{ik\cdot y}] \tag{18-A.38}$$

where $a^\mu(k)$ are complex vector functions of k in general. (We ignore positivity issues for the moment.) Substitution of eq. 18-A.38 in eq. 18-A.35 yields an expression with a form reminiscent of bosonic string expressions.[211] We will take up this point later in subsequent chapters.

18-A.4. The Composition of Extrema – Hamiltonian Formulation

The previous section established a Lagrangian formulation of dynamics based on the composition of extrema. In this section we will develop an equivalent hamiltonian formulation. We will assume a Minkowskian space-time with X^0 and y^0 playing the role of the time coordinates in the respective coordinate systems.

Initially, we will assume a scalar field ϕ with a Lagrangian of the form in eq. 18-A.15 and define canonical momenta with

$$\Pi_\phi = \partial\mathscr{L}/\partial\dot{\phi} \equiv \partial\mathscr{L}/\partial(\partial\phi/\partial X^\mu)\, \partial y^0/\partial X^\mu \tag{18-A.39}$$

$$\Pi_X{}^\mu = \partial\mathscr{L}/\partial\dot{X}_\mu \tag{18-A.40}$$

where

$$\dot{\phi} = \partial\phi/\partial y^0 \equiv \partial\phi/\partial X^\mu\, \partial X^\mu/\partial y^0 \tag{18-A.41}$$

$$\dot{X}^\mu = \partial X^\mu/\partial y^0 \tag{18-A.42}$$

Then we define the hamiltonian density as

$$\mathscr{H} = \Pi_\phi\, \dot{\phi} + \Pi_X{}^\mu\, \dot{X}_\mu - \mathscr{L}(\phi(X), \partial\phi/\partial X^\nu, X^\mu(y), \partial X^\mu(y)/\partial y^\nu, y) \tag{18-A.43}$$

and the hamiltonian

---

[211] See for example Polchinski (1998) and Bailin (1994).

$$H = \int \mathscr{H}\, d^3y \tag{18-A.44}$$

The hamiltonian density has the general form

$$\mathscr{H} = \mathscr{H}(\phi(X), \partial\phi/\partial X^i, \Pi_\phi, X^\mu(y), \partial X^\mu(y)/\partial y^j, \Pi_X{}^\mu, y^\nu) \tag{18-A.45}$$

for the case of one scalar field where the indices i and j represent space coordinates; time coordinates are assigned index value 0.

If we calculate the differential change in H using eq. 18-A.45 we obtain

$$dH = \int \{\partial\mathscr{H}/\partial\phi\, d\phi + \partial\mathscr{H}/\partial\Pi_\phi\, d\Pi_\phi - \partial/\partial y^\nu[\partial\mathscr{H}/\partial(\partial\phi/\partial X^j)\partial y^\nu/\partial X^j]d\phi + $$
$$+ \partial\mathscr{H}/\partial X^\mu\, dX^\mu + \partial\mathscr{H}/\partial\Pi_X{}^\mu\, d\Pi_X{}^\mu - \partial/\partial y^j\, [\partial\mathscr{H}/\partial(\partial X^\mu/\partial y^j)]\, dX^\mu\}\, d^3y \tag{18-A.46}$$

after some partial integrations. (Repeated indices indicate summations. Indices labeled i and j indicate space coordinates. Greek indices include all space-time components of a variable.)

Expressing the differential in H using eq. 18-A.43 we obtain

$$dH = \int dy\, \{\Pi_\phi\, d\dot{\phi} + \dot{\phi}\, d\Pi_\phi - \partial\mathscr{L}/\partial\phi\, d\phi - \partial\mathscr{L}/\partial(\partial\phi/\partial X^\mu)d(\partial\phi/\partial X^\mu) + $$
$$+ \Pi_X{}^\mu\, d\dot{X}^\mu + \dot{X}^\mu\, d\Pi_X{}^\mu - \partial\mathscr{L}/\partial X^\mu\, dX^\mu - \partial\mathscr{L}/\partial(\partial X^\mu/\partial y^j)d(\partial X^\mu/\partial y^j)\} \tag{18-A.47a}$$

After some manipulations we find

$$dH = \int \{\, \dot{\phi}\, d\Pi_\phi + \dot{X}_\mu d\Pi_X{}^\mu \; - \partial/\partial y^0\, \Pi_\phi\, d\phi - \partial/\partial y^0\, \Pi_X{}^\mu\, dX_\mu\}\, dy \tag{18-A.47b}$$

using the equations of motion eqs. 18-A.19a and 18-A.19b.

Comparing eqs 18-A.46 and 18-A.47 we obtain Hamilton's equations in the case of the composition of extrema:

$$\dot{\phi} = \partial\mathscr{H}/\partial\Pi_\phi \tag{18-A.48a}$$

$$\dot{\Pi}_\phi = -\partial\mathscr{H}/\partial\phi + \partial/\partial y^\nu\, [\partial\mathscr{H}/\partial(\partial\phi/\partial X^j)\, \partial y^\nu/\partial X^j] \tag{18-A.48b}$$

$$\dot{X}_\mu = \partial\mathscr{H}/\partial\Pi_X{}^\mu \tag{18-A.48c}$$

$$\dot{\Pi}_X{}^\mu = -\partial\mathscr{H}/\partial X^\mu + \partial/\partial y^j\, [\partial\mathscr{H}/\partial(\partial X^\mu/\partial y^j)]$$

where

$$\dot{\Pi}_\phi = \partial\,\Pi_\phi/\partial y^0 \tag{18-A.49a}$$

$$\dot{\Pi}_X{}^\mu = \partial\Pi_X{}^\mu/\partial y^0 \tag{18-A.49b}$$

18-A.5. Translational Invariance

If the Lagrangian of a field theory has no explicit dependence on the coordinates then one expects translational invariance accompanied by a conservation law for an energy-momentum stress tensor. We will show this is the case for Lagrangians implementing the composition of extrema. We assume a Lagrangian without an explicit dependence on the coordinates y^ν:

$$\mathscr{L} = \mathscr{L}(\phi(X), \partial\phi/\partial X^\nu, X^\mu(y), \partial X^\mu(y)/\partial y^\nu) \tag{18-A.50}$$

Under an infinitesimal displacement,

$$y'^\nu = y^\nu + \epsilon^\nu \tag{18-A.51a}$$

$$\delta\phi = \phi(X(y + \epsilon)) - \phi(X(y))$$

$$= \epsilon^a\, \partial\phi/\partial y^a \tag{18-A.51b}$$

$$\delta X^\mu = \epsilon^a\, \partial X^\mu/\partial y^a \tag{18-A.51c}$$

$$\delta(\partial\phi/\partial X^\mu) = \epsilon^a\, \partial(\partial\phi/\partial y^a)/\partial X^\mu \tag{18-A.51d}$$

$$\delta(\partial X^\mu/\partial y^\nu) = \epsilon^a\, \partial(\partial X^\mu/\partial y^a)/\partial y^\nu \tag{18-A.51e}$$

and the Lagrangian changes by

$$\delta\mathscr{L} = \epsilon^a\, \partial\mathscr{L}/\partial y^a \tag{18-A.52}$$

The change can also be expressed in terms of the changes in the fields, their derivatives and the mapping X^μ:

$$\delta\mathscr{L} = \partial\mathscr{L}/\partial\phi\, \delta\phi + \partial\mathscr{L}/\partial(\partial\phi/\partial X^\mu)\, \delta(\partial\phi/\partial X^\mu) + \partial\mathscr{L}/\partial X^\mu\, \delta X^\mu + \\ + \partial\mathscr{L}/\partial(\partial X^\mu/\partial y^\nu)\, \delta(\partial X^\mu/\partial y^\nu) \tag{18-A.53}$$

Combining eqs. 18-A.51, 18-A.52 and 18-A.53 we obtain (after some manipulations):

$$\in^{\nu} \partial/\partial y_{\mu} \mathscr{T}_{\mu\nu} = 0 \tag{18-A.54}$$

where

$$\mathscr{T}_{\mu\nu} = -g_{\mu\nu} \mathscr{L} + \partial\mathscr{L} / \partial(\partial\phi/\partial X^{\delta}) \, \partial y_{\mu}/\partial X^{\delta} \, \partial\phi/\partial y^{\nu} + \partial\mathscr{L} / \partial(\partial X^{\delta}/\partial y_{\mu}) \partial X^{\delta}/\partial y^{\nu} \tag{18-A.55a}$$

or, alternately using Lemma 2,

$$\mathscr{T}_{\mu\nu} = - g_{\mu\nu} \mathscr{L} + \partial\mathscr{L} / \partial(\partial\phi/\partial y_{\mu}) \, \partial\phi/\partial y^{\nu} + \partial\mathscr{L} / \partial(\partial X^{\delta}/\partial y_{\mu}) \, \partial X^{\delta}/\partial y^{\nu} \tag{18-A.55b}$$

Since $\in^{a}$ is an arbitrary displacement we obtain the conservation law:

$$\partial/\partial y_{\mu} \mathscr{T}_{\mu\nu} = 0 \tag{18-A.56}$$

Eq. 18-A.56 implies the energy-momentum vector

$$P_{\beta} = \int d^{3}y \, \mathscr{T}_{0\beta} \tag{18-A.57}$$

is conserved. We note

$$\partial/\partial y^{0} P_{\beta} = 0 \tag{18-A.58}$$

since eq. 18-A.56 and 18-A.57 can be used to obtain the integral of a divergence, which results in zero.
The hamiltonian (eqs. 18-A.43-44) is

$$H = P_{0} \tag{18-A.59}$$

We note for later use that the total energy, H, which is conserved, contains a term that represents the energy in the X^{μ} mapping. Thus energy can be exchanged in principle between the ϕ field sector and the X^{μ} sector.

18-A.6. Lorentz Invariance and Angular Momentum Conservation

We can also verify Lorentz invariance and obtain the form of the conserved angular momentum by considering the effect of an infinitesimal Lorentz transformation. We will consider the case of a scalar field ϕ.

Under an infinitesimal Lorentz transformation ($\in_{\mu\nu} = - \in_{\nu\mu}$):

$$y'_\mu = y_\mu + \delta y_\mu = y_\mu + \epsilon_{\mu\nu} y^\nu \qquad (18\text{-}A.60a)$$

$$\delta\phi = \phi(X(y')) - \phi(X(y))$$

$$= \epsilon^{\mu\nu}\, y_\nu\, \partial\phi/\partial X^a\, \partial X^a/\partial y^\mu \qquad (18\text{-}A.60b)$$

$$\delta X^\mu = S^\mu{}_a X^a(y') - X^\mu(y) \qquad (18\text{-}A.60c)$$

$$= \epsilon^\mu{}_a X^a(y) + \partial X^\mu/\partial y^\beta\, \delta y^\beta \qquad (18\text{-}A.60d)$$

where $S^\mu{}_a$ is the matrix for the Lorentz transformation of a vector. (If X^μ were a gauge field then an additional operator gauge term would have to be added to eq. 18-A.60d.)

The Lagrangian changes by

$$\delta\mathscr{L} = \epsilon^{\mu\nu}\, y_\nu\, \partial\mathscr{L}/\partial y^\mu \qquad (18\text{-}A.61)$$

under the infinitesimal Lorentz transformation. The change in the Lagrangian can also be expressed as:

$$\delta\mathscr{L} = \partial\mathscr{L}/\partial\phi\, \delta\phi + \partial\mathscr{L}/\partial(\partial\phi/\partial X^\mu)\, \delta(\partial\phi/\partial X^\mu) + \partial\mathscr{L}/\partial X^\mu\, \delta X^\mu +$$
$$+ \partial\mathscr{L}/\partial(\partial X^\mu/\partial y^\nu)\, \delta(\partial X^\mu/\partial y^\nu) \qquad (18\text{-}A.62)$$

Combining eqs. 18-A.61 and 18-A.62, and substituting and simplifying terms leads to:

$$\epsilon_{\mu\nu}\, \partial/\partial y^\sigma\, \mathscr{M}^{\sigma\mu\nu} = 0 \qquad (18\text{-}A.63)$$

where

$$\mathscr{M}^{\sigma\mu\nu} = (g^{\mu\sigma}y^\nu - g^{\nu\sigma}y^\mu)\mathscr{L} + \partial\mathscr{L}/\partial(\partial\phi/\partial X^\alpha)\, \partial y^\sigma/\partial X^a\, (y^\mu\partial\phi/\partial y_\nu - y^\nu\partial\phi/\partial y_\mu) +$$
$$+ \partial\mathscr{L}/\partial(\partial X^\delta/\partial y^\sigma)\, (g^{\delta\nu}X^\mu - g^{\delta\mu}X^\nu + y^\mu\, \partial X^\delta/\partial y_\nu - y^\nu\, \partial X^\delta/\partial y_\mu) \qquad (18\text{-}A.64)$$

The conserved angular momentum is:

$$M^{\mu\nu} = \int d^3y\, \mathscr{M}^{0\mu\nu} \qquad (18\text{-}A.65)$$

with

$$\partial M^{\mu\nu}/\partial y^0 = 0 \qquad (18\text{-}A.66)$$

The angular momentum density can be written in the familiar form:

$$\mathcal{M}^{\sigma\mu\nu} = y^\mu \mathcal{T}^{\sigma\nu} - y^\nu \mathcal{T}^{\sigma\mu} + \partial\mathcal{L}/\partial(\partial X^\delta/\partial y^\sigma)\,(g^{\delta\nu}X^\mu - g^{\delta\mu}X^\nu) \quad (18\text{-A}.67)$$

taking account of the vector nature of X^μ. The spatial part of $M^{\mu\nu}$ is the angular momentum.

18-A.7. Internal Symmetries

We will now consider the case of a set of scalar fields ϕ_r in a Lagrangian with an internal symmetry. Under a local transformation

$$\phi_r(X) \rightarrow \phi_r(X) - i\epsilon\lambda_{rs}\,\phi_s(X) \quad (18\text{-A}.68)$$

If the Lagrangian is invariant under this transformation, then

$$\delta\mathcal{L} = 0 = \partial\mathcal{L}/\partial\phi_r\delta\phi_r + \partial\mathcal{L}/\partial(\partial\phi_r/\partial X^\alpha)\,\delta(\partial\phi_r/\partial X^\alpha) \quad (18\text{-A}.69)$$

Using the equation of motion eq. 18-A.19a satisfied by all the components ϕ_r we obtain a conserved current:

$$\mathcal{J}^\nu = -i\,\partial\mathcal{L}/\partial(\partial\phi_r/\partial X^\delta)\,\partial y^\nu/\partial X^\delta\,\lambda_{rs}\,\phi_s \quad (18\text{-A}.70)$$

which satisfies

$$\partial\mathcal{J}^\nu/\partial y^\nu = 0 \quad (18\text{-A}.71)$$

The conserved charge is

$$Q = \int d^3y\,\mathcal{J}^0 \quad (18\text{-A}.72)$$

$$\partial Q/\partial y^0 = 0 \quad (18\text{-A}.73)$$

18-A.8. Separable Lagrangians

We now consider the case of a separable Lagrangian such as in eq. 18-A.26. Adopting the definitions:

$$\phi' = \partial\phi/\partial X^0 \quad (18\text{-A}.74)$$

$$X_\mu' = \partial X_\mu/\partial y^0 \quad (18\text{-A}.75)$$

we define canonical momenta as

$$\pi_\phi = \partial\mathcal{L}/\partial\phi' \equiv \partial\mathcal{L}/\partial(\partial\phi/\partial X^0) \quad (18\text{-A}.76)$$

$$\pi_X{}^\mu = \partial\mathcal{L}/\partial X_\mu' \equiv \partial\mathcal{L}/\partial(\partial X_\mu/\partial y^0) \tag{18-A.77}$$

We now define the separable hamiltonian density as

$$\mathcal{H}_s = J\pi_\phi\,\phi' + \pi_X{}^\mu\, X_\mu' - \mathcal{L}_s \tag{18-A.78}$$

where J is the Jacobian (eq. 18-A.21) and

$$H_s = \int \mathcal{H}_s\, d^3y \tag{18-A.79}$$

The separable Lagrangian (from eq. 18-A.26) is:

$$\mathcal{L}_s = \mathcal{L}_F(\phi(X), \partial\phi/\partial X^\mu)\, J + \mathcal{L}_C(X^\mu(y), \partial X^\mu(y)/\partial y^\nu, y) \tag{18-A.80}$$

In the case of one scalar field the separable hamiltonian density has the general form

$$\mathcal{H}_s = \mathcal{H}_S(\phi(X), \pi_\phi, \partial\phi/\partial X^i, X^\mu(y), \pi_X{}^\mu, \partial X^\mu(y)/\partial y^j, y^\nu) \tag{18-A.81}$$

where the indices i and j indicate spatial components. In particular, the terms in the separable hamiltonian are:

$$\mathcal{H}_s = \mathcal{H}_F J + \mathcal{H}_C \tag{18-A.82}$$

with

$$\mathcal{H}_F(\phi(X), \pi_\phi, \partial\phi/\partial X^i) = \pi_\phi\,\phi' - \mathcal{L}_F \tag{18-A.83}$$

$$\mathcal{H}_C(X^\mu(y), \pi_X{}^\mu, \partial X^\mu(y)/\partial y^j, y^\nu) = \pi_X{}^\mu\, X_\mu' - \mathcal{L}_C \tag{18-A.84}$$

where J is the absolute value of the Jacobian defined in 18-A.21.

We now define the time integral of H as we did in eq. 18-A.14 when considering the Lagrangian formulation:

$$G = \int dy^0\, H_s \tag{18-A.85}$$

Thus G is an integral over all space-time coordinates. Using G we can develop a hamiltonian formulation. First we calculate the differential change in G. Using eqs. 18-A.81-2 and 18-A.85 we obtain

$$dG = \int\{J\,\partial\mathcal{H}_F/\partial\phi\, d\phi + J\,\partial\mathcal{H}_F/\partial\pi_\phi\, d\pi_\phi +$$

$$+ \int \partial\mathscr{H}_F/\partial(\partial\phi/\partial X^j)\ d(\partial\phi/\partial X^j) + \partial\mathscr{H}_C/\partial X^\mu\ dX^\mu +$$
$$+ \partial\mathscr{H}_C/\partial\pi_X^{\ \mu}\ d\pi_X^{\ \mu} + \partial\mathscr{H}_C/\partial(\partial X^\mu/\partial y^j)\ d(\partial X^\mu/\partial y^j)\Big\}\, d^4y$$
(18-A.86)

with summations implied by repeated indices. (Index labels i and j label spatial coordinates only; Greek indices label space-time coordinates.) Rewriting dG as two integrals and performing partial integrations yields:

$$dG = \int d^4X\Big\{\partial\mathscr{H}_F/\partial\phi\ d\phi + \partial\mathscr{H}_F/\partial\pi_\phi\, d\pi_\phi - \partial/\partial X^j[\partial\mathscr{H}_F/\partial(\partial\phi/\partial X^j)]\ d\phi\Big\} +$$
$$+ \int d^4y\Big\{\partial\mathscr{H}_C/\partial X^\mu\ dX^\mu + \partial\mathscr{H}_C/\partial\pi_X^{\ \mu}\ d\pi_X^{\ \mu} - \partial/\partial y^j[\partial\mathscr{H}_C/\partial(\partial X^\mu/\partial y^j)]\ dX^\mu\Big\}$$
(18-A.87)

Alternately, expressing the differential in G using eqs. 18-A.82-4 we obtain

$$dG = \int d^4X\Big\{\pi_\phi\ d\phi' + \phi' d\pi_\phi - \partial\mathscr{L}_F/\partial\phi\ d\phi - \partial\mathscr{L}_F/\partial(\partial\phi/\partial X^\mu)d(\partial\phi/\partial X^\mu)\Big\} +$$
$$+ \int d^4y\Big\{\pi_{X\mu}\ dX^{\mu\prime} + X^{\mu\prime}\ d\pi_{X\mu} - \partial\mathscr{L}_C/\partial X^\mu\ dX^\mu - \partial\mathscr{L}_C/\partial(\partial X^\mu/\partial y^j)d(\partial X^\mu/\partial y^j)\Big\}$$
(18-A.88)

which becomes

$$dG = \int d^4X\Big\{-\pi_\phi'\ d\phi + \phi'\ d\pi_\phi\Big\} + \int d^4y\Big\{-\pi_{X\mu}'\ dX^\mu + X^{\mu\prime}\ d\pi_{X\mu}\Big\} \quad (18\text{-A}.89)$$

using the equations of motion eqs. 18-A.31-2.

Comparing eqs 18-A.87 and 18-A.89 we obtain Hamilton's equations for the case of the composition of extrema for a separable Lagrangian:

$$\phi' = \partial\mathscr{H}_F/\partial\pi_\phi \quad (18\text{-A}.90)$$

$$\pi_\phi' = -\,\partial\mathscr{H}_F/\partial\phi + \partial/\partial X^j\ [\partial\mathscr{H}_F/\partial(\partial\phi/\partial X^j)] \quad (18\text{-A}.91)$$

$$X_\mu' = \partial\mathscr{H}_C/\partial\pi_X^{\ \mu} \quad (18\text{-A}.92)$$

$$\pi_{X\mu}' = -\,\partial\mathscr{H}_C/\partial X^\mu + \partial/\partial y^j\ [\partial\mathscr{H}_C/\partial(\partial X^\mu/\partial y^j)] \quad (18\text{-A}.93)$$

where

$$\pi_\phi' = \partial\ \pi_\phi/\partial X^0 \quad (18\text{-A}.94)$$

$$\pi_{X\mu}{}' = \partial\, \pi_{X\mu}/\partial X^0 \qquad (18\text{-A}.95)$$

Notice that $\mathscr{L}_F$, $\mathscr{H}_F$ *and* π_ϕ *have precisely the same form, as a function of* X^μ, *as one sees in a conventional field theory formalism.* Yet X^μ is a mapping/function of the coordinates y. In reality, it can be viewed as a field as we shall see.

18-A.9. Separable Lagrangians and Translational Invariance

The general rule for conventional Lagrangians is: if a Lagrangian has no explicit dependence on the coordinates then translational invariance follows accompanied by a conservation law for an energy-momentum tensor. We will show that this rule needs modification for separable Lagrangians that implement the composition of extrema.

Consider the Lagrangian:

$$\mathscr{L}_s = J\, \mathscr{L}_F(\phi(X), \partial\phi/\partial X^\mu) + \mathscr{L}_C(X^\mu(y), \partial X^\mu(y)/\partial y^\nu) \qquad (18\text{-A}.96)$$

in which the X^μ play a dual role as both fields and coordinates. Let us consider a variation in X^μ:

$$X^\mu(y) \rightarrow X^\mu(y) + \delta X^\mu(y) \qquad (18\text{-A}.97)$$

where $\delta X^\mu(y)$ is an arbitrary function of y that vanishes at the endpoints of the integration region of the integral. The action is:

$$I = \int \mathscr{L}_s d^4y \qquad (18\text{-A}.98)$$

We will show that a variation in $X^\mu(y)$ leads to a conserved energy-momentum tensor. But we will use integrals of the Lagrangian density since it provides a simpler derivation of the result. Under the variation of eq. 18-A.97 we find

$$\delta\phi = \phi(X(y) + \delta X^\mu(y)) - \phi(X(y))$$

$$= \delta X^\mu\, \partial\phi/\partial X^\mu \qquad (18\text{-A}.99a)$$

$$\delta(\partial\phi/\partial X^\nu) = \delta X^\mu\, \partial(\partial\phi/\partial X^\mu)/\partial X^\nu \qquad (18\text{-A}.99b)$$

$$\delta(\partial X^\mu/\partial y^\nu) = \partial(\delta X^\mu)/\partial y^\nu \qquad (18\text{-A}.99c)$$

The integral in eq. 18-A.98 changes by

$$\delta I = \int d^4y\, \delta \mathscr{L}_s = \int d^4y\, [\, \delta(J \mathscr{L}_F) + \delta \mathscr{L}_C\,] \qquad (18\text{-A}.100a)$$

which becomes:

$$\delta I = \int d^4y\, [\delta X^\mu\, \partial(J \mathscr{L}_F)/\partial X^\mu + \partial(\delta X^\mu \partial \mathscr{L}_C / \partial(\partial X^\mu/\partial y^\nu))/\partial y^\nu] \quad (18\text{-A}.100b)$$

due to the equations of motion of X^μ (eq. 18-A.19b) in X^μ's role. Since the second term is a total divergence its contribution to δI is zero. Thus we can express eq. 18-A.100b as:

$$\delta I = \int d^4y\, [\, J\, \delta \mathscr{L}_F + \mathscr{L}_F\, \delta J\,] \qquad (18\text{-A}.101)$$

realizing that the Jacobian J depends on y and thus X:

$$\delta J = \delta X^\mu\, \partial J/\partial X^\mu \qquad (18\text{-A}.102)$$

A partial integration gives

$$\mathscr{L}_F\, \delta J = \delta X^\mu\, \partial(\, J \mathscr{L}_F)/\partial X^\mu - \delta X^\mu\, J\, \partial \mathscr{L}_F/\partial X^\mu \qquad (18\text{-A}.103)$$

Evaluating $\delta \mathscr{L}_F$ we find:

$$\delta \mathscr{L}_F = \partial \mathscr{L}_F / \partial\phi\, \delta\phi + \partial \mathscr{L}_F / \partial(\partial\phi/\partial X^\mu)\, \delta(\partial\phi/\partial X^\mu) \qquad (18\text{-A}.104)$$

which gives

$$\delta \mathscr{L}_F = \delta X^\nu\, \partial/\partial X^\mu\, [\partial \mathscr{L}_F / \partial(\partial\phi/\partial X^\mu)\, \partial\phi/\partial X^\nu\,] \qquad (18\text{-A}.105)$$

using the equations of motion eq. 18-A.31, and using eq. 18-A.99b. Combining eqs. 18-A.100, 18-A.101, 18-A.103 and 18-A.105 we obtain:

$$\int d^4y\, J\, \delta X^\nu\, \partial/\partial X_\mu\, \mathscr{T}_{F\mu\nu} = \int d^4X\, \delta X^\nu\, \partial/\partial X_\mu\, \mathscr{T}_{F\mu\nu} = 0 \qquad (18\text{-A}.106)$$

where

$$\mathscr{T}_{F\mu\nu} = -\, g_{\mu\nu}\, \mathscr{L}_F + \partial \mathscr{L}_F / \partial(\partial\phi/\partial X_\mu)\, \partial\phi/\partial X^\nu \qquad (18\text{-A}.107)$$

after some manipulations. Since δX^ν is an arbitrary function of y the differential conservation law follows:

$$\partial/\partial X_\mu \mathscr{T}_{F\mu\nu} = 0 \tag{18-A.108}$$

Eq. 18-A.108 implies the energy-momentum vector

$$P_{F\beta} = \int d^3X \, \mathscr{T}_{F0\beta} \tag{18-A.109}$$

is conserved:

$$\partial/\partial X^0 P_{F\beta} = 0 \tag{18-A.110}$$

The hamiltonian density (eq. 18-A.83) is

$$\mathscr{H}_F = \mathscr{T}_{F0\beta} \tag{18-A.111}$$

Thus the field energy

$$H_F = P_{F0} = \int d^3X \, \mathscr{T}_{F00} \tag{18-A.112}$$

is conserved with respect to the "time" X^0. Later we will see that H_F is trivially conserved in the Coulomb gauge of X_μ. (We will also establish an electromagnetic-like quantum field theory for X_μ with gauge invariance.) In other gauges the conservation of H_F is not trivial.

18-A.10. Separable Lagrangians and Angular Momentum Conservation

We can also verify Lorentz invariance and obtain the form of the conserved angular momentum for a separable Lagrangian by considering the effect of an infinitesimal Lorentz transformation. We will consider the case of a scalar field ϕ.

Under an infinitesimal Lorentz transformation as specified by eqs. 18-A.60a – 18-A.60d the separable Lagrangian changes by

$$\delta\mathscr{L}_s = \epsilon^{\mu\nu} y_\nu \partial\mathscr{L}_s/\partial y^\mu \tag{18-A.113}$$

which can also be expressed as

$$\delta\mathscr{L}_s = \partial\mathscr{L}_s/\partial\phi\, \delta\phi + \partial\mathscr{L}_s/\partial(\partial\phi/\partial X^\mu)\, \delta(\partial\phi/\partial X^\mu) + \partial\mathscr{L}_s/\partial X^\mu\, \delta X^\mu + $$
$$+ [\partial\mathscr{L}_s/\partial(\partial X^\mu/\partial y^\nu)]\, \delta(\partial X^\mu/\partial y^\nu) \tag{18-A.114}$$

Combining eqs. 18-A.113 and 18-A.114 leads to:

$$\epsilon_{\mu\nu}\, \partial/\partial y^{\sigma} \mathscr{M}_{s}^{\;\sigma\mu\nu} = 0 \tag{18-A.115}$$

where

$$\mathscr{M}_{s}^{\;\sigma\mu\nu} = J\, \mathscr{M}_{F}^{\;\sigma\mu\nu} + \mathscr{M}_{C}^{\;\sigma\mu\nu} + \mathscr{M}_{M}^{\;\sigma\mu\nu} \tag{18-A.116}$$

$$\mathscr{M}_{F}^{\;\sigma\mu\nu} = (g^{\mu\sigma}y^{\nu} - g^{\nu\sigma}y^{\mu})\mathscr{L}_{F} + \partial\mathscr{L}_{F}/\partial(\partial\phi/\partial y_{\sigma})\,(y^{\mu}\partial\phi/\partial y_{\nu} - y^{\nu}\partial\phi/\partial y_{\mu}) \tag{18-A.117}$$

$$\mathscr{M}_{C}^{\;\sigma\mu\nu} = (g^{\mu\sigma}y^{\nu} - g^{\nu\sigma}y^{\mu})\mathscr{L}_{C} + \\ + \partial\mathscr{L}_{C}/\partial(\partial X^{\delta}/\partial y^{\sigma})(g^{\delta\nu}X^{\mu} - g^{\delta\mu}X^{\nu} + y^{\mu}\,\partial X^{\delta}/\partial y_{\nu} - y^{\nu}\,\partial X^{\delta}/\partial y_{\mu}) \tag{18-A.118}$$

$$\mathscr{M}_{M}^{\;\sigma\mu\nu} = \mathscr{L}_{F}\,\partial J/\partial(\partial X^{\delta}/\partial y^{\sigma})(g^{\delta\nu}X^{\mu} - g^{\delta\mu}X^{\nu} + y^{\mu}\,\partial X^{\delta}/\partial y_{\nu} - y^{\nu}\,\partial X^{\delta}/\partial y_{\mu}) \tag{18-A.119}$$

where the third term originates in the dependence of J on derivatives of X^{μ}. Eq. 18-A.117 was obtained in part by using the identity:

$$\partial\mathscr{L}/\partial(\partial\phi/\partial y^{\sigma}) = \partial\mathscr{L}/\partial(\partial\phi/\partial X^{\alpha})\,\partial y^{\sigma}/\partial X^{a} \tag{18-A.120}$$

where $\mathscr{L}$ and ϕ have the form specified in eq. 18-A.15.

The conserved angular momentum is:

$$M_{s}^{\mu\nu} = \int dy\, \mathscr{M}_{s}^{0\mu\nu} \tag{18-A.121}$$

with

$$\partial M_{s}^{\mu\nu}/\partial y^{0} = 0 \tag{18-A.122}$$

18-A.10.1. Angular Momentum and $\mathscr{L}_F$

An alternate conserved angular momentum can be obtained by considering the "field" part of the Lagrangian $\mathscr{L}_F$ under an infinitesimal Lorentz transformation ($\epsilon_{\mu\nu} = -\epsilon_{\nu\mu}$):

$$X'_{\mu} = X_{\mu} + \delta X_{\mu} \tag{18-A.123a}$$

$$\delta\phi = \phi(X'(y)) - \phi(X(y))$$

$$= \delta X^\mu \, \partial\phi/\partial X^\mu \tag{18-A.123b}$$

$$\delta X^\mu = S^\mu{}_a X^a(y) - X^\mu(y) \tag{18-A.123c}$$

$$= \epsilon^\mu{}_a X^a(y) \tag{18-A.123d}$$

where $S^\mu{}_a$ is the Lorentz transformation matrix for a vector. (If X^μ is a gauge field then an additional operator gauge term would have to be added to eq. 18-A.123d.)

The Lagrangian changes by

$$\delta\mathscr{L}_F = \epsilon^{\mu\nu} X_\nu \, \partial\mathscr{L}_F/\partial X^\mu \tag{18-A.124}$$

under an infinitesimal Lorentz transformation. The change can also be expressed as:

$$\delta\mathscr{L}_F = \partial\mathscr{L}_F/\partial\phi \, \delta\phi + \partial\mathscr{L}_F/\partial(\partial\phi/\partial X^\mu) \, \delta(\partial\phi/\partial X^\mu) \tag{18-A.125}$$

Combining eqs. 18-A.124 and 18-A.125 leads to:

$$\epsilon_{\mu\nu}\partial/\partial X^\sigma \, \mathscr{M}_{FX}{}^{\sigma\mu\nu} = 0 \tag{18-A.126}$$

where

$$\mathscr{M}_{FX}{}^{\sigma\mu\nu} = (g^{\mu\sigma}X^\nu - g^{\nu\sigma}X^\mu)\mathscr{L}_F + \partial\mathscr{L}_F/\partial(\partial\phi/\partial X^\sigma)\,(X^\mu\partial\phi/\partial X_\nu - X^\nu\partial\phi/\partial X_\mu) \tag{18-A.127}$$

The conserved angular momentum associated with the X coordinates is:

$$M_{FX}{}^{\mu\nu} = \int d^3X \, \mathscr{M}_{FX}{}^{0\mu\nu} \tag{18-A.128}$$

with

$$\partial M_{FX}{}^{\mu\nu}/\partial X^0 = 0 \tag{18-A.129}$$

The angular momentum density can be written in the familiar form:

$$\mathscr{M}_{FX}{}^{\sigma\mu\nu} = X^\mu \, \mathscr{T}_F^{\sigma\nu} - X^\nu \, \mathscr{T}_F^{\sigma\mu} \tag{18-A.130}$$

using eq. 18-A.107.

18-A.11. Separable Lagrangians and Internal Symmetries

We will now consider the case of a set of scalar fields ϕ_r in a separable Lagrangian with an internal symmetry under a local transformation

$$\phi_r(X) \rightarrow \phi_r(X) - i\epsilon\lambda_{rs}\,\phi_s(X) \tag{18-A.131}$$

If the Lagrangian is invariant under this transformation, then

$$\delta\mathscr{L}_S \equiv \delta\mathscr{L}_F = 0 = \partial\mathscr{L}_F/\partial\phi_r\,\delta\phi_r + \partial\mathscr{L}_F/\partial(\partial\phi_r/\partial X^\alpha)\,\delta(\partial\phi_r/\partial X^\alpha) \tag{18-A.132}$$

Using the equation of motion eq. 18-A.31, which is satisfied by all components ϕ_r, we obtain a conserved current:

$$\mathscr{J}^\nu = -i\,\partial\mathscr{L}_F/\partial(\partial\phi_r/\partial X^\nu)\,\lambda_{rs}\,\phi_s \tag{18-A.133}$$

satisfying

$$\partial\mathscr{J}^\nu/\partial X^\nu = 0 \tag{18-A.134}$$

The conserved charge is

$$Q = \int d^3X\,\mathscr{J}^0 \tag{18-A.135}$$

$$\partial Q/\partial X^0 = 0 \tag{18-A.136}$$

We note eq. 18-A.71 provides a corresponding conservation law for the y coordinate system.

COSMOLOGY

19. Dark Energy and the Big Bang

With chapter 18 we finish the derivation/construction of the form of the Extended Standard Model based primarily on the use of complex space-time coordinates and the Reality group. It includes Dark Matter and a justification for fermion generations based on a "new" broken symmetry, the Generation group symmetry. We introduced a universal field Y^μ that eliminated the divergences that appear in perturbation theory without the use of a renormalization program. We now turn to the extension of our theory to the Cosmology of our universe, and the extension of Nature to include other universes, which together with our universe, reside in a sixteen dimensional Megaverse. Much of this chapter[212] on Dark Energy and the Big Bang first appeared in Blaha (2004).

19.1 Dark Energy

Dark Energy surfaced in Cosmology when it was noticed that the speed of expansion of our universe was expanding due to an unkown, and "undetectable" energy, dubbed Dark Energy.A theoretical framework was developed by A. Guth and others called *inflation* that provided a scenario for the expansion of the universe through an unidentified source called Dark Energy.

We shall now show that the cause of inflaton is the $Y^\mu(y)$ quantum field of our Extended Standard Model that also resolved QFT divergence problems. We will see $Y^\mu(y)$ also makes the Big Bang finite—no singularity; as well as freeing The Standard Model and Quantum Gravity[213] of infinities. *Thus $Y^\mu(y)$ has a remarkable triple role in our view – to eliminate the Big Bang singularity, to generate the explosive growth of the universe, and to remove infinities from The Standard Model and Quantum Gravity.* This happy coincidence of solutions reflects the use of Ockam's Razor to find the simplest general solution and Leibniz's Minimax Principle to find the most minimal solution that has maximal effects – two sides of the same coin in a sense.

Since the $Y^\mu(x)$ quantum gauge field is a free field (neglecting gravity) the initial state of the universe can be permeated with quanta of this field as well as particle quanta. The total energy of the free $Y^\mu(x)$ field within the universe is Dark Energy.

19.2 The Big Bang Experimentally

In the light of our progress since then we now provide a slightly revised version that is similar to Blaha (2004).

The current state of our knowledge of the evolution of the universe has now been extended back in time to about 350,000 years after the Big Bang through recent astrophysical research. While this progress is encouraging we still face major issues: the nature of Dark Matter (hopefully resolved by the Extended Standard Model), the nature and origin of Dark Energy (hopefully resolved in this chapter), and the events of those critical years before the 350,000 year point that we are slowly reaching

---

[212] While the calculation is unchanged we have added clarifications of the original that might have led to misunderstandings particularly in clarifying the constant a vs. the scale function a(t)..

[213] See Blaha (2011c) and Blaha (2005a) for the removal of infinities in Quantum Gravity.

experimentally. Those early years and the Big Bang itself remain mysteries. This situation is especially critical since the early years of the universe apparently contain an uncertain beginning and an explosive growth.

In this chapter we will attempt to understand that unknown period in the neighborhood of t = 0 where quantum effects we believe play a major role. We will suggest that the inflationary growth of the universe, which is attributed to an unknown "particle", actually is caused by the energy of the q-number part of coordinates – the quantum field $Y^\mu(x)$ that we saw in earlier chapters.

19.3 The State of the Universe at t = 0

If we simply extrapolate the currently popular Standard Cosmological Models (with or without inflations) back to the Big Bang t = 0, we find a universe beginning as a single "mathematical point" with infinite mass density and infinite temperature. The Robertson-Walker metric scale factor a(t), which is a solution of the Einstein equation

$$\dot{a}^2 - 8\pi G\rho a^2/3 = -k \qquad (19.1)$$

typically is solved for a perfect fluid under the assumption of a matter-dominated or a radiation-dominated phase of the universe. If we assume the universe is matter-dominated, then the energy density is

Matter-Dominated: $$\rho = \rho_0/a(t)^3 \qquad (19.2)$$

Under the alternate assumption that the universe is radiation-dominated we have

Radiation-Dominated: $$\rho = \rho_0/a(t)^4 \qquad (19.3)$$

With either assumption we find that the scale factor behaves as

$$a(t) \propto t^n \qquad (19.4)$$

where $0 < n < 1$. Thus a(0) = 0 and the universe reduces to a point with infinite density (eqns. 19.2 and 19.3), and with infinite temperature since

$$T \propto a^{-1}(t) \qquad (19.5)$$

There are evidently grave difficulties in extrapolating the Standard Cosmological Model, or its current variants, to t = 0. The difficulty is compounded by the inherently quantum mechanical aspects that are normally associated with gravitation at ultra-small distances.

Currently, the only viable complete theory of Quantum Gravity is the Two-Tier Quantum Gravity of Blaha (2003) and (2005a). Blaha's type of quantum field theory has the interesting feature that all forces (particle propagators) become zero at very short distances (presumably much less than the Planck mass). Thus a point universe could have an infinite density of essentially "non-interacting" matter as a quasi-stable state. Furthermore if one uses the *generalized* Robertson-Walker metric as described in Blaha (2004), and Appendix 19-A of this book, one finds a classical scale factor of the form:

$$A(t, ř) = a(t)b(ř) \tag{19.6}$$

where a(t) satisfies eq. 19.1.

Since quantum effects can be expected to play a role near t = 0 (the Big Bang) it is possible that the expectation value of a quantum scale factor operator, taking account of quantum effects, could have the form

$$\langle A(t, ř)\rangle = \langle a(t)\rangle\langle b(ř, t)\rangle \tag{19.7}$$

where

$$\langle b(ř, t)\rangle \to \beta(ř)/\langle a(t)\rangle \tag{19.8}$$

as $t \to 0$ so that the zero of $\langle a(t)\rangle$ might be cancelled with the result

$$\langle A(t, ř)\rangle \to \beta(ř) \neq 0 \tag{19.9}$$

Quantum effects would thus eliminate the singularities at t = 0. A quantized version of the generalized Robertson-Walker model[214] opens the possibility of a universe with a finite size, density, and temperature at the time of the Big Bang.

With that possibility in mind, we will use Blaha's (2004) Two-Tier theory to develop a version of his quantum model of the universe in the neighborhood of t = 0. Starting from eq. 8.7.1 of that book which is eq. 19-A.7.1 in the following Appendix 19-A:

$$A(t, ř) = a(t)b(ř) = 2\alpha k^{-1/2}a(t)[1 + \alpha^2 ř^2]^{-1} \tag{8.7.1}$$

where a is a constant given by eq. 19-A.6.3 and introducing a Two-Tier variable Y (as in chapter 7 of Blaha (2004)) with the identification[215]

$$ř \equiv M_cX = M_c(y + iY/M_c^2)$$

we see

$$b(ř) = b(y, t) = 2\alpha k^{-1/2}[1 + \alpha^2(M_cy + iY/M_c)^2]^{-1} \tag{19.10}$$

If

$$Y = -M_c[a_1(y, t) - a_2(y, t)a(t)]^{1/2}/\alpha + iM_c^2a_3(y, t) \tag{19.11}$$

and if, as $t \to 0$,

$$a_1(y, t) \to a_1(y, 0) = 1 \tag{19.12}$$

$$a_2(y, t) \to a_2(y, 0) \neq 0 \tag{19.13}$$

$$a_3(y, t) \to a_3(y, 0) = y \tag{19.14}$$

then we find

$$b(y, t) \to 2\alpha k^{-1/2}/[a_2(y, 0)a(t)] \tag{19.15}$$

and

$$A(t, ř) = a(t)b(ř) = 2\alpha k^{-1/2}/a_2(y, 0) + a(t)\beta_1(y) + a^2(t)\beta_2(y) + \ldots \tag{19.16}$$

---

[214] This model appears in Blaha (2004).

[215] α is a constant and not the fine structure constant.

$$\rightarrow 2ak^{-\frac{1}{2}}/a_2(y, 0) \qquad \text{as} \quad t \rightarrow 0$$

where we omit symbols indicating expectation values for the sake of clarity. Thus, under these circumstances, space does not collapse to a point and the density and temperature—as well as other parameters of interest—are finite; *and the features of the Standard Cosmological Model at larger times are still valid.*

In our model we will make the following assumptions about the universe near t = 0:

1. The particles in the universe consist of fundamental elementary particles – gravitons, photons, electrons, neutrinos, quarks, gluons and so on – and their corresponding anti-particles.

2. The particles are described by Two-Tier quantum field theory. In this type of quantum field theory all particle interactions become negligible at very short distances (as described in chapter 7 of Blaha (2004)) and so the forces between particles may be neglected near t = 0 when the universe is immensely *small.*

3. The energy of the universe can be viewed as consisting of particles – bosons and fermions – each species having blackbody energy distributions since the universe is the best of all possible black bodies.

4. The enormous energy of the universe even if confined to a small region makes the classical Einstein equations a good approximation *due to its macroscopic nature* with one proviso (item 5). Therefore we assume the Generalized Robertson-Walker metric of Blaha (2004) and appendix 19-A.

5. In the neighborhood of t = 0 when the universe is effectively confined to a region whose scale is set by the Planck mass or smaller the quantum nature of the Two-Tier coordinate X^μ becomes significant. In particular the Y^μ field causes a profound change in the behavior of the scale factor A(t, r) as t → 0. (Note: $X^\mu(y) = y^\mu + iY^\mu(y)/M^2$ defines the quantum coordinates where y^μ is a c-number coordinate and Y^μ a free q-number field similar to the electromagnetic field.)

6. The Y^μ quanta are assumed to have a black body spectrum[216] – just like elementary particles – reflecting their continuous emission and absorption by gravitons and other elementary

[216] The only reasonable choice for the spectrum is a black body spectrum given the confinement of the field to the ultimate black body – the universe.

particles. The Y^{μ} blackbody spectrum is implemented via a coherent state. Effectively the coherent state opens a small "bubble" into complex space-time changing the dynamics of the universe at t = 0.

7. We will calculate the expectation value of the quantum field operator Y^{μ} in a closed Robertson-Walker space. In principle we must use the generalized Robertson-Walker metric since the scale factor will depend on both r and t through its dependence on the expectation value of Y^{μ}.

19.4 Two-Tier Quantum Model for the Beginning of the Universe

In our approach in this, and the following, sections we will follow a modest program using the known theoretical foundations of elementary particle physics: the Extendsd Standard Model unified with Quantum Gravity in a Two-Tier quantum field theoretic framework. We will supplement this framework with natural assumptions about the initial conditions of the universe in order to develop a theory describing the evolution of the universe from its initial state.

19.4.1 Einstein Equations Near t = 0

There is no physical reason to believe that the universe at the beginning of time, t = 0, was a mathematical point of infinite temperature and density since the extrapolation of the scale factor of the Standard Cosmological Model to t = 0 is unwarranted for many reasons including quantum considerations.

Ideally we would use the Quantum Theory of gravity to establish the physical theory of the universe near t = 0. However a quantum calculation of the global structure of the universe near t = 0 is not feasible. In view of this situation we must find an approximation that captures the physics of the universe near the Big Bang. One approach is based on the macroscopic energy of the early universe. One can expect that a classical gravitation model theory with appropriate quantum corrections may be a reasonable approximation to the early state of the universe. After all, macroscopic bodies are described by classical physics in general. And the universe is a macroscopic body by virtue of its content at the point of the Big Bang despite its small size. Therefore we will assume that we may start with a classical gravitation model and then introduce quantum corrections.

The natural first choices – based on symmetry considerations – are a Robertson-Walker model and a generalized Robertson-Walker model of the type described in chapter 8 of Blaha (2004) and Appendix 19-A following. The quantum part, that we will shortly introduce, will require us to use a generalized Robertson-Walker model since the quantum corrections reduce the symmetry to a maximally symmetric *two-dimensional* subspace within a four-dimensional space-time. *The quantum part eliminates the equivalence of the classical Robertson-Walker and generalized Robertson-Walker models* that was described in section 8.7 of Blaha (2004). See appendix 19-A.

Therefore we begin with the classical c-number equation for the invariant interval defined by

$$d\tau^2 = dt^2 - A^2(t, \check{r})[d\check{r}^2 + \check{r}^2(d\theta^2 + \sin^2\theta\, d\varphi^2)] \tag{19.17}$$

where

$$A(t, \check{r}) = a(t)b(\check{r}) = 2\alpha k^{-1/2}a(t)[1 + \alpha^2 \check{r}^2]^{-1} \tag{19.18}$$

where a is a constant[217] given by eq. 19-A.6.3 and a(t) is the solution of the Einstein equation:

$$\dot{a}^2(t) - 8\pi G\rho a^2(t)/3 = -k \qquad (19.19)$$

Next we introduce quantum coordinates

$$X^\mu = y^\mu + i\, Y^\mu(y)/M_c^{\,2} \qquad (19.20)$$

We choose the same transverse gauge for Y^μ as we did in chapter 7 of Blaha (2004):

$$\partial Y^i/\partial y^i = 0 \qquad (19.21)$$

$$Y^0 = 0 \qquad (19.22)$$

As a result we make the identification (definition of coordinates)

$$X^0 = y^0 \equiv t \qquad (19.23)$$
$$X^j = y^j + i\, Y^j(y)/M_c^{\,2} \equiv M_c^{\,-1}\check{r}^j \qquad (19.24)$$

The mass factor on the right side of the equal sign in eq. 19.20 is required on dimensional grounds if y is to have the usual dimension of length (inverse mass). As a result, since $\check{r} \in [0, 1]$ by Blaha (2004), eq. 19.24 implies

$$y = |\mathbf{y}| \in [0, M_c^{\,-1}] \qquad (19.25)$$

There are two constants with the dimension of mass to a power: k and M_c. The constant k determines the curvature of space – a large-scale feature of Robertson-Walker models. The constant M_c is related to the very short distance behavior of the theory – high energy phenomena with energies of the order of the Planck mass or larger, and, as we will see, the origin of the universe – a short distance, high energy phenomena as well. Therefore we have also chosen to use M_c on the right side of eq. 19.24.

Since X^0 is a c-number and since the density $\rho(t)$ is a large c-number to very good approximation we will assume a(t) is the c-number solution of the classical c-number eq. 19.19 as $t \to 0$. Further we assume that quantum effects appear solely through $b(\check{r})$. We also assume that the q-number equivalent of $b(\check{r})$ is $b(M_cX)$. These assumptions are consistent with applying Ockham's Razor. The function $b(M_cX)$ satisfies the functional equation:[218]

$$k + (M_c^{\,2}Xb^2)^{-1}\, \partial(Xb'/b)/\partial X = 0 \qquad (19.26)$$

where

$$b' = \partial b/\partial X \qquad (19.27)$$

---

[217] We use a to denote a constant, later set equal to 1, and a(t) to be the solution of the Einstein equation. They are not connected to each other.

[218] See eq. 19-A.5.3 in Appendix 19-A.

and $X = (\vec{X}\cdot\vec{X})^{1/2}$. The formal solution of eq. 19.26 has the same functional form as the c-number solution $b(\check{r})$ in eq. 19.18. Therefore

$$b(M_cX) = :2ak^{-1/2}[1 + a^2M_c^2X^2]^{-1}: \qquad (19.28)$$

where a is a constant given by eq. 19-A.6.3 and where we have specified normal ordering with : ... : to avoid trivial divergences.

Since eq. 19.28 is a q-number expression we must find the scale factor as the expectation value of A(t, X) for a suitable state. We note that, at this point, the invariant interval is an operator expression of the form:

$$d\tau^2 = dt^2 - B^2(t, X)[dX^2 + X^2(d\theta^2 + \sin^2\theta\, d\varphi^2)] \qquad (19.29)$$

where

$$B(t, X) = M_cA(t, M_cX) = a(t)b_M(X) \qquad (19.30)$$

and

$$b_M(X) = M_cb(M_cX) = :2aM_ck^{-1/2}[1 + a^2M_c^2X^2]^{-1}: \qquad (19.31)$$

where a is a constant given by eq. 19-A.6.3. Thus the expectation value of $b_M(X)$ also must be calculated in order to determine the invariant interval's expectation value.

19.4.2 Y Black-Body Coherent States

The Y quanta are continuously being emitted and absorbed by the particles in the primeval universe. As such, they may be expected to have a blackbody energy spectrum that is similar to that of the particles from which they derive their existence. In particular one expects the temperature T associated with their blackbody energy distribution to be the same as that of the "real" particles in the universe. After all, the universe is a black body.

Thus the blackbody energy of Y-quanta as a function of frequency ν per unit volume per unit frequency is assumed to be:

$$u_\nu = 8\pi hc^{-2}\nu^3\, [e^{h\nu/\kappa T} - 1]^{-1} \qquad (19.32)$$

where c is the speed of light, h is Planck's constant, and κ is Boltzmann's constant. At this point we adopt units in which c = 1 and $\hbar = h/2\pi = 1$.

The Hamiltonian for the Y field has a form that is familiar from electrodynamics

$$H = \int d^3y\, \mathscr{H}_Y(y) = \tfrac{1}{2}\int d^3y :E_Y^2 + B_Y^2: = \int d^3p\, \omega \sum_\lambda a^\dagger(p,\lambda)a(p,\lambda) \qquad (19.33)$$

where E_Y and B_Y are the "electric" and "magnetic" fields of the Y field, $\omega = p^0 = |\vec{p}| = 2\pi\nu$ is the energy (in our units), and λ labels the polarization. Note that we are using "infinite volume" continuum quantization formulation.

We now define coherent Y field bra and ket states that yield a spherically symmetric blackbody distribution as the eigenvalue of the Hamiltonian H:

$$|BB, T> = N \ \exp[\int d^3p\, f(\omega,T)\sum_\lambda a^\dagger(p, \lambda)]|0> \tag{19.34}$$

$$<BB, T| = N^*<0|\exp[\int d^3p\, f^*(\omega,T)\sum_\lambda a(p, \lambda)] \tag{19.35}$$

where $\omega = |\vec{p}|$, and where N is a normalization factor. The expectation value of H is

$$<BB, T|H|BB, T> = \int d^3p\, 2\omega|f(\omega,T)|^2 \tag{19.36}$$

$$= \int d\omega\, 8\pi\omega^3|f(\omega,T)|^2 \tag{19.37}$$

$$= \int d\nu\, 16\pi^2\omega^3|f(\omega,T)|^2 \tag{19.38}$$

where $\omega = 2\pi\nu$ in our units ($c = 1$, $\hbar = 1$), and where the factor of two in eq. 19.36 is the number of polarizations.

The expectation value (eigenvalue) of the energy per unit frequency is

$$H_\nu = 16\pi^2\omega^3|f(\omega,T)|^2 \tag{19.39}$$

We relate H_ν to the blackbody energy *per unit volume* per unit frequency u_ν using

$$H_\nu = u_\nu(2\pi/\omega)^3 \tag{19.40}$$

where the factor of $(2\pi/\omega)^3$ makes the right side of eq. 19.40 the blackbody energy per unit frequency in the continuum case of a quantum field in a space of infinite volume. Thus we find

$$f(\omega,T) = \omega^{-3/2}[e^{\omega/\kappa T} - 1]^{-1/2} \tag{19.41}$$

with the phase of $f(\omega,T)$ set to zero.

19.4.3 Expectation Value of Y in Coherent States

As a preliminary to the evaluation of the operator scale factor in eq. 19.31 we will evaluate the expectation value of powers of the Y field between black body coherent states defined by eqns. 19.34-19.35. We will then determine the expectation value of $b_M(X)$ in combination with $a(t)$ to obtain the behavior of the overall scale factor near $t = 0$. It should be apparent to the reader that the expectation value of $b_M(X)$ is dependent on t as well as y due to the time dependence of the Y field. Thus the scale factor will exhibit a considerably more intricate behavior than simply its $a(t)$ dependence.

The Fourier expansion of the Y field is:

$$Y^i(z) = \int d^3p\, N_0(\omega) \sum_{\lambda=1}^{2} \varepsilon^i(p, \lambda)[a(p,\lambda)\, e^{-ip\cdot z} + a^\dagger(p,\lambda)\, e^{ip\cdot z}] \tag{19.42}$$

where z^μ will be set equal to y^μ later, and where

$$N_0(\omega) = [(2\pi)^3 2\omega]^{-\frac{1}{2}} \qquad (19.43)$$

and

$$\omega = (\mathbf{p}^2)^{\frac{1}{2}} = p^0 \qquad (19.44)$$

with $\vec{\boldsymbol{\varepsilon}}(p, \lambda)$ being the polarization unit vectors for $\lambda = 1, 2$ and $\eta_{\mu\nu}p^\mu p^\nu = 0$. The expectation value of Y between the |BB, T> states is:

$$\langle BB, T|Y^i(z)|BB, T\rangle = \int d^3p\, N_0(\omega)f(\omega,T)[e^{-ip\cdot z} + e^{ip\cdot z}]\, \Sigma\varepsilon^i(p, \lambda) \qquad (19.45)$$

The evaluation of eq. 19.45 (and spherical symmetry) gives

$$\langle BB, T|Y^i(z)|BB, T\rangle = \hat{y}^i\int d^3p\, N_0(\omega)f(\omega,T)[e^{-ip\cdot z} + e^{ip\cdot z}] \sum_\lambda \hat{\mathbf{z}}\cdot \boldsymbol{\varepsilon}(p, \lambda) \qquad (19.46)$$
$$\equiv \hat{\mathbf{z}}^i\, Y_{BB}(t, z)$$

where $\hat{\mathbf{z}} = \vec{\mathbf{z}}/|\vec{\mathbf{z}}|$ is the unit three-vector in the direction of $\vec{\mathbf{z}}$, $z = |\vec{z}|$, and $p\cdot z = \omega(t - z\cos\theta)$. We define a spatial coordinate system – choosing the z-axis parallel to $\vec{z}$. Then we have

$$\vec{z} = (0, 0, z) \qquad (19.47)$$
$$\vec{p} = (\sin\theta\cos\phi, \sin\theta\sin\phi, \cos\theta) \qquad (19.48)$$
$$\vec{\boldsymbol{\varepsilon}}(p,1) = (\cos\theta\cos\phi, \cos\theta\sin\phi, -\sin\theta) \qquad (19.49)$$
$$\vec{\boldsymbol{\varepsilon}}(p,2) = (-\sin\phi, \cos\phi, 0) \qquad (19.50)$$

with the result (taking account of eq. 19.46)

$$Y_{BB}(t, z) = \langle BB, T|\, \hat{\mathbf{z}}\cdot \mathbf{Y}(t, z)\, |BB, T\rangle$$
$$= 2\pi\int_0^\infty d\omega\, \omega^2 N_0(\omega)f(\omega,T)\int_0^\pi d\theta\, \sin^2\theta\, [e^{-ip\cdot z} + e^{ip\cdot z}] \qquad (19.51)$$

where $p\cdot z = \omega(t - z\cos\theta)$ with $z = |\vec{z}|$. We will develop integral representations and approximations to Y_{BB} in a later section.

19.4.4 Expectation Values of the Scale Factor A(t, X) and the Invariant Interval $d\tau^2$

The scale factor

$$b_M(X) = :2aM_ck^{-\frac{1}{2}}[1 + a^2M_c^2X^2]^{-1}: \qquad (19.52)$$

where a is a constant given by eq. 19-A.6.3, is a normal-ordered q-number expression. We can formally expand (define) this expression as a power series of normal-ordered powers of X^2 and then evaluate it between blackbody coherent states. First we note that

$$<BB, T|:Y^{i1}(z)Y^{i2}(z)Y^{i3}(z)Y^{i4}(z) \ldots Y^{in}(z):|BB, T> = \hat{z}^{i1}\hat{z}^{i2}\hat{z}^{i3}\hat{z}^{i4} \ldots \hat{z}^{in}(Y_{BB}(t, z))^n \quad (19.53)$$

We now set $z^\mu = y^\mu$, and use Y_{BB} to represent $Y_{BB}(t, \mathbf{y})$:

$$Y_{BB} \equiv Y_{BB}(t, \vec{y}) \quad (19.54)$$

Thus

$$b_{BB}(y, t) = <BB, T|b_M(X)|BB, T>$$

$$= 2\alpha M_c k^{-1/2}\{1 + \alpha^2[M_c^2y^2 + 2i\, yY_{BB} - Y_{BB}^2/M_c^2]\}^{-1} \quad (19.55)$$

and

$$B_{BB}(t, y) = <BB, T|B(t, X)|BB, T> = a(t)b_{BB}(y, t) \quad (19.56)$$

where a is a constant given by eq. 19-A.6.3. The expectation value of the q-number invariant interval (eq. 19.29) is the c-number expression:

$$d\tau_{BB}^2 = <BB, T|d\tau^2|BB, T>$$

$$= dt^2 - B_{BB}^2(t, y)[dX_{BB}^2 + X_{BB}^2(d\theta^2 + \sin^2\theta\, d\varphi^2)] \quad (19.57)$$

where

$$X_{BB} = y + iY_{BB}/M_c^2 \quad (19.58)$$

and

$$dX_{BB} = dy(1 + iM_c^{-2}\partial Y_{BB}/\partial y) \quad (19.59)$$

The appearance of Y_{BB} in the expression for the invariant interval (eq. 19.57) has two effects: it introduces complex space-time into the model and the generalized Robertson-Walker metric is no longer equivalent to the Robertson-Walker metric for all α as it would be in the classical case.

We will see that Y_{BB} approaches zero at large times thus yielding the conventional Robertson-Walker models. But at small times of the order of the Planck time near the Big Bang we enter a brave new world of complex space-time. We will investigate the nature of this new complex world in the succeeding sections.

19.4.5 Representation and Approximations for $Y_{BB}(t, z)$

The angle integral in eq. 19.51 can be performed to yield

$$Y_{BB}(t, z) = \pi^{1/2}z^{-1}\int_0^\infty d\omega\, \omega^{-1}(e^{\omega/\kappa T} - 1)^{-1/2}\cos(\omega t)J_1(\omega z) \quad (19.60)$$

where $J_1(\omega z)$ is a Bessel function using 3.915.5 of Gradshteyn (1965).

19.4.5.1 Some Representations of Y_{BB}

The integral in eq. 19.60 does not appear to be simply expressible in terms of standard transcendental functions. A series representation of the integral can be obtained by expanding the exponential factor due to the Planck distribution:

$$Y_{BB}(t, z) = \tfrac{1}{2}\pi^{1/2}\kappa T \sum_{n=0}^{\infty} (2n)![2^{2n}(n!)^2]^{-1}\{(2n+1+2i\kappa Tt)^{-1} F(\tfrac{1}{2}, 1; 2; -[\kappa Tz/(n+\tfrac{1}{2} + i\kappa Tt)]^2) + (2n+1-2i\kappa Tt)^{-1} F(\tfrac{1}{2}, 1; 2; -[\kappa Tz/(n+\tfrac{1}{2} - i\kappa Tt)]^2)\} \tag{19.61}$$

where $F(a, b; c; w)$ is a hypergeometric function.[219]

Using an integral representation[220]

$$F(a, b; c; w) = \Gamma(c)[\Gamma(b)\Gamma(c - b)]^{-1} \int_0^1 dt\, t^{b-1}(1-t)^{c-b-1}(1-tw)^{-a}$$

for $F(\tfrac{1}{2}, 1; 2; w)$ we see eq. 19.61 can be written in terms of simpler algebraic expressions:

$$Y_{BB}(t, z) = \tfrac{1}{2}\pi^{1/2}(\kappa Tz^2)^{-1} \sum_{n=0}^{\infty}(2n)![2^{2n}(n!)^2]^{-1}\{[(n+\tfrac{1}{2}+i\kappa Tt)^2 + (\kappa Tz)^2]^{1/2} + [(n+\tfrac{1}{2}-i\kappa Tt)^2 + (\kappa Tz)^2]^{1/2} - (2n+1)\} \tag{19.62}$$

Eq. 19.62 shows the limit of $Y_{BB}(t, z)$ for large t is

$$Y_{BB}(t, z) \to \pi^{1/2}\kappa T \sum_{n=0}^{\infty}(2n)![2^{2n+2}(n!)^2]^{-1}(2n+1)[(n+\tfrac{1}{2})^2 + (\kappa Tt)^2]^{-1} \to 0 \tag{19.63}$$

if $tT \to \infty$ as $t \to \infty$ as we see in cosmological models (see section 19.4).

Thus

$$(b_{BB}(y, t))^2 dX^2 \to 2\alpha M_c^2 k^{-1}[1 + \alpha^2 M_c^2 y^2]^{-2} dy^2 \equiv 2\alpha k^{-1}[1 + \alpha^2 \check{r}^2]^{-2} d\check{r}^2 \tag{19.64}$$

where a is a constant given by eq. 19-A.6.3, for large t showing the Two-Tier cosmological model becomes a Robertson-Walker model at large times. However, the Two-Tier standard cosmological model is very different at small times of the order of the Planck time near the Big Bang point.

19.4.5.2 Approximate Solution for Y_{BB}

The integral representation and power series representation of $Y_{BB}(t, y)$ do not reveal the physical behavior of the model for small times t and distances y. Therefore we will examine an approximation for $Y_{BB}(t, y)$ for ranges of y, t and T that are relevant for our considerations. We begin by scaling the integration variable in eq. 19.60 with the result:

---

219 Based on the integral 6.621.1 on p. 711 of Gradshteyn (1965).

220 Magnus (1949) p. 8.

$$Y_{BB}(t, y) = \pi^{½}y^{-1}\int_0^\infty d\omega\, \omega^{-1}(e^\omega - 1)^{-½}\cos(\omega\kappa Tt)J_1(\omega\kappa Ty) \qquad (19.65)$$

The blackbody exponential factor $(e^\omega - 1)^{-½}$ in the integrand of $Y_{BB}(t, y)$ enables the leading order approximate behavior of $Y_{BB}(t, y)$ to be determined for $0 \le t \lessapprox 10^{108}$ s – for all time, practically speaking. In a later section (section 19.3.4) we will see that our approximation to the integral in eq. 19.65 is consistent with the solution that we obtain for Y_{BB}, for the scale factor and thus for the temperature T. The approximations that we will make in eq. 19.65 are

$$\cos(\omega\kappa Tt) \approx 1 \qquad (19.66)$$

$$J_1(\omega\kappa Ty) \approx \omega\kappa Ty/2 \qquad (19.67)$$

They are based on $\kappa Tt \ll 1$ and $\kappa Ty \ll 1$ for all y ($0 \le y \le M_c^{-1}$). The exponential factor tends to limit contributions to the integral to small ω. After making these approximations we find

$$Y_{BB}(t, y) \simeq ½\, \pi^{½}\kappa T\int_0^\infty d\omega\, (e^\omega - 1)^{-½}$$

$$\simeq \pi^{3/2}\kappa T/2 \qquad (19.68)$$

The limit as t gets large can also be approximately determined from eq. 19.65. For large t such that κTy is small (and ω is small due to the exponential Planck distribution factor) we can again approximate the Bessel function with its leading power series expansion term and the exponential factor can again be approximated by $e^\omega - 1 \approx \omega$ so that eq. 19.65 becomes approximately

<u>As $t \to \infty$:</u>

$$Y_{BB}(t, y) \simeq \pi^{½}2^{-1}\,\kappa T\int_0^\infty d\omega\, \omega^{-½}\cos(\omega\kappa Tt)$$

$$= \pi 2^{-3/2}[\kappa T/t]^{½} \qquad (19.69)$$

using 3.751.2 of Gradshteyn (1965). We note that, while κTy is small, κTt could possibly have been large in either a matter-dominated or radiation-dominated universe since it grows as t to a positive power (see section 19.3.) *However, since $Y_{BB}(t, y)$ approaches zero for large times its impact can only be seen in the initial formative stages of the universe near t = 0.*

19.5 The Scale Factor a(t) Near t = 0

The "time factor" a(t) of the scale factor $B_{BB}(t, y)$ appears in

$$B_{BB}(t, y) = \langle BB, T|B(t, X)|BB, T\rangle = a(t)b_{BB}(y, t) \qquad (19.70)$$

and is determined by the classical Einstein equation:

$$\dot{a}^2(t) - 8\pi G\rho a^2(t)/3 = -k \qquad (19.71)$$

As we have argued earlier, the source determining a(t) for small times in the neighborhood of t = 0 (the time of the Big Bang) is a large, macroscopic, classical density $\rho(t)$ and thus a(t) may be considered to be a c-number quantity determined by the c-number Einstein equation to good approximation. This approximation should continue to hold even if this macroscopic density becomes enormous as $t \rightarrow 0$. The quantum effects near t = 0 in the Two-Tier model, that we have developed, appear in the factor $b_{BB}(y, t)$ that we evaluated in previous sections.

19.6 A Complex Blackbody Temperature Near t = 0

The blackbody temperature T for relativistic particles (presumably the dominant type of particles near t = 0) is inversely proportional to the scale factor. At large times the blackbody temperature has the form

$$T = T_0/a(t) \qquad (19.72)$$

where T_0 is a constant.

At times in the neighborhood of t = 0 (the Big Bang) space has three complex dimensions in the Two-Tier model. Temperature can be viewed as a measure of the root mean square speed (or the "average energy") of the components of the perfect fluid that we have assumed. In the case of a gas of particles of average energy E:

$$T = E/(3k/2) \qquad (19.73)$$

In a complex space it is quite natural for the root mean squared speed to be complex as well. As a result complex temperatures naturally follow. Thus we will define

$$T = T_0/B_{BB}(t, y) \qquad (19.74)$$

for all time since t = 0. Since $B_{BB}(t, y)$ approaches $M_c a(t) b(\check{r})$ at large times we find its large time behavior is consistent with those of standard Robertson-Walker models. At times near t = 0, the blackbody temperature T is complex since space is complex and complex kinetic energy is allowed. Then we apply a Reality group transformation to obtain the physical temperature.

In the case of the complex temperature T above, the corresponding physical temperature is its absolute value (obtained by multiplying T by a phase factor from the Reality group U(4))

$$T_{physical} = |T_0/B_{BB}(t, y)| \qquad (19.74a)$$

We shall use eq. 19.74 for the temperature, transforming the results below, afterwards, to real values using the 4-dimensional Reality group.

19.7 The Nature of the Universe Near t = 0

At this point we are ready to examine the Two-Tier model for the Big Bang period.

19.7.1 Behavior of the Complete Scale Factor B(t, y) Near t = 0

The behavior of the expectation value of the scale factor B(t, y), under the assumption that the Y quanta have a blackbody spectrum, is described by the equations:

$$b_{BB}(y, t) = 2\alpha M_c k^{-1/2}\{1 + \alpha^2[M_c^2 y^2 + 2i\, yY_{BB} - Y_{BB}^2/M_c^2]\}^{-1} \quad (19.75)$$

$$B_{BB}(t, y) = \langle BB, T|B(t, X)|BB, T\rangle = a(t)b_{BB}(y, t) \quad (19.76)$$

where a is a constant given by eq. 19-A.6.3, and

$$Y_{BB}(t, y) \simeq \pi^{3/2}\kappa T/2 \quad (19.77)$$

$$a(t) = [2\pi G\rho_0 n^2/3]^{1/n}\, t^{2/n} \quad (19.78)$$

$$T = T_0/B_{BB}(t, y) \quad (19.79)$$

as t → 0. We will set the constant $\alpha = 1$ in the interests of simplicity knowing that this value results in a metric fully equivalent to the Robertson-Walker metric at large times. (Other values of α[221] would also result in a metric equivalent to the Robertson-Walker metric at large times after a re-scaling of the radial coordinate.) Thus we may write

$$B_{BB}(t, y) \cong 2k^{-1/2}M_c a(t)\{1 + M_c^2 y^2 + iy\pi^{3/2}\kappa T_0/B_{BB} - \pi^3\kappa^2 T_0^2/(4M_c^2 B_{BB}^2)\}^{-1} \quad (19.80)$$

This quadratic algebraic equation for B_{BB} has the solutions:

$$B_{BB}(t, y) \cong (1 + M_c^2 y^2)^{-1}\{-i\chi M_c y + k^{-1/2}M_c a(t) \pm [\chi^2 - 2i\chi y k^{-1/2}M_c^2 a(t) + k^{-1}M_c^2 a^2(t)]^{1/2}\} \quad (19.81)$$

with

$$\chi = \pi^{3/2}\kappa T_0/(2M_c) \quad (19.82)$$

As t gets very large we obtain the equivalent of the Robertson-Walker metric scale factor in this approximation if we choose the plus sign in eq. 19.81 (assuming $a^2(t)$ becomes very large so other terms within the square root can be neglected):

$$B_{BB}(t, y) \to 2k^{-1/2}M_c a(t)/(1 + M_c^2 y^2) \quad (19.83)$$

Thus we must choose the plus sign in eq. 19.81:

$$B_{BB}(t, y) \cong (1 + M_c^2 y^2)^{-1}\{-i\chi M_c y + k^{-1/2}M_c a(t) + [\chi^2 - 2i\chi y k^{-1/2}M_c^2 a(t) + k^{-1}M_c^2 a^2(t)]^{1/2}\} \quad (19.84)$$

At t = 0 (the Big Bang) eq. 19.84 simplifies to (assuming a(0) = 0)

$$B_{BB}(0, y) \cong [\chi - i\chi\, M_c y]/(1 + M_c^2 y^2) \quad (19.85)$$

[221] It is **not** the fine structure constant.

For small y, the real part of $B_{BB}(0, y)$ is a constant and the imaginary part of $B_{BB}(0, y)$ is proportional to y.

19.7.2 The Expectation Value of the Scale Factor $A_{BB}(0, y)$ near t = 0

Eq. 19.85 gives the approximate behavior of the expectation value of the scale factor $B_{BB}(0, y)$ near t = 0 as a function of y. If we compare eq. 19.85 with the mechanism described in eqns. 19.1.6 – 19.1.9 of section 19.1 for cancelling the a(t) factor within the complete scale factor we see that we have found the blackbody spectrum of the Y quanta implements this mechanism. The solution can be written in the form:

$$A_{BB}(t, y) = M_c^{-1}B_{BB}(t, y) \cong \beta_0(y) + \beta_1(y)a(t) + \ldots \quad (19.86)$$
$$\beta_0(y) = \varkappa(1 - iM_cy)/[M_c(1 + M_c^2y^2)] \quad (19.87)$$
$$\beta_1(y) = k^{-1/2}(1 - iM_cy)/(1 + M_c^2y^2) \quad (19.88)$$

Eqns. 19.86–19.88 are expressed in terms of the y variable. They can be expressed in terms of ř as:

$$A(t, ř) = a(t)b(ř) = 2\alpha k^{-1/2}a(t)[1 + \alpha^2ř^2]^{-1} \quad (19.89)$$

where a is a constant given by eq. 19-A.6.3 set to $\alpha = 1$. Evidently, we have $M_cy \equiv ř$ at the level of approximation that we are using. Furthermore we can use

$$ř = \{[1 - (1 - kr^2)^{1/2}]/[1 + (1 - kr^2)^{1/2}]\}^{1/2} \quad (19.90)$$

to express ř in terms of the Roberson-Walker radial coordinate r (from Appendix 19-A.) Thus we find that the Robertson-Walker scale factor a(t) becomes

$$a(t) \rightarrow (1 + M_c^2y^2)(2k^{-1/2})^{-1}A_{BB}(t, y) \equiv a_{BBRW}(t, ř) \quad (19.91)$$
$$a_{BBRW}(t, ř) \cong \beta_{0RW}(ř) + \beta_{1RW}(ř)a(t) + \ldots \quad (19.92)$$

using the subscript "RW" to denote quantities scaled to the standard Robertson-Walker metric, with

$$\beta_{0RW}(ř) = \varkappa(1 - iř)/[2k^{-1/2}M_c] \quad (19.93)$$
$$\beta_{1RW}(ř) = (1 - iř)/2 \quad (19.94)$$

where ř is specified by eq. 19.90.

Using the Reality group we can "rotate" the complex scale factor, which is a factor in coordinate expressions, to a real value—its absolute value

$$a(t) \rightarrow (1 + M_c^2y^2)(2k^{-1/2})^{-1}|A_{BB}(t, y)| \equiv |a_{BBRW}(t, ř)| \quad (19.91a)$$

We will study the implications of this scale factor in the following chapters.

Appendix 19-A. Derivation of the Extended Robertson-Walker Model

This appendix provides the derivation of a generalization of the Roberson-Walker solution of General Relativity that we used in chapter 19 in our discussion of the Quantum Big Bang Theory originally presented in Blaha (2004). This appendix is extracted from chapter 8 of Blaha (2004) as necessary background information for chapter 19 and the following chapters.

The generalization derived here has not been derived before, to our knowledge, because at the level of c-number General Relativity it is fully equivalent to the known Robertson-Walker model. However when the theory has q-number parts introduced in a physically meaningful way, as we do, it is no longer equivalent to the c-number Robertson Walker model.

19-A.1 The Robertson-Walker Metric

Much of the current modeling of the evolution and properties of the universe is based on the assumption of a Robertson-Walker metric which is used in the Einstein equations to obtain a first order differential equation for the scale factor R(t):

$$\dot{R}^2 + k = 8\pi G\rho R^2/3 \qquad (19\text{-A.1.1})$$

where k is a factor in the three-dimensional spatial curvature of the Robertson-Walker metric:

$$K_3(t) = k/R^2(t) \qquad (19\text{-A.1.2})$$

Eq. 19-A.1.2 suggests that accurate measurements of the Hubble constant and other cosmological quantities could lead to an accurate determination of the curvature, the time dependence of the Hubble constant, and of R(t).

The form of the Robertson-Walker metric (We consider a real space-time only in this appendix) follows from the assumption of a maximally symmetric three-dimensional subspace whose metric has eigenvalues of the same sign (negative in our formalism) residing within a four-dimensional space-time with one positive eigenvalue and three negative eigenvalues. A maximally symmetric space is isotropic and homogeneous.[222]

Although the Robertson-Walker metric does not appear to embody the concept of an absolute space-time the general arguments presented in chapter 2 of Blaha (2004) indicate that it does in fact implicitly define an absolute reference frame.[223]

---

[222] See Chapter 13 of Weinberg (1972) for a detailed discussion.

[223223] This conclusion is now obvious due to the existence of the Flatverse. We note there can only be one absolute reference frame up to a Lorentz transformation since the sets of inertial reference frames of two absolute reference

Therefore it is sensible to inquire whether a more general metric – a generalization of the Robertson-Walker metric – might be worth investigating – particularly in view of the major unexplained mysteries of Dark Energy as well as other new data showing the existence of massive black holes and quite mature galaxies shortly after (two or three billion years) the origin of the universe. The pile-up of mysteries from WMAP, SDSS and other sources indicates a reconsideration of fundamental assumptions may be worthwhile.

Therefore we will examine the "simplest" generalization of the Robertson-Walker metric in this appendix with a view towards elucidating some of these mysteries. More importantly, we will use this generalization in chapter 19 and subsequent chapters to develop a non-singular, Two-Tier formulation of the dynamics of the universe "at the beginning of time" – the Big Bang – taking account of quantum effects.

From the point of view of the definition of maximally symmetric subspaces the most immediate generalization of the Robertson-Walker metric is to assume a maximally symmetric *two-dimensional* subspace within a four-dimensional space-time. The general form of the metric in this case is:

$$d\tau^2 = A_{tt}(r,t)\, dt^2 + 2A_{rt}(r,t)\, dt\, dr + A_{rr}(r,t)\, dr^2 + B(r,t)(d\theta^2 + \sin^2\theta d\varphi^2) \qquad (19\text{-A.1.3})$$

where A_{ik} is a 2×2 symmetric matrix with one positive and one negative eigenvalue, and B(r,t) is a negative function of r and t.

19-A.2 A Generalization of the Roberson-Walker Metric

We shall consider a generalization of the Robertson-Walker metric (eq. 5.5.1 of Blaha (2004)), which is a special case of eq. 19-A.1.3 that preserves the overall form of the Robertson-Walker metric but allows the scale factor a(t) to depend on r as well as t:

$$R(t) \rightarrow A_0(t, r) \qquad (19\text{-A.2.1})$$

The generalized metric that we will analyze is embodied in the invariant interval expression:

$$d\tau^2 = dt^2 - A_0^2(t, r)[dr^2/(1 - kr^2) + r^2(d\theta^2 + \sin^2\theta\, d\varphi^2)] \qquad (19\text{-A.2.2})$$

The introduction of a dependence on the radius r in $A_0(t, r)$ in eq. 19-A.2.2 eliminates the homogeneity of the three-dimensional spatial subspace reducing it to a two-dimensional maximally symmetric subspace.

We note that the general solution of the Einstein equations for the standard case of a perfect fluid lead to the usual view of the expansion of the universe (after the Big Bang Epoch), Hubble's law, the red shifts of radiation from distant sources of radiation, and the Cosmic Microwave Background (CMB) radiation.

19-A.3 The Einstein Equations for the Generalized Robertson-Walker Metric

The Einstein equations can be written

---

frames must be the same. This implies any two absolute reference frames must be related by a Lorentz transformation since absolute frames are necessarily flat inertial frames.

$$R_{\mu\nu} = -8\pi G S_{\mu\nu} \qquad (19\text{-A.3.1a})$$
$$S_{\mu\nu} = T_{\mu\nu} - \tfrac{1}{2} g_{\mu\nu} T^{\sigma}{}_{\sigma} \qquad (19\text{-A.3.1b})$$

where $R_{\mu\nu}$ is the Ricci tensor and $T_{\mu\nu}$ is the energy-momentum tensor. Assuming the energy-momentum tensor has the form of the energy-momentum tensor of a perfect fluid with the only non-zero components:

$$T_{tt} = \rho\, g_{tt} \qquad (19\text{-A.3.2})$$
$$T_{rr} = -\, p g_{rr} \qquad (19\text{-A.3.3})$$
$$T_{\theta\theta} = -\, p g_{\theta\theta} \qquad (19\text{-A.3.4})$$
$$T_{\phi\phi} = -\, p g_{\phi\phi} \qquad (19\text{-A.3.5})$$

where ρ is the density and p is the pressure, then the non-zero components of $S_{\mu\nu}$ are:

$$S_{tt} = \tfrac{1}{2} (\rho + 3p)\, g_{tt} \qquad (19\text{-A.3.6})$$
$$S_{rr} = -\tfrac{1}{2} (\rho - p) g_{rr} \qquad (19\text{-A.3.7})$$
$$S_{\theta\theta} = -\tfrac{1}{2} (\rho - p) g_{\theta\theta} \qquad (19\text{-A.3.8})$$
$$S_{\phi\phi} = -\tfrac{1}{2} (\rho - p) g_{\phi\phi} \qquad (19\text{-A.3.9})$$

In particular, the fact that

$$S_{tr} = 0 \qquad (19\text{-A.3.10})$$

for a perfect fluid results in an important simplification in the solution of the Einstein equations for this case.

The density $\rho = \rho(t)$ and the pressure $p = p(t)$ are assumed to be solely functions of time t as is usual in the case of a perfect fluid.

19-A.4 The Differential Equations for the Generalized Scale Factor A(t, r)

The dependence of the scale factor A_0 on both r and t leads to a significantly more complicated calculation of the Ricci tensor. We start by noting

$$g_{tt} = 1 \qquad g_{rr} = -A_0^2/(1 - kr^2) \qquad g_{\theta\theta} = -A_0^2 r^2 \qquad g_{\phi\phi} = -A_0^2 r^2 \sin^2\theta \qquad (19\text{-A.4.1})$$

Despite the dependence of A_0 on both t and r in the generalized case a direct calculation of the tt Ricci tensor component yields the familiar expression:

$$R_{tt} = g_{tt}\, 3\ddot{A}_0/A_0 \qquad (19\text{-A.4.2})$$

where we use dots over A_0 to indicate partial derivatives with respect to time:

$$\ddot{A}_0 \equiv \partial^2 A_0/\partial t^2 \qquad (19\text{-A.4.3})$$

However, the tr-component of the Ricci tensor R_{tr}, which is zero in the case of the ordinary Robertson-Walker metric, is non-zero in the more general case under consideration:

$$R_{tr} = 2\,\partial(\partial A_0/\partial t)/\partial r \qquad (19\text{-A.4.4})$$

The corresponding Einstein equation is

$$R_{tr} = 2\,\partial(A_0^{-1}\,\partial A_0/\partial t)/\partial r = -8\pi G S_{tr} = 0 \qquad (19\text{-A.4.5})$$

for a perfect fluid. Eq. 19-A.4.5 implies that $A_0(t, r)$ factorizes:

$$A_0(t, r) = a(t)b_0(r) \qquad (19\text{-A.4.6})$$

This factorization results in a substantial simplification in the non-linear Einstein equations considered next, which are shown to be separable in the radial and time variables.

Before proceeding to the consideration of the remaining Einstein equations it is convenient to redefine the radial coordinate using

$$\check{r}\, b(\check{r}) = r\, b_0(r) \qquad (19\text{-A.4.7})$$

and

$$d\check{r}/dr = b_0(r)[b(\check{r})(1 - kr^2)^{1/2}]^{-1} = \check{r}\,[r(1 - kr^2)^{1/2}]^{-1} \qquad (19\text{-A.4.8})$$

While this change of coordinates does not change the physical content of the theory it does lead to simpler Einstein equations. The change of radial coordinate results in a new form of the invariant interval:

$$d\tau^2 = dt^2 - a^2(t)b^2(\check{r})[d\check{r}^2 + \check{r}^2(d\theta^2 + \sin^2\theta\, d\varphi^2)] \qquad (19\text{-A.4.9})$$

The new radial coordinate $\check{r}$ is related to the old radial coordinate by

$$\check{r} = \{[1 - (1 - kr^2)^{1/2}]/[1 + (1 - kr^2)^{1/2}]\}^{1/2} \qquad (19\text{-A.4.10})$$

and

$$r = 2k^{-1/2}\check{r}(1 + \check{r}^2)^{-1} \qquad (19\text{-A.4.11})$$

Note the range of $\check{r}$ is [0, 1].

A direct calculation of the Ricci tensor for the metric in eq. 19-A.4.9 with $A(t, \check{r})$ defined as:

$$A \equiv A(t, \check{r}) = a(t)b(\check{r}) \qquad (19\text{-A.4.12})$$

leads to the following Einstein equations (remembering our flat space cartesian metric is $\eta_{tt} = +1$ and $\eta_{ij} = -\delta_{ij}$ for i, j = spatial indices):

$$R_{tt} = g_{tt}3\ddot{A}/A = -8\pi G S_{tt} = -4\pi G(\rho + 3p) \qquad (19\text{-A.4.13})$$

$$R_{t\check{r}} = 2\,\partial(A^{-1}\,\partial A/\partial t)/\partial\check{r} = -8\pi G\, S_{t\check{r}} = 0 \qquad (19\text{-A.4.14})$$

$$R_{\check{r}\check{r}} = g_{\check{r}\check{r}}[\ddot{A}/A + 2(\dot{A}/A)^2 - 2(\check{r}A^2)^{-1}\,\partial(\check{r}\, A'/A)/\partial\check{r}]$$

$$= -8\pi G\, S_{\check{r}\check{r}} = 4\pi G\, g_{\check{r}\check{r}}(\rho - p) \quad (19\text{-}A.4.15)$$

$$R_{\theta\theta} = g_{\theta\theta}[\ddot{A}/A + 2(\dot{A}/A)^2 - A''/A^3 - 3A'/(\check{r}A^3)^{-1}]$$

$$= -8\pi G\, S_{\theta\theta} = 4\pi G\, g_{\theta\theta}(\rho - p) \quad (19\text{-}A.4.16)$$

$$R_{\phi\phi} = g_{\phi\phi}[\ddot{A}/A + 2(\dot{A}/A)^2 - A''/A^3 - 3A'/(\check{r}A^3)^{-1}]$$

$$= -8\pi G\, S_{\phi\phi} = 4\pi G\, g_{\phi\phi}\,(\rho - p) \quad (19\text{-}A.4.17)$$

where

$$A' = \partial A/\partial\check{r} \quad (19\text{-}A.4.18)$$

and

$$A'' = \partial^2 A/\partial\check{r}^2 \quad (19\text{-}A.4.19)$$

19-A.5 The Solution for the Generalized Scale Factor A(t, r)

We begin by substituting eq. 19-A.4.13 in eq. 19-A.4.15, and then substituting the factorization of A (eq. 19-A.4.12). The result is a separable equation:

$$\dot{a}^2 - 8\pi G\rho a^2/3 - (\check{r}b^2)^{-1}\,\partial(\check{r}\, b'/b)/\partial\check{r} = 0 \quad (19\text{-}A.5.1)$$

Since the first two terms in eq. 19-A.5.1 are solely functions of t while the third term is solely a function of ř we obtain the separated equations:

$$\dot{a}^2(t) - 8\pi G\rho a^2(t)/3 = -k \quad (19\text{-}A.5.2)$$

and

$$k + (\check{r}b^2)^{-1}\,\partial(\check{r}\, b'/b)/\partial\check{r} = 0 \quad (19\text{-}A.5.3)$$

where k is a separation constant which we provisionally identify with the curvature parameter k in eq. 19-A.2.2.

Eq. 19-A.5.2 is precisely the equation used for the time dependent scale factor in current cosmological models (eq. 5A.3.2) using the Robertson-Walker metric. Therefore its solution, which depends on the time dependence of the energy density, will be the same as that of the corresponding conventional cosmological model for the same density.

Eq. 19-A.5.3 is a differential equation for the *spatial* expansion scale factor b(ř). Under the assumption of a perfect fluid it depends solely on the constant k and is independent of the details of the perfect fluid (i.e. its density and pressure). There are two solutions of the second order non-linear differential equation for b(ř). These solutions can be written as

$$b_1(\check{r}) = 2\gamma\delta k^{-1/2}[\delta^2 + \gamma^2\,\check{r}^2]^{-1} \quad (19\text{-}A.5.4)$$

where γ and δ are constants, and

$$b_2(\check{r}) = \sigma k^{-1/2}[\check{r}\,(\varsigma \pm i\sigma \ln \check{r})]^{-1} \qquad (19\text{-A.5.5})$$

where σ and ς are constants.

$b_1(\check{r})$ (in eq. 19-A.5.4) is the only physically acceptable solution for the case of a perfect fluid with energy-momentum tensor specified by eqns. 19-A.3.2 – 19-A.3.5. Reason: The solution $b_1(\check{r})$ satisfies eqns. 19-A.4.16 and 19-A.4.17 while the other solution $b_2(\check{r})$ does not satisfy these equations. A necessary and sufficient condition for $b(\check{r})$ to satisfy eqns. 19-A.4.16 and 19-A.4.17 is that

$$b'' - \check{r}^{-1}\, b' - 2\, b'^2/b = 0 \qquad (19\text{-A.5.6})$$

where ′ denotes a derivative with respect to ř. Eq. 19-A.5.6 follows from subtracting the coefficients of the metric tensor component factors in eq. 19-A.4.15 from the coefficients of the metric tensor component factors in eq. 19-A.4.16 (or eq. 19-A.4.17). The solution $b_1(\check{r})$ satisfies eq. 19-A.5.6 for all values of γ and δ. The solution $b_2(\check{r})$ does not satisfy eq. 19-A.5.6 for any choice of α and β except the trivial choice $\alpha = 0$ and is therefore physically irrelevant. It might have some relevance in the case of a non-perfect fluid. We will not investigate that possibility in the present work.

19-A.6 The Solution Expressed in the Original Radial Coordinate

The solution that we have obtained for the generalized Roberson-Walker case with radial coordinate ř can be related back to the generalized Robertson-Walker solution using the original radial coordinate r. Eq. 19-A.4.7 implies

$$b_0(r) = \check{r}\, b(\check{r})/r \qquad (19\text{-A.6.1})$$

$$= 2\alpha\, [1 + \alpha^2 + (1 - \alpha^2)(1 - kr^2)^{1/2}]^{-1} \qquad (19\text{-A.6.2})$$

using eqns. 19-A.4.10 and 19-A.5.4, and defining the constant a as

$$\alpha = \gamma/\delta \qquad (19\text{-A.6.3})$$

An important special case of eq. 19-A.6.2 is the case where $\alpha = 1$ (Ockham's Razor!). In this case we find eq. 19-A.6.2 becomes

$$b_0(r) = 1 \qquad (19\text{-A.6.4a})$$

and

$$A_0(t, r) = a(t) \qquad (19\text{-A.6.4b})$$

thus *recovering the normal Robertson-Walker solution exactly as a special case from eqns. 19-A.2.2, 19-A.4.6, and 19-A.5.2*. In this case we see

$$b(\check{r}) = 2k^{-1/2}(1 + \check{r}^2)^{-1} \qquad (19\text{-A.6.5})$$

from eq. 19-A.5.4 in the ř, θ, ϕ coordinate system. We considered this case within the expanded framework of a quantized model of the beginning of the universe in chapter 19 where it has non-trivial consequences.

19-A.7 Equivalence of the General Solution with the Original Robertson-Walker Solution

The general solution of the Einstein equations in the case of a perfect fluid (eqns. 19-A.4.13 – 19-A.4.17) for the scale factor A(t, ř) in the generalized metric specified by eq. 19-A.4.9

$$d\tau^2 = dt^2 - A^2(t, \check{r})[d\check{r}^2 + \check{r}^2(d\theta^2 + \sin^2\theta\, d\varphi^2)] \qquad (19\text{-A.4.9})$$

is given by

$$A(t, \check{r}) = a(t)b(\check{r}) = 2\alpha k^{-1/2} a(t)[1 + \alpha^2\check{r}^2]^{-1} \qquad (19\text{-A.7.1})$$

where a is a constant given by eq. 19-A.6.3 and where a(t) is the solution of the standard equation (eq. 19-A.5.2) for the time dependent scale factor in the case of the Robertson-Walker metric. We now note that if we define a new radial vector

$$\mathfrak{r} = \alpha\check{r} \qquad (19\text{-A.7.2})$$

then

$$d\tau^2 = dt^2 - a^2(t)4k^{-1}[1 + \mathfrak{r}^2]^{-1}\, [d\mathfrak{r}^2 + \mathfrak{r}^2(d\theta^2 + \sin^2\theta\, d\varphi^2)] \qquad (19\text{-A.7.3})$$

Comparing this invariant interval expression with the $\alpha = 1$ expression for b(ř) given in eq. 19-A.6.5 we conclude that we have proved the following theorem:

Theorem: The solution of the Einstein equations for the case of a perfect fluid for the generalized Robertson-Walker metric (eq. 19-A.2.2) is equivalent to a solution of the Einstein equations for the case of a perfect fluid in the case of the Robertson-Walker metric with a scale factor a(t) that is solely dependent on time.

In the case of *classical* gravitation theory the solutions are fully equivalent and related by a simple change of radial coordinates. Thus the homogeneity condition that we had relinquished at the beginning of our discussion is reinstated and the solution of the generalized case is consistent with a *maximally symmetric three-dimensional space*. The origin of the coordinate system can be chosen to be any point in space.

In the case of quantized versions of the Robertson-Walker model and our generalization of it we will see that the solutions are *generally not equivalent*. We explored a particular example of a quantized gravitational model that illustrates this point in chapter 19. The quantized model, which we defined, should be viewed as a first attempt to explore the quantum regime existing at the beginning of time – the Big Bang. The reasonableness of its results suggests that we are on the right track for understanding the Big Bang Epoch.

19-A.8 Hubble's Law in the Generalized Robertson-Walker Model

Our generalized Robertson-Walker metric assumes an inhomogeneous space with some fixed center at $\check{r} = 0$. Presumably this center was the point at which the Big Bang took place at the beginning of the universe. In this section see how Hubble's Law emerges in the generalized model.

Hubble's law is one of the cornerstones of modern cosmology. While one might think a scale factor that depended on the radius coordinate might not be consistent with Hubble's Law it is easy to show that Hubble's Law is satisfied provided that the scale factor factorizes as required by the tr-component of Einstein's equations (eq. 19-A.4.14) for our generalized Robertson-Walker model.

First we give a simple derivation of Hubble's law for the case of a separable scale factor:

$$A(t, \check{r}) = a(t)b(\check{r}) \qquad (19\text{-A.4.12})$$

under the assumption that some remote galaxy lies on the same radial line as the line from the origin of the space coordinates to our galaxy. The proper distance between the remote galaxy and our galaxy has the form

$$D(t) = D_0A(t, \check{r}) \qquad (19\text{-A.8.1})$$

The rate of recession of the remote galaxy is then

$$v = dD/dt = D_0b(\check{r})da(t)/dt = HD(t) \qquad (19\text{-A.8.2})$$

with

$$H = d \ln a(t)/dt \qquad (19\text{-A.8.3})$$

H is Hubble's constant. If the speed of recession is small then it determines the first-order Doppler shift. If we denote the wavelength of the received radiation as λ_r and the wavelength of the radiation at the source as λ_s then the shift z is

$$z = \lambda_r/\lambda_s - 1 = v/c = HD/c \qquad (19\text{-A.8.4})$$

Eq. 19-A.8.4 is Hubble's Law: the red shift of a galaxy is proportional to its distance. Note Hubble's Law in the generalized case follows from the factorization of $A(t, \check{r})$ with a(t) satisfying the same differential equation as the Robertson-Walker scale factor.

Since a change of radial coordinate reduces the classical generalized model to the Robertson-Walker model, Hubble's Law can be proven in the general case of the non-collinearity of the coordinate origin, source and reception points.

20. Big Bang Scale Factor from t = 0 to the Present

20.1 Introduction

This chapter develops a numerical model for the scale factor from the Big Bang to the present. Although the universe that we live in is almost flat according to recent WMAP experimental data, the small curvature of space closes the universe and has significant effects. Therefore we will not use a flat space approximation. Our goal is to obtain an order of magnitude understanding of the evolution of the universe from the beginning. Our calculated numerical quantities appear generally to be of the right order of magnitude. And the physical ideas appear to be consistent with a reasonable view of reality.

Chapter 21 describes the physical implications of our blackbody Y quanta Dark Energy stabilization and expansion mechanisms for the universe. An especially important result of chapter 7 of Blaha (2004) for the expansion of the universe is: **Gravity is a repulsive force (anti-gravity!) at distances less than 9.08 × 10^{-34} cm.** (See Fig. 18.6.2 and Fig. 7.3.9.3 of Blaha (2004).) Thus the expansion of the universe gets an additional boost from gravity at ultra-short distances.

The data that we use in this chapter and throughout are the combined results of the WMAP and SDSS data[224] based on the assumption of a non-flat space. In particular we use the following values:

$$\begin{aligned} h &= \text{Hubble parameter} = 0.660 \\ \rho_{cr} &= \text{Critical density} = 1.88h^2 \times 10^{-29}\ \text{g/cm}^3 \\ \Omega_\Lambda &= \text{Dark Energy density}/\rho_{cr} = \rho_{de}/\rho_{cr} = 0.695 \\ \Omega_d &= \text{Dark matter density}/\rho_{cr} = \rho_d/\rho_{cr} = 0.115 \\ \Omega_b &= \text{Baryon density}/\rho_{cr} = \rho_b/\rho_{cr} = 0.0230 \\ \Omega_{tot} &= \Omega_m + \Omega_\Lambda = 1.012 \\ \Omega_m &= \text{Matter density}/\rho_{cr} = \rho_m/\rho_{cr} = 0.317 \\ t_0 &= t_{now} = \text{Age of universe} = 14.1\ \text{Gyr} \end{aligned} \qquad (20.1.1)$$

The reader is directed to the original paper for other parameters, error bars and a detailed analysis of the data. *We use units where $\hbar = c = 1$ unless stated otherwise.*

In addition to the above input values we will use:[225]

$$\Omega_\gamma = \text{radiation density}/\rho_{cr} = \rho_\gamma/\rho_{cr} = 2.47h^{-2} \times 10^{-5} \qquad (20.1.2)$$

and, also, based on an analysis of WMAP[226] data

---

[224] M. Tegmark et al, Phys Rev. **D69**, 103501 (2004): Table IV, column 6. Certain parameters have changed in the past eight years of experiments and observations. However the changes do not significantly change the discussions and conclusions of this chapter and the following chapters.

[225] Dodelson(2003) p. 41.

[226] N. J. Cornish, D. N. Spergel, G. D. Starkman, and E. Komatsu, Phys. Rev. Lett. **92**, 201302-1 (2004).

$$r_{universe}(t_{now}) = \text{current radius of the universe} > 7.4 \times 10^{28}\ \text{cm} \qquad (20.1.3)$$

20.2 The Behavior of the Scale Factor a(t) after the Big Bang Period

The universe, as we know it today, contains a variety of forms of energy. The current densities of these forms of energy are listed in section 20.1. From them we can develop the form of the total energy density and project it back to the instants after the Big Bang. The Big Bang period is significantly different as we have seen in the preceding chapter.

The total energy density as a function of time is

$$\rho_{tot} = [\Omega_\gamma/a^4(t) + \Omega_m/a^3(t) + \Omega_\Lambda]\rho_{cr} \qquad (20.2.1)$$

based on well-known arguments.[227] The time dependence of the scale parameter is given by the Einstein equation:

$$\dot{a}^2 - 8\pi G\rho_{tot}a^2/3 = -k \qquad (19\text{-A.5.2})$$

where a(t) is the Robertson-Walker scale factor with $a(t_{now}) = 1$.

Before proceeding to the solution of eq. 9-A.5.2 we need to obtain a reasonable estimate of the curvature constant k. We can use the Robertson-Walker expression for the radius of the universe

$$r_{universe}(t) = a(t)/k^{½} \qquad (20.2.2)$$

evaluated for the present time where eq. 20.1.3 sets a lower bound on the radius to obtain

$$k^{-½} > 7.4 \times 10^{28}\ \text{cm} \qquad (20.2.3)$$

If we *assume* the actual radius of the universe is twice the lower bound, 1.48×10^{29} cm, we get

$$k = (1.48 \times 10^{29})^{-2}\ \text{cm}^{-2} = 4.57 \times 10^{-59}\ \text{cm}^{-2} \qquad (20.2.4)$$

We will use this value of k in the following sections.

20.2.1 General Form of a(t) Scale Factor Einstein Equation

We find the general form of the scale factor differential equation by combining eqns. 19-A.5.2 and 20.2.1

$$\dot{a}^2 - H_0^2a^2(t)[\Omega_\gamma/a^4(t) + \Omega_m/a^3(t) + \Omega_\Lambda] = -k \qquad (20.2.1.1)$$

where Hubble's constant H_0 satisfies

$$H_0^2 = [d(\ln a)/dt]|_{t = t_{now}} = 8\pi G\rho_{cr}/3 \equiv 1.17 \times 10^{-56}h^2\ \text{cm}^{-2} \qquad (20.2.1.2a)$$

or

[227] Weinberg(1972), Dodelson(2003).

$$H_0 = 1.08 \times 10^{-28} h \text{ cm}^{-1} \equiv 3.24 \times 10^{-18} h \text{ s}^{-1} \qquad (20.2.1.2b)$$

If we evaluate eq. 20.2.1.1 for the present time we find

$$k = [\Omega_\gamma + \Omega_m + \Omega_\Lambda - 1]H_0^2 \cong [\Omega_m + \Omega_\Lambda - 1]H_0^2 = (0.012^{+0.087}_{-0.082})H_0^2$$

$$= 6.10^{+44.2}_{-41.7} \times 10^{-59} \text{ cm}^{-2} \qquad (20.2.1.2c)$$

Notice the error "bars" make the value of k uncertain. They result from the difficulties in experimentally measuring the densities Ω_m and Ω_Λ.

The range of values for k in eq. 20.2.1.2c is $[-47.8 \times 10^{-59}, 50.3 \times 10^{-59}]$. Therefore we feel that the radius value determined from WMAP data in eq. 20.1.3, even though it is a lower bound, may be a better indication of the value of k assuming a closed Robertson-Walker universe. In any case we will use this value for k in eq. 20.2.4. This value is only 25% different from the estimate in eq. 20.2.1.2c and thus well within the order of magnitude goal of our calculations. The value of k that we have selected:

$$k = 4.57 \times 10^{-59} \text{ cm}^{-2} \qquad (20.2.4)$$

implies

$$\Omega_m + \Omega_\Lambda = 1.009$$

which is within the error bars of these quantities.

We now define

$$\xi = k/H_0^2 = 3.92 \times 10^{-3} h^{-2} \qquad (20.2.1.3)$$

for later use.

The solution of eq. 20.2.1.1 can be put into the form of integrals representing combinations of elliptic integrals:

$$\int_{a(t')}^{a(t)} da\, a\, H_0^{-1} [\Omega_\Lambda a^4 - \xi a^2 + \Omega_m a + \Omega_\gamma]^{-1/2} = \int_{t'}^{t} dt \qquad (20.2.1.4)$$

The result of these integrations is an implicit equation for a(t) that cannot be expressed in a simple closed form in terms of known functions. This equation can be easily solved numerically. A graph of a(t) is displayed in Fig. 20.2.1.1. Although it looks linear there are significant non-linearities in various parts of the plot of a(t). The radiation-dominated phase is not visible. It is a small slice of the plot since it amounts to less than 10^{13} s.

Because we know physically that approximations are possible for each of the various epochs: the matter-dominated epoch, the radiation-dominated epoch and so on, we can find physically meaningful approximations for each epoch. We therefore provisionally divide the life of the universe into two epochs: an explosive growth epoch, and an expanding epoch subdivided into matter-dominated and radiation-dominated phases. We would have subdivided the expanding epoch into three phases if the constant ξ were not so small.

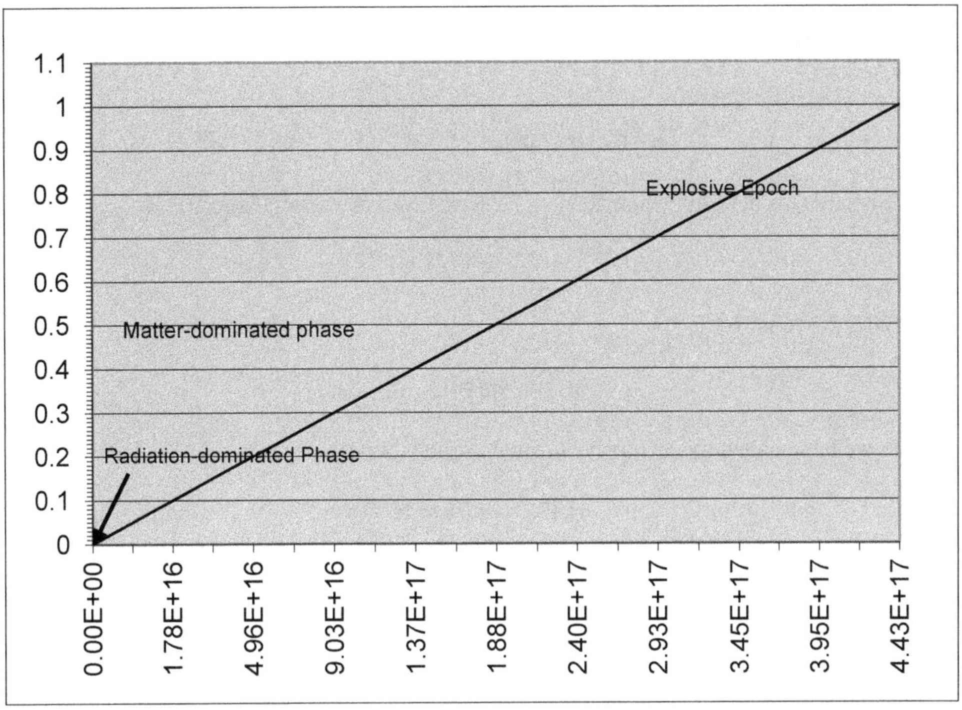

Figure 20.2.1.1. A plot of a(t) generated from eq. 20.2.1.4 through numerical integration.

Epoch	Type	Phases
I	**Explosive Growth**	
II	**Expanding**	**Matter-dominated** **Radiation-dominated**

Table 20.2.1.1. Epochs and phases of the universe after the Big Bang Epoch.

20.2.2 The Explosive Growth Epoch

The integral on the left in eq. 20.2.1.4 appears to support a simple approximation during the explosive growth phase. Notice that at $t = t_{now}$ we have $-\xi a^2 + \Omega_m a + \Omega_\gamma \cong .308$. The sum of these terms gets smaller as we proceed into the past. Therefore we approximate the left side with

$$\int_{a(t')}^{a(t)} da\, aH_0^{-1}[\Omega_\Lambda a^4]^{-\frac{1}{2}} = (H_0\Omega_\Lambda^{\frac{1}{2}})^{-1} \int_{a(t')}^{a(t)} da/a = (H_0\Omega_\Lambda^{\frac{1}{2}})^{-1}\ln[a(t)/a(t')] \cong t - t' \qquad (20.2.2.1)$$

or

$$a(t) = a(t')\exp[H_0\Omega_\Lambda^{1/2}(t - t')] \qquad (20.2.2.2)$$

Thus we have deSitter-like exponential growth. We would expect this growth phase to last approximately for the period where

$$\Omega_\Lambda a^4(t_E) \approx \Omega_m a(t_E) \quad \Rightarrow \quad a(t_E) = .77 \qquad (20.2.2.3)$$

until the present where t_E is the time when $a(t_E) = .77$. Since we have normalized a(t) by

$$a(t_{now}) = 1 \qquad (20.2.2.4)$$

we see that eq. 20.2.2.2 requires

$$a(t) = \exp[H_0\Omega_\Lambda^{1/2}(t - t_{now})] \qquad (20.2.2.5)$$

in the epoch that we have called the Explosive Growth Epoch. Note

$$H_0\Omega_\Lambda^{1/2} = 1.78 \times 10^{-18}\ s^{-1} \qquad (20.2.2.6)$$

The beginning of this epoch is set by eq. 20.2.2.3:

$$t_E = t_{now} + (H_0\Omega_\Lambda^{1/2})^{-1}\ln(.77) = 2.99 \times 10^{17}\ s \qquad (20.2.2.7)$$

giving the time interval of this epoch as 1.47×10^{17} s = 4.7 Gyr —much longer than the radiation-dominated era, and still expanding. The transition time t_E is close to the standard hypothesis for the appearance of Dark Energy of at a red-shift z ~ .5 or a(t) = 2/3. ($1 + z = a^{-1}$) The a(t) value of 2/3 corresponds to a time $t_E = 2.7 \times 10^{17}$ s by eq. 20.2.1.4 (numerical solution) which is to be compared to our estimate $t_E = 2.99 \times 10^{17}$ s – a roughly 10% difference.

20.2.3 The Expanding Epoch

The Expanding Epoch includes the matter-dominated and radiation-dominated phases. The solution for the scale factor in this epoch is obtained by neglecting the Ω_Λ term in eq. 20.2.1.4:

$$\int_{a(t')}^{a(t)} da\ a\ H_0^{-1}\,[-\boldsymbol{\xi a^2} + \boldsymbol{\Omega_m a} + \boldsymbol{\Omega_\gamma}]^{-1/2} = \int_{t'}^{t} dt \qquad (20.2.3.1)$$

This equation is easily integrated yielding:

$$[-\xi a^2(t) + \Omega_m a(t) + \Omega_\gamma]^{1/2} + \Omega_m(2\xi^{1/2})^{-1}\arcsin[(\Omega_m - 2\xi a(t))(\Omega_m^2 + 4\xi\Omega_\gamma)^{-1/2}] -$$

$$-[-\xi a^2(t') + \Omega_m a(t') + \Omega_\gamma]^{1/2} - \Omega_m(2\xi^{1/2})^{-1}\arcsin[(\Omega_m - 2\xi a(t'))(\Omega_m^2 + 4\xi\Omega_\gamma)^{-1/2}] = -\xi H_0(t - t') \qquad (20.2.3.2)$$

Assuming a(0) = 0 and letting $t_0 = 0$ we find

$$[-\xi a^2(t) + \Omega_m a(t) + \Omega_\gamma]^{1/2} + \Omega_m(2\xi^{1/2})^{-1}\arcsin[(\Omega_m - 2\xi a(t))(\Omega_m^{\ 2} + 4\xi\Omega_\gamma)^{-1/2}] -$$

$$- \Omega_\gamma^{\ 1/2} - \Omega_m(2\xi^{1/2})^{-1}\arcsin[\Omega_m(\Omega_m^{\ 2} + 4\xi\Omega_\gamma)^{-1/2}] = -\xi H_0 t \qquad (20.2.3.3)$$

Eq. 20.2.3.3 can be substantially simplified. Note the arguments of the arcsines are both near one in value due to the smallness of Ω_γ and ξ. Both arcsines can be approximated using

$$\arcsin(1 - \epsilon) \cong \pi/2 - (2\epsilon)^{1/2} \qquad (20.2.3.4)$$

for small ϵ.

First we approximate the arguments of the arcsines with

$$\arcsin[(\Omega_m - 2\xi a(t))(\Omega_m^{\ 2} + 4\xi\Omega_\gamma)^{-1/2}] \cong \arcsin(1 - 2\xi a(t)\Omega_m^{\ -1} - 2\xi\Omega_\gamma\Omega_m^{\ -2})$$

and

$$\arcsin[\Omega_m(\Omega_m^{\ 2} + 4\xi\Omega_\gamma)^{-1/2}] \cong \arcsin(1 - 2\xi\Omega_\gamma\Omega_m^{\ -2})$$

Then using eq. 20.2.3.4 we obtain

$$[-\xi a^2(t) + \Omega_m a(t) + \Omega_\gamma]^{1/2} - [\Omega_m a(t) + \Omega_\gamma]^{1/2} \cong -\xi H_0 t \qquad (20.2.3.5)$$

Noting that $\xi a^2(t)$ is much smaller than the other terms in the first square root in eq. 20.2.3.5 we can further approximate that equation by expanding the square root to obtain:

$$a^2(t)[\Omega_m a(t) + \Omega_\gamma]^{-1/2} \cong 2H_0 t \qquad (20.2.3.6)$$

Eq. 20.2.3.6 embodies the standard matter-dominated and radiation-dominated expressions for the scale factor.

In the case of the matter-dominated phase we have

$$\Omega_m a(t) > \Omega_\gamma$$

and can approximate eq. 20.2.3.6 accordingly

$$a^2(t)[\Omega_m a(t)]^{-1/2} \cong 2H_0 t \qquad (20.2.3.7)$$

or

<u>Matter-dominated Phase</u>

$$a(t) \cong [2\Omega_m^{\ 1/2} H_0 t]^{2/3} \qquad (20.2.3.8)$$

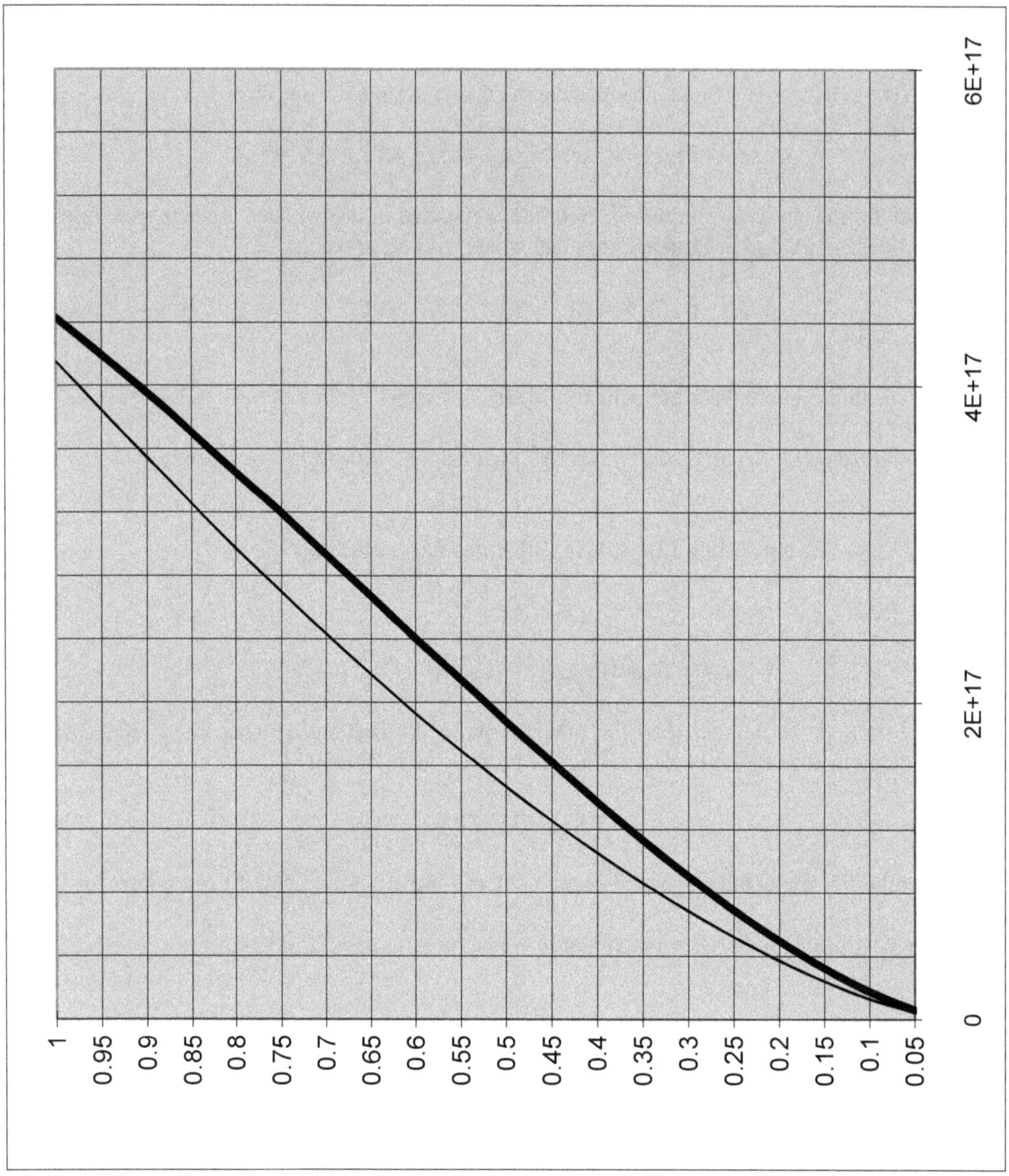

Figure 20.2.3.1. A plot of a(t) (horizontal axis) vs. time in seconds. The thick line is the plot of a(t) obtained by direct numerical integration of eq. 20.2.1.4 including the three density terms and the curvature constant term. The thin line is a plot of a(t) calculated directly from the approximation eq. 20.2.3.6. The approximation becomes increasingly better for small times as t → 0. (Reader: please rotate page 90 degrees clockwise.)

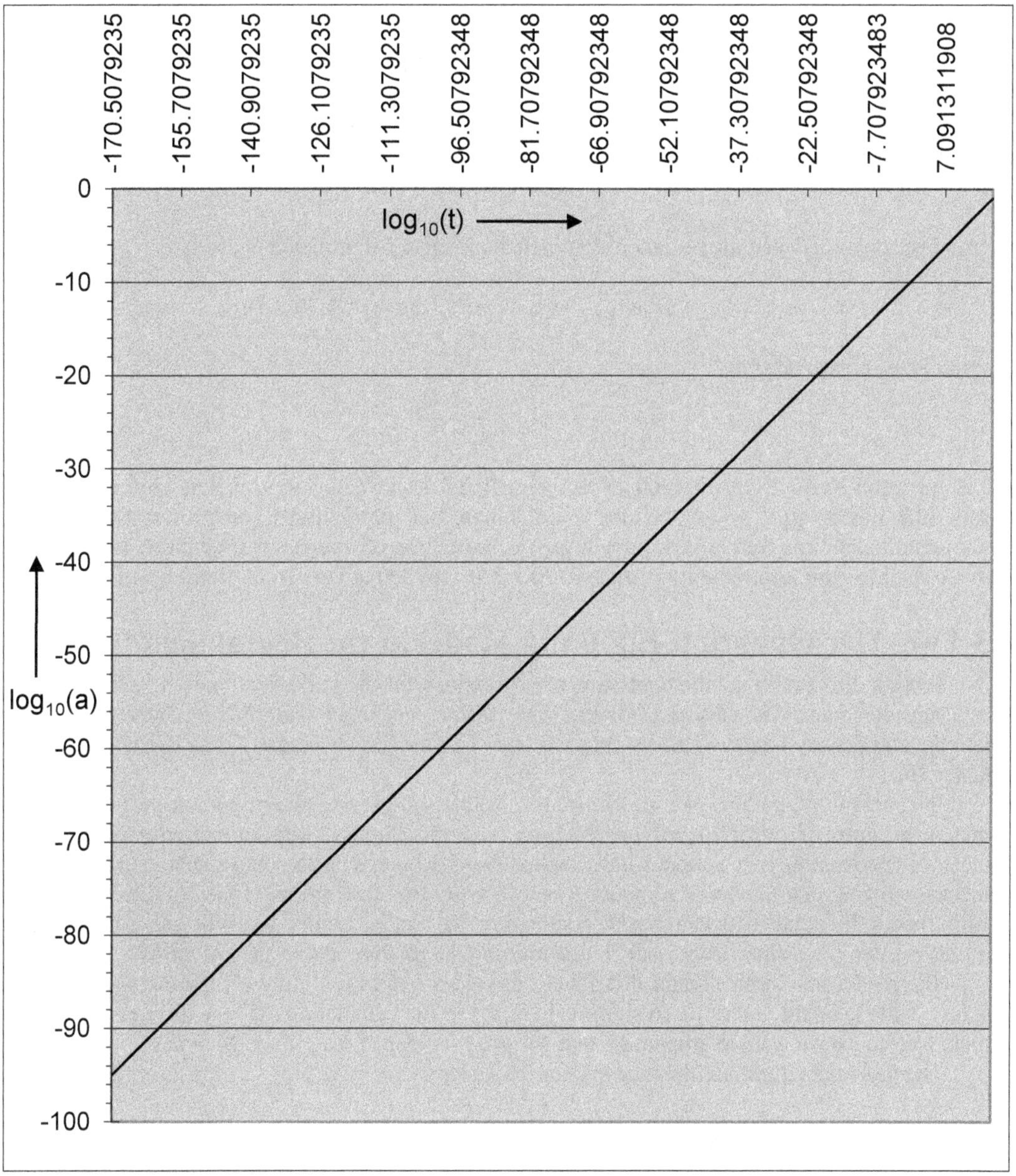

Figure 20.2.3.2. A log-log (base 10) plot of a(t) vs. t (in seconds) for small times calculated from eq. 20.2.3.6.

In the case of the radiation-dominated phase we have

$$\Omega_m a(t) < \Omega_\gamma$$

due to the smallness of the scale factor in that phase We approximate eq. 20.2.3.6 accordingly

$$a^2(t)[\Omega_\gamma]^{-1/2} \cong 2H_0t \tag{20.2.3.9}$$

Thus

Radiation-dominated Phase

$$a(t) \cong [2\Omega_\gamma^{1/2}H_0t]^{1/2} \tag{20.2.3.10}$$

The crossover point between the radiation-dominated and matter-dominated phase is at

$$\Omega_m a(t_{RM}) = \Omega_\gamma \qquad \text{or} \qquad a(t_{RM}) = 1.79 \times 10^{-4} \tag{20.2.3.11}$$

where t_{RM} is the crossover time:

$$t_{RM} = 7 \times 10^{11}\ \text{s} \tag{20.2.3.12}$$

Figs. 20.2.3.1 and 20.2.3.2 contain plots of a(t). Fig. 20.2.3.1 shows the approximate implicit equation for a(t) (eq. 20.2.3.6) is quite good over the entire range, and particularly good for small times in the radiation-dominated time frame. Since this region is the region of interest as it connects to the Big Bang Epoch we shall use this approximation, and eq. 20.2.3.10, for a(t) at very small times near t = 0.

20.3 Two-Tier Quantum Big Bang Model at the Beginning of Time

Section 20.2 presented the approximate expressions for the scale factor near t = 0. In this section we will consider numerical estimates for the scale factor, and other quantities of interest, such as the temperature and density near, and at, t = 0 in the Two-Tier model whose early time behavior is described in chapter 19.

In estimating quantities we are confronted with an imprecise determination of the needed input parameters because of experimental uncertainties and the impossibility of currently finding certain parameters experimentally in a model independent way. So we will use reasonable estimates for input parameters realizing that they may sometimes be off by up to a few orders of magnitude. Because of the vast differences in value between terms comprising the scale factors we will see a few orders of magnitude is often not a significant issue in determining the relative importance of terms.

The reader may notice slight differences in values due to rounding off numbers to three figures in the text while keeping values to 16 significant digits in the calculations. These differences can have a cumulative effect so we want to emphasize that our goal is order of magnitude accuracy.

The input data items are those of section 20.2 plus:

1. The current temperature of the Cosmic Microwave Background (CMB) radiation = 2.725 °K.

2. We assume $M_c = M_{Planck} = 1.22 \times 10^{28}$ ev since M_{Planck} is the only large mass intrinsic to the theory of gravitation and thus seemed to be a natural choice.

20.3.1 The CMB Temperature

From the current CMB temperature (T = 2.725 °K) we find

$$\kappa T = \kappa T_0/a(t_{now}) = \kappa T_0 = 2.34 \times 10^{-4} \text{ ev} \quad (20.3.1.1)$$

For later use we define

$$\chi = \pi^{3/2}\kappa T_0/(2M_c) \cong 5.34 \times 10^{-32} \quad (20.3.1.2)$$

20.3.2 The Generalized Robertson-Walker Scale Factor

A_{BB} is related to B_{BB} by eq. 19.86. Our approximation for the B_{BB} scale factor is:[228]

$$B_{BB}(t, y) \cong (1 + M_c^2y^2)^{-1}\{-i\chi M_c y + k^{-½}M_c a(t) + [\chi^2 - 2i\chi yk^{-½}M_c^2 a(t) + k^{-1}M_c^2a^2(t)]^{½}\} \quad (20.1.5)$$

If we let $y = M_c^{-1}$ then we can find B_{BB} "at the borders of the universe" where it simplifies to:

$$B_{BB}(t) = B_{BB}(t, M_c^{-1}) \cong \{-i\chi + \varpi a(t) + [\chi^2 - 2i\chi\varpi a(t) + \varpi^2a^2(t)]^{½}\}/2 \quad (20.3.2.1)$$

with

$$\varpi = k^{-½}M_c = 9.17 \times 10^{61} \quad (20.3.2.2)$$

If t and a(t) are small,

$$A_{BB} = A_{BB}(t, ř) = M_c^{-1}B_{BB}(t, ř) \cong \beta_0(ř) + \beta_1(ř)a(t) + \ldots \quad (20.3.2.3)$$

$$\beta_0(ř) = \chi(1 - iř)/[M_c(1 + ř^2)] \quad (20.3.2.4)$$

$$\beta_1(ř) = k^{-½}(1 - iř)/(1 + ř^2) \quad (20.3.2.5)$$

We noted earlier that if we transformed our results back to Robertson-Walker coordinates we would have a modified scale factor $a = a_{BBRW}$ which we can continue to express in terms of $ř = M_c y$.

$$a(t) \rightarrow (1 + M_c^2y^2)k^{½}A_{BB}(t, y)/2 \equiv a_{BBRW}(t, ř) \quad (19.5.2.4)$$

Thus

$$a_{BBRW}(t, ř) = (1 + M_c^2y^2)(2k^{-½}M_c)^{-1}B_{BB}(t, y) \quad (20.3.2.6)$$
$$= ½\{a(t) - i\chi\varpi^{-1}ř + [(\chi/\varpi)^2 + a^2(t) - 2i(\chi/\varpi)řa(t)]^{½}\} \quad (20.3.2.7)$$

by eq. 19.92. If we evaluate a_{BBRW} at $ř = M_c y = 1$ (the maximum value of y), since it determines the "size" of the universe, then

$$a_{BBRW}(t) \equiv a_{BBRW}(t, 1) = \{a(t) - i\gamma + [\gamma^2 + a^2(t) - 2i\gamma a(t)]^{½}\}/2 \quad (20.3.2.8)$$

---

[228] Please note that the physical value of scale factors is the absolute value of the complex scale factors (obtained by use of Reality group transformations) appearing here and in the following discussions.

with the dimensionless constant

$$\gamma = \chi/\varpi = 5.82 \times 10^{-94} \qquad (20.3.2.9)$$

This value reminds one of Eddington's famous remark that cosmological quantities often have orders of magnitude that are approximately multiples of 90.

The real and imaginary parts of $a_{BBRW}(t)$ are:

$$\text{Re } a_{BBRW}(t) = a(t)/2 + [R(t)(1 + \cos \psi(t))/2]^{½}/2 \qquad (20.3.2.10)$$

and

$$\text{Im } a_{BBRW}(t) = -\gamma/2 - [R(t)(1 - \cos \psi(t))/2]^{½}/2 \qquad (20.3.2.11)$$

where

$$R(t) = [(\gamma^2 + a^2(t))^2 + 4\gamma^2 a^2(t)]^{½} \qquad (20.3.2.12)$$

and

$$\cos \psi(t) = (\gamma^2 + a^2(t))/R \qquad (20.3.2.13)$$

There are 2 distinctly different periods specified by the time dependence of a_{BBRW}. The first period corresponds to a universe of "slowly" increasing size. The second period is the radiation and matter dominated phases with a fairly rapid increase in a(t). The boundary time t_c between these periods is specified by:

$$\gamma = a(t_c) \qquad (20.3.2.14)$$

yielding

$$t_c = 1.05 \times 10^{-167} \text{ s} \cong 10^{-167} \text{ s} \qquad (20.3.2.15)$$

In this period we find

$$\text{Re } a_{BBRW}(0) \cong \gamma/2 = 2.91 \times 10^{-94} \qquad (20.3.2.16a)$$

$$\text{Re } a_{BBRW}(t_c) \cong 1.28\gamma = 7.45 \times 10^{-94} \qquad (20.3.2.16b)$$

$$\text{Im } a_{BBRW}(0) = -\gamma/2 = -2.91 \times 10^{-94} \qquad (20.3.2.16c)$$

with the radius, and volume, of the universe "slowly" increasing during this period. The nature of this period directly reflects the effects of the blackbody Y quanta. This can be seen from its inverse square dependence on M_c. As $M_c \to \infty$ the constant $\gamma \to 0$ and thus $a_{BBRW} \to 0$ with the universe scaling down to zero size with the attendant catastrophes of infinite density and temperature that appear in the standard models. The blackbody quanta (Dark Energy) give the universe a meta-stable initial size thus avoiding catastrophic divergences.

Epoch	Type	Phases	Time Period
I	Explosive Growth	Dark Energy-dominated	2.99×10^{17} s – 4.46×10^{17} s
II	Expanding	Matter-dominated Radiation-dominated	7×10^{11} s – 2.99×10^{17} s 1.05×10^{-167} s – 7×10^{11} s
III	Metastable Big Bang	Blackbody Y quanta (Dark Energy) dominated	0 s – 1.05×10^{-167} s

Table 20.3.2.1 Epochs and phases of the Universe since t = 0.

The period after t_c is dominated by the usual scale factor a(t). This scale factor shows up directly in the real part of a_{BBRW}. Its behavior is:

$t < t_c$ (or $a(t) < \gamma$)

$$\text{Re } a_{BBRW}(t) \cong \gamma/2 + a(t)/2 \qquad (20.3.2.17a)$$

$t_c < t < t_{now}$

$$\text{Re } a_{BBRW}(t) \cong a(t) \qquad (20.3.2.17b)$$

The behavior of the imaginary part of a_{BBRW} is also indirectly dominated by a(t) but in a much less dramatic way. The gradual growth of the imaginary part of a_{BBRW} is more or less indicated by the following three values:

$$\text{Im } a_{BBRW}(0) \cong -\gamma/2 = -2.91 \times 10^{-94} \qquad (20.3.2.16c)$$
$$\text{Im } a_{BBRW}(t_c) \cong -0.822\gamma = -4.78 \times 10^{-94} \qquad (20.3.2.18a)$$
$$\lim_{t \to \infty} \text{Im } a_{BBRW}(t) = -\gamma = -5.82 \times 10^{-94} \qquad (20.3.2.18b)$$

and also the behavior

$a(t) \ll \gamma$ (or $t < t_c$)

$$\text{Im } a_{BBRW}(t) \cong -\gamma/2 - a(t)/2 \qquad (20.3.2.18c)$$

$\gamma \ll a(t)$ (or $t_c < t$)

$$\text{Im } a_{BBRW}(t) \cong -\gamma + O([\gamma/a(t)]^2) \cong -\gamma \qquad (20.3.2.18d)$$

Both the real and imaginary parts of a_{BBRW} roughly double in the time period $[0, t_c]$ and thereafter we see a gradual increase of Im a_{BBRW} in absolute value from $-.5\gamma$ to $-\gamma$ over the lifetime of the universe.

After t_c the real part grows dramatically (eq. 20.3.2.17) while the imaginary part remains minute. The details of the interpretation of the behavior of the scale factor a_{BBRW} in the Big Bang Epoch $[0, t_c]$ will be explored in chapter 21. It suffices, for now, to say the universe in the period before t_c is in a meta-stable state of "slowly" growing size due to the dynamics of the blackbody Y quanta. At t_c the epoch of the expanding universe as we know it begins!

In differentiating between the real and imaginary parts of a_{BBRW} we must realize that the Reality group combines them into a single real quantity when scaling coordinates. Happily the small size of the imaginary part it can be neglected in most situations.

Before proceeding to describe physically interesting features of the early universe we note the radial dependence of the scale factor $a_{BBRW}(t, ř)$ in general:

$$a_{BBRW}(t, ř) = \{-i\gamma ř + a(t) + [\gamma^2 - 2i\gamma ř a(t) + a^2(t)]^{½}\}/2 \quad (20.3.2.19)$$

$$\text{Re}\ a_{BBRW}(t, ř) = a(t)/2 + [R(t, ř)(1 + \cos \psi(t, ř)/2]^{½}/2 \quad (20.3.2.20)$$

$$\text{Im}\ a_{BBRW}(t, ř) = -\gamma ř/2 - [R(t, ř)(1 - \cos \psi(t, ř))/2]^{½}/2 \quad (20.3.2.21)$$

where

$$R(t, ř) = [(\gamma^2 + a^2(t))^2 + 4(\gamma ř a(t))^2]^{½} \quad (20.3.2.22)$$

and

$$\cos \psi(t, ř) = (\gamma^2 + a^2(t))/R(t, ř) \quad (20.3.2.23)$$

We find that the scale factor $a_{BBRW}(t, ř)$ is approximated in various time periods to well within an order of magnitude by

$\mathbf{0 \le t < t_c}$

$$\text{Re}\ a_{BBRW}(t, ř) \cong \gamma/2 + a(t)/2 \quad (20.3.2.24)$$

$$\text{Im}\ a_{BBRW}(t, ř) \cong -\gamma ř/2 - a(t)ř/2 \quad (20.3.2.25)$$

$\mathbf{t = t_c}$

$$\text{Re}\ a_{BBRW}(t_c, ř) = \{1 + [(1 + ř^2)^{½} + 1]^{½}\}\gamma/2 \le 1.28\gamma = 7.45 \times 10^{-94}$$

$$0 \ge \text{Im}\ a_{BBRW}(t_c, ř) = \{-ř - [(1 + ř^2)^{½} - 1]^{½}\}\gamma/2 \ge -0.822\gamma$$

$\mathbf{t_c < t}$

$$\text{Re}\ a_{BBRW}(t, ř) \cong a(t)\{1 + (1 + ř^2)\gamma^2/4\} \cong a(t) \quad (20.3.2.26)$$

$$\text{Im}\ a_{BBRW}(t, ř) \cong -\gamma ř/2 - [ř^2 + \gamma^2/(4a^2(t))]^{½}\gamma/2 \quad (20.3.2.27)$$

$\mathbf{t \to \infty}$

$$\text{Re}\ a_{BBRW}(t, ř) \cong a(t)\{1 + (1 + ř^2)\gamma^2/4\} \cong a(t) \quad (20.3.2.28)$$

$$\text{Im}\ a_{BBRW}(t, ř) \cong -\gamma ř = -5.82 \times 10^{-94} ř \quad (20.3.2.29)$$

20.3.3 The Temperature of the Early Universe in the Generalized Robertson-Walker Metric

We will now examine the temperature of the early universe based on the results of the preceding section. For t near t = 0, we note κT depends on the blackbody scale factor of the generalized Robertson-Walker metric:[229]

$$\kappa T = \kappa T_0/B_{BB}(t, y) \quad (20.3.3.1)$$

[229] Here again we remind the reader that complex values for radii, temperature and so on are transformed by the Reality group to real values – their absolute values.

and

$$B_{BB}(t, y) = 2k^{-½}M_c(1 + M_c^2y^2)^{-1}a_{BBRW}(t, ř) \quad (20.3.3.2)$$

by eq. 20.3.2.6 using the variable

$$ř \equiv M_cy \quad (20.3.3.3)$$

for convenience. Therefore from eqns. 20.3.2.24 – 20.3.2.27 we see

$0 \le t < t_c$

$$\kappa T_< \equiv \kappa T(0 \le t < t_c) \cong 2\kappa T_0(1 + iř)/\varkappa$$
$$= 8.76 \times 10^{27}(1 + iř) \text{ ev} \quad (20.3.3.4)$$

$t_c < t$

$$\kappa T_> \equiv \kappa T(t_c < t)$$

$$\cong \kappa T_0(a(t) + i\gamma ř)/[2k^{-½}M_c(1 + ř^2)^{-1}(a^2(t) + \gamma^2ř^2)]$$

$$\cong \kappa T_0(1 + ř^2)/[2k^{-½}M_ca(t)] = 1.28 \times 10^{-66}(1 + ř^2)/a(t) \text{ ev} \quad (20.3.3.5)$$

We note that the temperatures calculated in this section are in generalized Robertson-Walker coordinates (eq. 19.2.4.5) and not in Robertson-Walker coordinates.

Note also that the physical temperature is the absolute value of the complex temperature due to the use of the Reality group.

20.3.4 Consistency of Y_{BB} Approximation with the Resulting Temperature near t = 0

We now address the question of whether the values found for κT are consistent with the approximations made in chapter 19 in order to obtain an expression for Y_{BB}:

$$\cos(\omega\kappa Tt) \approx 1 \quad (19.66)$$
$$J_1(\omega\kappa Ty) \approx \omega\kappa Ty/2 \quad (19.67)$$

The first approximation is valid if $\kappa Tt \sim 0$ since the Planck distribution factor makes the largest contribution to the integral come from small ω. The values of $\kappa T_<$ and $\kappa T_>$ show that for larger times κTt is very small:

$$\kappa T_<t_c \le 1.98 \times 10^{-124}$$
$$\kappa T_>t \le (3.89 \times 10^{-51} \text{ s}^{-1})t/a(t) \le (3.89 \times 10^{-51} \text{ s}^{-1})t_{now}/a(t_{now}) = 1.74 \times 10^{-33} \quad (20.3.4.1)$$

The second approximation is valid if $\kappa Ty \ll 1$ since the Planck distribution factor again makes the largest contribution to the integral come from small ω. We find the maximum values of κTy in the two time periods from eqns. 20.3.3.4 and 20.3.3.5 to be

$0 \leq t < t_c$

$$MAX(|\omega\kappa T_< y|) = |\omega\kappa T_< M_c^{-1}| = 1.02\omega \quad (20.3.4.2)$$

which is small since the dominant part of the integration comes from small ω; and

$t_c < t$

$$MAX(|\omega\kappa T_> y|) = |\omega\kappa T_>(t_c) M_c^{-1}| = 2.10 \times 10^{-94}\omega/a(t)$$
$$\leq 2.10 \times 10^{-94}\omega/a(t_c) = 0.361\omega \quad (20.3.4.3)$$

which is also small. Thus the Bessel function power series expansion is well approximated by its first term:

$$J_1(\omega\kappa Ty) \approx (\omega\kappa Ty/2) \quad (20.3.4.4)$$

We conclude our approximate calculation of A(t, y) is valid for all time and for the complete range of y values: $0 \leq y \leq M_c^{-1}$.

20.3.5 Plots of the Scale Factor from $t = 0$ to the Present

We will create several plots of the scale factor from $t = 0$ to the present time $t = 4.46 \times 10^{17}$ s for the maximum value of $y = M_c^{-1}$, which corresponds to the maximum Robertson-Walker radius coordinate value $r = k^{-½}$. The approximation that we have developed for B_{BB} has been justified for times between the Big Bang and the present.

In Fig. 20.3.5.1 we show a log – log plot of the real and imaginary parts of $a_{BBRW}(t)$ vs. t using base 10 logarithms for $t \in [10^{-200}, 10^{20}]$ seconds. In Fig. 20.3.5.2 we plot the real and imaginary parts of a_{BBRW} vs. time from $t = 0$ to $t = 1.2 \times 10^{-246}$ s. In Fig. 20.3.5.3 we plot the real part of a_{BBRW}, and the Robertson-Walker scale factor a(t), vs. time in seconds in the period around 10^{-167} s. It shows the rapidity of the transition of Re a_{BBRW} from slowly rising to rapidly rising with a(t).

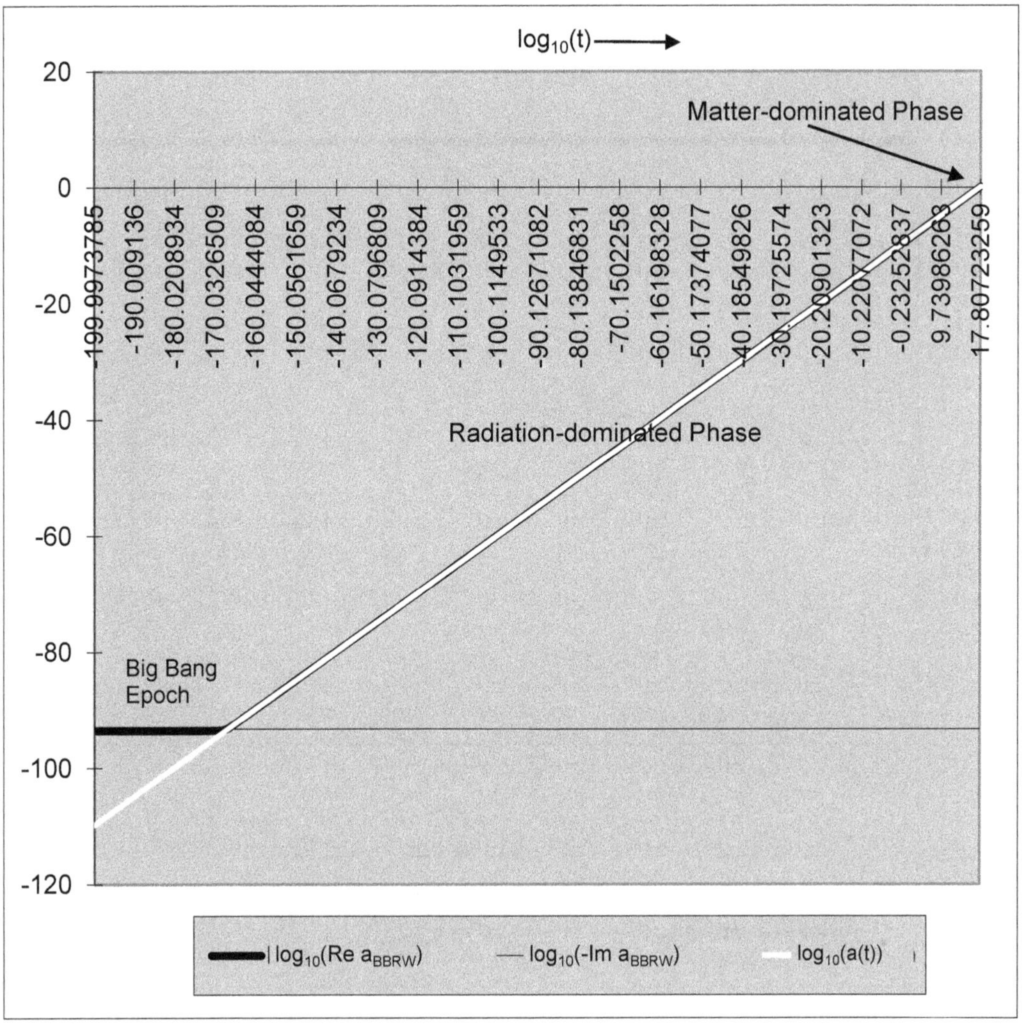

Figure 20.3.5.1. A log-log (base 10) plot of the real and imaginary parts of a_{BBRW} and the Robertson-Walker scale factor a(t) versus the log (base 10) of time in seconds. Note the imaginary part of a_{BBRW} is very slowly growing. The real part of a_{BBRW} is growing slowly until $t_c \approx$ 10–167 s and thereafter equals a(t) to good approximation.

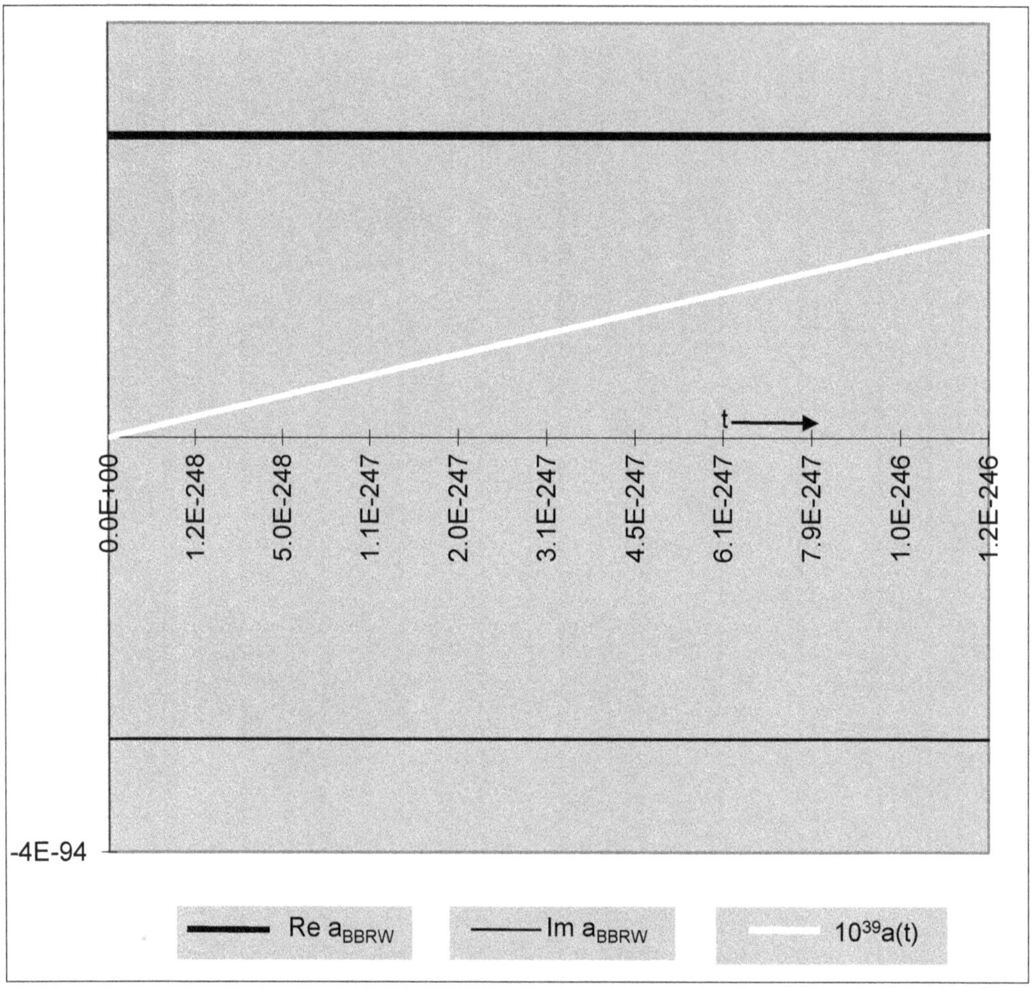

Figure 20.3.5.2. A plot of the real and imaginary parts of a_{BBRW}, and $10^{39} \times a(t)$, versus time from $t = 0$ to $t = 1.2 \times 10^{-246}$ s. Note they are slowly varying and well behaved in the neighborhood of $t = 0$ with only a(t) having the value of zero. "E" indicates a power of ten (for example: 2.0E-248 $= 2.0 \times 10^{-248}$ s).

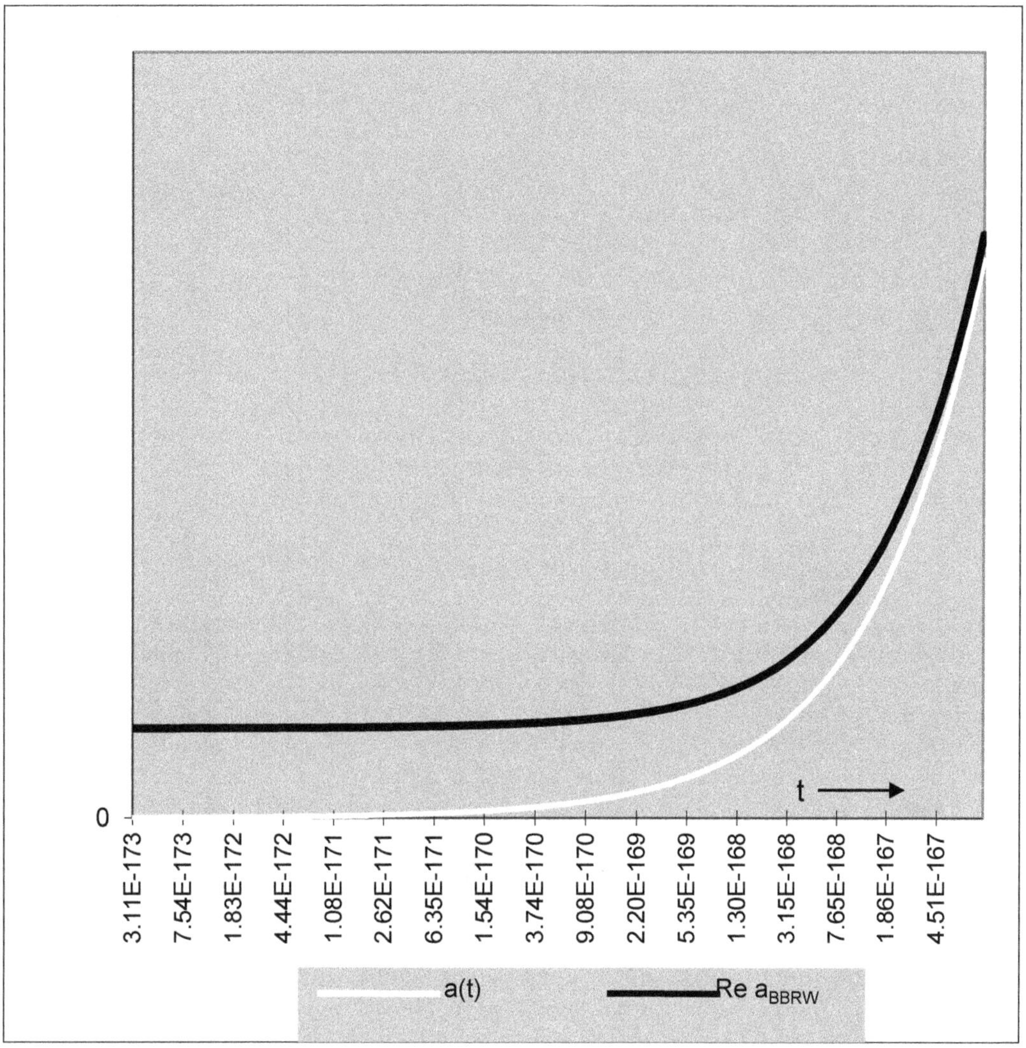

Figure 20.3.5.3. A plot of the real part of a_{BBRW} and a(t) vs. time in seconds around the time 10^{-167} s. Note Re a_{BBRW} quickly changes from slowly growing to growing like a(t).

20.4 The Interpretation of the Complex Scale Factor

The interpretation of the complex scale factor $a_{BBRW}(t, ř)$ hinges on its role in the expression for the proper interval. The expression for the proper interval in the Robertson-Walker metric is:

$$d\tau^2 = dt^2 - R^2(t)[dr^2/(1 - kr^2) + r^2(d\theta^2 + \sin^2\theta d\varphi^2)]$$

It was generalized to

$$d\tau^2 = dt^2 - A^2(t, ř)[dř^2 + ř^2(d\theta^2 + \sin^2\theta\, d\varphi^2)] \tag{19-A.4.9}$$

and an identification was made between the Robertson-Walker scale factor $R(t) \equiv a(t)$ and $a_{BBRW}(t, ř)$ through the following chain of equalities and correspondences

$$a(t) = a(t)b_0(r) = A(t, ř)/b(ř) \rightarrow A_{BB}(t, ř)(1 + ř^2)k^{1/2}/2 = a_{BBRW}(t, ř) \tag{20.4.1}$$

where the → relates the classical expressions on the left with the expressions on the right that embody quantum corrections due to Y-quanta blackbody radiation. Eq. 20.4.1 is based on eqns. 19-A.4.6, 19-A.4.7, 19-A.4.9, 19-A.4.11, 19-A.4.12, and 19-A.6.4a. Combining eqns. 20.4.1 we find

$$d\tau^2 = dt^2 - a_{BBRW}(t, ř(r))^2[dr^2/(1 - kr^2) + r^2(d\theta^2 + \sin^2\theta d\varphi^2)] \tag{20.4.2}$$

where the relation between ř and r, ř(r), is specified by eq. 19-A.4.20. We call the metric in eq. 20.4.2 a generalized Robertson-Walker metric. It is equivalent to eq. 19-A.4.9 – differing only in the definition of the radial coordinate.

Having determined the role of the complex scale factor in the metric tensor we can now physically interpret it by applying a Reality group transformation that effectively multiplies $a_{BBRW}(t, ř(r))$ by a phase making it real and equal to its absolute value $|a_{BBRW}(t, ř(r))|$.

We can simplify the physical interpretation without loss of generality by considering the radial coordinate of eq. 19-A.4.9 to be

$$r_{GRW} = A_{BB}ř \tag{20.4.3}$$

$$= r_r + ir_i \tag{20.4.4}$$

where

$$A_{BB}(t, y) = 2a_{BBRW}(t, ř)/[(1 + M_c^2y^2)k^{1/2}] \tag{19.5.2.4}$$

Thus

$$r_{GRW} = 2a_{BBRW}(t, ř)ř/[(1 + ř^2)k^{1/2}] \tag{20.4.5}$$

Applying a Reality group transformation we obtain the physically measurable, real-valued radius

$$r_{GRWphysical} = 2|a_{BBRW}(t, ř)ř/[(1 + ř^2)k^{1/2}]| \tag{20.4.5a}$$

20.5 Time Evolution of the Hubble Rate

The Hubble rate is one of the linchpins of modern cosmology. It is determined by Einstein's equation eq. 20.2.1.1 when written in the form:

$$H(t) = \dot{a}/a = [H_0^2(\Omega_\gamma/a^4(t) + \Omega_m/a^3(t) + \Omega_\Lambda) - k/a^2(t)]^{1/2} \qquad (20.5.1)$$

At small times, in the radiation-dominated phase, eq. 20.5.1 can be approximated by

$$H(t) \cong H_0\Omega_\gamma^{1/2}/a^2(t) \qquad (20.5.2)$$

If we define a Hubble rate H(t) using $a_{BBRW}(t, \check{r})$ then

$$H_{BBRW}(t, \check{r}) = |\dot{a}_{BBRW}(t, \check{r})/a_{BBRW}(t, \check{r}) \equiv \dot{A}_{BBRW}(t, \check{r})/A_{BBRW}(t, \check{r})|$$

$$= |[H_0^2(\Omega_\gamma/a_{BBRW}{}^4(t, \check{r}) + \Omega_m/a_{BBRW}{}^3(t, \check{r}) + \Omega_\Lambda) - k/a_{BBRW}{}^2(t, \check{r})]^{1/2}| \qquad (20.4.3)$$

$H_{BBRW}(t, \check{r})$ is the same as H(t) until we reach the first instants of the universe which we have called the Big Bag Epoch. Then we find

$0 \leq t < t_c$

$$H_{BBRW}(t, \check{r}) \cong H_0\Omega_\gamma^{1/2}/|a_{BBRW}(t, \check{r})|^2 \qquad (20.4.4)$$

where

$$\text{Re } a_{BBRW}(t, \check{r}) \cong \gamma/2 + a(t)/2 \qquad (20.3.2.24)$$

and

$$\text{Im } a_{BBRW}(t, \check{r}) \cong -\gamma\check{r}/2 - a(t)\check{r}/2 \qquad (20.3.2.25)$$

Substituting in eq. 20.4.4 we find

$$H_{BBRW}(t, \check{r}) \cong 4H_0\Omega_\gamma^{1/2}|[1 - \check{r}^2 + 2i\check{r}]/[(\gamma + a(t))^2(1 + \check{r}^2)^2]| \qquad (20.4.5)$$

$|H_{BBRW}(t, \check{r})|$ is a real physical number in this range. Thus space has a Hubble rate that is both space and time dependent in the Big Bang Epoch.

At t = 0 we find $H_{BBRW}(0, \check{r})$ is finite unlike the radiation-dominated Hubble rate (eq. 20.5.2):

$$H_{BBRW}(0, \check{r}) \cong 4H_0\Omega_\gamma^{1/2}|[1 - \check{r}^2 + 2i\check{r}]/[\gamma^2(1 + \check{r}^2)^2]| \qquad (20.4.6)$$

At the "edge" of the universe the Hubble rate is

$$H_{BBRW}(0, 1) \cong 2H_0\Omega_\gamma^{1/2}/\gamma^2 = 9.51 \times 10^{166}\ s^{-1} \qquad (20.4.7)$$

and is solely due to the imaginary part of $4H_0\Omega_\gamma^{1/2}[1 - \check{r}^2 + 2i\check{r}]/[(\gamma + a(t))^2(1 + \check{r}^2)^2]$ since the real part is zero.

Thus we consistently avoid the divergences that appear at t = 0 in the Standard Cosmological Model. Its radiation-dominated phase's Hubble rate is $(2t)^{-1}$ which diverges at t = 0.

21. The Big Bang Epoch

No great thing is created suddenly.
Discourses - *Epictetus*

21.1 The t = 0 Big Bang Scenario

The Two-Tier cosmological theory that we have developed in preceding chapters differs dramatically from the Standard Cosmological Model in the Big Bang Epoch and yet smoothly melds into the Standard Cosmological Model in the Expanding Universe and Exploding Universe epochs.

The universe has a finite size, temperature and density at the point of the Big Bang which we define to be at the time t = 0. The universe grows slowly for a period of time (until roughly 10^{-167} s) that we call the Big Bang Epoch. In this time period we see a very hot, very dense, macroscopic conglomeration of Y quanta, radiation and elementary particles that coexist with each other with non-singular interactions. These particles are not localized. Each particle can be said to occupy the entire universe since the size of the universe is infinitesimal compared to any particle's Compton radius (if it has one). The universe has an almost classical Robertson-Walker type of metric. In particular, it has a generalized Robertson-Walker metric with quantum corrections due to an effectively classical Y-quanta blackbody radiation field (the inflatons) that is both the source of the metastability of the universe at t = 0 and the source of infinitesimal imaginary spatial dimensions that are comparable in extent with the size of the real spatial dimensions of the universe during the Big Bang Epoch.

As the universe expands due to the Y quanta energy it appears that enormous amounts of gravitational energy are also released since the Two-Tier gravitational potential is zero at r = 0 and has a minimum around $r \approx 10^{-33}$ cm. Thus we view the universe in the Big Bang Epoch as in a "slowly" expanding metastable state. (This state is comparable to the metastable false vacuum state in inflation theories – but no scalar bosons are needed – the Y quanta play that role. The combination of gravitation and an effectively classical Y field serve to generate the metastable initial state of the Big Bang.

We can summarize our model's features (many of which are calculated later in this chapter) with:

1. The universe is a macroscopic object in terms of content and as such can be described by classical physics – modified by quantum effects.

2. Quantum fluctuations do not play any significant role in the Big Bang Epoch because of the nature of Two-Tier quantum field theory. For example, the quantum fluctuations of the quantized gravitational field were shown to be zero in chapter 7 of Blaha (2004):

$$<0|h_{\mu\nu}(X)h_{\alpha\beta}(X)|0> = \int d^3p\, b'_{\mu\nu\alpha\beta}(p)\, e^{-p^i p^j \Delta_{Tij}^{(0)}}/[(2\pi)^3 2\omega_p] = 0 \qquad (7.3.8.3.3)$$

3. All Two-Tier quantum fields also have zero quantum fluctuations. This behavior holds whether they are quantized in flat space or in a curved space. Thus quantum fluctuations (or foam) are not an issue in Two-Tier theories.

4. The radiation and matter in the Big Bang Epoch produces, in effect, a classical Y-quanta blackbody spectrum that modifies the nature of space making it complex with the real and imaginary parts of space being comparable. This effect appears in the spatial scale factor of the generalized Robertson-Walker metric described in previous chapters. A Reality group transformation transforms spatial coordinates to real physical values.

5. The usual scale factor a(t) is determined by the conventional Standard Cosmological Model classical Einstein equation since its source is the macroscopic energy density.

6. The radius of the universe (with both real and imaginary parts) at the beginning of the Big Bang Epoch is

$$r_{universe}(0) = a_{BBRW}(0)/k^{1/2} = \gamma(1 - i)/(2k^{1/2}) = 4.30(1 - i) \times 10^{-65} \text{ cm} \qquad (21.1.1)$$

The physical radius is the absolute value of $r_{universe}(0)$. The confinement of particles to a radius of this size means that they cannot be considered to be localized but, rather, they are spread over the entire volume of the universe.

7. The small size of the universe implies Two-Tier potentials between particles, and particle propagators, are effectively zero for all particles. Consequently the universe is in a metastable state. Some idea of the relative potential energy of Two-Tier QFT particles vs. ordinary QFT particles can be gleaned by comparing a standard Newton-Coulomb type of potential (knowing that it would be modified in strong gravitational fields if they were present)

$$V_{std} = 1/r \qquad (21.1.2)$$

with the Two-Tier potential at short distances (below the Planck scale):

$$V_{tt} = 2\sqrt{\pi}\, M_c^2 r \qquad (21.1.3)$$

obtained from eq. 7.3.9.3 of Blaha (2004). We have not displayed the coupling constant in eqns. 21.1.2 and 21.1.3. At $r = r_{universe}(0) = 4.30 \times 10^{-65}$ cm

$$V_{std} = 4.57 \times 10^{59} \text{ ev} \tag{21.1.4}$$

$$V_{tt} = 2\sqrt{\pi}\, M_c^2 r = 0.00115 \text{ ev} \tag{21.1.5}$$

Thus the Two-Tier potential between particles is negligible at distances up to the radius of the universe at t = 0 since the Heisenberg Uncertainty Principle, when applied using the "diameter" of the universe as the uncertainty in position, implies the uncertainty in a particle's energy, and thus the scale of particle energies, is (coincidentally) of the order of 10^{59} ev. (See the discussion in the following sections.)

8. In view of the above points it is reasonable, and self-consistent, to use the generalized Robertson-Walker metric that we have developed in the preceding chapters.

9. The t = 0 universe is very dense with an energy density of the order of 10^{339} g/cm^3 of particles with negligibly small, non-singular (as particle separation goes to zero) interactions. (See the discussion below.)

10. The Big Bang Epoch is a metastable state. Due to the form of the Two-Tier gravitational potential it has a much higher gravitational potential energy at t = 0 than when the radius of the universe is about 10^{-33} cm. **In fact, Gravity is a repulsive force (anti-gravity!) at distances less than 9.08 × 10^{-34} cm.** (Fig. 7.3.9.3 of Blaha (2004) is reproduced on the next page for the reader's convenience.)

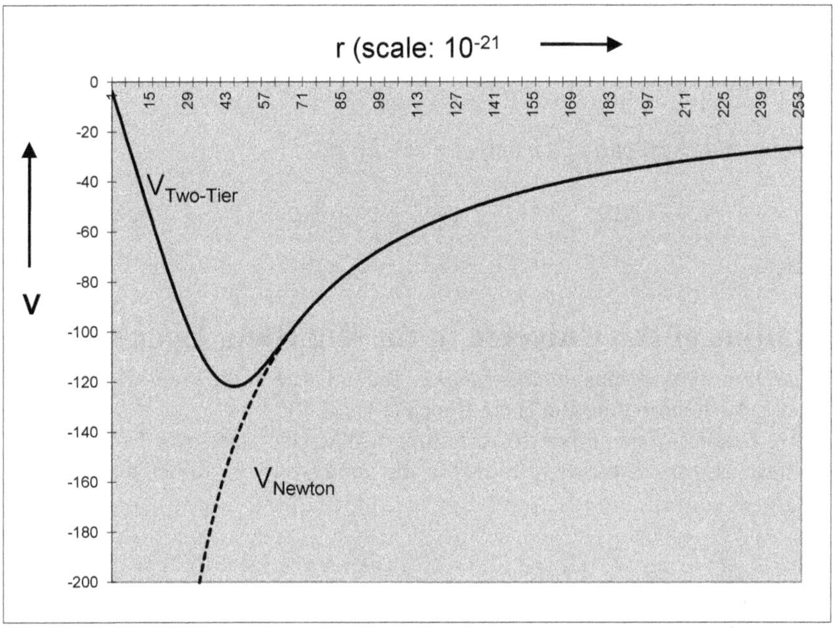

Figure 21.1.1. A plot of the Two-Tier gravitational potential (solid line) and the Newtonian gravitational potential (dashed line). Note anti-gravity at short distances. See Fig. 7.3.9.3 of Blaha (2004) for more details.

The Two-Tier gravitational potential between two particles has a minimum at (eq. 7.3.9.5 of Blaha (2004))

$$r_{MIN} = \pi^{-1/2} M_c^{-1} = 9.08 \times 10^{-34} \text{ cm} \tag{21.1.6}$$

At the minimum the gravitational $V_{\text{Two-Tier}}$ has the value:

$$V_{\text{Two-TierMIN}} = -\ .8427\sqrt{\pi}\ GM_c = -1.22 \times 10^{-28} \text{ ev}^{-1} \tag{21.1.7}$$

Since the energy of the universe is confined within a radius of 10^{-65} cm at t = 0 there is a tremendous release of gravitational potential energy as the universe expands to 10^{-33} cm and beyond. This energy is converted initially into kinetic energy and can be viewed as helping fuel the expansion. A crude approximation to this released energy is

$$\begin{aligned} \Delta E &\sim (\rho V_{\text{universe}})^2 V_{\text{Two-TierMIN}} \\ &\cong (1.6 \times 10^{212} \text{ g/cm}^3 \times 1.8 \times 10^{-100} \text{ cm}^3)^2 \times 1.22 \times 10^{-28} \text{ ev}^{-1} \\ &\cong 3 \times 10^{262} \text{ ev} \end{aligned} \tag{21.1.8}$$

The energy estimate (eq. 21.1.8) is far beyond the current total energy of the universe which is of the order of 10^{57} ev. Since energy is not conserved (It is considered somewhat undefined by many General Relativists.) as the universe evolves eq. 21.1.8 does not represent a problem in itself.

11. The temperature of the real part of the universe at t = 0 is

$$\kappa T \approx 2 \times 10^{89} \text{ ev} \qquad (21.1.9)$$

as shown later.

21.2 The Radius of the Universe in the Big Bang Epoch

In this section we will consider the radius of the universe at the point of the Big Bang (t = 0) and its relatively slow growth during the Big Bang Epoch ($t \leq t_c \cong 10^{-167}$ s).

The current radius of the universe (according to WMAP[230] data) is > 7.4×10^{28} cm. Motivated by that finding we have assumed the current radius of the universe in the Robertson-Walker model is twice that value:

Assumption: $$r_{universe}(t_{now}) = a(t_{now})/k^{1/2} \approx 2 \times 7.4 \times 10^{28} \text{ cm} \qquad (21.2.1)$$

and used it to determine k (eq. 20.2.4).

It is reasonable to define the radius of the universe at earlier times correspondingly. In our Two-Tier blackbody model for the universe it is

$$r_{universe}(t) = a_{BBRW}(t, \check{r} = 1)/k^{1/2} \qquad (21.2.2)$$

with

$$a_{BBRW}(t, \check{r}) = \{-i\gamma\check{r} + a(t) + [\gamma^2 - 2i\gamma\check{r}a(t) + a^2(t)]^{1/2}\}/2 \qquad (20.3.2.19)$$

Thus the radius of the universe is complex and given by

$$r_{universeBBRW}(t) = \{-i\gamma + a(t) + [\gamma^2 - 2i\gamma a(t) + a^2(t)]^{1/2}\}/(2k^{1/2}) \qquad (21.2.3)$$

The physical radius of the universe is the absolute value of eq. 21.2.3. It is calculated by applying a Reality group transformation.

The real and imaginary parts of $r_{universeBBRW}(t)$ are:

$$\text{Re } r_{universeBBRW}(t) = \{a(t) + [R(1 + \cos\psi)/2]^{1/2}\}/(2k^{1/2}) \qquad (21.2.4a)$$

and

$$\text{Im } r_{universeBBRW}(t) = \{-\gamma - [R(1 - \cos\psi)/2]^{1/2}\}/(2k^{1/2}) \qquad (21.2.4b)$$

where

$$R = [(\gamma^2 + a^2(t))^2 + 4\gamma^2 a^2(t)]^{1/2} \qquad (20.3.2.12)$$

---

[230] N. J. Cornish, D. N. Spergel, G. D. Starkman, and E. Komatsu, Phys. Rev. Lett. **92**, 201302-1 (2004).

and

$$\cos\psi = (\gamma^2 + a^2(t))/R \quad (20.3.2.13)$$

From the behavior of $a_{BBRW}(t)$ displayed in Figs. 20.3.5.1 – 20.3.5.3 we see that it is well approximated by eqns. 20.3.2.24 – 20.3.2.29 with ř = 1. Therefore for $t < t_c$

$t < t_c$

$$\text{Re } r_{universeBBRW}(0 \le t < t_c) \cong (\gamma + a(t))/(2k^{1/2}) \quad (21.2.5a)$$

$$\text{Re } r_{universeBBRW}(0) \cong \gamma/(2k^{1/2}) = 4.30 \times 10^{-65} \text{ cm} \quad (21.2.5b)$$

$$\text{Im } r_{universeBBRW}(0 \le t < t_c) \cong -(\gamma + a(t))/(2k^{1/2}) \quad (21.2.6a)$$

$$\text{Im } r_{universeBBRW}(0) \cong -\gamma/(2k^{1/2}) = -4.30 \times 10^{-65} \text{ cm} \quad (21.2.6b)$$

from eqns. 20.3.2.24 and 20.3.2.25. Since the Planck length is 1.61×10^{-33} cm we see the real part of the radius of the universe at $t = 0$ (until $t \approx t_c$) is over thirty orders of magnitude smaller than the Planck length.

If we use eqns. 20.3.2.28 and 20.3.2.29 then the radius at present is

$t = t_{now}$

$$r_{universeBBRW}(t_{now}) = a(t_{now})/k^{1/2} - i\gamma/k^{1/2} \quad (21.2.7)$$

with

$$\text{Re } r_{universeBBRW}(t_{now}) = a(t_{now})/k^{1/2} = 1.48 \times 10^{29} \text{ cm} \quad (21.2.8)$$

as above (eq. 21.2.1), and

$$\text{Im } r_{universeBBRW}(t_{now}) = -\gamma/k^{1/2} = -8.60 \times 10^{-65} \text{ cm} \quad (21.2.9)$$

The ratio of the current, and the $t = 0$ real parts of the, radius of the universe is huge:

$$r_{universe}(t_{now})/(\text{Re } r_{universeBBRW}(0)) = 3.44 \times 10^{93} \quad (21.2.10)$$

The real part of the universe has expanded dramatically while the imaginary part has remained almost constant in size. The physical radius of the universe for $t < t_c$ and $t = t_{now}$ are the absolute values of the complex radius values above.

The value of the Robertson-Walker radius coordinate at points *within* the universe is specified by

$$r_{RW}(t, ř) = a_{BBRW}(t, ř)r \quad (21.2.11)$$

The coordinate r is related to the ř coordinate by

$$r = 2k^{-1/2}ř(1 + ř^2)^{-1} \quad (19\text{-}A.4.11)$$

Thus

$$r_{RW}(t, \check{r}) = 2k^{-1/2}\check{r}a_{BBRW}(t, \check{r})(1 + \check{r}^2)^{-1} \qquad (21.2.12)$$

The real and imaginary parts of $a_{BBRW}(t, \check{r})$ are specified in eqns. 20.3.2.24 – 20.3.2.29. Again we note the absolute values of eqns. 21.2.11-21.2.12 are the physical values of the Robertson-Walker radius coordinate.

21.2.1 Localization of Particles at t = 0

The radius of the universe in the neighborhood of t = 0 is approximately 6.08×10^{-65} cm. The particles within that incredibly small universe are still the "wave- particles" that we are familiar with within the framework of quantum mechanics. As such, the position and momentum of the particles must satisfy the Heisenberg Uncertainty Condition:

$$\Delta p \Delta x \geq \hbar \qquad (21.2.1.1)$$

where $\hbar$ is Planck's constant divided by 2π. In view of the extraordinary small size of the universe – much smaller than the Compton wavelength of any known massive particle – the "spread" (uncertainty) in a particle's position is set by the radius of the universe:

$$\Delta x \approx 2\,|r_{universeBBRW}| \qquad (21.2.1.2)$$

Thus the "spread" in the momentum of a particle (the "size" of the region in momentum space where the Fourier transform of the particle's wave function is large) is

$$\Delta p \approx \hbar/\Delta x \approx \hbar/[2|\text{Re } r_{universeBBRW}|] = 1.62 \times 10^{59} \text{ ev} = 1.32 \times 10^{31} M_{Planck} \qquad (21.2.1.3)$$

One can only view the particles in the universe in the neighborhood of t = 0 as spread across the entire universe. They are entirely unlocalized within the universe from a quantum viewpoint. Since all forces are non-singular in the very small universe at the beginning of time we can view the particles as co-resident in the same spatial region that constitutes the universe. (Simply put, they interpenetrate each other.) Thus the question of particle horizons and the homogeneity of the universe in the Beginning are irrelevant. *The universe today does not scale down to a micro-universe of the same sort as our universe in the neighborhood of t = 0.*

21.3 A Quantum Big Bang and Evolutionary Theory

At this point we have shown that our Quantum Big Bang theory does not have a singularity at the beginning of the universe. It exhibits the known behavior of the universe since 350,000 years after the Big Bang epoch. Thus the long sought inflaton field turns out to be the quantum field, $Y^{\mu}(y)$, appearing in our definition of quantum coordinates. Remarkably our quantum coordinates also eliminate the divergences that have plagued quantum field theory for almost eighty years and enable calculations in The Extended Standard Model and Quantum Gravity to be divergence free without the cumbersome renormalization techniques that were needed for ElectroWeak theory and would not work for Quantum Gravity.

So our quantum coordinates free The Extwnded Standard Model and Quantum Gravity of infinities, both now, and at the beginning of the universe. Naturally this happy removal of infinities through one simple mechanism is a remarkable result that, we believe, reflects the simplicity of Nature when properly understood.Leibniz's Minimax Principle is realized by Two-Tier quantum theory that removes the infinities in The Extended Standard Model, Quantum Gravity, and at the beginning of the universe.

22. Quantum Gravity and the Wheeler-DeWitt Equation Extended to Complex Coordinates

22.1 Introduction

We earlier developed Quantum Gravity using the weak field expansion in sections 18.6 – 18.8 and showed that it gave finite results in perturbation theory calculations. There are other attenpts to develop Quantum Gravity without relying on the weak field expansion. These attempts have had some success. Our view is that the existence of one successful approach, in this case the weak field expansion, is proof that Quantum Gravity is viable. Thus we can claim a successful derivation of the unified Extended Standard Model and Quantum Gravity based on space-time geometry.

In this chapter we will consider Quantum Gravity as embodied in the Wheeler-DeWitt[231,232] equation. This equation has many noteworthy points that we will consider below – particularly its extension to complex space-time. It also raises important basic quantum questions: such as "Who is the Observer?" that we will address in chapter 23. Most of this chapter first appeared in Blaha (2014a).

22.2 Analytically Continued Wheeler-DeWitt Equation to Complex Metrics under a Faddeev-Popov Method Restriction

In this section we extend the Wheeler-DeWitt equation for Quantum Gravity to complex coordinates and metrics by analytic continuation (piece-wise if necessary) and impose the condition that metrics must be real-valued using the Faddeev-Popov Method. Our procedure will be to take the Wheeler-DeWitt equation for real-valued metrics (and coordinates), **analytically continue it to the case of complex metrics and coordinates,**[233] and then impose the condition that physically acceptable metrics must be real-valued through use of a Reality group transformation implemented via the Faddeev-Popov Method. Our motivation is two-fold: we have shown that the form of The Standard Model of Particles can be derived from complex space-time considerations demonstrating that we exist in a "masked" complex space-time; and we have shown the imposition of a restriction on a gauge theory such as gravitation[234] can be implemented using the Faddeev-Popov Method or equivalent.

We start by noting the canonical decomposition of a real-valued metric $g_{\mu\nu}$ is defined by:

$$g_{\mu\nu}(x) = \eta_{\alpha\beta}\, \partial\omega^\alpha/\partial x^\mu\ \partial\omega^\beta/\partial x^\nu \tag{22.1}$$

with

---

[231] DeWitt, B. S., Phys. Rev. **160**, 1113 (1987).

[232] Hartle and Hawking also derive the Wheeler-DeWitt equation from a path integral formalism for quantum gravity.

[233] The piecewise analytic continuation of general relativity to complex coordinates and metrics is described in some detail in Blaha (2004). The gist of the continuation is that all equations have the same form after analytic continuation due to a basic theorem of complex mathematics that the analytic extension of equations to complex values from real values is unique.

[234] Such as real-valuedness.

$$g_{\mu\nu} = g_{\nu\mu}$$

and with inverse

$$g^{\mu\nu} = \eta^{\alpha\beta}\, \partial x^{\mu}/\partial\omega^{\alpha}\, \partial x^{\nu}/\partial\omega^{\beta} \tag{22.2}$$

The decomposition of the real-valued metric is

$$g_{\mu\nu} = \begin{bmatrix} -\alpha^2\beta_k\beta^k & \beta_j \\ \beta_i & \gamma_{ij} \end{bmatrix} \tag{22.3}$$

$$g^{\mu\nu} = \begin{bmatrix} -\alpha^{-2} & \alpha^{-2}\beta^j \\ \alpha^{-2}\beta^i & \gamma^{ij} - \alpha^2\beta^i\beta^j \end{bmatrix} \tag{22.4}$$

where

$$\gamma_{ik}\,\gamma^{kj} = \delta_i^{\,j} \qquad\qquad \beta^i = \gamma^{ij}\,\beta_j \tag{22.5}$$

The Wheeler-DeWitt equation for real-valued metrics is

$$(G_{ijkl}\,\delta/\delta\gamma_{ij}\,\delta/\delta\gamma_{kl} + \gamma^{\frac{1}{2}\,(3)}R + 2\lambda\,\gamma^{\frac{1}{2}\,(3)})\Psi({}^{(3)}\mathcal{G}) = 0 \tag{22.6}$$

where λ is the cosmological constant, and where the Wheeler-DeWitt metric is

$$G_{ijkl} = \tfrac{1}{2}\,\gamma^{-\frac{1}{2}}(\,\gamma_{ik}\gamma_{jl} + \gamma_{il}\gamma_{jk} - \gamma_{ij}\gamma_{kl}) \tag{22.7}$$

The functional derivatives $\delta/\delta\gamma_{ij}$ have several interpretations that are presumably equivalent. DeWitt characterizes them as coordinate independent specifications of the 3-metric. The wave function $\Psi({}^{(3)}\mathcal{G}) = \Psi(\gamma_{ij})$, where ${}^{(3)}\mathcal{G}$ is a geometry, is *not* coordinate dependent. It is invariant under coordinate changes. ${}^{(3)}\mathcal{G}$ is a discrete infinity of independent invariants constructed from products of the Riemann tensor and its covariant derivatives.

Hartle and Hawking[235] derive the Wheeler-DeWitt equation from a path integral formalism for quantum gravity. Their path integral can be represented as

$$Z = N\int \delta g(x)\,\exp(iS_E[g]) \tag{22.8}$$

where S_E is the classical action for gravity and the functional integral is an integral over all 4-geometries. Changing to DeWitt's notation based on the spatial metric γ_{ij} and expressing eq. 22.8 in a more explicit form for use in conjunction with the Faddeev-Popov Method we have

$$Z = N\int \sum_{i,j}\prod_{x} d\gamma_{ij}(x)\,\exp(iS_E[\gamma]) = N\int D\gamma\,\exp(iS_E[\gamma]) \tag{22.9}$$

The integrand, being a functional integral over all space, is independent of the coordinates.

---

[235] Hartle, J. B. and Hawking, S. W., Phys. Rev. D **28**, 2960 (1983).

The Wheeler-DeWitt equation applies to real-valued metrics $\gamma^{ij}(x)$. We now extend this equation to apply to complex-valued metrics using the local Reality group. In doing this we realize that there are an infinite number of complex-valued metrics in the orbit corresponding to each real-valued metric.

This redundancy can be resolved by realizing that the physical measurement of an invariant interval, and the coordinates from which it is derived, are always real-valued. Yardsticks and clocks can only measure real-valued numbers. Based on this physical principle we can generalize quantum gravity to complex coordinates and metrics by using the Faddeev-Popov method to constrain the set of paths in the quantum gravity path integral eq. 22.9. Using the Faddeev-Popov Method the constraint can be expressed in terms of an infinitesimal transformation of the metric to a complex value. Using an infinitesimal Reality group transformation V:

$$[\exp(ia_j(x'',x')U_j)]^{\alpha}{}_{\mu} = S(x'',x')^{\alpha}{}_{\mu} = \partial x''^{\alpha}/\partial x'^{\mu} \qquad (22.10)$$

$$V = \exp(ia_k(x'',x')U_k) \cong I + ia_k(x'',x')U_k \qquad (22.11)$$

$$V^{\dagger} = [\exp(ia_k(x'',x')U_k)]^{\dagger} \cong I - ia_k(x'',x')U_k \qquad (22.12)$$

where $a_j(x'',x')$ is the jth real-valued local infinitesimal parameter, and U_k is one of the 16 hermitean generators of the Reality group. (We treat the index on a_k and U_k as lower case and sum on k.) We find the condition fixing the metric to a physical *real* value is:

$$F^{ij}(\gamma(x)) = \text{Im}\, \gamma^{ij}(x) \equiv -\tfrac{1}{2}\, i(\gamma^{ij}(x) + \gamma^{ij}(x)^*) = 0 \qquad (22.13)$$

where

$$F^{aij}(\gamma^{a}(x)) = \text{Im}\{(\delta^{i}{}_{m} + ia_k(x',x)U_k{}^{i}{}_{m})\,(\delta^{j}{}_{p} + ia_k(x',x)U_k{}^{j}{}_{p})\gamma^{mp}(x)\} \qquad (22.14)$$

using the infinitesimal form.

The Reality condition eq. 22.13 is implemented within the path integral formalism with the Faddeev-Popov Method identity

$$1 = \int D\gamma\, \Delta(F(\gamma))\, \delta(\delta F^{aij}(\gamma^{a}(x))/\delta a_n(x',x)) = \int D\gamma\, \Delta(F(\gamma))\, \delta(F(\gamma^{ij})) \qquad (22.15)$$

Eq. 22.14 yields

$$\begin{aligned}\delta F^{aij}(\gamma^{a}(x))/\delta a_n(x',x)|_{a=0} &= \text{Im}\{\delta_{kn} i U_k{}^{i}{}_{m}\gamma^{mp}(x)\delta^{j}{}_{p} + \delta_{kn}\delta^{i}{}_{m} i\gamma^{mp}(x)U_k{}^{j}{}_{p}]\} \\ &= \text{Re}\{\gamma^{mp}[[U_n{}^{i}{}_{m}\delta^{j}{}_{p} + \delta^{i}{}_{p}U_n{}^{j}{}_{m}]\} \\ &= \gamma^{mp}(x)\text{Re}\{U_n{}^{i}{}_{m}\delta^{j}{}_{p} + \delta^{i}{}_{p}U_n{}^{j}{}_{m}\} = \gamma^{mp}(x)\xi^{ij}{}_{nmp}\end{aligned} \qquad (22.16)$$

since $\gamma^{mp}(x)$ is made real by the $\delta(\text{Im}\,\gamma^{mp}(x))$ where

$$\xi^{ij}{}_{nmp} = \text{Re}\{U_n{}^{i}{}_{m}\delta^{j}{}_{p} + \delta^{i}{}_{p}U_n{}^{j}{}_{m}\} \qquad (22.17)$$

Note $\xi^{ij}{}_{nmp}$ is symmetric in i and j, and becomes effectively symmetric in m and p when combined with $\gamma^{mp}(x)$ in eq. 22.16. Calculating $\Delta(F(\gamma))$ we obtain

$$\Delta(F(\gamma)) = [\det \delta F^{aij}(\gamma^{a}(x))/\delta a_n(x',x)|_{a=0}]^{-1} \qquad (22.18)$$

We can rewrite this Faddeev-Popov determinant as a path integral over an anti-commuting c-number scalar field χ with a ghost Lagrangian:

$$\Delta(F(\gamma)) = \int D\chi^* D\chi \exp[i\int d^4x\, \gamma^{½} \mathscr{L}_\gamma^{ghost}(x)] \tag{22.19}$$

where

$$\mathscr{L}_\gamma^{ghost}(x) = \chi^*(x)[U_{nij}\xi^{ij}{}_{nmp}\gamma^{mp}(x)]\chi(x)$$
$$= \chi^*(x)(\text{Re } U_{nij})\xi^{ij}{}_{nmp}\gamma^{mp}(x)]\chi(x) \tag{22.20}$$
$$= \chi^*(x)\xi_{mp}\gamma^{mp}(x)\chi(x) \tag{22.21}$$

since $\xi_{mp}\gamma^{mp}(x)$ is a c-number:

$$\xi_{mp} = (\text{Re } U_{nij})\xi^{ij}{}_{nmp} = 2 \text{ Re } U_{npi} \text{ Re } U_n{}^i{}_m \tag{22.22}$$

using $\text{Re } U_n{}^i{}_m = \text{Re } U_n{}^m{}_i$ for the U(4) generators in its fundamental representation $\underline{4}$.

We then find ξ_{mp} is a diagonal matrix and has the value

$$\xi_{mp} = \xi_m \delta_{mp} \tag{22.22a}$$

where the diagonal elements are

$$\xi_0 = 8$$
$$\xi_1 = 8$$
$$\xi_2 = 8$$
$$\xi_3 = 8 \tag{22.22b}$$

and the non-diagonal elements are zero making ξ_{mp} considered as a matrix a multiple of the Identity matrix..

The Faddeev-Popov generated terms when added to the Einstein Action appear to have important ramifications – particularly with respect to the Cosmological Constant. We explore that issue next.

22.3 Possible Source of the Cosmological Constant in the Complex Space-time – Faddeev-Popov Constraint Term

The Wheeler-DeWitt equation

$$(G_{ijkl}\, \delta/\delta\gamma_{ij}\, \delta/\delta\gamma_{kl} + \gamma^{½\,(3)} R + 2\lambda\, \gamma^{½\,(3)})\Psi({}^{(3)}\mathcal{G}) = 0 \tag{22.23}$$

has the cosmological constant, Λ, as one of its terms. This equation is derived from the Hamiltonian constraint that ultimately follows from the Einstein action. Inserting the Faddeev-Popov term of the previous section in the Einstein lagrangian yields

$$S_E(\gamma) = -(16\pi G)^{-1} \int d^4x\, \gamma^{½}\{ R(x) - 2\lambda + \chi^*(x)\chi(x)\xi_{mp}\gamma^{mp}(x)\} \tag{22.24}$$

The constant matrix ξ_{mp} is a product of parts of the generators of U(4) given by eq. 22.22a:

$$\xi_{mp} = 2\,(\mathrm{Re}\,U_{npi})\,(\mathrm{Re}\,U_{n\ m}^{\ i}) \tag{22.25}$$

We will now show that the term

$$\Pi = \gamma^{½}\chi^{*}(x)\chi(x)\xi_{mp}\gamma^{mp}(x) \tag{22.26}$$

upon variation of the metric $\delta\gamma_{\mu\nu}$, gives a term which is approximately a constant cosmological term assuming $\chi(x)$ is approximately constant (with its implied divergence eliminated by renormalization of the path integral), and assuming an almost flat space-time $\gamma_{\mu\nu} \cong \eta_{\mu\nu}$. Varying the metric for Π yields

$$\delta\Pi = \delta\gamma_{\mu\nu}\,\{½\,\gamma^{½}\chi^{*}(x)\chi(x)\xi_{mp}\gamma^{mp}(x)\gamma^{\mu\nu} - \gamma^{½}\chi^{*}(x)\chi(x)\xi_{mp}\gamma^{m\mu}\gamma^{p\nu}\} \tag{22.27}$$

The resulting modified Einstein field equation is

$$R^{\mu\nu} - ½\, g^{\mu\nu}R + \lambda\,\gamma^{\mu\nu} + ½[\gamma^{mp}(x)\gamma^{\mu\nu} - \gamma^{m\mu}\gamma^{p\nu}]\xi_{mp}\chi^{*}(x)\chi(x) = -8\pi T^{\mu\nu} \tag{22.28}$$

The terms $½[\gamma^{mp}(x)\gamma^{\mu\nu} - \gamma^{m\mu}\gamma^{p\nu}]\xi_{mp}\chi^{*}(x)\chi(x)$ appearing above is, or is a contribution to, the cosmological constant assuming a nearly flat universe as our universe seems to be. Thus $\gamma_{\mu\nu} \cong \eta_{\mu\nu}$ to good approximation and $\chi^{*}(x)\chi(x)$ can be taken to be constant since the time derivative of $\chi(x)$ does not appear in the lagrangian. Consequently the total cosmological constant term is

$$\lambda_{tot}^{\ \mu\nu} \cong \lambda g^{\mu\nu} + 4\,\chi^{*}(x)\chi(x)g^{\mu\nu} = (\lambda + \lambda_{F\text{-}P})\, g^{\mu\nu} \tag{22.29}$$

by eq. 22.22b.

Given the somewhat problematic state of our understanding of the cosmological constant it is not impossible that the complexity of space-time leading to $\lambda_{F\text{-}P}$ may be the sole origin of the cosmological constant. In evaluating eq. 22.29 we may normalize $\chi^{*}(x)\chi(x) = 1/4$ by adjusting the overall normalization of the path integral since $\chi(x)$ is time independent. Then if the "bare" cosmological constant is zero we obtain

$$\lambda_{tot} = \lambda_{F\text{-}P} = 1 \tag{22.30}$$

by eq. 22.29.

If this is true then we have achieved a space-time origin for the cosmological constant rather than an ad hoc origin. The modified Einstein field equation is then

$$R^{\mu\nu} - ½\, g^{\mu\nu}R + \lambda_{tot}\gamma^{\mu\nu} = -8\pi T^{\mu\nu} \tag{22.31}$$

by eqs. 22.28 and 22.30.

22.4 Impact of the Faddeev-Popov Complexity Term on the Wheeler-DeWitt Equation

The Faddeev-Popov term that arises because of the restriction of the metrics and coordinates to real values also impacts on the Wheeler-DeWitt equation since it is derived from the lagrangian via the Hamiltonian it generates. The Wheeler-DeWitt equation changes to

$$(G_{ijkl}\ \delta/\delta\gamma_{ij}\ \delta/\delta\gamma_{kl} + \gamma^{\frac{1}{2}\,(3)}\,R + 2\lambda\,\gamma^{\frac{1}{2}\,(3)})\Psi({}^{(3)}\mathcal{G}) = 0 \tag{22.32}$$

The functional Wheeler-DeWitt equation of eq. 22.32 resembles a Klein-Gordon equation.[236] Solutions of this equation can be expressed as path integrals:

$$\Psi({}^{(3)}\mathcal{G},\ \mathcal{L}_F) = N \int_{\mathcal{C}} \delta g(x)\ \exp(-I(g,\ \mathcal{L}_F)) \tag{22.33}$$

where $I(g, \mathcal{L}_F)$ is the effective total Euclidean action for the open universe case. See Hartle and Hawking for a detailed study in the case of a symmetric cosmological constant.

It does not appear that the issues of the Wheeler-DeWitt equation are resolved by the extended Extended Wheeler-DeWitt equation presented here:

- Divergences in integrals in inner products, thus requiring renormalization.
- Negative probabilities in inner products,
- Issues with the requirement of space-like surfaces,
- The frontier divergence singularity.

Later (in chapter 23) we consider another form of the Wheeler-DeWitt equation expressed in terms of Megaverse geometry. We will reconsider the issues of the above formulation again in the Megaverse.

---

[236] Hartle, J. B. and Hawking, S. W., Phys. Rev. D **28**, 2960 (1983).

23. The Megaverse: a 16-Dimensional Space

23.1 Introduction to the Megaverse

We introduce the Megaverse in this chapter and discuss its features in some detail. The Megaverse[237] is a 16-dimensional complex space that contains 4-dimensional "island universes" floating within it. We view the universes as capable of interaction and, in fact, having the properties we usually attribute to particles. We develop a formalism for quantum universe particles interacting via Baryonic and Dark Baryonic long range, but ultraweak, forces using the 16-dimensional Baryonic force presented in chapter 15. We suggest that "universe-antiuniverse pairs" may be created in a double "Big Bang event" that may well be the source of our universe (and an antiuniverse). We also consider the generalization of the Wheeler-DeWitt equation to sixteen dimensions.

We plan to present reasons in this chapter (based on Leibniz's Principle and Ockham's Razor as well as the observed unity[238] of Nature in its manifestations in many disparate areas) for the above brief summary of our Megaverse theory with its somewhat "speculative" nature.

23.2 Reasons for a Megaverse

In the first part of this book we constructed an Extended Standard Model and Quantum Gravity theory from the fundamental basis of our universe in Asynchronous Logic, complex 4-dimensional space-time, the Reality group, the Generation group, General Relativity, and Two-Tier Quantum Field Theory. We now will proceed to construct an extension of our theory to multiple universe residing in a Megaverse.

In this book and earlier books we have suggested that there are weighty reasons to believe that other universes exist.[239] The existence of other universes is suggested as the solution to a number of problems. They provide:

1. The ultimate source of mass and inertia in our universe.
2. The "clock" that sets the time in our universe.
3. The "Observer" that makes a quantum Megaaverse of our universe.

---

[237] In earlier books such as Blaha (2014a) we called the Megaverse the Multiverse. We now use the name Megaverse to avoid confusion with the multiverse of Everitt and others. Giordano Bruno suggested that earth was not alone and proposed that many worlds existed. In the early 20th century astronomers found nebulae were "island universes". We now know them to be galaxies. In the 19th century the possibility of extra dimensions and parallel universes was much discussed. William James[237] appears to have first coined the word "Megaverse" to describe parallel universes. This possibility was explored further theoretically in the latter part of the 20th century.

[238] The appearance of similar mechanisms in very different areas of Nature has been noted by many obserers since the 18th century. Nature repeats its successful mechanisms.

[239] In Blaha (2013a), before the Higgs particle was discovered at CERN we suggested an alternate mechanism was possible if a sister universe existed (making the existence of other universes a reasonable possibility. The Higgs discovery makes the sister universe mechanism unlikely.

These problems have a source in Quantum Gravity and the interpretation of the Wheeler-DeWitt equation in particular.

23.2.1 Universe Clocks

Asynchronous Logic provides the equivalent of a clock for the synchronization of processes within large electrical systems such as VLSI chips. Similarly there is a need for a clock for a Megaverse. For quantum gravity DeWitt[240] points out,

'"The variables ... [of the quantized Friedmann model] because of their lack of hermiticity, are not rigorously observable and hence cannot yield a measure of proper time which is valid under all circumstances. ... . It is for this reason that we may say that "time" is only a phenomenological concept ... If the principle of general covariance is truly valid then the quantum mechanics of everyday usage with its dependence on the Schrödinger equations ... is only a phenomenological theory. For the only "time" which a covariant theory can admit is an intrinsic time defined by the contents of the universe itself. Any intrinsically defined time is necessarily non-Hermitean, which is equivalent to saying that there exists no clock, whether geometrical or material, which can yield a measure of time which is operationally valid under *all* circumstances, and hence there exists no operational method for determining the Schrödinger state function with arbitrarily high precision."

The lack of a clock within our universe invalidates quantum mechanics in principle and Quantum Gravity in particular. DeWitt concludes, "Thus [quantum gravity] will say nothing about time unless a clock to measure time is provided."

Unruh[241] also has an issue with the question of time:

"One of the key problems is that of time. We see and experience the world in terms of time. We see things grow, develop, and change. However, time does not enter into the Euclidean formulation of quantum gravity directly. In the usual Hamiltonian formulation, the Hamiltonian for quantum gravity is made up of densities which are the generators, not only of spatial coordinate transformations, but also of temporal coordinate transformations. The content of four of Einstein's equations, namely, the 6 „components, is that these generators are zero. Thus all wave functions are invariant under all spatial and all temporal coordinate transformations. There is nothing in the wave function or the amplitudes which refers to the coordinate t, or the corresponding points of the manifold in any way. How then do we recover the indubitable and ubiquitous experience we have of time? The standard answer is that our experience of time is actually an experience of different correlations between physical quantities in the world. Time is replaced by the readings of clocks. I know that time has changed, not through any direct experience with time, but because the hands of my watch have changed.

Although the implementation of this idea is actually extremely difficult in practice, and although I personally believe that one should formulate one's quantum theory of gravity so as to contain time explicitly, let us nevertheless pursue the consequences of this idea of time as defined internally, as the "reading" of a dynamic variable. For an observer inside the theory, his "time" is not the coordinate t. Rather his time is some one of the given dynamic variables of the theory: y or P. Thus although the coupling to the baby universes via the effective action S,. is independent of the coordinates t or x, that does not mean that the observer inside the theory will experience the interactions as being independent of

[240] DeWitt, B. S., Phys. Rev. **160**, 1113 (1987).
[241] Unruh, W. G., Phys. Rev. D **40**, 1053 (1989).

time. For him and/or her, time is one of the dynamic variables and so it can depend on the various dynamic variables of the theory, even if it does not depend on the time coordinate t. In general one would expect the observer to see what looks to him like a time-dependent interaction with the baby universes. At one time, some one of the baby universes may couple strongly to the large universe, while at some other time, another of the baby universes will couple more strongly."

We suggest the existence of other universes plays the role of a clock for our universe. And being universes, they are an excellent clock. DeWitt points out, 'Because every clock has a "one-sided" energy spectrum, its ultimate accuracy must necessarily be inversely proportional to its rest mass. When the whole universe is cast in the role of a clock, the concept of time can of course be made fantastically accurate (at least in principle) … ' Setting a mass scale using other universes, also sets[242] a time scale and resolves the issue of a clock for our universe. *In principle the existence of other universes validates the role of time in the Copenhagen interpretation of quantum mechanics.*

23.2.2 Quantum Observer

Attempts to create a quantum gravity theory have to confront the need for an *Observer* in any quantum theory in the Copenhagen interpretation. DeWitt points out,

"The Copenhagen view depends on the assumed a priori existence of a classical level to which all questions of observation may ultimately be referred. Here, however, the whole universe is the object of inspection; there is no classical vantage point, and hence the interpretation question must be re-argued from the beginning. While we do not wish to stress this point unduly, since, after all, the Friedmann model ignores the vast complexities of the real universe, it is nevertheless clear that the quantum theory of space-time must ultimately force a deviation from the traditional Copenhagen doctrine."

And Unruh states

"One of the key features in the interpretation of such transition amplitudes, or wave functions, is the idea that we, as observers are also a part of the Universe as a whole. We, as physical observers, must be describable from within the theory and not as observers external to the theory as in usual quantum mechanics. In usual quantum mechanics, the interpretation is usually given in terms of observers that are outside of the theory. There one makes a split, with the quantum world at one side of the split, and the observer on the other. von Neumann argued that the predictions of quantum mechanics, at least under certain assumptions, are independent of the exact location of that split, but Bohr argued adamantly for the necessity of such a split (classical observers and quantum world). *There is a great difficulty in setting up such a split for physical observers contained within and influenced by a quantum universe,* [itallics added] and for the Universe as a whole, especially including gravity, one cannot argue that the predictions will be independent of where one puts the split. Since all energies interact gravitationally, and our observations are surely energetic phenomenon, the treatment of the energetics of observation as classical would lead to different predictions than if they were treated quantum mechanically. One is therefore forced to devise an interpretation of quantum mechanics in which the observer is part of the quantum system, rather than outside the quantum system.

---

[242] For example the Planck time value is set by the Planck mass.

This means that the interpretation of these transition amplitudes becomes somewhat non-intuitive. One must ask what the system looks like from within, from the viewpoint of an observer who is part of that world, rather than being able to interpret them directly in terms of probabilities for observations made by an external observer."

While the *Observer* question is addressed by a number of authors, the proposed answers are not entirely convincing. *The existence of other universes provides macroscopic quantum Observers for our universe.* And our universe provides a macroscopic quantum observer for other universes. Thus the quantum observer issue is resolved.

These considerations lead us to view the existence of other universes as a critical solution to the above problems.

23.2.3 Asynchronous Logic is a Requirement of Universes

By establishing Asynchronous Logic principles as the basis for the existence of universes and for setting the number of dimensions in each universe – four; and basis of fermion particles - iotas – we have found deeper principles of organization for the foundations of physics. The principles built on this foundation serve to enable the coordination of complex physical processes.

Usually we look at particle processes primarily from a space-time perspective: particles collide and produce new particles. We primarily think of the incoming and outgoing particles in a collision. However, considering the set of fundamental particles – and the particle transforming interactions in themselves – neglecting space-time and momentum considerations – leads us to view particles as constituting an alphabet and their interactions as a type of computer grammar.[243] Then the Asynchronicity Principles enable us to bring in space-time in a way that gives us the maximum complexity with the most minimal assumptions. As Leibniz[244] points out our universe has maximal complexity with minimal assumptions.

23.3 General Nature of the Megaverse

Having seen the rationale for a Megaverse – to make a theory of Quantum Gravity physically meanful – we now turn to questins about the nature of the Megaverse. Our procedure is to use Leibniz's Minimax Principle and Ockham's Razor for guidance as well as to assume that our universe is not unique but is representative in its qualitative features of all universes in the Megaverse – "Nature repeats its successful features and mechanisms."

23.3.1 The Universes Within the Megaverse

So we will assume all universes are 4-dimensional due to the need for parallel physical processes from our earlier considerations on Asynchronous Logic. Although larger dimension universes are not ruled out we will use Ockham's Principle of Parsimony, and the assumed representative nature of our universe, to rule them out of consideration in this book.

Since our universe has complex coordinates we assume all universes have complex coordinates. The Reality group of our universe suggests that other universes have a Reality group. These obervations about our universe which originate in geometry imply the form of The Extended Standard Model. *So we*

---

[243] This conceptual approach was first described in Blaha (1998) who went on to characterize our universe as one enormous word evolving in time.

[244] See Rescher (1967).

take other universes to have the physics of The Extended Standard Model – perhaps modified by gravity if a given universe has exceptionally large curvature. Other universes have General Relativity and Quantum Gravity since these theories are also based on geometry. So in sum, we expect other universes to have the same fundamental physical form as our universe although attributes such as the Dark Matter proportion and the size of the universe will possibly be points of difference. The physics is the same but the physical attributes such as constants can be expected to vary.

23.3.2 Megaverse characteristics

We now have a view of the general nature of universes in the Megaverse. The next question we examine is the general nature of the Megaverse itself. *Ockham's Razor implies that complex coordinates universes require embedding in a complex Megaverse.* If the Megaverse had real-valued coordinates an embedding of our universe, and other universes, in a flat Megaverse would require the Megaverse to have a minimum of ten dimensions. This type of embedding has been known since the early 20th century. The embeddings then were usually within a real-valued 10-dimensional flat space since a 4-dimensional real-valued metric $g_{\mu\nu}$ has 10 independent components.

In the case of a *complex* 4-dimensional space-time the resulting $g_{\mu\nu}$ would require, at minimum, embedding in a 10-dimensional *complex* space or a 20-dimensional real space. But we have an additional requirement: we wish to extend the Reality group, R = SU(3)⊗SU(2)⊗U(1)⊗SU(2)⊗U(1), we found in four dimensions to be the Reality group of the complex Megaverse.[245] This group has 16 generators and a 16-dimensional complex regular representation composed of the direct sum of the regular representations of the factors comprising the Reality group. *Consequently, we choose to use a complex 16-dimensional space*[246] *for the Megaverse (significantly exceeding the minimally required 10 complex dimensions.)* An additional point of importance is that the Megaverse must also be Euclidean in order to support Quantum Gravity[247] as we will discuss later in more detail.

The complex 16-dimensional Megaverse will have an analytic, complex, symmetric metric g_{ik} which satisfies

$$g_{ij} = g_{ji} \tag{23.1}$$

for i, j = 1, 2, … , 16.

The choice of 16 dimensions is consistent with Leibniz's Minimax Principle giving multiple benefits (as we will see) from a minimal assumption.

The embedding equations for our curved universe within the complex, 16-dimensional Megaverse are:

$$z_i = f_i(x) \tag{23.2}$$

$$ds^2 = g_{ik}dz^i dz^k \tag{23.3}$$

---

245 We require the Reality group to be the same for the Megaverse and universes because we wish Reality transformations in the Megaverse to apply to the Megaverse space occupied by universes. This will become apparent later.

246 We will refine this assumption about the Megaverse's complex coordinates later in section 23.8.

247 Hawking, S. W., Phys. Rev. D **37** 904 (1988).

where x is a complex 16-vector in the Megaverse, g_{ik} is the complex Megaverse metric, ds is the 16-dimensional invariant distance,[248] and the complex, 4-dimensional metric of a universe is

$$g_{\mu\nu} = \partial f_j/\partial x^\mu \, \partial f_j/\partial x^\nu \qquad (23.4)$$

with an implied sum over the subscript j.

The picture we paint of the Megaverse is a complex, 16-dimensional space containing (perhaps) countless universes, some of which may be flat or almost flat like ours, and some of which may be curved, closed 4-surfaces. Part of the Megaverse is outsideof universes (since gravity does not penetrate the surface of universes.) It does have gravitation due to the Two-Tier field Y^μ, the Baryonic Field B^μ and the Dark baryonic field $B_D{}^\mu$. Y^μ, and these long range forces between universes, give it a dynamic aspect which we describe later.

A universe occupies a region within the Megaverse. Its gravitational field and mass-energy are confined to within the universe's region – rather like quarks and gluons are confined within hadrons.[249] We will consider the confinement of gravitation in more detail later.

From the viewpoint of Megaverse coordinates a universe can change its location and/or size. Universes can expand or contract. Within a universe, its size and mass-energy distributions can change.

We thus arrive at a view of the Megaverse as an ocean of island universes. This view is in accord with the observation that the total mass-energy contained within a universe is zero (the mass-energy of matter and radiation being cancelled by the universe's gravitational energy), and the total 3-momentum of a universe is also zero. But baryonic long range forces, the presence of Y^μ in coordinates and gravitation effects due to them, are the source of dynamic interactions between universes.

Although the Megaverse appears to be non-quantum, in chapter 24 we will introduce quantum interactions between universes that can also probe the contents of universes. We will be guided by an analogy with electromagnetic probes of protons: for low energies the probes treat the proton as an elementary particle with a spatial extension; for very high energies the probes "penetrate" the proton revealing its quark-parton constituents. Thus the quantum "observer" will use interactions to observe quantum universe phenomena.

Chapter 24 will consider gauge boson interactions somewhat like QED but dependent on baryon number, and Dark Baryon number. Their existence is due to the excess of baryons and Dark baryons in our universe and other universes.

23.4 Summary of Features of the Megaverse

There are a number of features of the Megaverse that follow from previous discussions which we may believe to be true:

- The Megaverse has non-zero gravitation and curvature outside of universes, not due to the mass of a universe but due to the energy of the Baryonic field, Dark baryonic field,

[248] We note that ds^2 is complex and might be found troubling. A Reality group transformation would yield the absolute value of ds^2 and thus $ds_{physical} = |ds^2|^{1/2}$ would be the physical real-valued invariant distance. This remark applies to 4-dimensional complex General Relativity also.

[249] The possibility that the gravitational field and the strong interaction gauge field might have similarities such as confinement and a linear potential was first considered by this author in "Quantum Gravity and Quark Confinement" by Stephen Blaha, Lett. Nuovo Cim. 18:60, 1977. Honorable Mention in the Gravity Research Foundation Essay Competition in 1978.

and the Y^i(y field), and quantized universe particles representing the universes (described later).

- There are baryonic and Dark baryonic long range gauge fields in universes and the Megaverse. They exert forces between a universe and the baryons within the same universe, and between a universe and the baryons within another universe. At low energies the forces are between universes. At high energies baryons in different universes interact directly.

- We can define a mass for a universe (see below) that is time dependent and proportional to the area of the universe in a manner analogous to black holes.

- Universes can be treated as second quantized particles with either spin ½ or and spin zero – universe particles. Spin ½ universe particles are implied by the "handedness" that we see in our universe.

- Each universe occupies a region of the Megaverse that is a closed set in Megaverse coordinates and an open set in the curved coordinates of the universe.

- Within the boundaries of a universe one can use either Megaverse coordinates or the curved coordinates of the universe.

- The Baryon and Dark baryon gauge fields can create universe-antiuniverse pairs. Our universe may have been created through such a Big Bang process.

- Due to the size of universes, the creation or interaction of universes via the baryonic fields will have form factors or structure constants analogous to similar features in the electromagnetic behavior of hadrons.

- It is possible that universes can be tachyonic. This is explicitly seen in chapter 24 in the solutions of the Wheeler-DeWitt equation.

- We assume that other universes have the same physics as our universe with the possible difference that they may have differing coupling constants and particle masses.

23.5 External Properties of Universes Within the Megaverse

Major non-null characteristics of our universe are its baryon number and its surface area. In the case of our universe the preponderance of matter over anti-matter yields an enormous baryon number. In chapter 15 we considered the possibility of a baryonic gauge field that would produce a force between universes, between a universe and baryons in another universe, and between baryons within a universe. This field, if it exists, and there is some evidence for it, would account for *baryon number conservation* and also would have significant consequences on the galactic level within our universe. Naturally it must be significantly weaker than the force of gravity. But the plenitude of baryons in our universe gives it importance on large scales.

There is also the possibility that a baryonic force could induce movement, collisions, universe amalgamations, and other effects between universes in the Megaverse making universes a peculiar form of particle on a Megaverse scale. We consider this possibility in chapter 24..

Another important characteristic of universes is their area. While a universe is not believed to be a black hole (although Hawking has recently jokingly suggested that our universe may be a black hole, and even more recently suggested black holes are not quite black holes – grey?), there are general qualitative similarities that lead us to consider the possibility that the four laws of black holes[250] may apply in part (or their entirety) to universes. In particular the 2nd law states

$$dM = \kappa dA/8\pi + \Omega dJ \tag{23.5}$$

where dM is the change in "mass/energy," A is the area of the Black Hole (universe), Ω is its angular velocity (zero for universes) and J is the angular momentum (also zero for universes). Since $J = \Omega = 0$ for universes we can define a "mass" for a universe particle[251] in terms of a universe's area with

$$M = \kappa A/8\pi \tag{23.6}$$

This definition seems to capture the physics of universes that could be used in developing a dynamics of universes as we do in chapter 24. It also allows us to escape the dilemma of zero total energy of universes that would preclude treating universes as particles in the Megaverse and developing a Megaverse dynamics of universe-particles.

Thus we have universes with (large) baryon numbers and areas and most likely large Dark Baryon numbers that gives rise to a Dark baryon force.

23.6 The Surface of a Universe – A Type of Horizon

The surface of a universe may play a significant role in a fashion similar to the role of the surface of a black hole. Consider the non-quantum definition of an embedded universe defined by eqs. 23.2 – 23.4.

A universe has a surface. Assuming rectangular coordinates one would expect that the surface[252] is approached when one or more of the coordinates x^μ approaches infinity where there is no matter and the metric approaches the flat space metric $\eta_{\mu\nu}$. The surface itself is defined by a set of Megaverse equations:

$$z_k = g_k(z) \tag{23.7}$$

Near the surface of a universe the Megaverse coordinates can be expressed in a simple rectangular form

$$\begin{aligned} z_1 &= f_1(x) \cong ix_0 \\ z_2 &= f_2(x) \cong x_1 \\ z_3 &= f_3(x) \cong x_2 \\ z_4 &= f_4(x) \cong x_3 \end{aligned} \tag{23.8}$$

which yields the metric

---

250 Wald, R. M., "The Thermodynamics of Black Holes", *Living Reviews in Relativity* **4** (6): 12119 (2001).

251 *We limit all subsequent discussions to spatially closed universes.*

252 Barring singularities within the universe and topological complications.

$$g_{\mu\nu} \cong \eta_{\mu\nu} \qquad (23.9)$$

of a flat space Lorentz metric.

The surface (horizon) of a universe thus encloses the universe in a manner that is analogous to quark confinement in hadrons where the strong interaction goes to zero "outside" the hadron. At the surface of a universe the universe has zero total energy and zero total value for all quantities except baryon number, Dark baryon number and surface area. An interesting feature of a universe surface, in comparison to quark confinement, is that leakage of energy/matter through the surface cannot happen. A universe has confinement. However it achieves confinement in a radically different way than quark confinement. A universe expands if energy/matter penetrates its horizon. So "confinement" is achieved by expansion. This confinement mechanism is not present in quark confinement.[253]

Turning to Quantum Gravity: If we consider the behavior of the wave function of a universe at its horizon then we can only expect that it goes to zero as well. If universes collide then a quantum dynamical event occurs. The study of these phenomena experimentally is not technically feasible currently or in the near future.

23.7 Fully Reducible Representation of 16-Dimensional Reality Group

The Reality group which was so important in determining the form of The Extended Standard Model is also important in determining the form of the dynamics of universes within the Megaverse. We now discuss some features of the Reality group representations in 16-dimensional space.

The 16-dimensional Reality group R of the Megaverse can be represented in a fully reduced form in terms of the regular representations of SU(3) – 8 by 8 matrices, SU(2) – 3 by 3 matrices, and U(1) 1 by 1 matrices. (See Fig. 2.2 of Blaha (2012b), which is reproduced here for the reader's convenience as Fig. 23.2.)

Denote the 16 matrix generators of R as V_i. Corresponding to each generator is a connection or field that we used to define covariant derivatives earlier. We denote the connections as A_{ki} where k labels the coordinate and i labels the connection. Then we divide these connections into sets of connections for each of the interactions that will ultimately become the Extended Standard Model gauge fields:

$$\begin{aligned} &SU(3) - A_{ik} \text{ for } i = 1, 2, \ldots, 8 \\ &SU(2) - W_{ik} \text{ for } i = 1, 2, 3 \\ &U(1) - W_{0k} \\ &SU(2) - W'_{ik} \text{ for } i = 1, 2, 3 \\ &U(1) - W'_{0k} \end{aligned} \qquad (23.10)$$

where k = 1, 2, …, 16 is the coordinate index.

Each matrix of each subgroup's generators has its usual matrix regular representation in its block. It has 1's along the diagonal for all the other groups' blocks and zeroes otherwise. Thus each of the sets of generators has their conventional commutation relations although they are embedded in 16×16 matrices. For example the SU(3) generators would "fill" the top 8 by 8 block in Fig. 23.1. The rest of the

[253] Recently discrepancies have been found in the size of the proton in various experiments. These discrepancies are very small, and subject to dispute, but they are interesting when we consider "universe confinement."

generator matrices would consist of 1's along the diagonal following the block. All other components of its matrices would be zeros. See Fig. 23.2.

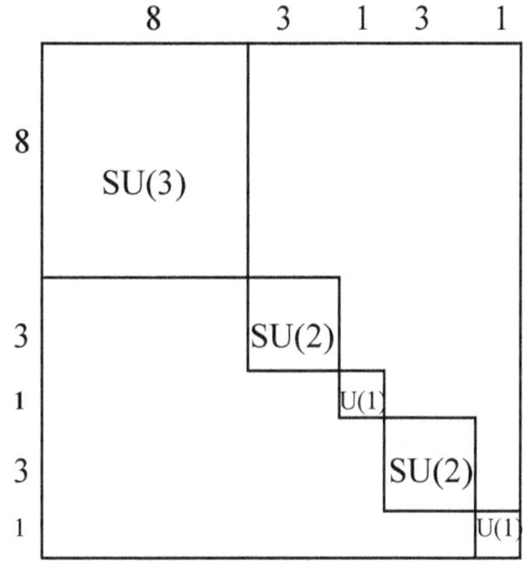

Figure 23.1. The fully reduced block structure of R matrix representation.

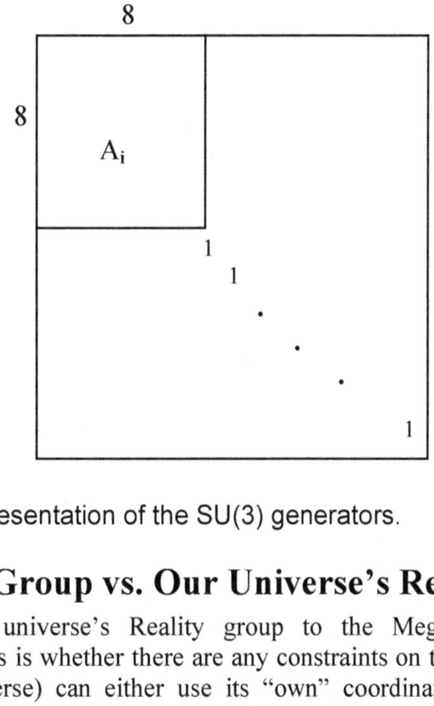

Figure 23.2. 16-dimensional representation of the SU(3) generators.

23.8 Megaverse Reality Group vs. Our Universe's Reality Group

We have extended our universe's Reality group to the Megaverse with its differing dimenionality. A question that arises is whether there are any constraints on this extension. In particular, since our universe (and any universe) can either use its "own" coordinate system or a Megaverse coordinate system to describe events within it, is the use of both Reality groups always consistent, or are there consistency restrictions on the use of the Megaverse Reality group within a universe?

The answer to this question is that there are required consistency conditions. The proof begins with a Reality group transformation denoted G_{4x} on 4-dimensional coordinates in our universe. Then we assume that there is a corresponding Reality group transformation G_{16z} of the Megaverse coordinates of our universe. Consequently if

$$z = f(x) \tag{23.11}$$

then, if we apply corresponding Reality group transformations of our universe's coordinates and its Megaverse coordinates, we find

$$y = G_{16z}z = f(G_{4x}x) \tag{23.12}$$

where y is a real 16-vector, G_{16z} is an element of the 16-dimensional Reality group R_{16}, G_{4x} is an element of our universe's Reality group G_{4x}, z is a 16-dimensional vector and f is a vector composed of the 16 f_i functions. The 16-dimensional group R_{16} has 16 generators that we will denote V_i. The 4-dimensional Reality group R_4 has 16 generators that we will denote U_i. If we make an infinitesimal transformation G_{4b}

$$G_{4b} = I + \beta_i U_i \tag{23.13}$$

then there must be a corresponding infinitesimal transformation G_{16a}

$$G_{16a} = I + \alpha_i V_i \tag{23.14}$$

where α_i and β_i are constants[254] for i = 1, 2, …, 16. Substituting in eq. 23.12 and expanding to first order we find

$$\alpha_i V_{ijk} f_k = \beta_i U_i{}^{\mu}{}_{\nu}\, x^{\nu}\, \partial f_j / \partial x^{\mu} \tag{23.15}$$

for j = 1, 2, …, 16 with summations over i, k, μ, and ν. Eq. 23.15 is 16 equations that determine the β_i parameters in terms of the α_i parameters. Thus a G_{4x} transformation uniquely determines a G_{16z} transformation.

In the case of a flat 4 dimensional space-time where we can limit the Megaverse to 4 dimensions also ($z_i = x_i$ for i = 1, 2, …, 4), then eq. 23.15 simplifies to[255]

$$\alpha_i V_{ijk} = \beta_i U_{ijk} \tag{23.16}$$

implying $\alpha_i = \beta_i$ and $V_{ijk} = U_{ijk}$. In this case the Megaverse's and our universe's Reality transformations coincide. In the case of a curved universe eq. 23.15 requires a more complex calculation. In any case, the equality of the number of Reality group generators in our universe and the Megaverse is crucial. Otherwise a solution is either ambiguous or non-existent in general.

23.8.1 Restriction of Complexity of the 16-dimensional Megaverse

The above discussion shows that compatibility of the Reality groups of a universe and the Megaverse part within the universe is required to have both sets of coordinates compatible within a universe. Continuity requires the Reality group of the Megaverse within a universe be the same outside

---

254 They could be functions of z and x respectively in the case of a curved 4-dimensional space-time or for accelerating reference frame transformations.

255 We will not distinguish between raised and lowered indices for the sake of simplicity as they do not have physical import in these considerations.

the universe. Therefore the Reality group G_{16z} throughout the Megaverse must be SU(3)⊗SU(2)⊗U(1)⊗SU(2)⊗U(1). Consequently we must restrict the complex coordinate systems of the Megaverse to those coordinate systems that can be transformed to real-valued coordinate systems by a Reality group transformation. We note the Reality group transformations can be local, position-dependent transformations (Yang-Mills transformations). Stated differently, all allowed *complex* Megaverse coordinate systems (in a flat or curved Megaverse space) can be generated from a coordinate system in the set of *real-valued* Megaverse coordinate systems by Reality group transformations.

The reason for this clarification will become apparent when we consider the types of *universe particles*[256] that can be created by 16-dimensional Lorentzian boosts which will, in some cases, create tachyonic universe particles with complex coordinates. The complex coordinates of a tachyonic universe particle must be transformable to real-valued Megaverse coordinates by a Reality group transformation.

We make these points because complex vectors and tensors in 16-dimensional space require the full set of U(16) transformations to be made real-valued. The Reality group only has 16 generators as opposed to 256 generators for U(16). Thus only "special" classes of complex coordinate systems can be made real-valued by the Reality group.

23.9 Covariant Derivatives and Connections in the Megaverse

The introduction of the Reality group in the Megaverse leads to a need for covariant derivatives for quantities that are subject to Reality group transformations. This question has been addressed in Blaha (2012a). In this section we will introduce covariant derivatives necessitated by the Reality group.

Consider a 16 component vector function of the coordinates of the Megaverse **F**(z). If we apply a Reality group transformation G_{16z} to it (to make all its components real-valued) then the partial derivative of $G_{16z}\mathbf{F}$ changes in a non-covariant way.

$$\partial(G_{16z}\mathbf{F})_j/\partial z^i \neq G_{16zjk}\partial(\mathbf{F})_k/\partial z^i \qquad (23.17)$$

In order to have a covariant derivative expression we must take the 16 generators of G_{16z}, which we denote V_i for i = 1, 2, … , 16, and define the covariant derivative

$$D_k = \partial/\partial z^k - iV_iZ_{ki} \qquad (23.18)$$

where the quantities Z_{ki} are connections (using the terminology of General Relativity), and where there is an implicit sum over i. Using the covariant derivative we find

$$D_k(G_{16z}\mathbf{F}) = G_{16z}\, D_k\mathbf{F} \qquad (23.19)$$

23.10 Covariant Derivatives in 4-Dimensional Lorentzian Universes such as Our Universe

A similar need for covariant derivatives appears in a 4-dimensional Lorentzian universe. In a universe the Reality connections lead to the gauge fields that embody the interactions of The Extended Standard Model.

In our complex 4-dimensional Lorentzian universe

---

[256] A universe particle is a universe treated as a "quantum" particle. We are led to universe particles by an extension of Quantum Gravity to the Megaverse. See chapter 24.

$$ds^2 = {}_4g_{\mu\nu}dx^\mu dx^\nu \tag{23.20}$$

where ${}_4g_{\mu\nu}$ is the complex conventional Lorentz metric with the metric satisfying

$${}_4g_{\mu\nu} = {}_4g_{\nu\mu} \tag{23.21}$$

Again we define a covariant derivative similar to that of eq. 23.18. Now the 4-dimensional space-time Reality group of the Extended Standard Model is R = SU(3)⊗SU(2)⊗U(1)⊗SU(2)⊗U(1), a direct product group. It performs a role similar to the 16-dimensional case:

$$w = R_1x \tag{23.22}$$

where x is a complex 4-vector, w is a real-valued 4-vector and R_1 is an R transformation. The *specific* form of the Reality group is derived in chapters 18, 18A, and 19 of Blaha (2011c) and chapter 1 of Blaha (2013) from a consideration of complex Lorentz group transformations. As expected the 4-dimensional Reality group has 16 generators, and so it can transform any complex 4-vector into a real-valued vector. Since it is independent of the complex Lorentz group the "noGo" theorems of Coleman and others do not apply within this framework – there is no unification of the Reality group and the Lorentz group such as appeared, for example, in the case of SU(6).

Due to the Reality group, the covariant derivative of the Extended Standard Model must have the form

$$D_\mu = \partial/\partial x^\mu - iV_{8i}\, A^i_{\ \mu} \tag{23.23}$$

where the $A^i_{\ \mu}$ are 16 gauge fields – connections (i = 1, …, 16), and the V_{8i} are the 16 generators of SU(3)⊗SU(2)⊗U(1)⊗SU(2)⊗U(1) as shown earlier.

23.11 Corresponding Lorentzian Transformations in a Universe and the Megaverse

We have embedded our complex universe in a 16-dimensional complex space, the Megaverse,

$$y_i = h_i(x) \tag{23.24}$$

23.11.1 Special Case of Flat Universes

In the special case of a flat universe in the Megaverse, we can easily relate a Lorentz transformation in the universe to a transformation in the Megaverse. Suppose we define the Megaverse coordinates of our universe as

$$\begin{aligned} y^0 &= x^0 \\ y^1 &= x^1 \\ y^2 &= x^2 \\ y^3 &= x^3 \\ y^4 = y^5 = y^6 = \ldots &= y^{15} = 0. \end{aligned} \tag{23.25}$$

If we now perform a (real or complex) Lorentz transformation:

$$x'^{\mu} = \Lambda^{\mu}{}_{\nu}x^{\nu} \tag{23.26}$$

then by eq. 23.25

$$y'^{\mu} = \Lambda^{\mu}{}_{\nu}y^{\nu} \tag{23.27}$$

for μ, ν = 0, 1, 2, 3, is the corresponding transformation that yields

$$\begin{gathered} y'^{\mu} = x'^{\mu} \\ y'^{4} = y'^{5} = y'^{6} = \ldots = y'^{15} = 0. \end{gathered} \tag{23.28}$$

for μ, ν = 0, 1, 2, 3.

23.11.2 Corresponding General Coordinate Transformations in a Universe and the Megaverse

If we perform a general coordinate transformation such as that in eq. 23.24 then

$$y_i = h_i(x) = h_i(f(x')) \tag{23.29}$$

where we surpress displaying the index on the coordinate transformation f(). Now x' are the new coordinates of a point in our universe. Therefore, assuming well-behaved functions, we can write

$$y_i = h'_i(x') \tag{23.30}$$

where h'_i is the composition of the $h_i()$ and f() functions, which we symbolize by

$$h'_i = h_i \circ f \tag{23.31}$$

Suppose we define "new" y coordinates y_i' by

$$y_i' = h_i(x') \tag{23.32}$$

It is clear that a set of functions $g_i()$ exists, that relate the y_i coordinates to the y_i' coordinates:

$$y_i = g_i(y') \tag{23.33}$$

where the argument y' represents the set of 16 y'_i coordinates. From eq. 23.29 we see that

$$y_i = g_i(h(x')) \tag{23.34}$$

where h() represents the set of 16 $h_i()$ functions.

This development enables us to relate a general coordinate transformation in our universe to a transformation in the 16-dimensional Megaverse

$$\begin{aligned} y_i &= g_i(h(x')) = g_i(h(f^{-1}(x))) \\ &= h_i(x) \end{aligned} \tag{23.35}$$

Therefore we have the composition

$$h = g \circ h \circ f^{-1} \tag{23.36}$$

as illustrated in Fig. 23.3. Eq. 23.36 implies the transformation g() between the y and y' coordinates is symbolized by

$$g = h \circ f \circ h^{-1} \tag{23.37}$$

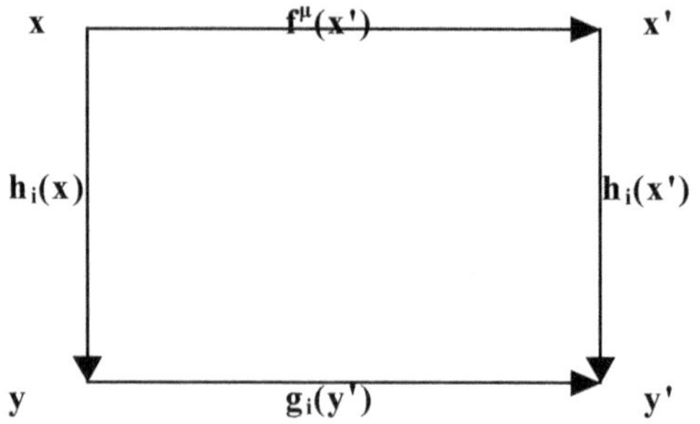

Figure 23.3. A diagram illustrating the various functions relating the universes, and representing the coordinate transformations.

Fig. 23.3 illustrates eq. 23.4 as well. The h^{-1} function takes the y' coordinates to x' coordinates. Then f takes the x' coordinates to x coordinates. Lastly h takes the x coordinates to the y coordinates. The net result is the y' coordinates are transformed to the y coordinates as specified by eq. 23.33. Eq. 23.37 is equivalent to

$$g_i(y') = h(f(h^{-1}(y'))) \tag{23.38}$$

We assume the above functions are all well behaved, non-singular, and have inverses.

Finally we note

$$\begin{aligned} \partial y'_i/\partial y_j &= \partial h_i(x')/\partial x'^{\mu}\, \partial x^{\nu}/\partial h_j(x)\, \partial x'^{\mu}/\partial x^{\nu} \\ &= \partial h_i(x')/\partial h_j(x) \end{aligned} \tag{23.39}$$

23.12 Quantum Gravity in the Megaverse and its Universes

Since our universe is described by Quantum Gravity, and other universes are also it is reasonable to assume the Megaverse is described by classical and Quantum Gravity. The source of Quantum Gravity in a universe is the mass-energy within the universe including particle masses, the Baryonic and Dark Baryonic gauge fields and the Two-Tier Y^{μ} field. The sources of classical Gravitation and Quantum Gravity in the Megaverse are the "masses" of universe particles (described in chapter 24), the Baryonic and Dark Baryonic gauge fields and the Megaverse Two-Tier Y^{μ} field, but not the mass-energy *within* universes.

Classical and Quantum Gravity smoothly transition through a universes boundary as their sources change from in-universe sources (particularly the mass-energy within a universe) to extra-universe sources (which exclude the mass-energy within universes and include the "masses" of universe particles). The key to a smooth transition is a transition through the surface of a universe where the mass-energy of the universe has become negligible. It is incongruous for a universe to have a surface with large mass-energy concntrations.

The 16-dimensional surface of a universe in Quantum Gravity is fuzzy and determined by the quantum universe's wave function. This wave function is a solution of the Wheeler-Dewitt equation. The Wheeler-DeWitt equation assumes a single 4-dimensional universe with real-valued coordinates and metrics.

Since our theory requires complex coordinates and thus complex-valued metrics, we extended the Wheeler-DeWitt equation to complex metrics in chapter 22. Now we must also extend it to 16-dimensional complex-valued coordinates and metrics. The Wheeler-DeWitt equation then assumes a Megaverse form that we can apply uniformly to the universes within the Megaverse and to the 16-dimensional space between them.

23.13 Megaverse Wheeler-Dewitt Equation

The Megaverse Wheeler-DeWitt equation inside a universe (based on that for a universe eq. 22.23) is

$$\Big(G_{ijkl}\Big\{\prod_x\sum_m\{[\delta^m{}_j\,\partial f_n/\partial x^i + \delta^m{}_i\,\partial f_m/\partial x^j\,](\partial^2 f_n/\partial x^{m2})\}^{-1}\partial/\partial x^m\Big\}\Big\{\prod_x\sum_m\{[\delta^m{}_k\,\partial f_n/\partial x^l +$$

$$+ \delta^m{}_l\,\partial f_n/\partial x^k\,](\partial^2 f_n/\partial x^{m2})\}^{-1}\partial/\partial x^m\Big\} + \gamma^{1/2\,(3)}R + 2\lambda\gamma^{1/2\,(3)}\Big)\Psi({}^{(3)}G, L_F) = 0 \qquad (23.40)$$

where f_n is given by eqs. 23.2-23.4. In terms of Megaverse coordinates y_n

$$0 = \Big(G_{ijkl}\Big\{\prod_x\sum_m\{[\delta^m{}_j\,\partial y_n/\partial x^i + \delta^m{}_i\,\partial y_m/\partial x^j\,](\partial^2 y_n/\partial x^{m2})\}^{-1}\partial/\partial x^m\Big\}\Big\{\prod_x\sum_m\{[\delta^m{}_k\,\partial y_n/\partial x^l +$$

$$+ \delta^m{}_l\,\partial y_n/\partial x^k\,](\partial^2 y_n/\partial x^{m2})\}^{-1}\partial/\partial x^m\Big\} + \gamma^{1/2\,(3)}R + 2\lambda\,\gamma^{1/2\,(3)}\Big)\Psi({}^{(3)}G, L_F) \qquad (23.41)$$

where the sums over n and m in each pair of {} are done independently. All references to the metric are expressed in terms of the Megaverse using eq. 23.4.

Due to the products over all coordinates, the Megaverse expression for $\delta/\delta\gamma_{ij}$ is independent of x and Megaverse coordinates in accordance with the space-time independence of the original Wheeler-DeWitt equation.

The Megaverse form of the Wheeler-DeWitt equation also directly relates the metric of a universe within the universe to the Megaverse. Every universe has two sets of coordinates: universe coordinates, usually labeled x, embodying the curvature of the universe induced by gravitation, and Megaverse coordinates usually labeled y.

Outside of universes the Megaverse Wheeler-DeWitt equation is

$$(G_{ijkl}\,\delta/\delta\gamma_{ij}\,\delta/\delta\gamma_{kl} + \gamma^{1/2\,(15)}\,R + 2\lambda_M\,\gamma^{1/2\,(15)})\Psi({}^{(15)}\mathfrak{G}) = 0 \qquad (23.42)$$

with a cosmological constant, λ_M which is assumed to be present based on its presence in the universe case above and the uniformity of Nature. We assume a representation of the 16-dimensional metric of a form analogous to eqs. 22.3 and 22.4 with a "spatial"sub-metric part γ_{ij} in eq. 22.3. $\gamma^{1/2\,(15)}$ is the determinant of γ_{ik}.

Due to the products over all coordinates, the Megaverse expression for $\delta/\delta\gamma_{ik}$ is independent of Megaverse coordinates in analogy with the space-time independence of the original Wheeler-DeWitt equation. But the solutions of the Wheeler-DeWitt equations for a universe must be related to the solutions of the Wheeler-DeWitt equations of the Megaverse within the universe. This relationship must constrain the universe solutions at the boundary of a universe.

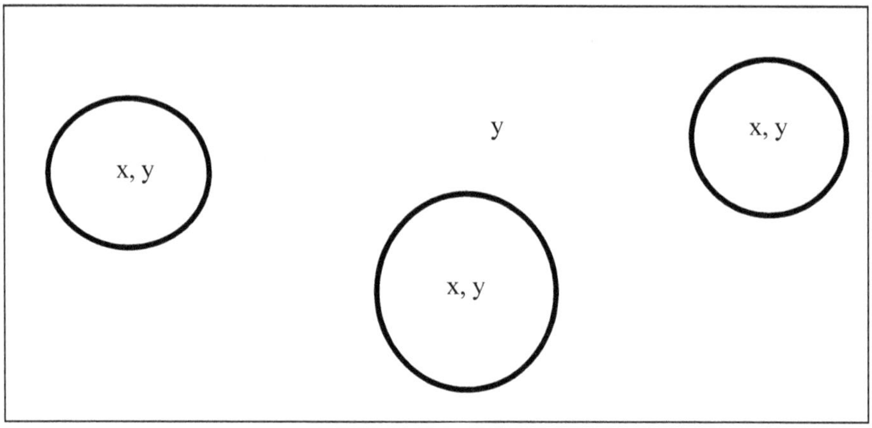

Figure 23.4. A symbolic view of part of the Megaverse with universes depicted as circles. Each universe has its own curvilinear coordinate 4-vector denoted x. The Megaverse coordinates are a 16-vector y. A Wheeler-DeWitt equation applies within each universe and a comprehensive equation describes the entire Megaverse transitioning to each universe's wave function within its domain.

23.15 Megaverse Metric Functional Integral in a Universe

Corresponding to the functional derivative in eq. 23.41 is an explicit Megaverse form of the functional integral for a universe

$$\int D\gamma \equiv \int \prod_x \sum_{i,j} d\gamma_{ij}(x) = \int \prod_x \sum_{i,j,k} d(\partial f_k/\partial x^i\, \partial f_k/\partial x^j) \tag{23.43}$$

$$= \int \prod_x \sum_{i,j,k} [d(\partial f_k/\partial x^i)\, \partial f_k/\partial x^j + d(\partial f_k/\partial x^j)\, \partial f_k/\partial x^i]$$

$$= 2\int \prod_x \sum_{i,j,k} d(\partial f_k/\partial x^i)\, \partial f_k/\partial x^j$$

$$= 2\int \prod_x \sum_{i,j,k} dx(\partial^2 f_k/\partial x^{i2})\, \partial f_k/\partial x^j \equiv 2\int \delta f(x) \tag{23.44}$$

$$= 2\int \prod_x \sum_{i,j,k} dx(\partial^2 y_k/\partial x^{i2})\, \partial y_k/\partial x^j \equiv 2\int \delta f(x) \tag{23.45}$$

The third line is due to the summation over i and j. The factor of 2 can be absorbed in the normalization factor. Eqs. 23.45-23.45, like the Wheeler-DeWitt equation, are independent of the coordinates due to the product over coordinates x. Thus the solution of the Wheeler-DeWitt equation takes the form

$$\Psi({}^{(3)}\mathcal{G},\ \mathcal{L}_F) = N \int \delta f(x)\ \exp(-I(g,\ \mathcal{L}_F)) \qquad (23.46)$$

upon absorption of the factor of 2 into the normalization constant N

The Wheeler-DeWitt equation depends on the metric γ_{ij} which is a 15 by 15 = 225 component construct with 120 independent components due to its symmetry.

23.16 Megaverse Wheeler-DeWitt Solutions

23.16.1 Tachyonic Solutions of Wheeler-DeWitt Equation

The original Wheeler-DeWitt equation and its Megaverse equivalents have a form that is similar in many respects to the Klein-Gordon equation. In particular as DeWitt[257] noted it resembles "a Klein-Gordon equation with $-\gamma^{½}\ {}^{(3)}R$ (our notation) playing the role of a mass-squared term. An important difference, however, is that ${}^{(3)}R$ can be either positive or negative, and hence the wave propagation of the state functional is not confined to time-like directions." DeWitt proceeds later in his paper to exclude consideration of negative "mass" squared terms.

In our view the negative "mass" squared cases represent tachyonic solutions of the Wheeler-DeWitt equation and should be considered as having the same validity as the positive mass squared solutions.

In a universe G_{ijkl} can be regarded as the contravariant metric of a 6-dimensional Riemannian manifold M with hyperbolic signature (-1, 1, 1, 1, 1, 1) with a time-like coordinate.[258] Tachyonic solutions are part of the set of solutions of the Wheeler-DeWitt equation. In the Megaverse formulation the tachyonic solutions are indicators of tachyonic universes in the Megaverse.

If we use the Megaverse form of the Wheeler-DeWitt equation then the changes in time (dilations of γ_{ij} or equivalently dilations of y_n) can be viewed as specifying the overall motion of entire universes. We see now that we can have "normal" or tachyonic motion of universes.

Clearly we are building towards a particle view of universes.[259]

23.16.2 Problems in the Solutions of the Wheeler-DeWitt Equation

There are a number of problem areas associated with the original and our Megaverse Wheeler-DeWitt equations. DeWitt identified most of them in his paper, referenced below. While one could view these apparent problems as negatives, we will take the view that they are indicators of a deeper structure of universes below the Wheeler-DeWitt solutions just as the Dirac equation resolves analogous difficulties with the Klein-Gordon equation.

---

[257] DeWitt, B. S., Phys. Rev. **160**, 1113 (1987).p. 1124.

[258] In the case of complex-valued metrics the Riemannian manifold would be 9-dimensional with signature (-1, 1,1,1,1,1,1,1,1).

[259] The Reality groups of 4-dimensional universes, and of the Megaverse, play a role in the physical interpretation of Megaverse phenomena.

23.17.2.1 Negative Frequencies and Probabilities – Anti-Universes

DeWitt noted the existence of negative frequencies and negative probabilities associated with the solutions of the Wheeler-DeWitt equation. The rather close analogy to the Klein-Gordon equation in whose solutions similar issues appear is suggestive.

It appears that two general types of universes are embodied in the Wheeler-DeWitt equation. One type, which we call "normal", consists of universes like ours which have an excess of baryons (and consequently electrons to make an overall charge neutral universe). The other type of universe we will call *anti-universes*. These universes have an excess of anti-baryons (and positrons). We suggest such pairs can result from vacuum fluctuations in the Megaverse. (They are described in some detail in chapter 24.)

In addition we have "normal" universes and tachyonic universes whose motions in the Megaverse are analogous to the motions of fermions within a universe. In direct analogy with the Klein-Gordon equation the negative frequency and negative probability issues are thereby resolved.

One can speculate that the Wheeler-Dewitt equation, which is effectively second order in the "time" derivative, can be factored into first derivative equations – perhaps through the introduction of more degrees of freedom in a fashion similar to Dirac's introduction of spinors – to achieve first order equations in "time." Then we would have to face the issue of interpreting "Dirac-like" Wheeler-DeWitt equations.[260] We could again fall back on the rationale for introducing spinors for Standard Model fermions – Asynchronous Logic – and suggest that universes embody a 4-valued Logic – although the concept of universes as logic values is somewhat strange. It is strange to view universes as having logic values – "spin." However the "left handedness" of our universe is suggestive of other universes with "right handedness." The quality of handedness is usually an attribute of spin. Therefore we can see a meaning for spin in universe internal properties.

We will be content, for the present, to assume both universes and anti-universes exist. An anti-universe will be presumed to be a universe in the Megaverse where anti-particles predominate. We will postulate that universes, and anti-universes like protons and anti-protons, are always bodies with extension and not point-like. Blaha (2013) and chapters 19-21 provide an example of a Big Bang where a universe begins with extension and is not point-like.

Chapter 24 describe a proposed dynamics of universes and anti-universes based on Megaverse gauge fields associated with baryon number, dark baryon number, lepton number and Dark lepton number. (The lepton fields are short range and massive within universes due to a Higgs Mechanism in a universe.) However universes can have a total lepton number if lepton number is conserved. In the external Megaverse, since Higgs fields are absent, lepton fields with sources that are the total lepton numbers of universes, would exist and would be massless long range fields.

23.17 General Types of Universes within the Megaverse

We determined the dimensionality of our universe based on principles of Asynchronous Logic that suggested a 4-valued logic that could be embodied in a 4-dimensional spinor matrix formulation. This 4-dimensional spinor formulation led to a 4-dimensional space-time. The requirement that the speed of light is the same in all inertial reference frames, and that transformations between reference frames in faster than light relative motion are physical, led to the requirement of complex coordinates and the complex Lorentz transformation group. The reality of all physical time and distance measurements led to

260 We will not consider factoring the Wheeler-DeWitt equation in this volume.

the introduction of the Reality group that mapped complex quantities to real physical values. This chain of logic is in accord with Leibniz's minimax principle: nature uses the simplest means to create complex physical phenomena.

While the preceding paragraph applies very nicely to our universe, and other universes, the question of other possible types of universes within the Megaverse naturally arises. Stars and galaxies have many varieties. Why should all universes be similar?

Having developed the fundamental nature of our universe from Logic (the only sure requirement of any physical theory) it seems reasonable to classify possible universes based on their fundamental logic. In Blaha (2011c) we developed matrix formulations for many-valued logics. Assuming no separate clock mechanism to synchronize parts of complex processes, we developed an n × n matrix formalism for n-valued logic.

Therefore we could develop a principal sequence of types of universes based on n-valued logic. We can summarize the small n-valued cases as follows:

n-Valued Logic	Matrix Representation Size	Spinor Components	Space-Time Dimensionality[261]
1	1×1	1	1
2	2×2	2	2
3	3×3	2	3
4	4×4	4	4
5	5×5	4	5
6	6×6	8	6

Table 4.1. Space-time dimensionality and number of spinor components corresponding to various n-valued logics.

Fant (2005) points out that VLSI circuits with spatially separated parts, which require time synchronization of activity without clocks, need a 4-valued logic at minimum. Thus for a complex universe such as ours the minimum space-time dimensionality is 4. For a smaller number of dimensions the complexity of physical processes is much diminished as the many solvable models of low space-time dimensionality show. Easily solved – not very complex phenomena!

Smaller dimensioned universes may well exist – but not with the richness of complexity that leads to our type of universe's phenomena such as life.

Larger dimension universes may well also exist. They would have an excess of phenomena that might preclude life as we know it.

The general tendency of physical phenomena to be largely based on extrema suggests that 4-dimensional space-time based on Leibniz's minimax principle is the "logical" choice for all universes.

The above classification scheme for universes is based on logic. Another important consideration is size. It appears that universes can have differing sizes and in fact can also grow or diminish in size (expansion or contraction). We will consider the size issue in chapter 24. Other possible differentiating factors between universes will also be considered.

---

[261] Weinberg (1995) p. 216 exhibits an equation that relates the number of components of a spinor to the dimension of its space-time.

23.18 Quantum Aspects of the Megaverse

The Megaverse contains quantized universes. Being quantized, the horizon (surface) of a universe is not precisely defined but undergoes presumably mild quantum smearing. In particular the surface of a quantum universe in the Megaverse must be defined, as particle positions are defined in quantum mechanics, by a wave function whose "square" at each Megaverse point is the probability of a part of the universe being there. Physically we would expect the probability to fall sharply shortly beyond the classical horizon (surface) of the universe.

So in a Megaverse with a low density of universes most of the Megaverse will have zero probability of a universe being present.

There are however three sources of quantum phenomena in the Megaverse: 1) gauge fields that provides interactions between universes as well as quantum fluctuations that create universes; 2) universe particles; and 3) the Y^{μ} field that appears within universes and in the exterior Megaverse that quantizes the coordinates of each universe. Y^{μ}'s existence in Megaverse makes gauge field (chapter 24) perturbation theory computations finite to all orders.

24. The Particle Interpretation of Universes

The Wheeler-DeWitt equation, because of its similarity to the Klein-Gordon equation, has led to numerous proposals to view universes as particles.[262]

In this chapter we will consider a possible particle interpretation of universes that, while consistent with the spirit of the Wheeler-DeWitt equation and the Megaverse, goes far beyond our current experimental knowledge, although some recent astronomical data tends to support it. It can only be justified in this century by its generality and simplicity. It just looks right.

The Wheeler-DeWitt equation specifies the internal dynamics of universes. The Megaverse, and its universal Baryonic, Dark Baryonic, Leptonic, and Dark Leptonic gauge fields, embody the dynamics of universes. There is also a 16-dimensional Two-Tier Y^μ field that eliminates potential infinities in perturbation theory calculations.

We view universes as extended particles in the 16-dimensional Megaverse rather like hadrons in particle physics. As we did in the low energy days of particle physics we will first quantize universes as point-like particles. We then take account of their internal structure in interactions (between universes) using solutions of the Wheeler-DeWitt equation, spectral representations of vacuum expectation values, form factors, "deep inelastic" structure functions, and so on. The interaction between universes, and between a universe and an elementary baryon particle in another universe, can be similarly treated.

The sole interactions between universes are assumed to be the aforementioned gauge interactions. Gravity is present in the Megaverse outside of universes due to universe particles and the gauge fields. The gauge bosons are cloaked with Y^μ fields through the use of Megaverse quantum coordinates similar to those discussed in chapter 18.

24.1 The Hierarchy of the Cosmos

In our universe we have seen that natural phenomena form a hierarchy ranging from the simplest to the largest/most complex phenomena. One current view of the hierarchy of levels of physical phenomena is:

Elementary particles: leptons, quarks, gluons, gauge bosons, and Higgs particles
Hadrons: protons, neutrons, …
Molecules
Agglomerations of molecules
Macroscopic objects
Planets
Stars
Galaxies
Clusters of galaxies
Supergalaxies
The Universe

---

[262] Some suggestions of this interpretation are: DeWitt, B. S., Phys. Rev. **160**, 1113 (1967); Robles-Perez, S. J., arXiv:1212.4598 (2012); and references therein.

Each level generally has a set of "simplified" physical laws that describe its phenomena. For example molecules have quantum mechanical laws and regularities that help to understand the phenomena at the molecular level.

Interestingly, while all phenomena at each level should be explainable by the laws at lower levels, and ultimately, all phenomena should be explainable at the level of elementary particles, connecting phenomena at different levels is often quite difficult and, in many cases, impossible.

Consequently, while we believe physical phenomena are ultimately reducible to the lowest level, the problem of relating phenomena at different levels is largely unresolved.

In this book we introduce new levels in the hierarchy of nature: the level of multiple universes, and the level of the all-encompassing Megaverse. In doing this, we seek to maintain what we know of our universe, as embodied in The New Standard Model and Quantum Gravity, while introducing one feature – particle number gauge fields, that appear to be needed to theoretically give the four laws of particle number conservation as well as to provide a dynamics for universes in the Megaverse.[263]

We will now turn to a discussion of the universes level and a portrayal of universes as extended particles.

24.2 The Particle Interpretation of Extended Wheeler-DeWitt Equation Solutions

In earlier sections we described features of the Wheeler-DeWitt equation that suggested that universes could be viewed as particles or anti-particles, or tachyons. The solutions of this equation are scalar wave functions on a manifold that are analogous to the solutions of the Klein-Gordon equation. Yet the issues of negative probabilities, possible tachyonic solutions, and negative frequency solutions suggest a need for an appropriate particle interpretation of universes that can possibly resolve these problems.

Some physicists have taken the Wheeler-DeWitt equation as the starting point for a theory of a universe as a particle. The Wheeler-DeWitt equation describes the interior of a universe in a quantum framework. We will take a different approach using the Megaverse as the environment of universe particles that internally have Quantum Gravity, and externally have Megaverse Quantum Gravity.

We view a universe as an extended particle and begin by ignoring the detailed inner structure of universes. This approach is similar to the historical treatment of hadrons such as the proton as particles and developing a theory of them as fundamental particles using form factors, structure functions and so on to approximate their inner structure. Afterwards, as detailed data became available, the detailed investigation of the internal structure of hadrons using quark-parton models followed. We will pursue a similar theoretical development beginning with a theory of universes as extended particles in the 16-dimension Megaverse. The internal structure of the particle universes will eventually be specified by the Wheeler-DeWitt equation expressed in Megaverse coordinates.

The two simplest choices for the nature of universes are "spin 0" *bosonic universes* and *fermionic universes* with odd half integer spin.[264] In this section we use Logic and spinor mathematics to establish a structural dimensional framework for universes. We begin with the minimalist implication of

---

[263] See Sakurai (1964) pp. 185-186 for early discussions. We also argue for lepton number conservation and for Dark baryon and lepton number conservation as well. (Sakurai (1964) pp. 190-194 provides an early discussion.) Both baryon number and lepton number conservation laws are still undergoing periodic testing with very low limits on possible violations.

[264] Since the Megaverse is 16-dimensional, the spin of fermionic universe particles will be shown to be 127/2.

Asynchronous Logic of (complex-valued) 4-dimensional universes resulting from a 4-valued logic.[265] *We will assume universes are 4-dimensional constructs applying Leibniz's Minimax Principle.*

We will first consider the possibility of fermionic universes, and then briefly consider "spin 0" *bosonic* universes.

The first issue of fermionic universes (reminiscent of the discussions of spin in the 1920's) is the interpretation of spin states. We suggest that the upper (128) components (64 "spin up" and 64 "spin down") of a universe wave function represent a left-handed universe with an excess number of baryons. The lower (128) components lead to right-handed anti-universes where there is an excess of anti-baryons. These associations are analogous to the interpretations of the Dirac electron wave function.[266]

The universe particle "spin up" and "spin down" states are distinguished by their interactions with gauge fields in a manner analogous to quantum electrodynamics.

24.3 "Free Field" Dynamics of Fermionic Universe Particles

We now view universes as extended particles with an odd half integer spin – *fermionic universe particles* - in the 16-dimensional Megaverse. In the Megaverse there are sixteen 256×256 matrices that are the equivalent of the four Dirac matrices in four dimensions. We will denote these sixteen matrices as γ^i for i = 1, 2, … , 16. They satisfy the anti-commutation relations:

$$\{\gamma^i, \gamma^j\} = 2\,\delta^{ij} \tag{24.1}$$

and thus form a Clifford algebra. We will choose y^{16} to be the time coordinate and thus make it imaginary with a Reality group transformation. (The 16-dimensional Megaverse space is complex.) Therefore γ^{16} will be hermitean ($(\gamma^{16})^2 = 1$) and γ^i for i = 1, … , 15 will be anti-hermitean with $(\gamma^i)^2 = -1$.
The number of linearly independent matrices in 16 dimensions is $2^{16} = 65{,}536$.

The Megaverse metric is (by use of the Reality group) chosen to be

$$g^{ij} = -\delta^{ij}, \qquad g^{16,16} = 1 \tag{24.2}$$

for i, j = 1, 2, … , 15; and zero otherwise.

Except for the additional dimensions, fermion dynamics is quite similar to the 4-dimensional case. The free universe particle Dirac equation is

$$(i\gamma^i\partial_i + m)\psi(y) = 0 \tag{24.3}$$

summed over i = 1, 2, … , 16 where the mass is temporarily assumed to be constant, and set by eq. 24.119 below. The derivative operator, is based on the use of quantum coordinates[267]

$$Y^i(y) = y^i + i\,Y_u^{\,i}(y)/M_u^{\,2} \tag{24.4}$$

and is defined to be

$$\partial_i = \partial/\partial Y^i(y) = \partial/\partial(y_i - Y_{ui}(y)/M_u^{\,2}) \tag{24.5}$$

---

[265] Larger dimensioned universes are possible in the Megaverse. We will not consider these cases here.

[266] It is known that phenomena in our universe tend to be left-handed. If this feature of our universe's phenomena is also a property of the universe itself, then, since handedness is an attribute of spin, the treatment of a universe as having spin is not unreasonable.

[267] Giving Two Tier renormalization. See chapter 18.

where *we assume* $M_u = M_c$ *with* M_c *being a very large mass scale of perhaps the order of the Planck mass.*

$Y_u^{\,i}$ is a 16-dimensional Megaverse gauge field equivalent of the $Y^\mu(x)$ used in Two-Tier renormalization (discussed in chapter 18):

$$Y^\mu(z) = z^\mu + i\, Y^\mu(z)/M_u^{\,2}$$

where $Y^\mu(z)$ is a free QED-like field. The $Y^i(y)$ quantum coordinates will be used in the Megaverse to eliminate potential divergences, in a manner similar to the case of our universe as outlined in chapter 18 when universe particle interactions are introduced later.

24.3.1 Four Types of Fermionic Universe Particles

There are four possible types of fermionic universe particles in the Megaverse that are analogous to the four species of fermion described in chapter 15 (and Blaha (2010b)) for The Extended Standard Model. Two of these types are tachyonic. It is important to note that DeWitt points out that the Wheeler-DeWitt equation has tachyonic solutions since the mass-like term dependent on ${}^{(3)}R$ can be positive or negative.[268] A negative mass is an indication of tachyonic behavior wherein the wave propagation of the state functional is not necessarily in time-like directions and is thus tachyonic.

Eq. 24.3 is a Dirac-type 16-dimensional Dirac equation. There are three other general types of universe particle equations. (Fermionic universes come in four species like fermions.) The derivation of the four types of universe particles is similar to the derivation of fermion types in the Extended Standard Model in 4-dimensional complex space-time given in Blaha (2010b). We will now consider the 16-dimensional equivalent for universe particles in the Megaverse.

The general form of a pure 16-dimensional complex Lorentz group[269] boost can be expressed in terms of a complex relative 15-velocity $\mathbf{v}_c$ between inertial reference frames. A 16-dimensional coordinate boost has the form

$$\Lambda_C(\mathbf{v}_c) \equiv \Lambda_C(\omega, \mathbf{v}_c) = \exp[i\omega\hat{\mathbf{w}}\cdot\mathbf{K}] \tag{24.6}$$

where

$$\omega = (\omega_r^{\,2} - \omega_i^{\,2} + 2i\omega_r\omega_i\, \hat{\mathbf{u}}_r\cdot\hat{\mathbf{u}}_i)^{1/2} \tag{24.7}$$

and

$$\hat{\mathbf{w}} = (\omega_r\hat{\mathbf{u}}_r + i\omega_i\hat{\mathbf{u}}_i)/\omega \tag{24.8}$$

with all vectors being 15-dimensional spatial vectors. We define the real and imaginary unit vectors $\hat{\mathbf{u}}_r\cdot\hat{\mathbf{u}}_r = 1 = \hat{\mathbf{u}}_i\cdot\hat{\mathbf{u}}_i$ with the result

$$\hat{\mathbf{w}}\cdot\hat{\mathbf{w}} = 1 \tag{24.9}$$

The complex relative velocity is

$$\mathbf{v}_c = \hat{\mathbf{w}}\tanh(\omega) \tag{24.10}$$

---

[268] DeWitt, B. S., Phys. Rev. **160**, 1113 (1967) p. 1124.

[269] The 16-dimensional complex Lorentz group has similar features to the 4-dimensional complex Lorentz group. We shall only discuss it to the extent needed for our universe particle type's derivation. See Weinberg (1995) for the 4-dimensional Lorentz group – the 16-dimensional Lorentz group generalizes directly from the features of the 4-dimensional Lorentz group.

The free dynamical equations of the four universe particle species will be generated by 16-dimensional Lorentz boosts of the free Dirac equation of a universe particle at rest with the *requirement that the time variable* ($t = y^{16}$) *and energy are real in the resulting field equations.*[270] The procedure can most easily be performed in 16-dimensional momentum space with the Megaverse coordinate space version of the generated equation determined from the momentum space version.

24.3.1.1 Dirac-like Equation – Type I universe Particle

A positive energy plane wave solution of the Dirac equation eq. 24.3 for a universe particle at rest is

$$\psi(y) = \exp[-imt]w(0) \tag{24.11}$$

where we set $\partial_t = \partial/\partial y_{16}$ while temporarily ignoring the $Y_u^i(y)/M_u^2$ term. w(0) is a 256 component spinor column vector. The solution $\psi(y)$ satisfies the momentum space Dirac equation for a particle at rest:

$$(m\gamma^{16} - m)\psi(y) = 0 \tag{24.12}$$

The 256 x 256 spinor matrix form of a 16-dimensional Lorentz boost with relative real velocity **v** of the Dirac matrices is[271]

$$S^{-1}(\Lambda(\mathbf{v}))\gamma^{\nu}S(\Lambda(\mathbf{v})) = \Lambda^{\nu}{}_{\mu}(\mathbf{v})\gamma^{\mu} \tag{24.13}$$

where $\Lambda^{\nu}{}_{\mu}(\mathbf{v})$ is a 16-dimensional Lorentz boost. $S(\Lambda(\mathbf{v}))$ has the form

$$\begin{aligned} S(\Lambda(\mathbf{v})) &= \exp(-\omega\gamma^{16}\boldsymbol{\gamma}\cdot\mathbf{v}/(2|\mathbf{v}|)) \\ &= \cosh(\omega/2)I + \sinh(\omega/2)\gamma^{16}\boldsymbol{\gamma}\cdot\mathbf{p}/|\mathbf{p}| \end{aligned} \tag{24.14}$$

with *real* $\omega = \text{arctanh}(|\mathbf{v}|)$ and *real* **v**. $|\mathbf{p}|$ is the magnitude of the spatial 15-vector. Also

$$\begin{aligned} S^{-1}(\Lambda(\mathbf{v})) &= \gamma^{16}S^{\dagger}(\Lambda(\mathbf{v}))\gamma^{16} = \exp(\omega\gamma^{16}\boldsymbol{\gamma}\cdot\mathbf{v}/(2|\mathbf{v}|)) \\ &= \cosh(\omega/2)I - \sinh(\omega/2)\gamma^{16}\boldsymbol{\gamma}\cdot\mathbf{p}/|\mathbf{p}| \end{aligned} \tag{24.15}$$

If we now apply $S(\Lambda(\mathbf{v}))$ to the momentum space Dirac equation of a particle at rest (eq. 24.12) we find

$$\begin{aligned} 0 &= S(\Lambda(\mathbf{v}))(m\gamma^{16} - m)\,\psi(y) \\ &= [mS(\Lambda(\mathbf{v}))\gamma^{16}S^{-1}(\Lambda(\mathbf{v})) - m]S(\Lambda(\mathbf{v}))w(0) \end{aligned}$$

A straightforward evaluation shows

$$mS(\Lambda(v))\gamma^{16}S^{-1}(\Lambda(v)) = g_{16\mu\nu}p^{\mu}\gamma^{\nu} = \not{p} \tag{24.16}$$

---

[270] The 16-dimensional "energy" must be real since it relates to the area of the universe – a real number.

[271] **The indices ν and μ from this point in this chapter have values: 1, 2, … , 16.**

where p is a momentum 16-vector. In addition we define the 16-dimension spinor (256 components)

$$S(\Lambda(v))w(0) = w(p) \tag{24.17}$$

which can be viewed as a "positive energy Dirac spinor". The Dirac equation in momentum space has the familiar form:

$$(\not{p} - m)\exp[-ip\cdot y]w(p) = 0 \tag{24.18}$$

Eq. 24.18 implies the free, coordinate space Dirac equation:

$$(i\gamma^{\mu}\partial/\partial y^{\mu} - m)\psi(y) = 0 \tag{24.19}$$

We identify this equation as the dynamical equation of a type 1 universe particle. It corresponds to the free charged lepton elementary particle species Dirac equation in particle physics.

24.3.1.2 Complex Boosts

The form of the 16-dimensional spinor boost transformation corresponding to the coordinate transformation eq. 24.6 is:

$$\begin{aligned} S_C(\omega, \mathbf{v}_c) \equiv S_C &= \exp(-\omega\gamma^{16}\boldsymbol{\gamma}\cdot\hat{\mathbf{w}}/2) \\ &= \cosh(\omega/2)I + \sinh(\omega/2)\gamma^{16}\boldsymbol{\gamma}\cdot\hat{\mathbf{w}} \end{aligned} \tag{24.20}$$

with *complex* $\mathbf{v}_c$ and $\hat{\mathbf{w}}$ defined by eqs. 24.10 and 24.8 respectively. The inverse transformation is

$$\begin{aligned} S_C^{-1}(\omega, \mathbf{v}_c) &= \exp(\omega\gamma^{16}\boldsymbol{\gamma}\cdot\hat{\mathbf{w}}/2) \\ &= \cosh(\omega/2)I - \sinh(\omega/2)\gamma^{16}\boldsymbol{\gamma}\cdot\hat{\mathbf{w}} \end{aligned} \tag{24.21}$$

Note that S_C is not unitary just as in the 4-dimensional case.

We now apply a spinor boost to the Dirac equation for a particle at rest in this more general case of complex ω and $\hat{\mathbf{w}}$.

$$\begin{aligned} 0 &= S_C(\omega, \mathbf{v}_c))(m\gamma^{16} - m)\exp[-imt]w(0) \\ &= [mS_C\gamma^{16}S_C^{-1} - m]\exp[-imt]S_C w(0) \end{aligned} \tag{24.22}$$

where $S_C = S_C(\omega, \mathbf{v}_c)$. After some algebra we find

$$mS_C\gamma^{16}S_C^{-1} = m[\cosh(\omega)\gamma^{16} - \sinh(\omega)\boldsymbol{\gamma}\cdot\hat{\mathbf{w}}] \tag{24.23}$$

We will use these *complex* boosts to generate the other species' Dirac-like equations.

24.3.1.3 Tachyon Universe particle Dirac Equation

The development of the complex spinor boost transformation (subsection 24.3.1.2 above) leads to two possible forms of the tachyon Dirac-like equation. One form will lead to a lagrangian dynamics for left-handed universe particles. The other form leads to a lagrangian dynamics for right-handed universe particles.

24.3.1.4 Type IIa Case: Left-Handed Tachyonic Universe Particles

If the real and imaginary relative vectors parts of $\hat{\mathbf{w}}$, namely $\hat{\mathbf{u}}_r$ and $\hat{\mathbf{u}}_i$, are parallel, then $\hat{\mathbf{u}}_r \cdot \hat{\mathbf{u}}_i = 1$ and

$$\omega = \omega_r + i\omega_i \tag{24.24}$$

Eqs. 24.23 and 24.24 then imply

$$\begin{aligned} mS_C\gamma^{16}S_C^{-1} = m[\cosh(\omega_r)\cos(\omega_i) + i\sinh(\omega_r)\sin(\omega_i)]\gamma^{16} - \\ - m[\sinh(\omega_r)\cos(\omega_i) + i\cosh(\omega_r)\sin(\omega_i)]\boldsymbol{\gamma}\cdot\hat{\mathbf{u}}_r \end{aligned} \tag{24.25}$$

or

$$mS_C\gamma^{16}S_C^{-1} = \cos(\omega_i)\gamma\cdot p_r + i\sin(\omega_i)\gamma\cdot p_i \tag{24.26}$$

where

$$p_r^{\ 0} = m\cosh(\omega_r) \qquad p_i^{\ 0} = m\sinh(\omega_r) \tag{24.27}$$

and

$$\mathbf{p}_r = m\hat{\mathbf{u}}_r\sinh(\omega_r) \quad \mathbf{p}_i = m\hat{\mathbf{u}}_r\cosh(\omega_r) \tag{24.28}$$

If $\omega_i = 0$, then we recover the momentum space Dirac-like equation. If $\omega_i = \pi/2$, then we obtain the left-handed momentum space tachyon equation:

$$mS_C\gamma^{16}S_C^{-1} = i\gamma\cdot p_i \tag{24.29}$$

and the tachyon energy and momentum expressions

$$\mathbf{p} = m\mathbf{v}\gamma_s \qquad\qquad E = m\gamma_s \tag{24.30}$$

where $\sinh(\omega) = \gamma_s = (\beta^2 - 1)^{-1/2}$ with $\beta = v/c > 1$. v is the absolute value of the 15 component spatial velocity. Also

$$S_C w(0) = w_C(p) \tag{24.31}$$

is a tachyon spinor.

The momentum space tachyonic Dirac-like equation is

$$(i\not{p} - m)\exp[-ip\cdot y]w_T(p) = 0 \tag{24.32}$$

where $p\cdot y = p^{16}y^{16} - \mathbf{p}\cdot\mathbf{y}$ after performing a corresponding boost in the exponential factor. If we apply $i\not{p}$ to eq. 24.32 we find the tachyon mass condition is satisfied

$$-E^2 + \mathbf{p}^2 = m^2 \tag{24.33}$$

Transforming back to coordinate space we obtain the "left-handed" *tachyonic Dirac-like equation*:

$$(\gamma^\mu \partial/\partial y^\mu - m)\psi_T(y) = 0 \tag{24.34}$$

24.3.1.5 Type IIb Case: Right-Handed Tachyonic Universe Particles

If the real and imaginary relative vectors parts of $\hat{\mathbf{w}}$, $\hat{\mathbf{u}}_r$ and $\hat{\mathbf{u}}_i$, are anti-parallel $\hat{\mathbf{u}}_r = -\hat{\mathbf{u}}_i$, then $\hat{\mathbf{u}}_r \cdot \hat{\mathbf{u}}_i = -1$ and

$$\omega = \omega_r - i\omega_i \tag{24.35}$$

then

$$mS_C\gamma^{16}S_C^{-1} = m[\cosh(\omega_r)\cos(\omega_i) - i\sinh(\omega_r)\sin(\omega_i)]\gamma^{16} - \\ - m[\sinh(\omega_r)\cos(\omega_i) - i\cosh(\omega_r)\sin(\omega_i)]\boldsymbol{\gamma}\cdot\hat{\mathbf{u}}_r \tag{24.36}$$

or

$$mS_C\gamma^{16}S_C^{-1} = \cos(\omega_i)\gamma\cdot p_r - i\sin(\omega_i)\gamma\cdot p_i \tag{24.37}$$

where

$$p_r^{16} = m\cosh(\omega_r) \qquad p_i^{16} = m\sinh(\omega_r) \tag{24.38}$$

and

$$\mathbf{p}_r = m\hat{\mathbf{u}}_r\sinh(\omega_r) \qquad \mathbf{p}_i = m\hat{\mathbf{u}}_r\cosh(\omega_r) \tag{24.39}$$

If $\omega_i = \pi/2$, then we obtain the right-handed momentum space tachyon equation.[272]

$$(-\gamma^\mu \partial/\partial y^\mu - m)\psi_T(y) = 0 \tag{24.40}$$

24.3.1.6 Type III Case: "Up-Quark-like" Universe Particles

There are two other cases where we can obtain fermion dynamical equations with a *real* time variable and real energy. In one case we set $\hat{\mathbf{u}}_r \cdot \hat{\mathbf{u}}_i = 0$ and have a real ω.

If the real and imaginary relative vectors parts of $\hat{\mathbf{w}}$, namely $\hat{\mathbf{u}}_r$ and $\hat{\mathbf{u}}_i$, are perpendicular, $\hat{\mathbf{u}}_r \cdot \hat{\mathbf{u}}_i = 0$, then

$$\omega = (\omega_r^2 - \omega_i^2)^{1/2} \tag{24.41}$$

Thus ω is either pure real ($\omega_r \geq \omega_i$) or pure imaginary ($\omega_r < \omega_i$).

The momentum space equation generated by the corresponding spinor boost is

$$\{m\cosh(\omega)\gamma^{16} - m\sinh(\omega)\boldsymbol{\gamma}\cdot(\omega_r\hat{\mathbf{u}}_r + i\omega_i\hat{\mathbf{u}}_i)/\omega - m\}\exp[-imt]w_c(p) = 0 \tag{24.42}$$

Defining the momentum 4-vector

$$p = (p^{16}, \mathbf{p}) \tag{24.43}$$

where

$$p^{16} = m\cosh(\omega) \qquad \mathbf{p} = \mathbf{p}_r + i\mathbf{p}_i \tag{24.44}$$

---

[272] We note that $\gamma_s = (\beta^2 - 1)^{-1/2}$, *if expressed in terms of* ω, *has a branch cut extending from* <-∞, +∞> *in the complex* ω *plane. Thus values of* ω *with positive imaginary parts are physically different from values of* ω *with negative imaginary parts.*

with

$$\mathbf{p}_r = m\omega_r\hat{\mathbf{u}}_r \sinh(\omega)/\omega \quad \mathbf{p}_i = m\omega_i\hat{\mathbf{u}}_i \sinh(\omega)/\omega \tag{24.45}$$

$$\mathbf{p}_r\cdot\mathbf{p}_i = 0 \tag{24.46}$$

then we obtain a positive energy Dirac-like equation

$$[p\cdot\gamma - m]\exp[-imt]w_c(p) = 0$$

or

$$[p^{16}\gamma^{16} - (\mathbf{p}_r + i\mathbf{p}_i)\cdot\boldsymbol{\gamma} - m]\exp[-ip\cdot y]w_c(p) = 0 \tag{24.47}$$

with a complex 3-momentum **p** and the 4-momentum mass shell condition:

$$p^2 = p^{16\,2} - \mathbf{p}_r\cdot\mathbf{p}_r + \mathbf{p}_i\cdot\mathbf{p}_i = m^2 \tag{24.48}$$

Note

$$|\mathbf{v}_c| = |\mathbf{p}|/p^{16} = [(\mathbf{p}_r + i\mathbf{p}_i)\cdot(\mathbf{p}_r + i\mathbf{p}_i)]^{1/2}/p^{16} = \tanh(\omega) \tag{24.49}$$

and so the Lorentz factor is

$$\gamma = \cosh(\omega) \tag{24.50}$$

Eq. 24.47 is the momentum space equivalent of the wave equation[273]

$$[i\gamma^{16}\partial/\partial t + i\boldsymbol{\gamma}\cdot(\nabla_r + i\nabla_i) - m]\psi_u(t, \mathbf{y}_r, \mathbf{y}_i) = 0 \tag{24.51}$$

where $\mathbf{y} = \mathbf{y}_r - i\mathbf{y}_i$, and where the grad operators ∇_r and ∇_i are with respect to $\mathbf{y}_r$ and $\mathbf{y}_i$ respectively. Since $\hat{\mathbf{u}}_r\cdot\hat{\mathbf{u}}_i = 0$ we see that there is a subsidiary condition on the wave function

$$\nabla_r\cdot\nabla_i\, \psi_u(t, \mathbf{y}_r, \mathbf{y}_i) = 0 \tag{24.52}$$

We note eq. 24.52 can be put into covariant form as the difference of two vectors squared (which is a real 16-dimensional Lorentz group invariant):

$$[\gamma^{16}\partial/\partial t + i\boldsymbol{\gamma}\cdot(\nabla_r + i\nabla_i)]^2 - [\gamma^{16}\partial/\partial t + i\boldsymbol{\gamma}\cdot(\nabla_r - i\nabla_i)]^2 = 4\nabla_r\cdot\nabla_i.$$

We identify eq. 24.51 as the dynamical equation of an "up-quark-like" universe particle.

24.3.1.7 Type IVa Case: Left-Handed "Down-Quark-like" Tachyonic Universe Particles

In this case we set $\hat{\mathbf{u}}_r\cdot\hat{\mathbf{u}}_i = 0$. Then by eq. 24.7

$$\omega = (\omega_r^2 - \omega_i^2)^{1/2}$$

Thus ω again starts out either pure real (if $\omega_r \geq \omega_i$) or pure imaginary (if $\omega_r < \omega_i$). In this case we also choose ω real, and then change ω to

---

[273] The gradient operators ∇_r and ∇_i are 15-dimensional spatial gradient operators.

$$\omega = (\omega_r^2 - \omega_i^2)^{½} \rightarrow \omega' = (\omega_r^2 - \omega_i^2)^{½} + i\pi/2 = \omega + i\pi/2$$

by adding $i\pi/2$ to ω since ω is a free parameter. We then proceed as we did in the prior tachyon case.[274]. The resulting Lorentz boost

$$\Lambda_C = \exp[i((\omega_r^2 - \omega_i^2)^{½} + i\pi/2)(\omega_r\hat{\mathbf{u}}_r + i\omega_i\hat{\mathbf{u}}_i)\cdot\mathbf{K}/\omega] \qquad (24.53)$$

becomes a left-handed "quark-like" boost. The tachyon dynamical equation is[275]

$$[\gamma^{16}\partial/\partial t + \boldsymbol{\gamma}\cdot(\boldsymbol{\nabla}_r + i\boldsymbol{\nabla}_i) - m]\psi_d(y) = 0 \qquad (24.54)$$

with the constraint equation

$$\boldsymbol{\nabla}_r\cdot\boldsymbol{\nabla}_i\,\psi_d(t, \mathbf{y}_r, \mathbf{y}_i) = 0 \qquad (24.55)$$

We will call the universe particles satisfying eqs. 24.54 and 24.55 left-handed *tachyonic quark-like universe particles.*

24.3.1.8 Type IVb Case: Right-Handed Down-Quark-like Tachyonic Universe Particles

In this case we set $\hat{\mathbf{u}}_r\cdot\hat{\mathbf{u}}_i = 0$. Then by eq. 24.7

$$\omega = (\omega_r^2 - \omega_i^2)^{½}$$

Thus ω again starts out either pure real (if $\omega_r \geq \omega_i$) or pure imaginary (if $\omega_r < \omega_i$). In this case we also choose ω real, and then change ω to

$$\omega = (\omega_r^2 - \omega_i^2)^{½} \rightarrow \omega' = (\omega_r^2 - \omega_i^2)^{½} - i\pi/2 = \omega - i\pi/2$$

since ω is a free parameter and proceed as we did in the prior case. The resulting Lorentz boost

$$\Lambda_C = \exp[i((\omega_r^2 - \omega_i^2)^{½} - i\pi/2)(\omega_r\hat{\mathbf{u}}_r + i\omega_i\hat{\mathbf{u}}_i)\cdot\mathbf{K}/\omega] \qquad (24.56)$$

becomes a right-handed quark-like boost. The resulting tachyon dynamical equation is

$$[-\gamma^{16}\partial/\partial t - \boldsymbol{\gamma}\cdot(\boldsymbol{\nabla}_r + i\boldsymbol{\nabla}_i) - m]\psi_d(y) = 0 \qquad (24.57)$$

with the constraint equation

$$\boldsymbol{\nabla}_r\cdot\boldsymbol{\nabla}_i\,\psi_d(t, \mathbf{y}_r, \mathbf{y}_i) = 0 \qquad (24.58)$$

---

[274] Here again the choice of ω in eq. 24.53 leads to a "left-handed" universe particle while the choice $\omega' = \omega - i\pi/2$ leads to a right-handed one.

[275] The gradient operators $\boldsymbol{\nabla}_r$ and $\boldsymbol{\nabla}_i$ are 15-dimensional spatial gradient operators.

We will call the universe particles satisfying eqs. 24.57 and 24.58 right-handed *tachyonic quark-like universe particles.*

24.3.2 Lagrangians

In this section we will develop a lagrangian formalism for each of the four types of universe particles noting that a tachyonic universe particles have two forms: left-handed and right-handed (discussed later in section 24.3.5).

The various types of universe particles described in section 24.3.1 correspond to universes with differing internal characteristics and motion in the Megaverse. The equations are all free field equations. Internal potentials and interactions must be introduced in these equations to complete the universe dynamical equations. A connection to the Wheeler-DeWitt description of their internal quantum structure also remains to be established (section 24.3.6).

In defining the lagrangians for the four universe types that yield their dynamical equations in a canonical manner, we require the conventional quantum field theory feature that the hamiltonian derived from the lagrangian is hermitean. We will develop a separate lagrangian for each type.

24.3.2.1 Type I Universe Particle Lagrangian

The Universe particle Dirac equation lagrangian is

$$\mathcal{L}_u = \bar{\psi}(i\gamma^\mu \partial/\partial y^\mu - m)\psi(y) \tag{24.59}$$

where

$$\bar{\psi} = \psi^\dagger \gamma^{16}$$

and $\psi^\dagger$ is the hermitean conjugate of ψ.

24.3.2.2 Type II Tachyon Universe Particle Lagrangian

This lagrangian includes both left-handed and right-handed cases. It can be separated into lagrangian terms for each case using parity projection operators.

$$\mathcal{L}_{uT} = \psi_T^{\,S}(\gamma^\mu \partial/\partial y^\mu - m)\psi_T(y) \tag{24.60}$$

where

$$\psi_T^{\,S} = \psi_T^{\,\dagger}\, i\gamma^{16}\gamma^5 \tag{24.61}$$

with γ^5 being the 16-dimensional equivalent for γ^5 in 4 dimensions. The peculiar form of the tachyon universe lagrangian is necessitated by the hermiticity of the hamiltonian calculated from it.

24.3.2.3 Type III "Up-Quark-like" Universe Particle Lagrangian

The lagrangian density of a free "up-quark-like" universe particle is

$$\mathcal{L}_u = \bar{\psi}_u(i\gamma^\mu D_\mu - m)\psi_u(y) \tag{24.62}$$

where $\bar{\psi}_u = \psi_u^{\,\dagger}\gamma^{16}$ and

$$\psi_u^{\dagger} = [\psi_u(\mathbf{y}_r, \mathbf{y}_i)]^{\dagger}\,|_{\mathbf{y}_i = -\mathbf{y}_i} \tag{24.63}$$

$$D_{16} = \partial/\partial y^{16}$$
$$D_k = \partial/\partial y_r^{k} + i\, \partial/\partial y_i^{k} \tag{24.64}$$

for k = 1, 2, … , 15. The action

$$I = \int d^{15}y \mathcal{L}_u \tag{24.65}$$

It is easy to show that this action is also real.

24.3.2.4 Type IV "Down-Quark-like" Tachyon Universe Particle Lagrangian

The lagrangian density of a free "down-quark-like" universe particle is

$$\mathcal{L}_d = \psi_d^{C}(y)(\gamma^{16}\partial/\partial t + \boldsymbol{\gamma}\cdot(\boldsymbol{\nabla}_r + i\boldsymbol{\nabla}_i) - m)\psi_d(y) \tag{24.66}$$

where

$$\psi_d^{C}(y) = [\psi_d(y)]^{\dagger}|_{\mathbf{y}_i = -\mathbf{y}_i}\, i\gamma^{16}\gamma^{5} \tag{24.67}$$

In words, eq. 24.67 states: take the hermitean conjugate of $\psi_d(y)$; change $\mathbf{y}_i$ to $-\mathbf{y}_i$; and then post-multiply by the indicated factors.

The action is

$$I = \int d^{15}y \mathcal{L}_d \tag{24.68}$$

The action is real. The lagrangian can also be separated into left-handed and right-handed parts using projection operators.

24.3.3 Form of The Megaverse Quantum Coordinates Gauge Field

The discussions of sections 24.3.1 and 24.3.2 assumed the coordinates were Megaverse coordinates and their derivatives. Prior to those discussions we indicated we would use quantum coordinates in the Megaverse of the form[276]

$$Y^{i}(y) = y^{i} + i\, Y_u^{i}(y)/M_u^{8} \tag{24.4}$$

and their derivatives

$$\partial_i = \partial/\partial Y^{i}(y) = \partial/\partial(y^{i} - Y_u^{i}(y)/M_u^{8}) \tag{24.5}$$

for i = 1, 2, … , 16 to eliminate divergences in quantum field theory. The subscript "u" signifies universes. The mass constant for the Megaverse M_u may be the same as the mass constant M_c appearing

[276] The denominator M_u^{8} is necessitated by the dimension of $Y_u^{i}(y)$ which is $[m]^{7}$. Eqs. 24.78 and 24.81 below imply this conclusion.

in the Two Tier mechanism for our universe. (See chapter 18 for a discussion of eliminating infinities with this mechanism.)

In this section we define the gauge fields $Y_u^i(y)$ of the Megaverse.[277] They are similar to the $Y^\mu(y)$ fields of our New Standard Model.[278] The $Y_u(y)$ 16-dimensional vector gauge field, in the absence of external sources, will be defined in a 16-dimensional Coulomb gauge:

$$Y_u^{16}(y) = 0 \qquad (24.69)$$
$$\partial Y_u^j(y)/\partial y^j = 0$$

where the sum over j is over the 15 spatial y coordinates. We follow a procedure similar to Blaha (2003) but for 16-dimensional space. The lagrangian density for the free $Y_u^j(y)$ fields is

$$\mathscr{L}_u = -\tfrac{1}{4} F_u^{\mu\nu} F_{u\mu\nu} \qquad (24.70)$$

and the lagrangian is

$$L_u = \int d^{15}y \mathscr{L}_u \qquad (24.71a)$$

with

$$F_{u\mu\nu} = \partial Y_{u\mu}/\partial y^\nu - \partial Y_{u\nu}/\partial y^\mu \qquad (24.71b)$$

The equal time commutation relations, derived in the usual way, are:

$$[Y_u^\mu(\mathbf{y}, y^0), Y_u^\nu(\mathbf{y}', y^0)] = [\pi_u^\mu(\mathbf{y}, y^0), \pi_u^\nu(\mathbf{y}', y^0)] = 0 \qquad (24.72)$$
$$[\pi_u^j(\mathbf{y}, y^0), Y_{uk}(\mathbf{y}', y^0)] = -i\,\delta^{15tr}{}_{jk}(\mathbf{y} - \mathbf{y}') \qquad (24.73)$$

for μ, ν, j, k = 1, 2, … , 15 where

$$\pi_u^k = \partial \mathscr{L}_u / \partial Y_{uk}' \qquad (24.74)$$
$$\pi_u^0 = 0 \qquad (24.75)$$

and

$$\delta^{tr}{}_{jk}(\mathbf{y} - \mathbf{y}') = \int d^{15}k\, e^{i\,\mathbf{k}\bullet(\mathbf{y}-\mathbf{y}')} (\delta_{jk} - k_j k_k/\mathbf{k}^2)/(2\pi)^{15} \qquad (24.76)$$

$$Y_{uk}' = \partial Y_{uk}/\partial y^{16} \qquad (24.77)$$

The Coulomb gauge indicates fourteen degrees of freedom are present in the vector potential. The Fourier expansion of the vector potential is:

$$Y_u^i(y) = \int d^{15}k\, N_0(k) \sum_{\lambda=1}^{14} \varepsilon^i(k, \lambda)[a(k,\lambda)\, e^{-ik\cdot y} + a^\dagger(k,\lambda)\, e^{ik\cdot y}] \qquad (24.78)$$

where

$$N_0(k) = [(2\pi)^{15} 2\omega_k]^{-1/2} \qquad (24.79)$$

---

277 This choice implies that the Megaverse Y mass $M_u = M_C$, its universe mass.
278 See Blaha (2005a) for details.

and (since the field is massless)

$$k^{16} = \omega_k = (\mathbf{k}^2)^{1/2} \tag{24.80}$$

where k^{16} is the energy, and where the $\varepsilon^i(k, \lambda)$ are the polarization unit vectors for $\lambda = 1, \ldots, 14$ and $k^\mu k_\mu = k^{16\,2} - \mathbf{k}^2 = 0$.

The commutation relations of the Fourier coefficient operators are:

$$[a(k,\lambda), a^\dagger(k',\lambda')] = \delta_{\lambda\lambda'}\delta^{15}(\mathbf{k} - \mathbf{k'}) \tag{24.81}$$
$$[a^\dagger(k,\lambda), a^\dagger(k',\lambda')] = [a(k,\lambda), a(k',\lambda')] = 0 \tag{24.82}$$

and the polarization vectors satisfy

$$\sum_{\lambda=1}^{14} \varepsilon_i(k, \lambda)\varepsilon_j(k, \lambda) = (\delta_{ij} - k_i k_j/\mathbf{k}^2) \tag{24.83}$$

It will be convenient to divide the Y field into positive and negative frequency parts:

$$Y_u{}^+{}_i(y) = \int d^{15}k\, N_0(k) \sum_{\lambda=1}^{14} \varepsilon_i(k, \lambda)\, a(k,\lambda)\, e^{-ik\cdot y} \tag{24.84}$$

and

$$Y_u{}^-{}_i(y) = \int d^{15}k\, N_0(k) \sum_{\lambda=1}^{14} \varepsilon_i(k, \lambda)\, a^\dagger(k,\lambda)\, e^{ik\cdot y} \tag{24.85}$$

For later use we note the commutator between the positive and negative frequency parts is:

$$[\, Y_u{}^-{}_j(y_1), Y_u{}^+{}_k(y_2)] = -\int d^{15}k\, e^{ik\cdot(y_1 - y_2)}\, (\delta_{jk} - k_j k_k/\mathbf{k}^2)/[(2\pi)^{15} 2\omega_k] \tag{24.86}$$

24.3.3.1 Y^μ Fock Space Imaginary Coordinate States

States can also be defines for the quantized Y^μ field. These states will be similar in form to electromagnetic photon states but play a different role in our approach since they are in fact coordinate excitation states for the imaginary part of $Y^i(y)$ (eq. 24.4). Thus universe particles (and other fields) will exist in a real 16-dimensional space with quantum excitations into imaginary Quantum Dimensions. These excitations become significant at high energies. At the low energies space appears as c-number complex; at very high energies space becomes slightly q-number complex.

There are two types of imaginary coordinate excitations: 1.) Quantum excitations into Fock states consisting of a superposition of states with a definite finite number of Y_u "particles" and 2.) Imaginary coordinate excitations into coherent Y_u states with an "infinite" number of particles. Coherent states can be viewed as representing "classical" fields.

In this section we will consider Y_u field states with a definite number of excitations ("particles"). The raising and lowering operators of the Y_u field can be used to define free particle states. For example a one particle state can be defined by

$$|k, \lambda\rangle = a^\dagger(k, \lambda)|0\rangle \tag{24.87}$$

with corresponding bra state

$$<k, \lambda| = <0|a(k, \lambda) \tag{24.88}$$

where the "coordinate vacuum" is defined as usual:

$$a(k, \lambda)|0> = 0 \tag{24.89}$$

$$<0|a^\dagger(k, \lambda) = 0 \tag{24.90}$$

Multi-particle states can also be defined in the conventional way with products of the raising and lowering operators applied to the vacuum. The set of all states containing a finite number of "particles" constitutes a Fock space.

A state with a finite number of Y_u "particles" represents a quantum fluctuation into imaginary Quantum Dimensions.

24.3.3.2 Y_u Coherent Imaginary Coordinate States

Coherent Y_u states bring us closer what we might consider to be "classical" imaginary dimensions – dimensions that we can, in principle, experience as we do normal dimensions. Let us define the coherent state[279]

$$| y, p> = e^{-\mathbf{p}\cdot Yu^-(y)/M_u^8}|0> \tag{24.91}$$

This state is an eigenstate of the coordinate operator $Y_u^+(y')$:

$$Y_u{}^+{}_j(y_1)\ |y_2, p> = -[Y_u{}^+{}_j(y_1), \mathbf{p}\cdot\mathbf{Y}^-(y_2)]/M_u{}^8|y, p> \tag{24.92}$$

$$= -\int d^{15}k\ [N_0(k)]^2\ e^{ik\cdot(y_2 - y_1)}\ (p_j - k_j\mathbf{p}\cdot\mathbf{k}/\mathbf{k}^2)/M_u{}^8|y, p>$$

$$= p^i\Delta_{Tij}(y_1 - y_2)/M_u{}^8|y, p> \tag{24.93}$$

where $p^i\Delta_{Tij}(y_1 - y_2)/M_u^8$ is the eigenvalue of $Y_u{}^+{}_j(y_1)$. As we will see later, the eigenvalue of Y_u^+ becomes large as $(y_1 - y_2)^2 \rightarrow 0$. Thus the imaginary Quantum Dimensions become significant at very short distances, and significantly modify the high-energy behavior of quantum field theories. In particular, Quantum Dimensions have a significant effect when

$$(y_1 - y_2)^2 \lessapprox (2^{14}\pi^{14}M_u{}^2)^{-1} \tag{24.94}$$

We assume the mass scale $M_u = M_C$ is very large – perhaps of the order of the Planck mass (1.221×10^{19} GeV/c^2).

---

279 Coherent states are well known in the physics literature. See for example T. W. B. Kibble, J. Math. Phys. **9**, 315 (1968) and references therein; V. Chung, Phys. Rev. **140**, B1110 (1965); J. R. Klauder, J. McKenna, and E. J. Woods, J. Math. Phys. 7, 822 (1966) and references therein.

24.3.3.3 Quantization of the Type I Free Universe Particle Dirac Field

The quantization procedure is formally identical to that of a conventional Dirac particle. The standard equal time anti-commutation relations for a 16-dimensional fermion field are:

$$\{\psi_\alpha(\mathrm{Y}), \psi_\beta(\mathrm{Y}')\} = \{\pi_{\psi\alpha}(\mathrm{Y}), \pi_{\psi\beta}(\mathrm{Y}')\} = 0 \qquad (24.95)$$
$$\{\pi_{\psi\alpha}(\mathrm{Y}), \psi_\beta(\mathrm{Y}')\} = \mathrm{i}\,\delta_{\alpha\beta}\,\delta^{15}(\mathbf{Y} - \mathbf{Y}') \qquad (24.96)$$

where α and β are the spinor indices ranging from 1 to 256 and where

$$\pi_{\psi\alpha}(\mathrm{Y}) = \mathrm{i}\,\psi_\alpha^\dagger(\mathrm{Y}) \qquad (24.97)$$

The field can be expanded in a fourier series:

$$\psi(\mathrm{Y(y)}) = \sum_s \int d^{15}p\, N^d_m(p)\, [b(p,s)u(p,s) :e^{-ip\cdot(y + iYu/M_u^8)}: + d^\dagger(p,s)v(p,s) :e^{ip\cdot(y + iYu/M_u^8)}:] \qquad (24.98)$$

$$\psi^\dagger(\mathrm{Y(y)}) = \sum_s \int d^{15}p\, N^d_m(p)\, [b^\dagger(p,s)\bar{u}(p,s)\gamma^0 :e^{+ip\cdot(y + iYu/M_u^8)}: + d(p,s)\bar{v}(p,s)\gamma^0 :e^{-ip\cdot(y + iYu/M_u^8)}:] \qquad (24.99)$$

where

$$N^d_m(p) = [m/((2\pi)^{15}E_p)]^{1/2} \qquad (24.100)$$

and

$$E_p = p^{16} = (\mathbf{p}^2 + m^2)^{1/2} \qquad (24.101)$$

The commutation relations of the Fourier coefficient operators are:

$$\{b(p,s), b^\dagger(p',s')\} = \delta_{ss'}\delta^{15}(\mathbf{p} - \mathbf{p}') \qquad (24.102)$$
$$\{d(p,s), d^\dagger(p',s')\} = \delta_{ss'}\delta^{15}(\mathbf{p} - \mathbf{p}') \qquad (24.103)$$
$$\{b(p,s), b(p',s')\} = \{d(p,s), d(p',s')\} = 0 \qquad (24.104)$$
$$\{b^\dagger(p,s), b^\dagger(p',s')\} = \{d^\dagger(p,s), d^\dagger(p',s')\} = 0 \qquad (24.105)$$
$$\{b(p,s), d^\dagger(p',s')\} = \{d(p,s), b^\dagger(p',s')\} = 0 \qquad (24.106)$$
$$\{b^\dagger(p,s), d^\dagger(p',s')\} = \{d(p,s), b(p',s')\} = 0 \qquad (24.107)$$

The spinors u(p,s) and v(p,s) are defined in a conventional way (as in Bjorken and Drell). However their form is different from the 4-dimensional case. If one takes the 256×256 $\gamma\cdot p$ matrix, then the first 128 columns give u(p,s) up to a normalization for the free particle case, the remaining 128 columns give v(p,s) up to a normalization.

Since there are 256 possible spin values, using the equation 2s + 1 = total number of spin values we see that the spin of a fermionic universe particle is s = 127/2. The possible universe particle spin values are:

Up spin values: +1/128, +2/128, … , +64/128
Down spin values: -64/128 = -½, -63/128, … , -1/128

24.3.3.4 Feynman Propagators for the Type I Free Universe Particle Dirac Field

The form of the fermionic universe particle Feynman propagator differs from a conventional fermion propagator by having a Gaussian factor R(**p**, z) in its fourier expansions. This follows from using quantum Megaverse coordinates (eq. 24.4).

$$iS_F^{TT}(y_1 - y_2) = <0|T(\bar{\psi}(Y(y_1))\psi(Y(y_2)))|0> \quad (24.108)$$

where the time ordering is with respect to y_1^{16} and y_2^{16}. Expanding the free fields leads to the fourier representation:

$$iS_F^{TT}(y_1 - y_2) = i\int \frac{d^{16}p\, e^{-ip\cdot(y_1 - y_2)}\,(\not{p} + m)\, R(\mathbf{p}, y_1 - y_2)}{(2\pi)^{16}\,(p^2 - m^2 + i\varepsilon)} \quad (24.109)$$

where

$$R(\mathbf{p}, y_1 - y_2) = \exp[-p^ip^j\Delta_{Tij}(y_1 - y_2)/M_u^{16}] \quad (24.110)$$
$$= \exp\{-p^2[A(v) + B(v)\cos^2\theta] / [(2\pi)^{14}M_c^4z^2]\} \quad (24.111)$$

(Note p^2 is the square of the spatial 15-vector.) with

$$z^\mu = y_1^\mu - y_2^\mu \quad (24.112)$$
$$z = |\mathbf{z}| = |\mathbf{y}_1 - \mathbf{y}_2| \quad (24.113)$$
$$p = |\mathbf{p}| \quad (24.114)$$
$$v = |z^0|/z \quad (24.115)$$
$$A(v) = (1 - v^2)^{-1} + .5v\ \ln[(v - 1)/(v + 1)] \quad (24.116)$$
$$B(v) = v^2(1 - v^2)^{-1} - 1.5v\ \ln[(v - 1)/(v + 1)] \quad (24.117)$$
$$\mathbf{p}\cdot\mathbf{z} = pz\cos\theta \quad (24.118)$$

and $|\mathbf{p}|$ denoting the length of a spatial 15-vector **p** while $|z^0|$ is the absolute value of $z^0 \equiv z^{16}$.

As eq. 24.109 indicates, the Gaussian damping factor R(p, z) for large spatial momentum p is the same for both the positive and negative frequency parts of the Two-Tier Feynman propagator. We are assuming the spatial momentum is real-valued in this discussion. It is also important to note that R(p, z) does not depend on $p^0 = p^{16}$ (in the Y Coulomb gauge) and thus the integration over p^0 proceeds in the usual way to produce time-ordered positive and negative frequency parts.

24.3.3.5 Feynman Propagators for the Types II, III, and IV Free Universe Particle Dirac Fields

These propagators differ in details from the Type I propagator. The differences modulo the change in dimension appear in Blaha (2011c). See also Blaha (2005a) for a detailed discussion of 4-dimensional spin ½ particle propagators.

24.3.4 Expanding and Contracting Universes: Impact of Time Dependent Universe Particle Masses

Our discussions of the dynamics of universe particles assumed their masses were constant. However the definition of mass in terms of the area of a universe based on the physics of black holes is

$$M = \kappa A/8\pi \qquad (24.119)$$

where A is the area of the black hole shows that *the mass of a universe particle is time dependent* because the area of a universe is time dependent. For example, our universe is expanding and its surface area is thus growing with time.

Eqs. 24.11 (and subsequent fermionic dynamic equations) must then be modified from

$$\psi(y) = \exp[-imt]w(0) \qquad (24.11)$$

to a covariant form:

$$\psi(y) = \exp[-i\int_0^{w\cdot y} m(t')dt']w(0) \qquad (24.120)$$

where w is a unit 16-vector in the time (y^{16}) direction ($w^2 = 1$). The lower bound on the integral, 0, is the time of the beginning of the universe particle – its Big Bang. Thus the cumulative change in the mass of the universe particle is significant. It is interesting to note that the Wheeler-Dewitt equation also has a variable value mass term R that also depends on the evolution of the universe.

Eq. 24.120 satisfies the free covariant Dirac-like universe particle field dynamic equation

$$[i\gamma^i\partial/\partial y^i - m(w\cdot y)]\psi(y) = 0 \qquad (24.121)$$

In contrast to the constant mass equation eq. 24.19. Substituting eq. 24.120 in eq. 24.121 we find

$$(\gamma^i w_i\, m(w\cdot y) - m(w\cdot y))\psi(y) = 0 \qquad (24.122)$$

or

$$(\gamma^i w_i - 1)\psi(y) = 0 \qquad (24.123)$$

Upon performing a 16-dimensional Lorentz boost (of the type of eqs. 24.13 – 24.16) on eq. 24.123 we obtain

$$(\gamma_i p^i/m_0 - 1)\psi(y) = 0$$

or

$$(\gamma_i p^i - m_0)\psi(y) = 0 \qquad (24.124)$$

where p^i is a momentum 16-vector with $p^2 = m_0^2$. Eq. 24.123 is the constant mass momentum space dynamic equation. It determines the spinor in $\psi(y)$. After taking account of the quantum coordinates the quantum Dirac-like universe particle wave function has the form

$$\psi(Y(y)) = \sum_s\int d^{15}p\, N^d_m(p)\, [b(p,s)u(p,s) : \exp[-iG(p, Y(y))]: + d^\dagger(p,s)v(p,s) :\exp[+iG(p, Y(y))]:\} \qquad (24.125)$$

$$\psi^\dagger(Y(y)) = \sum_s\int d^{15}p\, N^d_m(p)\, \{b^\dagger(p,s)\bar{u}(p,s)\gamma^0 :\exp[+iG(p, Y(y))]: + d(p,s)\bar{v}(p,s)\gamma^0 :\exp[-iG(p, Y(y))]:\} \qquad (24.126)$$

where : … : denotes normal ordering and

$$G(p, Y(y)) = \int_0^{p\cdot Y(y)/\lambda} m(t')dt' \qquad (24.127)$$

with $\lambda = m_0$, and $N^d_m(p)$ a normalization constant. Contrast eqs. 24.125-24.126 to the constant mass case eqs. 24.98-24.101. The *constant mass case* simply sets $m(t') = m_0$.

If we examine the integral eq. 24.127 for a short time interval δt in the particle's rest frame then $G(p, Y(y)) \approx m(0)\delta t$ and so we define $m(0) = m_0$. Based on the formula for universe particle mass (eq. 24.119) we anticipate that m_0 might be as large as the Planck mass or larger – thus an extremely short radius. Blaha (2013) describes a quantum Big Bang model in which the initial radius of the universe is $O(EM_{Planck}^{-2})$ where E is of the order of 1 and has the dimensions of [mass].

The spinors u(p,s) and v(p,s) are defined in a conventional way. However their form is different from the 4-dimensional case. If one takes the 256×256 $\gamma\cdot p$ matrix, then the first 128 columns give u(p,s) up to a normalization for the free particle case, the remaining 128 columns give v(p,s) up to a normalization.

Thus we have a closed form definition of a quantum universe particle wave function for universe particles of type I. A similar procedure can be followed for universe particles of types II, III, and IV.

The Feynman propagator for type I quantum fields is *not* eq. 24.109 but now has a form reflecting the Y(y) dependence of the quantum fields in eqs. 24.125 and 24.126:

$$iS_F^{TT}(y_1, y_2) = i\int \frac{d^{16}p\ \{ <0|\theta(y_{1 16} - y_{2 16})G(y_1, y_2) + \theta(y_{2 16} - y_{1 16})G(y_2, y_1)\}0>}{(2\pi)^{16}(\not p - m_0)} \qquad (24.128)$$

where p^{16} is the energy and

$$G(y_1, y_2) = :\exp[-iG(p, Y(y_1))]: :\exp[+iG(p, Y(y_2))]: \qquad (24.129)$$

Let

$$G_{tot}(y_1, y_2) = <0|\theta(y_{1 16} - y_{2 16})G(y_1, y_2) + \theta(y_{2 16} - y_{1 16})G(y_2, y_1)\}0> \qquad (24.130)$$

$$= <0|\theta(y_{1 16} - y_{2 16}):\exp[-iG(p, Y(y_1))]::\exp[+iG(p, Y(y_2))]: +$$
$$+\ \theta(y_{2 16} - y_{1 16}) :\exp[-iG(p, Y(y_2))]::\exp[+iG(p, Y(y_1)):]|0>$$

$$= <0|\theta(y^{16}_1 - y^{16}_2): \exp[-i\int_0^{p\cdot Y(y1)/\lambda} m(t')dt']::\exp[+i\int_0^{p\cdot Y(y2)/\lambda} m(t')dt']: +$$

$$+\ \theta(y^{16}_2 - y^{16}_1):\exp[-i\ \exp[+i\int_0^{p\cdot Y(y2)/\lambda} m(t')dt']::\exp[+i\int_0^{p\cdot Y(y1)/\lambda} m(t')dt']:|0>$$

with $\lambda = m_0$ then

$$iS_F^{TT}(y_1, y_2) = i\int \frac{d^{16}p\ G_{tot}(y_1, y_2)}{(2\pi)^{16}\,(\not{p} - m_0)} \qquad (24.131)$$

Except for the case of a constant mass, where $m(t) = m_0$, the Feynman propagator is not a function of $y_1 - y_2$. The evaluation of eq. 24.130 in the general case of a variable mass is straightforward but cumbersome. For the special case of a linear time dependence of the mass, $m(t) = at$, we find eq. 24.130 gives

$$G_{tot}(y_1, y_2) = \langle 0|\theta(y^{16}{}_1 - y^{16}{}_2){:}\exp[-ia(p\cdot Y(y_1)/m_0)^2/2]{:}{:}\exp[+ia(p\cdot Y(y_2)/m_0)^2/2]{:} + \\ + \theta(y^{16}{}_1 - y^{16}{}_2){:}\exp[-ia(p\cdot Y(y_2)/m_0)^2/2]{:}{:}\exp[+ia(p\cdot Y(y_1)/m_0)^2/2]{:}|0\rangle \qquad (24.132)$$

yielding a complex function of p, y_1, and y_2. Note that the lower bound of the integrals in the Feynman propagator cancel and thus the need for an understanding of the beginning of a universe is removed in this case.

We have shown that universe particle theory can handle the case of a variable universe mass m(t). Expanding or contracting (or oscillating) universe particles correspond to expanding and contracting (or oscillating) universes.

24.3.5 Left-Handed and Right-handed Universe Particles

In sections 24.3.1 and 24.3.2 we found that left-handed and right-handed tachyonic universe particles existed. The tachyonic nature of the universe particles indicates that their speed in the universe exceeds the "speed of light" of the Megaverse. The physical meaning of the handedness of these types of universes is an interesting issue. When we consider our universe we see left-handedness in the weak interactions of elementary particles. In addition it appears that organic molecules overwhelmingly favor left-handedness on earth although right-handed molecules exist in outer space and can be created in the laboratory. Right-handed molecules transform into left-handed molecules in watery media through electromagnetic effects.

Why nature favors left-handedness is an open question. It has given rise to speculations that gravitation, especially quantum gravitons, may be left-handed. The European Space Agency's Planck telescope will study polarization effects in the cosmos and may well be able to show that the gravitons starting from the beginning of the universe, and magnified by inflation in the universe's expansion, may be left-handed.

If handedness of gravitation is verified experimentally, then our theory of left-handed/right-handed universe particles would be supported. *Our universe would then be tachyonic and probably left-handed. We, in the universe, would of course not know of the velocity of the universe in the Megaverse.*

24.3.6 Internal Structure of Universe Particles

We have treated universes as particles in the preceding discussion taking an extremely large view of Megaverse particles just as elementary particle theory viewed nucleons at low energies (large distances). Now we develop a detailed view of universe particles in a manner analogous to the high energy view of the internal dynamics of nucleons that led to the quark-parton model of nucleons. In the present case we shall see that high energy Baryonic and other field probes of universe particles can yield a model of the internal baryonic structure of universes.

We know that universes are composed of matter and radiation. We believe that there is at least one possibly accessible interaction between universes dependent on baryon number – a baryonic, 16-dimensional gauge field. There are also other particle number gauge fields – but these are less likely to be significant[280] since Dark matter is yet to be found except through its gravitational effects. In this section we will discuss the use of a baryonic gauge field to probe the baryon structure of universe particles.

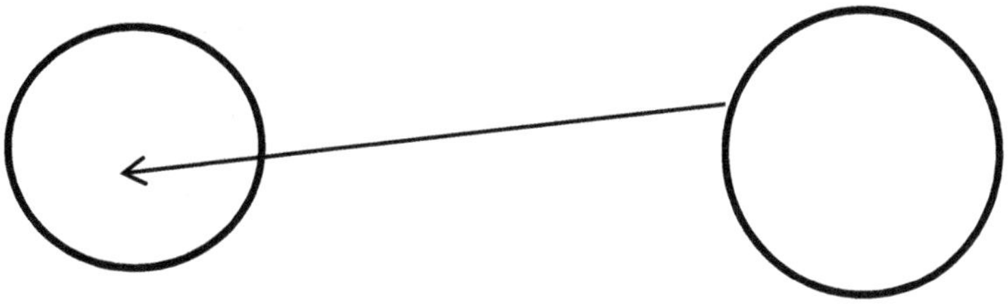

Figure 24.1. A symbolic view of a high resolution (high energy) probe from a universe to a specific baryonic part of another universe.

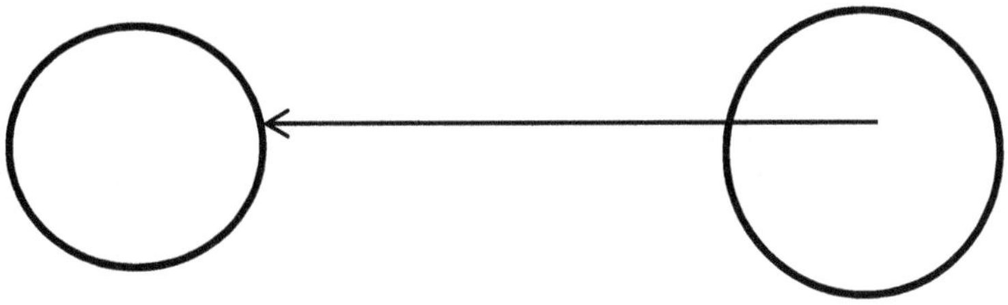

Figure 24.2. A symbolic view of a low resolution (lower energy) probe from a universe to an entire universe.

There appears to be two types of baryonic probes of a universe: 1) a series of high energy probes to specific small regions inside another universe for the purpose of mapping its internal structure; 2) a low energy probe of another universe to get a global view of its baryonic structure. The first type of probe corresponds to deep inelastic (high energy) electron-nucleon scattering which led to the quark-parton model of nucleons. The second type of probe corresponds to low energy electron-nucleon scattering to get a "global" view of a nucleon. In both case an electromagnetic (gauge) field particle (photon) was the probe particle.

Besides the inherent scientific interest in such experiments it is possible that they may be of use in the very distant future in Mankind is able to develop Megaverse starships that can travel in the Megaverse to other universes. Then the baryonic gauge field becomes the "eyes" of the starship just as electromagnetic fields (light) are the eyes of current spaceships. We considered the possibility of universe starships in the book entitled, *All the Multiverse! II* in detail, and briefly in the next chapter.

---

[280] In our universe for purposes such as starship entry into the exterior Megaverse as described in the following chapter.

24.4 When Universes Collide: Interactions and Collisions of Universe Particles

24.4.1 Gravitation and the Fifth Force

As we saw in chapter 15 the primary forces involved in the interactions and collisions of universe particles are the force of gravity, and a fifth set of forces which we take to be the four forces associated with particle number conservation: the baryonic force, Dark baryonic force, lepton force, and the Dark lepton force. We describe the baryonic force insections 24.11-24.15. The other forces are similar so we will not describe them in any detail. One interesting point of interest is the two baryonic forces are long range forces both inside universes and in the Megaverse because their gauge particles are massless.

The two leptonic forces are short range forces within universes because their gauge particles are massive - probably very massive inside universes due to the Higgs Mechanism. However outside of universes in the Megaverse they are long range forces because their gauge fields are massless. (No Higgs particles exist in the external Megaverse.) Since their sources are the total number of leptons and Dark leptons in each universe they will exist in the Megaverse. Thus four long range forces exist in the external Megaverse.

24.4.2 Universes in Collision

We assume that the dynamics of universes in collision will be analogous to that of galaxies in collision since gravity is a dominant force in both cases. Colliding galaxies have been observed often. Their dynamics should provide guidance for the case of universes in collision.[281]

It is clear in the case of colliding galaxies and of colliding large nuclei (gold and lead typically) that there are several types of collisions with differing results. These types of universe collisions can be qualitatively classified as

1. Clean collisions in which universes nudge each other but retain their identity. These are extreme peripheral collisions. If the universes overlap slightly then the typically spherical symmetry of the universes may become distorted and they may become lopsided.[282]

2. Peripheral collisions in which the universes retain their identity but are connected by a trailing string of mass-energy. Eventually the string breaks and the universes separate. Subsequently the pieces of trailing string in each universe contract due to their universe's gravitational effects.

3. Two universes can collide and produce multiple universes.

4. Two universes can collide in a "central" collision and amalgamate into one universe.

---

[281] The high energy collision of atomic nuclei at Brookhaven, CERN and other laboratories also is analogous in overall detail with universes in collision.

[282] The Wilkinson Microwave Anisotropy Probe (WMAP) and the Planck European Space Agency satellite has been accumulating data since 2001 that suggests the universe may be lopsided with hot and cold spots on opposite sides of the universe differing from those on the other side being hotter and colder respectively. Perhaps the result of a collision when the universe was young.

We will discuss universe interactions in more detail later.

24.5 Bosonic Universe Particles

The previous section has described fermionic universe particles. In this section we will briefly describe aspects of bosonic (spin 0) universe particles. First it is important to note that the Wheeler-DeWitt equation being second order like the Klein-Gordon equation seems to suggest that universe particles can be bosonic – like Klein-Gordon equation particles.[283] The Wheeler-DeWitt equation has a mass-like term R that can be positive or negative. If the mass term is negative then the wave-like propagation of the state functional (wave equation solution) can be in space-like directions implying a tachyonic solution. Thus the Wheeler-DeWitt equation supports "normal" state functionals that propagate in time-like directions as well as tachyonic propagation.

For this reason we suggest that bosonic universe particles can be either normal or tachyonic. Tachyonic bosonic universe particles can fission in a manner similar to tachyonic fermionic universe particles. The fission equations of section 24.5.2 also apply to tachyonic bosonic universe particles.

The quantum field theory of normal and tachyonic bosonic universe particles is similar to that of ordinary bosons. See Blaha (2005a) for the boson case discussion that is paralleled by our universe particle formalism.

24.6 Physical Meaning of Universe Particle Spin

The physical meaning of spin is a continuing discussion topic. We have suggested that spin states are in essence logic states with changes in spin an analogous to changes in logical values in a discourse or computer program. Since the matrix formalism for spin ½ and higher spin states is formally similar to the formalism for angular momentum, one can combine spin and angular momentum as we do in quantum theory.

In the case of universe particles, one can also associate universe particles with "true" and "false" values. Fermionic universe states have 128 truth values and correspond to a multi-valued logic. The numerousness of truth values is due to the 16-dimensional space within which universe particles reside.

Naturally one would like to know the physical differences between these 128 types of universe particles. Does the difference reside in different shapes of the universe particles? Or is the difference somehow a consequence of the global mass-energy distribution of the universe that we have not been able to discern since we only know of one universe?

The physical meaning of spin for elementary particles is also somewhat elusive. It does not reflect the flow of charge within a particle. For if it did reflect physical spinning of a particle the outer edges of a particle such as an electron would be traveling at a speed faster than light. So spin is not a mechanical property of the internal structure of an elementary particle. We have suggested that it is a truth value in the matrix formulation of a 4-valued logic called Asynchronous Logic. Thus it has no certain tangible physical basis.

In the case of universe particles the situation is unclear at present. It could be taken to be an indirect reflection of the structure of mass-energy within a universe. This view would be contrary to our proposed view of elementary particle spin as truth values. So we can only assert that a logic interpretation is the only sensible one (based on our present knowledge or our lack thereof). The physical role of

[283] One should remember that the Wheeler-DeWitt equation is not in space but in a 6-dimensional manifold, denoted M, of metrics with one "time" dimension – having hyperbolic signature – + + + + + when the metric is positive definite. See DeWitt's paper.

universe particle spin is only evident in interactions between universe particles via the baryonic gauge field proposed later. Thus one must simply view it as a construct for the present.

Other than our mapping of spin values to logic values in Asynchronous Logic there is little anyone can say about the physical origin of elementary particle spin. Specifying a symmetry group as the origin, such as SuperSymmetry, is not sufficient.

24.7 Elementary Particles with Time Dependent Masses

The discussions of this chapter were presented for universe particles. However they could also apply to elementary particles or condensed matter excitations with some changes – primarily changes due to different dimensions.

Elementary particles with time dependent masses do not exist as far as we know. And that is a good thing. The idea of an elementary particle expanding indefinitely, like a universe, would have disastrous consequences for life as would particles contracting indefinitely. However, particles with oscillating masses would seem to be physically acceptable. The development of an elementary theory with oscillating masses would be an interesting exercise that might have applications in condensed matter physics.

24.8 Impact of Universe Particle Acceleration – Lopsided Internal Structure of Universe

We are developing a theory of universe particle interactions. Such interactions would cause universe particles to accelerate and should be detectable within the universe as a "lopsidedness" – there would be a shift of parts of the universe away from the direction of acceleration resulting in a difference in the features of the universe "in front" compared to those "in back" – an acceleration effect just as one sees when a jet accelerates.

Interestingly new data from the Planck observatory of the European Space Agency confirms and extends earlier data from NASA's WMAP observatory that one side of the universe appears different from the other side. There are temperature differences and mass distribution differences – just as one might expect if the universe were accelerating as a unit.

Thus we see the beginning of data suggesting our universe may be moving – "indeed accelerating" – through a Megaverse. Some Planck observatory scientists have suggested their data is a preliminary indication of the Megaverse.

24.9 Megaverse Baryonic Gauge Field - Plancktons

The conservation of baryon number has been repeatedly investigated by experimenters and found to be true to extremely high accuracy. For decades theorists have suggested that the conservation law follows from the existence of a gauge field in a manner much like electric charge conservation follows from the properties of the electromagnetic abelian gauge field.[284]

We will therefore assume a baryonic gauge field exists that is similar to the electromagnetic field except for features due to its definition and existence in the 16-dimensional Megaverse. This field will couple extremely weakly to individual baryons as well as universe particles with non-zero baryon

[284] See Gell-Mann, M. and Levy, M. *Nuovo Cimento* 16, 705 (1960) for a proof.

number. We will call the baryonic gauge field particle a *planckton*. Its electromagnetic analogue is the photon. We described 16-dimensional baryonic gauge field quantization in chapter 15.

Plancktons propagate in the Megaverse, both within universes, and in the Megaverse external to universes. So we will define the planckton field 16-dimensional Megaverse coordinates. They will interact with baryons within a universe with Megaverse coordinates mapped to the curvilinear coordinates in the universe. (This mapping was discussed earlier.)

Since a planckton field in 16-dimensional conventional coordinates would lead to divergences we will use quantum coordinates:

$$Y^i(y) = y^i + i\, Y_u^{\,i}(y)/M_u^{\,8}$$

with quantum coordinate derivatives defined by

$$\partial_i = \partial/\partial Y^i(y) = \partial/\partial(y^i - Y_u^{\,i}(y)/M_u^{\,8})$$

to obtain a completely finite theory of planckton interactions with elementary particles and universe particles.

Plancktons, other particle number fields, universe fields, gravitation, and the $Y_u^{\,i}(y)$ field of quantum coordinates are the only fields in the space between universes in the Megaverse. Since the mass-energy and charge of universes is zero, Standard Model fields are zero in the space between universes.[285]

24.10 Beyond the Planckton

The analogy between plancktons and the electromagnetic field photons raises the possibility that the baryonic gauge field may be one of a set of gauge fields. We showed that the four particle number interactions embody a broken U(4) symmetry within universes in chapter 15. Their U(4) symmetry gives rise to the four fermion generations.

24.11 Planckton Interactions with Universe Particles and Individual Baryons

Section 15,1 describes the second quantization of plancktons. In this section we will develop an interacting theory of universe particles and plancktons from the lagrangian terms of universe particles, plancktons and quantum coordinates. We will only consider the case of type I universe particles since the other cases differ from it only in details. The universe particle – planckton lagrangian terms are:

$$\mathcal{L} = \bar{\psi}(Y(y))[i\gamma^\mu\partial/\partial y^\mu - e_B\gamma^\mu B_{u\mu}(Y(y)) - m(t)]\psi(Y(y)) - \tfrac{1}{4} F_{Bu}^{\;\mu\nu}(Y(y))F_{Bu\mu\nu}(Y(y)) - \tfrac{1}{4} F_u^{\;\mu\nu}(y)F_{u\mu\nu}(y) \qquad (24.133)$$

where $\mu, \nu = 1, 2, \ldots, 16$ and where

$$\bar{\psi} = \psi^\dagger\gamma^{16}$$

$$F_{Bu\mu\nu} = \partial B_{u\mu}(Y(y))/\partial Y^\nu(y) - \partial B_{u\nu}(Y(y))/\partial Y^\mu(y) \qquad (24.134)$$
$$F_{u\mu\nu} = \partial Y_\mu/\partial y^\nu - \partial Y_\nu/\partial y^\mu$$

[285] The vacuum energy of the baryonic field and the $Y_u^{\,i}(y)$ fields being uniform throughout the Megaverse do not exert forces or cause gravitational effects except possibly through baryonic Casimir forces between universes.

$$Y^i(y) = y^i + i\, Y_u^{\,i}(y)/M_u^{\,8}$$
$$e_B = e_{B0}/M_u^{\,6}$$

with e_{B0} a dimensionless coupling constant, and with μ and ν ranging from 1 through 16.

The corresponding lagrangian is

$$L = \int d^{15}y \mathscr{L} \tag{24.135}$$

Note the dimensions of the fields differ in the 16 dimensional space:

$$Y^\mu \sim [\text{mass}]^{7-7}$$
$$B_{u\mu} \sim [\text{mass}]^{-7}$$
$$\psi \sim [\text{mass}]^{-15/2}$$

as can be seen from the above lagrangian as well as earlier equations. Note also that the mass and thus the size of universe particles is time dependent. They can expand or contract with time depending on their internal characteristics (gravitation and effects of elementary particle interactions) which are not embodied in this lagrangian. As a result this theory, incomplete as it is, does not conserve energy unless m(t) is constant.

The lagrangian generates the baryonic interactions of universe particles using Two Tier quantum coordinates which prevent infinities in perturbation theory calculations.

The interaction of baryon elementary particles with the baryonic field requires terms in The Extended Standard Model covariant derivatives specifying the baryon field interaction baryons with the form

$$e_B\gamma^\mu B_{u\mu}(Y(y))$$

The following sections describe some of the physically significant interactions that the lagrangian (eq. 6.19) implies.

24.12 Creation of Universes through Baryonic Gauge Field Fluctuations

One of the most exciting questions in Cosmology is the origin of our universe. The conventional view is that it originated in a Big Bang from an infinitesimal point in space. The source of the Big Bang and the prior state of the Cosmos, if there was one, is the subject of much speculation. Based on the particle interpretation of the Wheeler-DeWitt equation, the possibility of a baryonic force strongly supported by conservation of baryon number, and the Megaverse concept it is reasonable to consider the possibility that the universe originated in a vacuum fluctuation.

In this case there would be two Big Bangs one for our universe and one for an anti-universe. One would expect that they would have reverse corresponding features: one with baryon dominance – one with anti-baryon dominance, and one left-handed – one right-handed.

Our formulation of universe particle theory provides for the generation of a universe particle and anti-particle as a vacuum fluctuation. We view a universe particle as having a substantial excess of baryons, N, as we see in our universe. Its anti-universe at the time of creation (the Big Bang point) is its "mirror image" having the "same" number of anti-baryons (baryon number –N) so that baryon number is conserved by the fluctuation event. Thus the excesses one is compensated by the excesses of the other.

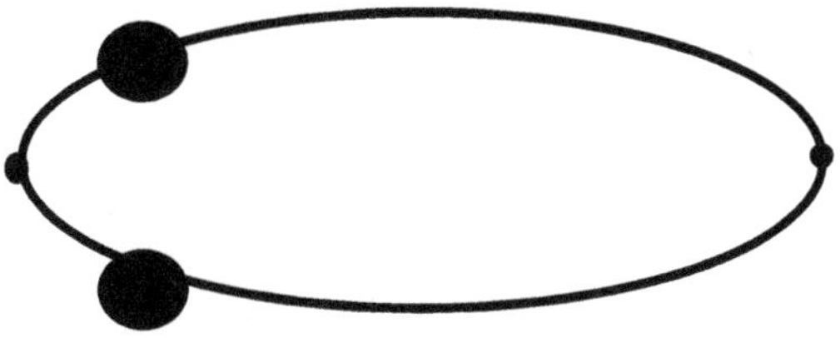

Figure 24.3. Generation of a universe – anti-universe pair as a vacuum fluctuation.

The small value of the coupling constant should lead to an extremely long lifetime for the universes generated by the fluctuation. Thus the 13.7 billion year life of our universe is not unreasonable. The probability of the creation of universes by vacuum fluctuations should be correspondingly small.

24.13 When Universes Collide: Coalescence of Universes

Universes moving in the Megaverse can collide through chance, or due to the planckton field which can cause universes with excess baryons to attract universes with excess anti-baryons.

When universes collide several possibilities present themselves:

1. They can graze each other distorting each other's shape and internal baryon distribution through the baryonic force while maintain their individual identity.

2. They can intermix with both the baryonic and gravitational forces causing a redistribution of their masses. They may separate afterwards or may coalesce into a single universe. One result of this may be lopsided universes. Our universe appears to be lopsided. Some cosmologists believe this is due to a near collision of our universe with another shortly after the Big Bang.

In our discussion we have been referring only to the planckton baryon field for the sake of concreteness. But the three other massless particle number fields will also play a role in universe interactions. The relative strength of these interactions is not known since we do not know the size of their coupling constants.

24.14 Fission of Universes

Under certain circumstances the distribution of matter in the universe may lead to the fission of the universe into two separate universes. Our model lagrangian supports this possibility for universe particles. The detailed mechanism of the fission process is not specified by the model.

24.14.1 Fission of Normal universes

The fission of universe particles in our universe particle model is depicted in the Feynman diagram in Fig. 24.4.

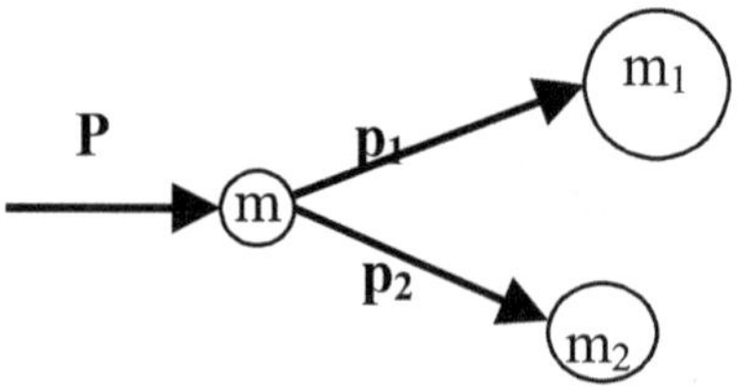

Figure 24.4. Fission of a universe particle into two universe particles.

The sum of the masses of the output universe particles is usually less than the original universe particle mass. However if the fission takes a long time and the masses are time dependent then the produced universe particles combined masses may exceed the original universe's mass.

24.14.2 Tachyon Universe Particle Fission to More Massive Universe Particles

In Blaha (2007a) we showed that a tachyonic (faster than light) particle could fission into particles of larger mass. In this section we will show that a tachyonic universe particle may fission into two more massive universe particles. This phenomenon is of particular interest because it enables tachyonic universes to spawn in a new novel way not previously considered in discussions of the origin of universes.

The lagrangian for a tachyonic universe particle is

$$\mathcal{L}_{\|} = \psi_T^{S}(Y(y))[\gamma^\mu\partial/\partial y^\mu - e_B\gamma^\mu B_{u\mu}(Y(y)) - m(t)]\psi(Y(y)) - \tfrac{1}{4}F_{Bu}^{\mu\nu}(Y(y))F_{Bu\mu\nu}(Y(y)) - \tfrac{1}{4}F_u^{\mu\nu}(y)F_{u\mu\nu}(y) \tag{24.136}$$

We assume m(t) is constant.

When a particle or a universe particle fissions (decays) one normally expects that the masses of the particles or universe particles produced by the decay to be smaller than the mass of the original particle or nucleus. In the case of tachyonic (faster-than-light) elementary particles or universe particles a much different possibility is present: a tachyon can decay into heavier tachyons. We will consider the specific case of a tachyon universe particle decaying into two universe particles whose total mass is greater than the original. (See Fig. 24.5.)

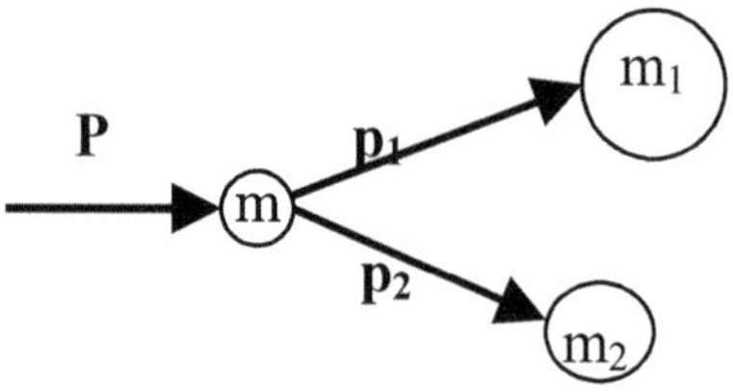

Figure 24.5. Two universe particle decay of a tachyon universe particle.

We will assume the initial tachyon universe particle has zero energy ($p^{16} = 0$) and thus the tachyons universe particles emerging from the decay also have total universe particle energy zero. The analysis is based on conservation of total universe energy and momentum in Megaverse space outside of universes. The below discussion applies to 16-dimensional space with 15-dimensional spatial coordinates.

Momentum conservation implies

$$\mathbf{P} = \mathbf{p}_1 + \mathbf{p}_2 \tag{24.137}$$

Since all energies are zero

$$\begin{aligned}(cP)^2 &= (c\mathbf{P})^2 = m^2 \\ (cp_1)^2 &= (c\mathbf{p}_1)^2 = m_1^2 \\ (cp_2)^2 &= (c\mathbf{p}_2)^2 = m_2^2\end{aligned} \tag{24.138}$$

where $P = |\mathbf{P}|$, $p_1 = |\mathbf{p}_1|$, and $p_2 = |\mathbf{p}_2|$. If we now square eq. 24.137 and then use eqs. 24.138 we obtain

$$m^2 = m_1^2 + m_2^2 + 2m_1m_2 \cos\theta \tag{24.139}$$

where θ is the opening angle between the emerging universe particles momenta $\mathbf{p}_1$ and $\mathbf{p}_2$. Eq. 24.139 has a number of interesting cases:

Case $\theta = 0$:

$$m = m_1 + m_2 \tag{24.140}$$

The masses of the outgoing universe particles sum to the mass of the original tachyon universe particle.

Case $\theta = \pi/2$:

$$m^2 = m_1^2 + m_2^2 \tag{24.141}$$

The masses of each outgoing universe particle tachyon is less than the mass of the original tachyon universe particle.

Case $\theta = \pi$:

$$m^2 = (m_1 - m_2)^2 \tag{24.142}$$

In this case either $m_1 > m$ or $m_2 > m$. Thus one of the outgoing tachyon universe particles has a greater mass than the original tachyon universe particle. Mass is effectively created from the spatial momentum of the initial universe particle. This process is the inverse of normal particle and universe particle fission where the sum of the outgoing masses is always less than the original particle's mass and the difference is mass converted into energy in the form of additional photons.

This last case, where one of the outgoing universe particles is more massive than the original universe particle, is not just for $\theta = \pi$. Since

$$\cos\theta = (m^2 - m_1^2 - m_2^2)/(2m_1m_2) \tag{24.143}$$

we see that the sum of the outgoing universe particle masses is always greater than the original tachyon universe particle *mass (except when $\theta = 0$)* since

$$\cos\theta = 1 + [m^2 - (m_1 + m_2)^2]/(2m_1m_2) \leq 1 \quad (24.144)$$

and thus

$$[m^2 - (m_1 + m_2)^2]/(2m_1m_2) \leq 0 \quad (24.145)$$

Note $m = m_1 + m_2$ only if $\theta = 0$.

Since we can transform the above discussion to the case of universe particle tachyons having non-zero Megaverse energy using an ordinary 16-dimensional Lorentz transformation the discussion in this subsection is general.

We therefore conclude that when a tachyon universe particle decays into two tachyon universe particles the sum of the masses of the produced tachyon universe particles is greater than the mass of the original tachyon universe particle except if the angle between the momenta of the produced tachyon universe particles is zero. In that case the sum of the masses of the produced tachyon equals the mass of the original tachyon universe particle and the produced universe particles overlap.

24.15 Universe Particle – Planckton Interactions

These interactions are quite similar to Two-Tier electromagnetic interactions except that universe particles have time-dependent masses, and that the space is 16-dimensional.

The interactions have a new aspect due to the time dependence of the universe particle masses. This feature is illustrated by Fig. 24.6: the mass of a universe particle after a baryonic interaction vertex is the same as it was before the interaction assuming the point-like interaction specified in the lagrangian.

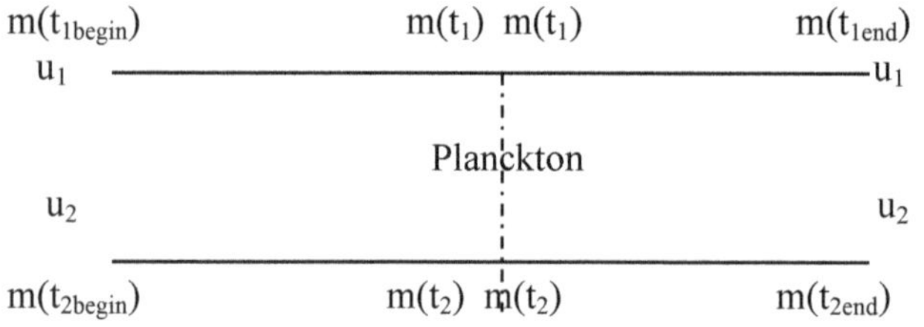

Figure 24.6. A Feynman diagram illustrating the continuity of a universe particle mass through a Planckton interaction.

The reader may verify this by writing the perturbation theory equivalent. A universe particle vertex corresponds to

$$iS_F^{TT}(y_1, y_2)\gamma^\mu iS_F^{TT}(y_2, y_3) \quad (24.146)$$

By eqs. 5.131 and 5.132 the universe particle mass is the same on either side of the interaction vertex.

24.16 Internal Structure of Universe Particles

We have developed a beginning in planckton field theory that gives interactions between baryons. This theory is applicable to universe-universe interactions. It also yields baryon particle – baryon particle interactions as well as baryon particle – universe particle interactions.

It is possible for a planckton to be emitted in one universe and interact with a baryon elementary particle in another universe. This type of "probe" must be a high energy probe just as a photon probe of the internal structure of a nucleon[286] must be a high energy photon to bring out the nucleon's internal structure (parton model).

In this section we will discuss planckton probes of other universes, and the internal structure of a universe as a mass distribution governed by gravitation as it relates to universe particles.

24.16.1 Planckton Probes

Plancktons can be generated in one universe and be used to probe the baryon distribution of another universe. Since the planckton propagator is expressed in Megaverse coordinates the baryon distribution in the target universe will be a distribution in Megaverse coordinates. Megaverse coordinates can be expressed in terms of the curved space-time coordinates of a universe x^μ. However the inversion of the map between universe coordinates and Megaverse coordinates

$$x^\mu = f^{-1\mu}(y) \tag{24.147}$$

is not 1:1 since x^μ is 4-dimensional and y is a 16-dimensional vector. The universe coordinates x^μ are each individually determined up to a subspace. One might be concerned about this situation but the determination of the distribution in Megaverse coordinates gives a more direct picture not convoluted by the curvature of the target universe.

The detailed probing of a target universe requires high energy plancktons. The similarity of this procedure to deep inelastic electro-nucleon scattering is obvious to the high energy physicist. But in doing this planckton probe experiment one obtains a picture of a different universe – something that is not possible to do with electromagnetic or graviton probes.

24.16.2 Internal Structure of a Universe Particle

The development of the theory of universe particles which resulted in the lagrangian of eq. 24.136 does not fully describe universe particles since it neglects the internal structure of a universe particle. The internal structure of a universe particle is primarily determined by gravitation, electromagnetic effects and nuclear physics.

Consequently the full lagrangian of a universe particle has the form

$$\mathcal{L}_{tot} = \mathcal{L}_{internal} + \mathcal{L} \tag{24.148}$$

where $\mathcal{L}$ is determined above. As a result the complete quantum wave function of a universe particle has the form

$$\psi_{tot} = \psi_{internal}(Y)\psi_{ext}(Y) \tag{24.149}$$

where $\psi_{internal}(Y)$ is the internal wave function and $\psi_{ext}(Y)$ is external wave function. It seems reasonable to have a separable equation except when universes collide. In that situation a perturbative mixing of the universes and their wave functions applies and it may be possible to calculate the collision output

---

[286] Deep inelastic electron-nucleon scattering.

universes by introducing a further interaction between the internal and external aspects of the universe particles.

24.17 Central Role of the Baryonic Force for Travel into the Megaverse

We have focused on the baryonic force in this chapter although the Dark baryonic force and the lepton forces may well play a role in universe particle dynamics. The reason is the long range possibility of travel into the Megaverse to other universes using the baryonic force to exit our universe into 16-dimensional space using a slingshot orbit around a neutron star. We briefly discuss this possibility in chapter 25 and refer the interested reader to Blaha (2014c) for a detailed discussion.

25. Starships in the Universe and Megaverse

The proof of success in new Science is the benefits that humanity derives and the doorways it opens to a better understanding of Nature. The development of an Extended Standard Model has shed light on the fundamental nature of reality – the nature of space and time, and the basis of elementary particle Physics in Logic – bringing us closer to Plato's vision of reality as essentially a mirror of a World of Ideas.

In this chapter we consider the potential benefits of our work for the future exploration of our universe and the Megaverse in starships capable of enormous speeds and "short" travel times thoughout the Megaverse.

Currently we are exploring the Solar System using relatively slow unmanned rockets powered by chemical fuels. Soon we will be using nuclear (of fusion) power to explore and colonize selected parts of the Solar System. These approaches to spaceship power are not sufficient to successfully explore and colonize nearby stars. At best they only offer the possibility of multi-generation voyages to even the clsest stars. Multi-generation voyages will be few, and insufficient for colonization and the expansion of humanity to the stars. The fault is basically in the the limitation of travel to speeds below the speed of light. This limitation cannot be overcome with advanced conventional rocketry.

We have proposed the development of a new type of power configuration that will enable starships to far exceed the speed of light using a quark-gluon "ion" drive that *evades* the speed of light limit[287] by accelerating starships with complex-valued thrust.[288] In our derivation of The Extended Standard Model we showed that quarks and gluons have complex speeds in general that are usually masked by color confinement of quarks within hadrons. When heavy atomic nuclei collide in a very high energy accelerator such as CERN in Geneva, Switzerland or the Brookhaven National Laboratory in the United States, a quark-gluon plasma is created that contains free quarks and gluons. These particles, if extracted by magnetic fields before the plasma collapses ("freezes out") and used to provide a complex-valued thrust to a starship, would not be limited by the speed of light and could power a starship to large speeds that could reduce travel time to the stars to days rather than hundreds of years.

The quark-gluon drive approach requires very powerful accelerators and magnet technology that have yet to be developed.[289] But our Extended Standard Model theory indicates that it is possible in principle.

We envision a two stage development and exploration effort for the conquest of the universe and the Megaverse realizing that this development effort will take many hundreds of years, and this exploration effort will take tens of thousands of years. Our purpose then is to set a great goal for the world to make a beginning.

The first stage of the effort is to develop accelerator and magnet technology, and other needed technology such as suspended animation, to the point that they can be used to create starship prototypes

[287] The speed of light is a real number. By using complex-valued speeds a vehicle can "go around" the speed of light to faster than light speeds. Thus Einstein's limit remains true for real velocities. But complex-valued faster than light velocities are possible.

[288] This chapter is based on the author's books listed in chapter 29.

[289] There are also significant other problems such as collisions with space "dust" and debris, and a need for the development of long time hibernation/suspended animation for humans. These problems are solvable.

with faster than light capability and the ability to travel to nearby stars. Subsequently an exploration effort to nearby stars should begin followed by colonization and the exploration of the galaxy. Then jouneys to other far galaxies with an exploration program for interesting parts of the universe.

The second phase of the space initiative would be to travel into the Megaverse and perhaps travel to nearby universes. This phase can be pursued in parallel with phase one after quark-gluon starships become a reality because the doorway to the Megaverse is "just around the corner." In section 25.2 we will a design for a starship that can "slingshot" around a "nearby" neutron star into the Megaverse. Once in the Megaverse the starship, which would have a powerful energy source such as fuel tanks of protons and antiprotons for proton-antiproton annihilation energy and computer support for visualizing and navigation in the sixteen dimensions of the Megaverse, would seek and find a nearby universe for exploration. The mechanism for slingshoting out of a universe would be based on the baryonic force described earlier. Seeing in the Megaverse would also be based on baryonic planckton "optics" since universes do not emit light.

After initial exploration, the possibilities of colonization and trade might be pursued if justified. An exploration program for the Megaverse could begin with a planckton observatory searching for other universes just as we now search for galaxies with earth-bound and satellite observatories.

After accomplishing this two phase program Mankind would have achieved its destiny to see all of Creation and to grow to maturity. And yet that will not be the end. For the human mind is an infinite space itself and its exploration will have just begun. In the past ten thousand years, since civilizations began, Mankind has only slightly progressed in its mental abilities and understanding. Much time is needed to grow.

25.1 Phase One: Starship Engine Development

In this section we will consider a constant, propulsive force in a starship's rest frame that drives the starship from a sublight velocity to a superluminal velocity. The key factor in achieving a superluminal speed is evading the singularity in γ at $v/c = 1$. We accomplish this goal by having a complex force – a force with a real and imaginary part – that generates a complex acceleration, and thus a complex velocity, that "goes around" the singularity in γ in the complex velocity plane. We assume that an "instantaneous" Lorentz transformation relates the earth reference frame and the starship reference frame.

We assume a constant, complex force exists in the rest frame of the starship due to the starship's thrust in the direction of the positive x' (and x) axis. The starship (primed coordinates) and earth (unprimed coordinates) coordinates have parallel axes. The spatial force in the positive x direction is

$$\mathbf{F'} = g\hat{\mathbf{x}} \tag{25.1}$$

where g is assumed to now be a complex constant.

The fourth component of the force (since force is a Lorentz 4-vector) is zero in the rocket's rest frame:

$$F'^0 = 0 \tag{25.2}$$

Applying an inverse Lorentz transformation we find the force in the earth rest frame is

$$F^0 = \gamma(F'^0 + \beta F'^x/c) = \gamma\beta F'^x/c = \gamma vg/c^2 \qquad (25.3)$$
$$F^x = \gamma(F'^x + \beta c F'^0) = \gamma F'^x = \gamma g$$
$$F^y = F^z = 0$$

where $\beta = v/c$, c is the speed of light, and $\gamma = (1 - \beta^2)^{-1/2}$ as before. We again use the superscripts x, y, and z to identify the components of the spatial force. The spatial momentum of an object of mass m is

$$\mathbf{p} = \gamma m\mathbf{v} \qquad (25.4)$$

and the dynamical equation of motion is

$$d\mathbf{p}/dt = \mathbf{F} \qquad (25.5)$$

in the "earth" coordinate system resulting in

$$dp^x/dt = \gamma g \qquad (25.6)$$

with[290]

$$dp^y/dt = dp^z/dt = 0 \qquad (25.7)$$

The differential equation resulting from eq. 25.5 is

$$d(\gamma v)/dt = \gamma g/m \qquad (25.8)$$

Assuming initially that g is real we must use $\gamma = (1 - \beta^2)^{-1/2}$ for $v < c$ and $\gamma = (\beta^2 - 1)^{-1/2}$ for $v > c$ based on the need for real coordinates for faster than light travel. The solutions for real v are[291]

<u>$v < c$, Re $v_0 < c$</u>

$$v = c\{1 - 2/(1 + ((c + v_0)/(c - v_0))\exp[2g(t - t_0)/(mc)])\} \qquad (25.9a)$$

<u>$v > c$, Re $\acute{\upsilon}_0 \geqslant c$</u>

$$v = c\{1 - 2/(1 + ((c + \acute{\upsilon}_0)/(c - \acute{\upsilon}_0))\exp[2\breve{g}(t - t_0)/(mc)])\} \qquad (25.9b)$$

where the velocity is v_0 at time t_0, $\acute{\upsilon}_0$ is the velocity[292] at $t = t_0$ and $\breve{g}$ is the acceleration for Re $v \geqslant c$.[293]

Analytically continuing eqs. 25.9 to complex v with a complex force constant g we obtain the starship equation of motion. We require continuity when the real part of $v = c$ by requiring that when Re v(t) of eq. 25.9a equal c, that $t_0 = t$ and $v(t_0)$ of eq. 25.9a equal $\acute{\upsilon}_0$. These conditions fix t_0 and $\acute{\upsilon}_0$.

<u>Note:</u>

---

[290] There is thrust in the y and z direction as well. To avoid getting distracted by the details of an exact calculation we approximate the force in those directions as zero.

[291] The velocity is entirely in the x-direction in this calculation. It can, and does, have complex values in this example.

[292] It is greater than c by assumption in the calculation of eq. 25.9b.

[293] Although eqs. 25.9a and 25.9b have the same form, the acceleration for Re $v < c$ can be changed to a new value $\breve{g}$ after Re v exceeds the speed of light in order to approach the singularity discussed later.

Eqs. 25.9 can easily be integrated to give the distance traveled in the x direction.

$v < c$, Re $v_0 < c$

$$x = x_0 + (mc^2/g)\ln((1 - v_0/c + (1 + v_0/c)\exp[2g(t - t_0)/(mc)])/2) - c(t - t_0) \quad (25.10a)$$

$v \geqslant c$, Re $\acute{\upsilon}_0 \geqslant c$

$$x = x_0 - (mc^2/g)\ln((1 - \acute{\upsilon}_0/c + (1 + \acute{\upsilon}_0/c)\exp[2\breve{g}(t - t_0)/(mc)])/2) - c(t - t_0) \quad (25.10b)$$

or, correspondingly,

$v < c$, Re $v_0 < c$

$$x = x_0 + (mc^2/g)\ln[(1 - v_0/c)/(1 - v/c)] - c(t - t_0) \quad (25.11a)$$

$v \geqslant c$, Re $\acute{\upsilon}_0 \geqslant c$

$$x = x_0 + (mc^2/\breve{g})\ln[(1 - \acute{\upsilon}_0/c)/(1 - v/c)] - c(t - t_0) \quad (25.11b)$$

The complexity of g and thus of the velocity causes x to be complex. The starship is thus generally at a point x in complex space which can be mapped to real space by a Reality grou transformation.

25.1.1 High Speed

To achieve the type of motion we desire the constant force value $\breve{g}$ required after Re $v \geqslant c$ must satisfy a special set of conditions. These conditions emerge from a consideration of the denominator of eq. 25.9b:

$$1 + ((c + \acute{\upsilon}_0)/(c - \acute{\upsilon}_0))\exp[2\breve{g}(t - t_0)/(mc)] \quad (25.12)$$

where $\acute{\upsilon}_0 \geqslant c$. If this denominator approaches zero then the speed v becomes infinite if g has an appropriate complex value. Let

$$\breve{g} = g_1 + ig_2 \quad (25.13)$$

If we wish the velocity to get very large (approach infinity) after some acceleration time interval $\Delta t = t_1 - t_0$ we set

$$1 + ((c + \acute{\upsilon}_0)/(c - \acute{\upsilon}_0))\exp[2\breve{g}\Delta t/(mc)] = 0 \quad (25.14)$$

with the result

$$g_2 = (mc/(2\Delta t))\{n\pi + \text{Im}\ln[(c - \acute{\upsilon}_0)/(c + \acute{\upsilon}_0)]\} \quad (25.15)$$

and

$$g_1 = (mc/(2\Delta t))\,\text{Re}\ln[(c - \acute{\upsilon}_0)/(c + \acute{\upsilon}_0)] \quad (25.16)$$

where n is an odd, positive integer, since $\acute{\upsilon}_0$ is complex in general. Eqs. 25.15 and 25.16 enable the real part of the velocity to become infinite in the time interval Δt. We assume n = 1 in the following discussions. Substituting in eq. 25.9b we obtain

$$v = c\{1 - 2/[1 + ((c + \acute{\upsilon}_0)/(c - \acute{\upsilon}_0))^{1-(t-t_0)/\triangle t} e^{in\pi(t-t_0)/\triangle t}]\} \quad (25.17)$$

25.1.2 From Light Speed to Enormous Speeds

In the example given in section 25.4 we considered an illustrative example of a starship accelerating to light speed. In this section we consider the second part of the acceleration: from light speed to enormous speeds taking advantage of the mechanism described in section 25.1.1. We will use an approximation to eq. 25.9b as its denominator approaches zero. Letting $t = t_1 + \tau$ where τ is small, and letting $\triangle t = t_1 - t_0$ then eq. 25.9b becomes

$$\begin{aligned} v &= c\{1 - 2/(1 + ((c + \acute{\upsilon}_0)/(c - \acute{\upsilon}_0))\exp[2\breve{g}(\triangle t + \tau)/(mc)])\} \\ &= c\{1 - 2/(1 - \exp[2\breve{g}\tau/(mc)])\} \\ &\simeq c\{1 - 2/(1 - (1 + 2\breve{g}\tau/(mc))\} \\ &\simeq c\{1 + (mc/\breve{g})(1/\tau)\} \\ &\simeq (\breve{g}^*mc^2/|\breve{g}|^2)(1/\tau) \end{aligned} \quad (25.23a)$$

Continuing the preceding illustrative example with $\acute{\upsilon}_0 = c + ic$, and m = 10,000 metric tons, and choosing $\triangle t$ = 30 days we find

$$\breve{g} = -4.66\times10^{13} + i6.43\times10^{13} \text{ gm-cm/sec}^2 \quad (25.23b)$$

Given the signs of g_1 and g_2 we see that

- For small negative τ both the real and imaginary parts of v approach $+\infty$ as $\tau \to 0$ from below.
- For small positive τ both the real and imaginary parts of v approach $-\infty$ as $\tau \to 0$ from above.

as displayed in Figs. 25.5 and 25.6. A starship can decide to switch off engines and "coast" at high speed towards the destination at some time close to the singularity point.

At $t = 30 - 1\times10^{-13}$ days ($\tau = 8.64\times10^{-9}$ sec) we find using eq. 25.23a that the starship velocity in the earth's reference frame is

$$v = 8547c + i11802c \quad (25.23b)$$

which becomes after a Reality group transformation

$$v_r = 14572c \quad (25.23c)$$

At 14572c any of the 100 or so known stars within 21 light years can be reached in a few hours. There is also time needed to decelerate the starship so the actual travel time would be longer. At 30,396c any point in the galaxy could be reached in about 4 years. *Thus Milky Way travel times become comparable to 16th century oceanic travel times via ships to various parts of the world!*

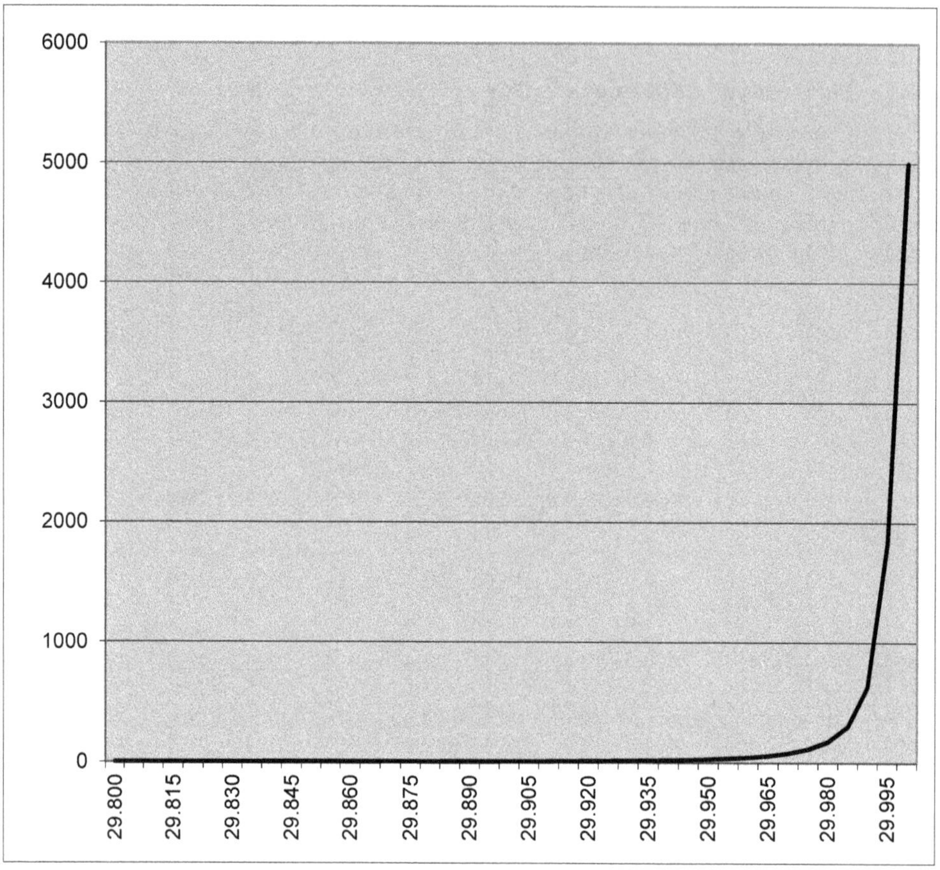

Figure 25.1. A plot of the real part of the velocity of a starship on its 29th and 30th earth day of travel up to 5,000c. The dynamics of this case are described in the text where the real speed reaches 14572c and beyond. Time is measured in earth days. Note: as the speed of the starship increases rapidly near the singularity point, time on the starship also passes more quickly so that the starship occupants do not experience very high acceleration. Starship time $t' \approx \beta t$ when $\beta \gg 1$ where t is earth time.

25.1.3 The Acceleration Experienced on the Starship

The rapid acceleration, particularly in the neighborhood of $\tau = 0$ raises the question of the inertial forces that would be experienced by passengers on the starship.

The calculation of the maximum acceleration begins with the inverse of the relativistic transformation from earth coordinates to starship coordinates (eq. 25.3):

$$F'^0 = \gamma(F^0 - \beta F^x/c)$$
$$F'^x = \gamma(F^x - \beta c F^0) \quad (25.24)$$
$$F'^y = F'^z = 0$$

which implies the apparent acceleration of the starship calculated in the starship's reference frame is

$$a' = F'^x/m = \breve{g}/m \quad (25.25)$$

whereas in the earth's reference frame, the acceleration a is given by the derivative of eq. 25.9b.

$$a = dv/dt$$
$$= 4(\breve{g}/m)((c + \acute{\upsilon}_0)/(c - \acute{\upsilon}_0))\exp[2\breve{g}(t - t_0)/(mc)]/\{1 +$$
$$+ ((c + \acute{\upsilon}_0)/(c - \acute{\upsilon}_0))\exp[2\breve{g}(t - t_0)/(mc)]\}^2 \quad (25.26)$$

At $t = t_1 + \tau$ we see

$$a \simeq -mc^2/(\breve{g}\tau^2)[1 + 2\breve{g}\tau/mc] \approx -mc^2/(\breve{g}\tau^2) \quad (25.26a)$$

in the earth's reference frame.

The acceleration experienced by the starship occupants, the relevant acceleration for the occupants, is

$$a' = \breve{g}/m = -4.66\times10^3 + i6.43\times10^3 \text{ gm-cm/sec}^2$$
$$= (-4.75 + i6.75)g_E \quad (25.27)$$
$$\equiv 8.25 g_E$$

by eq. 25.25 and 25.23b using a Reality group transformation in the last step. This acceleration (experienced by the starship occupants) is acceptable for short periods of time according to US Air Force studies.

25.1.4 Travel Time on the Starship – Suspended Animation

Another issue is the travel time experienced by the starship occupants. It will appear much, much longer than that measured on earth (See appendix 5-C.) For example if v = 5,000c then it will be approximately 5000 times longer. A 2 month trip from earth's view would take around 1,000 years from the view of the occupants of the starship. *Therefore a practical method of suspended animation must be found for long distance travel. A 4 month round trip to a star would require the starship occupants to be in suspended animation for approximately 1,700 years – starship time. With suspended animation they could be kept biologically roughly "in sync" with the earth measured travel time of 4 months (plus time spent at the destination) despite the starship elapsed time of about 1,700+ years.*

25.1.5 Constant Superluminal Starship Travel

Having reached an enormous *real* speed such as a speed between 5000c and 30,000c we can turn off the superluminal engines.

The starship then moves at this constant speed in the absence of forces (and neglecting gravity and other minor perturbative forces). At a real speed of 5000c any place in the galaxy is a short travel time away. And nearby galaxies are reachable as well. Figure 25.2 shows the time required to reach various interesting destinations at 30,000c.

Destination	Distance (ly)	Approximate Travel Time (years)
To the other end of the Milky Way Galaxy	**100,000**	**3**
To the Center of the Milky Way	**30,000**	**1**
Large Magellenic Galaxy	**150,000**	**5**
Small Magellenic Galaxy	**200,000**	**7**
Andromeda Galaxy	**2,000,000**	**70**

Figure 25.2. "Coasting" part of travel time to various destinations at a real velocity of 30,000c.

Since much, much higher "coasting" velocities are also possible, almost the entire visible universe becomes accessible to Mankind. Mankind then has an incredible future if it has the will to seize it.

Figure 25.3. A visualization of a starship. The outer disk contains the colliding hadron ring(s) which generate quark-gluon fireballs in a "combustion chamber." The fireball expands through a "rocket nozzle" generating a complex-valued thrust that enables the starship to exceed the speed of light. The four nuclear engines depicted on the underside of the ship provide "intra-solar system" speeds.

25.2 Phase Two: Starship Slingshot into the Megaverse

In this section we will consider a baryonic force mechanism for trips into the Megaverse using a slingshot around a neutron star.

25.2.1 The Characteristics of a Rapidly Spinning Neutron Star

It is estimated that there are 100 million neutron stars in our galaxy. It is likely that other galaxies of comparable size have a similar plethora of neutron stars. Neutron stars represent the end point of the evolution of main sequence stars with initial masses above ten solar masses. If such a star evolves and declines to a mass between the Chandrasekhar Limit (1.44 solar masses) and three solar masses, then it will become a neutron star – a star composed primarily of neutrons compacted to a radius of approximately ten kilometers with an enormous density of the order of 10^{17} kg/m^3. We expect neutron stars to be plentiful in our universe and other mature universes of the Megaverse.

Neutron stars are thought to have a series of layers: an outer crust composed of a crust of iron with an estimated density of the order of 10^9 kg/m^3 and an innermost purely neutron (possibly a quark-gluon plasma) core region with a density of order of 10^{18} kg/m^3 or more.

An important factor in the nature of neutron stars is their unusually large rotation rates ranging up to over 700 rotations per second. The rotation of a neutron star (assuming a baryonic gauge force exists as we do) generates a baryonic magnetic field as well as a baryonic electric field. However we shall see that the neutron star spin-generated baryonic magnetic and electric fields only impart a force to a uniship within our universe. Thus the Coulomb baryonic force is the key to slingshoting a uniship into the Flatverse as we pointed out in earlier books.

The important neutron star parameters for the determination of the slingshot trajectory into the Flatverse are its mass, its density as function of radius, and its spin. From these parameters we can determine its baryon number, its baryonic current density, and its baryonic electric and magnetic fields. Then the trajectory of a uniship into the Megaverse can be determined.

25.2.2 The Baryonic Fields of a Rapidly Spinning Neutron Star

Spinning neutron stars, and they all spin at varying rates except in extreme old age, generate baryonic fields which *might* have provided a mechanism to escape from our universe into the Megaverse.[294] In fact, we shall see that *the spin generated force does not* have a component in the direction of the Megaverse spatial coordinates. Thus they do cannot be used to exit into the Megaverse.

The baryonic Coulomb force does have components in Megaverse directions outside our universe and can slingshot uniships (starships designed for travel in the Megaverse) into the Megaverse.

This section calculates the spin-dependent baryonic fields for a neutron star of mass M, internal mass density ρ(r), radius R, volume V, and rotation rate Ω measured in rotations per second. We assume the neutron star is rigid and rotates uniformly. We believe this is a reasonable assumption in view of the close packing of the neutron star throughout most of its body.

---

[294] It is possible that other types of stars such as white dwarfs and perhaps even ordinary stars might be used by uniships to escape from a universe into the Megaverse. The enormous mass, and extremely small size, of neutron stars make them more favorable for slingshot escape from the universe.

25.2.3 Baryonic Current of a Neutron Star

We assume that a neutron star is spherical to good approximation although a rapidly spinning neutron star will deviate slightly from a sphere. We also assume that the neutron star can be treated as point-like since its radius of the order of 10 km is very small compared to the closest point of approach of a uniship which will be, at minimum, tens of thousands of kilometers distant.

The baryonic current of a spinning neutron star, if the coordinate system is oriented so that it spins around the z-axis, yields a current J_φ in the φ direction of spherical coordinates (r, θ, φ). The current is to good approximation

$$J_\varphi = \int drd\theta d\varphi r^2 \beta_B \Omega \rho(r)/m_n = (\beta_B\Omega/m_n)\int drd\theta d\varphi r^2 \rho(r) = \beta_B \Omega M/m_n \qquad (25.28)$$

where β_B is the baryonic charge (analogous to e in electromagnetism), where m_n is the mass of a neutron, $\rho(r)/m_n$ is the neutron (nucleon) number density at radius r, and where the current at a radial distance r is $\beta_B\Omega\rho(r)/m_n$.

As discussed in Blaha (2014) the baryonic "Coulomb" potential in Megaverse coordinates is

$$\phi(y_1, y_2, \ldots, y_{15}) = (\beta_B^2/4\pi)N_1N_2/(y_1^2 + y_2^2 + \ldots + y_{15}^2)^{1/2} \qquad (25.29)$$

where $\alpha_B = (\beta_B^2/4\pi)$ is the equivalent of the electromagnetic fine structure constant α. Earlier (chapter 15) we estimated the order of magnitude of α_B using the not very well understood differences[295] between various experiments to determine the gravitational constant G. We found the order of magnitude to be

$$\alpha_B = \beta_B^2/4\pi \simeq .118\ Gm_H^2 \qquad (25.30)$$

where m_H is the mass of a hydrogen atom.

We conclude this subsection by determining the baryonic current from the above equations:

$$\begin{aligned} J_\varphi &= (4\pi \cdot 0.118G)^{1/2}\Omega M \\ &= 1.22\ G^{1/2}\Omega M \end{aligned} \qquad (25.31)$$

using the approximation $m_H = m_n$.

25.2.4 The Sixteen Component Baryonic Vector Potential

In chapter 15 we described some of the features of the baryonic vector potential, which we recapitulate here for the reader's convenience. As in electromagnetism there is an antisymmetric tensor of the second rank that appears in the free part of the baryonic field $F_{Bu\mu\nu}(y)$ lagrangian:

$$\mathscr{L}_{Bu} = -\tfrac{1}{4} F_{Bu}{}^{ij}(y)F_{Buij}(y) \qquad (25.32)$$

where

$$F_{Buij}(y) = \partial B_{ui}(y)/\partial y^j - \partial B_{uj}(y)/\partial y^i \qquad (25.33)$$

---

[295] The recent experiment by T. Quinn et al, Phys. Rev. Lett. **111**, 101102 (2013) differs significantly from the 2010 CODATA world average of previous experiments. See P. J. Mohr, B.N. Taylor, and D. B. Newell, Rev. Mod. Phys. **84**, 1527 (2012).. We attribute the difference to the baryonic force between masses.

and i, j = 1, 2, … , 16. The 16th coordinate corresponds to the time coordinate. While the coordinates are complex in general we will treat the 15 spatial coordinates as real and the 16th coordinate as pure imaginary with the resulting invariant interval

$$ds^2 = dy_1^{\ 2} + dy_2^{\ 2} + \ldots + dy_{15}^{\ 2} - c^2 dy_{16}^{\ 2} \qquad (25.34)$$

which is invariant under 16 dimensional Lorentz transformations. The coordinates can be transformed into complex-valued coordinates using the Reality group defined earlier.

The tensor F_{Buij} is conveniently separated into an baryon electric part and a baryon magnetic part in a manner similar to the separation of the electromagnetic fields into electric and magnetic fields. However the 15 spatial dimensions change the forms of the baryon fields. Analogously to electromagnetism the baryonic force is given by

$$f_i = F_{Buij}(y) J_B^{\ j}/c \qquad (25.35)$$

where $J_B^{\ j}$ is the jth baryonic current.

The baryon "electric" field is

$$E_{Bui} = -F_{Bui0}(y)/c \qquad (25.36)$$

while the baryon "magnetic" field is

$$B_{Bui} = \varepsilon_{ijk} F_{Bu}^{\ jk}(y) \qquad (25.37)$$

where i, j, k = 1, 2, … , 15 and where ε_{ijk} is a totally anti-symmetric tensor with component values ±1. If i < j < k then ε_{ijk} is +1. Even permutations of these three indices yield a value of +1 for the tensor components. Odd permutations of these three indices yield a value of –1. For example, $\varepsilon_{246} = +1$, $\varepsilon_{426} = -1$, $\varepsilon_{642} = -1$, $\varepsilon_{264} = -1$, $\varepsilon_{462} = +1$, $\varepsilon_{624} = +1$.

With these definitions of the $\mathbf{E_{Bu}}$ and $\mathbf{B_{Bu}}$ fields we derive the 16-dimensional generalization of the *Lorentz force law* for a baryon of charge q and 15-velocity v_j:

$$F_i = qE_{Bui} + q\varepsilon_{ijk} v_j B_{Buk}/c \qquad (25.38)$$

for i = 1, 2, … , 15. One important difference from the 4-dimensional case is the forms of the $\mathbf{E_{Bu}}$ and $\mathbf{B_{Bu}}$ fields

$$E_{Bui} = -F_{Bui0}(y)/c = [-\partial B_{u16}(y)/\partial y^i - \partial B_{ui}(y)/\partial y^{16}]/c \qquad (25.39)$$

or, expressed as a 15-vector,

$$\mathbf{E_{Bu}} = [-\nabla_{15}\phi(y) - \partial \mathbf{B_u}(y)/\partial y^{16}]/c \qquad (25.40)$$

where ϕ is the baryonic Coulomb potential $B_{u16}(y)$, ∇_{15} is the 15-dimensional grad operator, and $\mathbf{B_u}(y)$ is the baryonic 15-vector potential.

The 15-dimensional baryon magnetic field has the form of eqn. 25.37. The baryon magnetic field exhibits more complexity than the 3-dimensional magnetic field of electromagnetism:

$$B_{Bu1} = \varepsilon_{1jk}F_{Bu}{}^{jk}(y)/c = [F_{Bu}{}^{23}(y) + F_{Bu}{}^{24}(y) + \ldots + F_{Bu}{}^{2,15}(y) + F_{Bu}{}^{34}(y) + F_{Bu}{}^{35}(y) + \ldots + F_{Bu}{}^{3,15}(y) + F_{Bu}{}^{45}(y) + \ldots + F_{Bu}{}^{14,15}(y)]/c \qquad (25.41)$$

25.2.5 The Baryonic Electric and Magnetic Field Strengths

In this section we will calculate the baryonic electric and magnetic field strengths for a neutron star due to its spin. The spatial field strengths are determined by the dynamical equation:

$$\partial F_{Bu}{}^{ij}(y)/\partial y^i = J^j \qquad (25.42)$$

The current for a neutron star is well approximated by the constant current in the φ direction (in the spherical coordinates)

$$J_\varphi = \beta_B \Omega M/m_n \qquad (25.43)$$

due to the neutron star's small size. Expressing J_φ in rectangular coordinates assuming the rotation is along the z-axis we find

$$J_x = -\sin\varphi\ J_\varphi \qquad (25.44)$$

$$J_y = \cos\varphi\ J_\varphi \qquad (25.45)$$

We will use the relative flatness of space, and the small size of the neutron star neighborhood, to identify the x, y, and z of our universe with the Megaverse coordinates y^1, y^1, and y^3. Inserting eqns. 25.44 and 25.43 in eq. 25.43 yields fifteen equations:

$$\begin{aligned} \partial F_{Bu}{}^{ix}(y)/\partial y^i &= -\sin\varphi\ J_\varphi \\ \partial F_{Bu}{}^{iy}(y)/\partial y^i &= \cos\varphi\ J_\varphi \\ \partial F_{Bu}{}^{ij}(y)/\partial y^i &= 0 \end{aligned} \qquad (25.46)$$

for j = 4, … , 15. These 15 equations have a solution that gives a magnetic force that is solely within the three spatial dimensions of our universe. *Thus they cannot participate in a slingshot into the Megaverse.*

25.2.6 The Baryonic Coulomb Force Slingshot into the Megaverse

The 16-dimensional version of the Lorentz force is

$$F_i = qE_{Bui} + q\varepsilon_{ijk}v_jB_{Buk}/c \qquad (25.47)$$

where

$$\mathbf{E_{Bu}} = [-\nabla_{15}\phi(y) - \partial \mathbf{B_u}(y)/\partial y^{16}]/c \qquad (25.48)$$

We have seen that the baryonic force slingshot is wholly derived from the baryonic Coulomb force:

$$\phi(y_1, y_2, \ldots, y_{15}) = (\beta_B{}^2/4\pi)N_1N_2/(y_1{}^2 + y_2{}^2 + \ldots + y_{15}{}^2)^{1/2} \qquad (25.49)$$

between two baryon masses with baryon numbers N_1 and N_2. The baryonic Coulomb force is:

$$F_i = \nabla_{15i}\phi(y) \tag{25.50}$$

where ∇_{15i} is the ith component of the 15-dimensional grad operator $\boldsymbol{\nabla}_{15}$.

The part of the Lorentz force that slingshots a uniship into the Megaverse is

$$\begin{aligned} F_{isling} &= \partial\phi(y)/\partial y^i \\ &= (\beta_B{}^2/4\pi)N_1N_2\, y_i\,/(y_1{}^2 + y_2{}^2 + \ldots + y_{15}{}^2)^{3/2} \end{aligned} \tag{25.51}$$

for i = 4, 5, … , 15.

A uniship will have a baryonic "Coulomb" force directing it out of our universe into the Megaverse. The baryonic force between the uniship and the neutron star is repulsive since they are both composed of a majority of baryons.

This force will undoubtedly be small compared to the gravitational forces during a slingshot maneuver. Thus we can see that the uniship will slowly ease out of our universe. For a time it will be partly in and partly out of the universe creating a physical situation not hitherto encountered in physics. It has some advantages since, for example, it allows an "umbrella" of thrust tubes that originally are in 3-dimensional space to widen into a 15-dimensional umbrella of thrust tubes that would enable the uniship to travel in any direction in the Megaverse – Megaverse maneuverability. We shall consider this possibility later.

Since the force of gravity is confined to within our universe it will have no effect on directions outside our universe. Baryonic forces in directions within our universe will be of little consequence compared to gravitation.

In directions into the Megaverse, gravitation forces being absent, the baryonic force will be the sole force.

Thus we find a clear division: gravitation dominates in directions within the universe; baryonic force dominates in directions out of our universe. Fig. 25.4 depicts the trajectory of a uniship in a slingshot maneuver. Note the attractive hyperbolic motion due to the dominance of gravity in our universe.

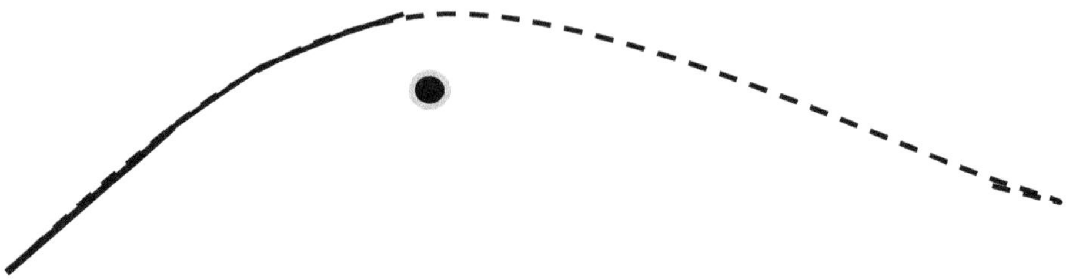

Figure 25.4. The trajectory of a uniship in a slingshot maneuver with a neutron star (the dark circle). The repulsive baryonic force causes the "turn" away from the star into the Megaverse. The solid line corresponds to the time in which the uniship is wholly within the universe and dominated by gravity. The dotted line reflects the transition of the uniship into the Megaverse.

25.2.7 Point-like Uniship Slingshots

The major problems of a close approach of perhaps 100,000 km to a neutron star are the large gravity of the neutron star, its strong tidal gravitation effects (stresses) on the uniship's structure, its strong magnetic field, and its emission of large amount of primarily x-ray/gamm ray radiation. These properties of a neutron star neighborhood would appear to significantly affect the structural integrity of the uniship, and, more importantly, seriously impact on the safety and life of its human crew.

Fortunately there is a saving grace in this physical environment. A uniship approaching the neutron star could resolve these issues by traveling at extremely high speed[296] so that the time spent in the "danger" zone of the neutron star would be very small thus sharply reducing its deleterious effects.

25.2.8 The Uniship Slingshot Trajectory

The slingshot trajectory of a uniship is approximately a hyperbola in our universe due to the dominance of gravity. As it approaches a neutron star, with a distance of closest approach of perhaps 100,000 km, a uniship could be programed to take perhaps 3 seconds or so to circle around the neutron star. Consequently the uniship need only spend a minimal time near the neutron star with little gravitational tidal stress, magnetic field exposure, and radiation exposure issues.

Since the difficulties of a neutron star slingshot are surmountable, we will now turn to the issue of a uniship escape from our universe. We are fortunate that the neutron star's force on the uniship can be conveniently divided into two parts: gravitation which only influences the spatial motion of the uniship in our universe's coordinates, and the baryonic force which only significantly influences the uniship's other 12 spatial Megaverse coordinates. The weakness of the baryonic force compared to gravity makes its impact on motion in our universe's spatial coordinates negligible.

The baryonic force generated by the interaction of a uniship's baryons with the baryons within the neutron universe causes the uniship's course to be deflected into the Megaverse.

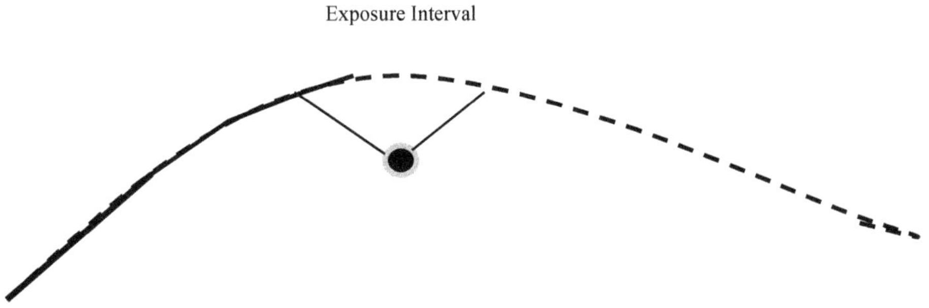

Figure 25.5. Depiction of the uniship exposure interval of closest approach to a neutron star. The time interval spent in this region might be about three second or so. The uniship speed as it approaches might be about 0.7c or so to avoid capture by the neutron star. When the uniship totally exits our universe it disappears from view since electromagnetic radiation (light) from the

---

[296] The starship would approach at perhaps c/2 and acquire an additional speed due to gravity of c/3. Combining these speeds using the rules of special relativity for the addition of velocities yields an approach speed of 0.7c near the neutron star.

uniship (or any object) "cannot penetrate" the boundary of our universe.[297] The dashed line indicates the partial exit of the uniship from our universe into the Megaverse with Megaverse velocity and momentum components.

25.2.9 Uniship Neutron Star Slingshot Dynamics

In this subsection we will describe the dynamics of the neutron star slingshot. We will assume a flat space-time in our universe with the understanding that space-time may be significantly curved in the immediate vicinity of the neutron star. We assume the uniship will not enter that region. We will define the neutron star to be at the origin of the Megaverse coordinates. Thus $y_i = 0$ for i = 1, … , 15. We will assume the universe spatial coordinates x_i to be equal to the first three Megaverse coordinates:

$$y_i = x_i \qquad (25.52)$$

for i = 1, 2, 3. We are allowed to do this by the flat space-time assumption for the universe.

The total potential energy of the uniship in the neutron star's reference frame is[298]

$$V_{tot} = -GM_1M_2/r_u + (\beta_B^2/4\pi)\, N_1N_2/r_F \qquad (25.53)$$

where M_1 and M_2 are the masses of the neutron star and uniship, N_1 and N_2 are their baryon numbers, and where

$$r_u^2 = x_1^2 + x_2^2 + x_3^2 = y_1^2 + y_2^2 + y_3^2 \qquad (25.54)$$
$$r_F^2 = y_1^2 + \ldots + y_{15}^2 = r_u^2 + y_4^2 + \ldots + y_{15}^2 \qquad (25.55)$$

The force is the gradient of the potential

$$\begin{aligned} F^i_{slingshot} &= \partial V_{tot}/\partial y_i \\ &= GM_1M_2\,(\delta^{i1} + \delta^{i2} + \delta^{i3}) y^i/r_u^{3/2} - (\beta_B^2/4\pi) N_1N_2\, y^i/r_F^{3/2} \end{aligned} \qquad (25.56)$$

where the Kronecker delta functions restrict the gravity force to the spatial coordinates of the universe.

The dynamic equation of the uniship motion[299] is

$$dp^i/d\tau = F^i_{slingshot} \qquad (25.57)$$

where τ is the invariant interval. We now consider the initial phase of the escape to the Megaverse in which

$$r_u >> r_F - r_u \qquad (25.58)$$

---

[297] It cannot penetrate in the sense that it would have to travel an enormous distance from the edge of the universe in universe coordinates.

[298] Again we note that we are assuming a uniship trajectory in flat space-time so that we may use special relativistic potentials and dynamic equations.

[299] The neutron star is assumed to be stationary due to the largeness of its mass relative to the uniship.

The universe spatial distance is much greater than the purely Megaverse spatial distance $r_F - r_u$. In this case eq. 25.56 has different forms to good approximation for i = 1, 2, 3 and the remaining Megaverse spatial coordinates:

$$dp^i/d\tau \simeq GM_1M_2y^i/r_u^{3/2} \quad (25.59)$$

for i = 1, 2, 3 and

$$dp^i/d\tau = -(\beta_B^2/4\pi)N_1N_2\, y^i/r_F^{3/2} \quad (25.60)$$

for i = 4, … , 15. Eq. 25.59 yields a solution of the central force problem which in the present case is an approximately hyperbolic trajectory of the form depicted in Fig. 25.5.

Due to our well-justified assumption that the distance into the Megaverse will be small compared to the distance of the uniship from the neutron star we can approximate

$$r_F^{-3/2} \simeq r_u^{-3/2}[1 - (3/2)(r_F^2 - r_u^2)/r_u^2] \quad (25.61)$$

Eq. 5.9 then becomes approximately

$$dp^i/d\tau \simeq -(\beta_B^2/4\pi)N_1N_2\, y^i[1 - (3/2)(r_F^2 - r_u^2)/r_u^2]/r_u^{3/2} \quad (25.62)$$

or

$$dp^i/d\tau \simeq -(\beta_B^2/4\pi)N_1N_2\, y^i[1 - (3/2)\Sigma y_k^2/r_u^2]/r_u^{3/2} \quad (25.63)$$

where the sum in eq. 5.12 is from k = 4, …, 15. The above equations yield an initially exponential-like trajectory into the Megaverse to leading order.

Thus we have shown that the neutron star slingshot clearly drives the uniship out of the universe into the Megaverse. The uniship passes into the Megaverse. For a small time it is partly in and partly out of the universe.

25.2.10 Umbrella-Shaped Uniship Slingshots

We take it for granted that we can move in one spatial direction or another with ease. However when one enters a higher dimensional space from a space of lower dimension, movement in the additional dimensions, which requires the expenditure of force in those dimensions, becomes an issue. A lower dimension object does not automatically have forces within it in the additional directions and so it cannot move itself, or part of itself in those directions.

In the previous subsections we saw how to use the baryonic force to escape from our universe to the higher dimension Megaverse. In this subsection we will show how to give "wings" to a uniship so that it will have the ability to maneuver in any direction in the Megaverse. To achieve this capability we will have to design the uniship so that it will expand in all Megaverse directions as it enters the Megaverse. This will require an umbrella-like configuration that will use the baryonic force to open the umbrella in all Megaverse directions. The spokes of the umbrella will be long thrust tubes through which uniship thrust can be directed to move the uniship in a desired direction. Fig. 25.6 is a depiction of the simplest form of umbrella uniship.

25.2.11 Scenario for the Opening of a Uniship Umbrella

As an umbrella uniship slingshots around a neutron star the body of the ship including fuel tanks is rigid and moves as a unit. The spokes of the umbrella, the thrust tubes, are moveable and will each move differently because of their differing average distance from the neutron star. As a result they will point in different directions in the Megaverse and can be further moved within the Megaverse relative to each other to deliver thrust in any Megaverse direction. After positioning they can be locked in place and used to maneuver the uniship towards any universe.

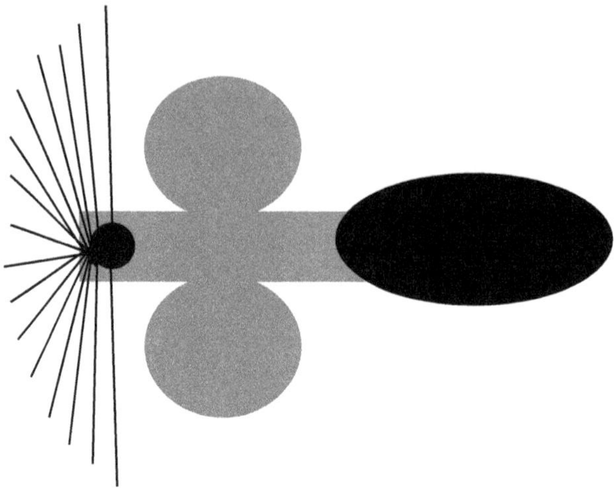

Figure 25.6. Tentative umbrella-like uniship design with the spokes of the umbrella forming a fan. The thrust tubes (umbrella spokes) extend kilometers from the thrust power generator(s) core to enable the baryonic force to maneuver them in the 15 different Megaverse directions. The thrust tubes are able to swivel into all 15 spatial directions in response to the baryonic force as the uniship enters the Megaverse. Two fuel spheres are depicted under the assumption that one holds hydrogen and the other holds anti-hydrogen since they are presently the most powerful known possible energy source. The black forward part is for crew, supplies, and cargo. The rear gray part holds the engine apparatus and other related engine components.

25.2.12 Equations for the Motion of the Thrust Tubes Entering the Megaverse

Earlier we developed the equations for the slingshot mechanism for a compact rigid uniship. In this subsection we will extend the equations to the case of a rigid uniship with an umbrella with a fan shape (a flat array of spokes as picture in Fig. 25.6. The array is flat with each spoke in the plane of the ship's trajectory.) A uniship with a true umbrella of spokes (thrust tubes) is another possibility that may be of importance. This is a technical question that we will not address.

We define radial distances from the neutron star center for a "fan-shaped" umbrella with the nth spoke end[300] at radial distance r_{un}. Then eq. 25.61 becomes

[300] The spoke center of mass actually.

$$r_{Fn}^{-3/2} \simeq r_{un}^{-3/2}[1 - (3/2)(r_{Fn}^{2} - r_{un}^{2})/r_{un}^{2}] \quad (25.64)$$

for the end of each spoke and eq. 25.62 then becomes

$$dp_n^{i}/d\tau \simeq -(\beta_B^{2}/4\pi)N_1N_2\, y^{i}[1 - (3/2)(r_{Fn}^{2} - r_{un}^{2})/r_{un}^{2}]/r_{un}^{3/2} \quad (25.65)$$

or

$$dp_n^{i}/d\tau \simeq -(\beta_B^{2}/4\pi)N_1N_2\, y^{i}[1 - (3/2)\Sigma y_k^{2}/r_{un}^{2}]/r_{un}^{3/2} \quad (25.66)$$

for the nth spoke where the sum in eq. 25.66 is from k = 4, …, 15. For each spoke radial distance r_{un} these equations yield different initial trajectories into the Megaverse to leading order.

Thus the spokes will enter the Megaverse in different Megaverse directions since they are not rigidly attached to the uniship body. Upon entry they can change each other's direction giving a final configuration with full Megaverse maneuverability. Uniships can now move in any Megaverse spatial directions.

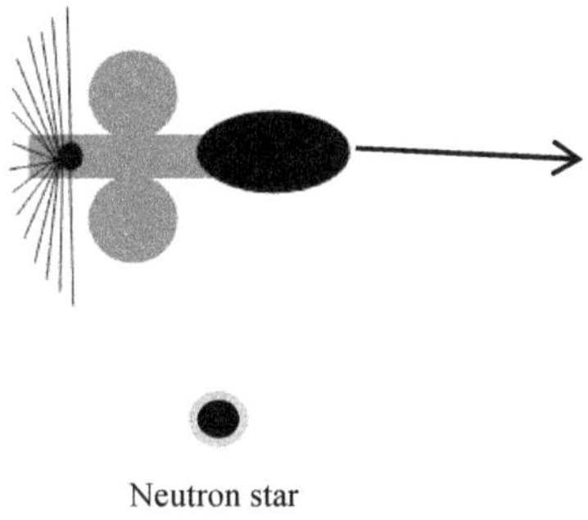

Figure 25.7. Uniship slingshot past neutron star. Note the umbrella spokes which are attached to the uniship but moveable will respond differently to the baryonic force since they are at different radial distances from the neutron star and thus feel differing amounts of force. The baryonic force will twist them in different directions. They then can be re-oriented by the uniship computer to provide mobility in all Megaverse directions.

26. Some Experimental Tests of Our Extended Standard ModelTheory

There are many tests of the Extended Standard Model that could be made in the near future to verify its correctness. This chapter lists possible experimental tests.

26.1 Are Neutrinos Tachyons?

Although neutrino experiments have tested this hypothesis in the past. The results have been mixed as we discussed earlier in this book. (Appendix 5-B) The problem is the small mass of neutrinos which results, at best, in velocities only slightly greater than the speed of light. Sensitive, careful tests are needed to decide this question.

26.2 Are Quarks Complexons?

We suggest both up-type and down-type quarks are complexions – particles with complex-valued velocities. Their confinement within hadrons makes velocity measurements difficult. Deep inelastic electron-nucleon scattering experiments have yielded structure functions that might show complex velocity partons (quarks). However curve fitting can mask any complexity of the parton velocity. Anomalies in the parton momentum spectrum suggest there is a possibility that complexon behavior might be detected. Another area where complex velocity quarks might be detected is in quark-gluon plasmas generated by the collision of heavy ions such as gold or lead. The models that have been created for the limited data currently available are not mature and further experimental and theoretical work is needed to elucidate this question. One might inquire whether parts of the quark-gluon plasma are moving faster than the speed of light.

26.3 Dark Matter Structure

Dark Matter is currently a cosmological phenomena. There is some evidence of a force existing between Dark Matter blobs in studies of colliding galaxies. However the force may well be gravity. If it is more than gravitation then it is possible that the Dark baryon force described in chapter 15 might also be partly the cause. This issue will probably ned to be resolved at accelerators. Some recent experiments suggest that Dark particles have been created. If a sufficiently large number are created and they can be led to interact – a difficult task – then the properties of Dark Matter could be found. We have suggested in chapter 8 that they have a Dark SU(2)⊗U(1) (probably broken) symmetry and an interaction with ElectroWeak gauge bosons. Sections 13.2.1 and 13.2.2 provides the detailed lagrangian terms.

26.4 Dark Energy – What is it? Two-Tier Y Particles?

Cosmological studies have suggested the existence of a Dark Energy powering the expansion of the universe. In chapters 19 - 21 we have provided a detailed Big Bang model that embodies inflation based on the role of the Two-Tier Y field (chapter 18) which not only yields finite results at the beginning of the Big Bang but also causes perturbation theory calculations to be finite – no divergences. An improved knowledge of Big Bang Cosmology near the beginning may help confirm our model or a variant thereof.

26.5 A Baryonic Force?

In chapter 15 we have suggested that recent experiments on the gravitational constant G can be reconciled if a baryonic force exists. We have even done a tentative estimate of its coupling constant. More experiments on G's value should be done (section 15.1.1) to obtain an accurate value for G removing current disagreements and isolating a baryonic force term in the total force between two lumps of matter.

26.6 A 4th Generation?

We have suggested (chapter 16) there is good theoretical reason to believe a 4th generation of fermions and Dark fermions existed. If it exists the generations trend of lepton and quark masses suggests that 4th generation masses would be of the order of hundreds, or more, of times the 3rd generation masses. Consequently it appears that the CERN LHC may not have enough energy to discover the next generation. A higher energy accelerator proposal is under consideration. This issue is a strong motivation for its construction as is further studies of quark-gluon plasmas.

27. Summary of the Author's Particle Physics Books (2001- 2014)

Below is a list of books that contain some additional information on our derivation/construction of The Extended Standard Model, which was a fourteen year effort with some deviations from the path along the way.

Blaha, S., 1998, *Cosmos and Consciousness* (Pingree-Hill Publishing, Auburn, NH, 1998).

_______2002, *A Finite Unified Quantum Field Theory of the Elementary Particle Standard Model and Quantum Gravity Based on New Quantum Dimensions™ & a New Paradigm in the Calculus of Variations* (Pingree-Hill Publishing, Auburn, NH, 2002).

_______2004, *Quantum Big Bang Cosmology: Complex Space-time General Relativity, Quantum Coordinates™ Dodecahedral Universe, Inflation, and New Spin 0, ½, 1 & 2 Tachyons & Imagyons* (Pingree-Hill Publishing, Auburn, NH, 2004).

______ *2005a, Quantum Theory of the Third Kind: A New Type of Divergence-free Quantum Field Theory Supporting a Unified Standard Model of Elementary Particles and Quantum Gravity based on a New Method in the Calculus of Variations* (Pingree-Hill Publishing, Auburn, NH, 2005).

______, 2005b, *The Metatheory of Physics Theories, and the Theory of Everything as a Quantum Computer Language* (Pingree-Hill Publishing, Auburn, NH, 2005).

______, 2005c, *The Equivalence of Elementary Particle Theories and Computer Languages: Quantum Computers, Turing Machines, Standard Model, Superstring Theory, and a Proof that Gödel's Theorem Implies Nature Must Be Quantum* (Pingree-Hill Publishing, Auburn, NH, 2005).

______, 2006a, *The Foundation of the Forces of Nature* (Pingree-Hill Publishing, Auburn, NH, 2006).

______, 2006b, *A Derivation of ElectroWeak Theory based on an Extension of Special Relativity; Black Hole Tachyons; & Tachyons of Any Spin.* (Pingree-Hill Publishing, Auburn, NH, 2006).

_______, 2007a, *Physics Beyond the Light Barrier: The Source of Parity Violation, Tachyons, and A Derivation of Standard Model Features* (Pingree-Hill Publishing, Auburn, NH, 2007).

______, 2007b, *The Origin of the Standard Model: The Genesis of Four Quark and Lepton Species, Parity Violation, the ElectroWeak Sector, Color SU(3), Three Visible Generations of Fermions, and One Generation of Dark Matter with Dark Energy* (Pingree-Hill Publishing, Auburn, NH, 2007).

______, *2008a, A Direct Derivation of the Form of the Standard Model From GL(16) (Pingree-Hill Publishing, Auburn, NH, 2008).*

______, 2008b, *A Complete Derivation of the Form of the Standard Model With a New Method to Generate Particle Masses Second Edition* (Pingree-Hill Publishing, Auburn, NH, 2008)

______, 2009, *The Algebra of Thought & Reality: The Mathematical Basis for Plato's Theory of Ideas, and Reality Extended to Include A Priori Observers and Space-Time Second Edition* (Pingree-Hill Publishing, Auburn, NH, 2009).

______, 2010a, *Operator Metaphysics: A New Metaphysics Based on a New Operator Logic and a New Quantum Operator Logic that Lead to a Mathematical Basis for Plato's Theory of Ideas and Reality* (Pingree-Hill Publishing, Auburn, NH, 2010).

______, 2010b, *The Standard Model's Form Derived from Operator Logic, Superluminal Transformations and GL(16)* (Pingree-Hill Publishing, Auburn, NH, 2010).

______, 2011a, *21st Century Natural Philosophy Of Ultimate Physical Reality* (McMann-Fisher Publishing, Auburn, NH, 2011).

______, 2011b, *All the Universe! Faster Than Light Tachyon Quark Starships & Particle Accelerators with the LHC as a Prototype Starship Drive Scientific Edition* (Pingree-Hill Publishing, Auburn, NH, 2011).

______, 2011c, *From Asynchronous Logic to The Standard Model to Superflight to the Stars* (Blaha Research, Auburn, NH, 2011).

______, 2011d, *The Standard Model Symmetries and U(4) and U(16) Complex General Relativity* (Blaha Research, Auburn, NH, 2011).

______, 2012a, *From Asynchronous Logic to The Standard Model to Superflight to the Stars volume 2: Superluminal CP and CPT, U(4) Complex General Relativity and The Standard Model, Complex Vierbein General Relativity, Kinetic Theory, Thermodynamics* (Blaha Research, Auburn, NH, 2012).

______, 2012b, *Standard Model Symmetries, And Four And Sixteen Dimension Complex Relativity; The Origin Of Higgs Mass Terms* (Blaha Reasearch, Auburn, NH, 2012).

______, 2013b, *The Bridge to Dark Matter; A New Sister Universe; Dark Energy; Inflatons; Quantum Big Bang; Superluminal Physics; An Extended Standard Model Based on Geometry* (Blaha Reasearch, Auburn, NH, 2013).

______, 2014a, *Universes and Megaverses: From a New Standard Model to a Physical Megaverse; The Big Bang; Our Sister Universe's Wormhole; Origin of the Cosmological Constant, Spatial Asymmetry of the Universe, and its Web of Galaxies; A Baryonic Field between Universes and Particles; Flatverse Extended Wheeler-DeWitt Equation* (Blaha Reasearch, Auburn, NH, 2014).

28. Summary of the Author's Books on Space travel, and Starship Travel

Below is a list of the author's books on space travel, starships, and travel in the universe and the Megaverse (which we called the Multiverse in 2014.)

______, 2009a, *Bright Stars, Bright Universe Advancing Civilization by Colonization of the Solar System and the Stars using a Fast Quark Drive* (Pingree-Hill Publishing, Auburn, NH, 2009)

______, 2009b, *To Far Stars and Galaxies: Second Edition of Bright Stars, Bright Universe* (Pingree-Hill Publishing, Auburn, NH, 2009).

______, 2011b, *All the Universe! Faster Than Light Tachyon Quark Starships & Particle Accelerators with the LHC as a Prototype Starship Drive Scientific Edition* (Pingree-Hill Publishing, Auburn, NH, 2011).

______, 2011c, *From Asynchronous Logic to the Standard Model to Superflight to the Stars* (Blaha Research, Auburn, NH, 2011).

______, 2013a, *Multi-Stage Space Guns, Micro-Pulse Nuclear Rockets, and Faster-Than-Light Quark-Gluon Ion Drive Starships* (Blaha Research, Auburn, NH, 2013).

______, 2014b, *All the Multiverse! Starships Exploring the Endless Universes of the Cosmos Using the Baryonic Force* (Blaha Research, Auburn, NH, 2014).

______, 2014c, *All the Multiverse! II Between Multiverse Universes: Quantum Entanglement Explained by the Multiverse Coherent Baryonic Radiation Devices – PHASERs Neutron Star Multiverse Slingshot Dynamics Spiritual and UFO Events, and the Multiverse Microscopic Entry into the Multiverse* (Blaha Research, Auburn, NH, 2014).

Appendix A. C, P and T for Tachyons and Dirac Fermion Fields

A.1 Parity for Dirac and Tachyonic Fermions

In this section we will describe the parity transformation for tachyons. We start with the usual Dirac parity transformation to elucidate the similarities and differences.

A.1.1 Parity for Dirac Fermions

We begin by recapitulating the effect of the parity transformation on Dirac fermions to illustrate the similarities and differences with parity transformations of tachyons. The parity transformation $\mathcal{P}$ transforms a Dirac field by[301]

$$\mathcal{P}\psi(\mathbf{x}, t)\mathcal{P}^{-1} = \gamma^0\psi(-\mathbf{x}, t) \tag{A.1}$$

Upon substituting the fourier expansion of the Dirac field (eq. 3.2) we find after some manipulations and using

$$\gamma^0 u(-p, s) = u(p, s) \qquad \gamma^0 v(-p, s) = -v(p, s)$$

the following effect on the creation and annihilation operators

$$\mathcal{P}b(p, s)\mathcal{P}^{-1} = b(-p, s) \qquad \mathcal{P}d^\dagger(p, s)\mathcal{P}^{-1} = -d^\dagger(-p, s) \tag{A.2}$$

plus their hermitean conjugates noting that $\mathcal{P}^\dagger = \mathcal{P}^{-1}$. Note $p = (p^0, -\mathbf{p})$.

A.1.2 Parity for Tachyonic Fermions

There are significant differences between tachyon fermions and conventional fermions. In defining the parity transformation for spin ½ tachyons we try to retain as much similarity as possible to the Dirac spin ½ fermion parity transformation. By definition the parity transformation changes $\mathbf{x} \to -\mathbf{x}$. In the case of a Dirac field the transformation is defined by eq. A.1.

If we now consider a spin ½ tachyon field and assume the same general form for the transformation:

$$\mathcal{P}\psi_T(\mathbf{x}, t)\mathcal{P}^{-1} = \gamma^0\psi_T(-\mathbf{x}, t) \tag{A.3}$$

where we use the free tachyon fourier expansion

[301] We use the conventional definition of Dirac matrices here.

$$\psi_T(x) = \sum_{\pm s} \int d^2p dp^+ N_T(p)\theta(p^+)[b_T(p, s)u_T(p, s)e^{-ip\cdot x} + d_T^\dagger(p, s)v_T(p, s)e^{+ip\cdot x}] \tag{A.4}$$

where the projection operators C and R are defined in chapter 5 for the spinors. Then we find

$$\gamma^0 u_T(-p, s) = u_T(p, s) \qquad \gamma^0 v_T(-p, s) = -v_T(p, s) \tag{A.5}$$

and the creation and annihilation operators satisfy

$$\mathcal{P} b_T(p, s)\mathcal{P}^{-1} = b_T(-p, s) \qquad \mathcal{P} d_T^\dagger(p, s)\mathcal{P}^{-1} = -d_T^\dagger(-p, s) \tag{A.6}$$

plus their hermitean conjugates. (Note $\mathcal{P}^\dagger = \mathcal{P}^{-1}$.) The free tachyon lagrangian is

$$\mathcal{L}_T = \psi_T^S(\gamma^\mu \partial/\partial x^\mu - m)\psi_T(x) \tag{A.7}$$

where

$$\psi_T^S = \psi_T^\dagger i\gamma^0\gamma^5 \tag{A.8}$$

Applying the parity transformation

$$\begin{aligned}\mathcal{P}\mathcal{L}(\mathbf{x}, t)\mathcal{P}^{-1} &= \mathcal{P}\psi_T^\dagger(x)\mathcal{P}^{-1}\mathcal{P}\, i\gamma^0\gamma^5(\gamma^\mu\partial/\partial x^\mu - m)\mathcal{P}^{-1}\mathcal{P}\psi_T(x)\mathcal{P}^{-1} \\ &= \psi_T^\dagger(-x)\gamma^0 i\gamma^0\gamma^5(\gamma^\mu\partial/\partial x^\mu - m)\gamma^0\psi_T(-x) \\ &= -\psi_T^\dagger(-x)i\gamma^0\gamma^5(\gamma^\mu\partial/\partial x'^\mu - m)\psi_T(-x) = -\mathcal{L}(-\mathbf{x}, t)\end{aligned} \tag{A.9}$$

where x' = –x = (x^0, –**x**). The tachyon anti-commutation relations are

$$\{\psi_{T\,a}^\dagger(x), \psi_{Tb}(y)\} = -[\gamma^5]_{ab}\,\delta^3(x - y) \tag{A.10}$$

Under the parity transformation the anti-commutation relations become

$$\{\psi_{T\,a}^\dagger(x'), \psi_{Tb}(y')\} = [\gamma^5]_{ab}\,\delta^3(x' - y') \tag{A.11}$$

where x' = (–**x**, t) and y' = (–**y**, t).

Thus the lagrangian and the anti-commutation relations change sign under the parity transformation. Therefore the physics of tachyons is not invariant under parity. This fact is directly evidenced in chapter 5 by the expression of the lagrangian in terms of left-handed and right-handed fields shows the lagrangian changes sign under the interchange of left and right handed fields (an effect of the parity transformation). Thus spin ½ tachyon theory, like nature, violates parity.

Note that parity violation is intrinsic to tachyons – even free tachyons. The discussion of the Standard Model in the following chapters will associate tachyon parity violation with Standard Model parity violation.

A.2 Charge Conjugation for Dirac and Tachyonic Fermions

In this section we will describe the charge conjugation transformation for tachyons. We start with the usual Dirac charge conjugation transformation to elucidate the similarities and differences.

A.2.1 Charge Conjugation for Dirac Fermions

We begin by describing the effect of the charge conjugation transformation on Dirac fermions. The charge conjugation transformation $\mathcal{C}$ transforms a free Dirac field by[302]

$$\mathcal{C}\psi(x)\mathcal{C}^{-1} = i\gamma^2\psi^*(x) \quad (A.12)$$

$$\mathcal{C}\psi^\dagger(x)\mathcal{C}^{-1} = i\psi^T(x)\gamma^2 \quad (A.13)$$

Defining the current with

$$j^\mu(x) = [\psi^\dagger(x)\gamma^0, \gamma^\mu\psi(x)] \quad (A.14)$$

we find

$$\mathcal{C}j^\mu(x)\mathcal{C}^{-1} = -j^\mu(x) \quad (A.15)$$

as required.

Substituting the fourier expansion of the Dirac field (chapter 5) we find after some manipulations

$$iu^\dagger(p, s)\gamma^2 = \varphi(p, s)v(p, s) \qquad iv^\dagger(p, s)\gamma^2 = \varphi(p, s)u(p, s) \quad (A.16)$$

where $\varphi(p, s)$ is a phase factor that we will set to unity in the interests of simplicity.

The creation and annihilation operators satisfy

$$\mathcal{C}b(p, s)\mathcal{C}^{-1} = \varphi(p, s)d(p, s) \qquad \mathcal{C}d^\dagger(p, s)\mathcal{C}^{-1} = \varphi(p, s)b^\dagger(p, s) \quad (A.17)$$

plus their hermitean conjugates (noting that $\mathcal{C}^\dagger = \mathcal{C}^{-1}$.)

A.2.2 Charge Conjugation for Tachyonic Fermions

There is a difference in the charge conjugation properties of tachyon fermions and conventional fermions. In defining the charge conjugation transformation for spin ½ tachyons we again try to retain as much similarity as possible to the Dirac spin ½ fermion charge conjugation transformation.

We now consider a tachyon field and assume the same general form as conventional fermions for the charge conjugation transformation:[303]

---

302 We use the conventional definition of Dirac matrices.

303 The charge conjugation operation agrees with that of the Dirac charge conjugation. We feel this definition is appropriate since it has the same form as that of Dirac fields.

$$\mathcal{C}\psi_T(x)\mathcal{C}^{-1} = i\gamma^2\psi_T{}^*(x) \quad (A.18)$$

$$\mathcal{C}\psi_T{}^\dagger(x)\mathcal{C}^{-1} = i\psi_T{}^T(x)\gamma^2 \quad (A.19)$$

If we use the free tachyon fourier expansion of chapter 5 then the tachyon spinors satisfy

$$iu_T{}^\dagger(p, s)\gamma^2 = \varphi_T(p, s)v_T(p, s) \qquad iv_T{}^\dagger(p, s)\gamma^2 = \varphi_T(p, s)u_T(p, s) \quad (A.20)$$

where $\varphi_T(p, s)$ is a phase factor that we will set to unity in the interests of simplicity.

The tachyon creation and annihilation operators satisfy

$$\mathcal{C}b_T(p, s)\mathcal{C}^{-1} = \varphi_T(p, s)d_T(p, s) \qquad \mathcal{C}d_T{}^\dagger(p, s)\mathcal{C}^{-1} = \varphi_T(p, s)b_T{}^\dagger(p, s) \quad (A.21)$$

plus their hermitean conjugates.

Charge conjugating the free tachyon lagrangian

$$\mathcal{L}_T = \psi_T{}^S(\gamma^\mu\partial/\partial x^\mu - m)\psi_T(x) \quad (A.22)$$

where

$$\psi_T{}^S = \psi_T{}^\dagger i\gamma^0\gamma^5 \quad (A.23)$$

we find

$$\begin{aligned}\mathcal{C}\mathcal{L}(\mathbf{x}, t)\mathcal{C}^{-1} &= \mathcal{C}\psi_T{}^\dagger(x)\mathcal{C}^{-1}\mathcal{C}\, i\gamma^0\gamma^5(\gamma^\mu\partial/\partial x^\mu - m)\mathcal{C}^{-1}\mathcal{C}\psi_T(x)\mathcal{C}^{-1} \\ &= -\psi_T(x)\gamma^2 i\gamma^0\gamma^5(\gamma^\mu\partial/\partial x^\mu - m)\gamma^2\psi_T{}^*(x) \\ &= \psi_T{}^\dagger(x)i\gamma^0\gamma^5(\gamma^\mu\partial/\partial x'^\mu - m)\psi_T(x) = \mathcal{L}(\mathbf{x}, t)\end{aligned} \quad (A.26)$$

The tachyon anti-commutation relations are

$$\{\psi_T{}^\dagger{}_a(x), \psi_{Tb}(y)\} = -[\gamma^5]_{ab}\,\delta^3(x - y) \quad (A.27)$$

Under the charge conjugation transformation the anti-commutation relations remain the same. Thus free tachyon theory is charge conjugation invariant.

A.3 Time Reversal for Dirac and Tachyonic Fermions

Time reversal invariance is also a significant theoretical and experimental issue.

A.3.1 Time Reversal for Dirac Fermions

The standard Dirac fermion time reversal transformation is:

$$\mathcal{T}\psi(\mathbf{x}, t)\mathcal{T}^{-1} = T\psi(\mathbf{x}, -t) \quad (A.28)$$

where $\mathfrak{T} = \mathcal{U}K$, $\mathcal{U}$ is a unitary operator, and K is the operator that takes the complex conjugate of all c-numbers, and where

$$T = i\gamma^1\gamma^3 \tag{A.29}$$

A.3.2 Time Reversal for Tachyonic Fermions

We will assume the *tachyon* time reversal transformation is the same as that of Dirac fermions:

$$\mathfrak{T}_T\psi_T(\mathbf{x}, t)\mathfrak{T}_T^{-1} = T_T\psi_T(\mathbf{x}, -t) \tag{A.30}$$

where $\mathfrak{T}_T = \mathcal{U}_T K$, $\mathcal{U}_T$ is a unitary operator defined for tachyons, and K is the complex conjugation operator for all c-numbers, and where

$$T_T = i\gamma^1\gamma^3 = T \tag{A.31}$$

The matrix T_T satisfies:

$$T_T^{-1} = -i\gamma^3\gamma^1 = T_T = T_T^\dagger \tag{A.32}$$

$$T_T\gamma^\mu T_T^{-1} = \gamma^{\mu T} \tag{A.33}$$

Under time reversal the current becomes

$$\mathfrak{T}_T j_{T\mu}(\mathbf{x}, t)\mathfrak{T}_T^{-1} = j_T^{\mu}(\mathbf{x}, -t) \tag{A.34}$$

where

$$j_T^{\mu}(\mathbf{x}, t) = \psi_T^{S}\gamma^\mu\psi_T(x) \tag{A.35}$$

If we assume the electromagnetic field satisfies

$$\mathfrak{T}\mathbf{A}(\mathbf{x}, t)\mathfrak{T}^{-1} = -\mathbf{A}(\mathbf{x}, -t)$$

under time reversal, then the electromagnetic tachyon lagrangian

$$\mathfrak{L} = \psi_T^{S}[\gamma^\mu(\partial/\partial x^\mu + ieA_\mu) - m]\psi_T(x) \tag{A.36}$$

satisfies

$$\mathfrak{T}_T\mathfrak{L}(\mathbf{x}, t)\mathfrak{T}_T^{-1} = \mathfrak{L}(\mathbf{x}, -t) \tag{A.37}$$

under time reversal. Although the action changes by a translation in time, Poincaré translation invariance implies the action is invariant. The equation of motion derived from the lagrangian is covariant under the tachyon time reversal transformation.

Thus we find the dynamics of the tachyon lagrangian theory to be invariant under the tachyon time reversal transformation.

Substituting the tachyon field fourier expansion of chapter 5 in eq. A.30 we find the tachyon creation and annihilation operators satisfy

$$\mathfrak{T}_T b_T(p, s)\mathfrak{T}_T^{-1} = -\varphi_{Tt}(p, s)b_T(-p, -s) \tag{A.38}$$

$$\mathfrak{T}_T d_T^{\dagger}(p, s)\mathfrak{T}_T^{-1} = -\varphi_{Tt}(p, s)d_T^{\dagger}(-p, -s) \tag{A.39}$$

plus their hermitean conjugates where $\varphi_{Tt}(p, s)$ is a phase factor. The momentum is $-p = (-\mathbf{p}, t)$ and the spin is $-s = (-\mathbf{s}, s^0)$.

Tachyon fermion dynamics is invariant under time reversal.

Appendix B. CP and CPT for Neutrino and Quark Tachyons

B.1 CP and CPT in Tachyon Quantum Field Theory

In chapter 5 we described our theory of second quantized tachyons. By choosing to second quantize on the light front, and separating the left-handed and right-handed parts of spin ½ tachyon fields, we were able to develop a canonical second quantized field embodying a complete set of solutions to our tachyon Dirac equation.

With this basis we were able to construct the form of The Extended Standard Model with an $SU(3)_{color}\otimes SU(2)\otimes U(1)\otimes SU(2)\otimes U(1)$ symmetry and parity violation. These features followed from the enlargement of the set of transformations between reference frames from Lorentz transformations to include superluminal transformations. Since superluminal transformations from the "lab" frame to a frame moving at a velocity greater than the speed of light necessarily produce coordinates in the "faster than light" reference frame that are complex-valued numbers in general, and since clocks and rulers always measure real numbers, we were led to introduce $SU(2)\otimes U(1)\otimes SU(2)\otimes U(1)$ rotations in the superluminal frame to "rotate" complex-valued coordinates to real-valued coordinates.

In this chapter we will consider CP and CPT. We will show that we can introduce CP violation in the dynamic terms of the Extended Standard Model lagrangian if experimental results should require it (in addition to the known CP violation due to certain complex values in the Cabibbo-Kobayashi-Maskawa (CKM) mass matrix.)

In addition we will consider CPT symmetry in tachyon quantum field theory and show that microcausality is violated. We will then show CP eigenfields and eigenstates enable us to redefine the Extended Standard Model lagrangian in such a way as to support, or not support, CPT violation and microcausality as future experiments dictate.

B.2 CP Symmetry, Tachyon Particle States, and the Lorentz Invariance of Tachyon Particle Number

B.2.1 Tachyon Quantum Fields

In this section we will discuss the effect of Lorentz transformations on quantum tachyon fields and their states. We will consider a left-handed neutrino tachyon field that is quantized on the light front to avoid completeness problems that would appear if it were quantized on a time-like surface.

The free, "+" light-front,[304] *left-handed* tachyon wave function Fourier expansion appropriate for neutrinos is:[305]

$$\psi_{TL}{}^{+}(x) = \sum_{\pm s} \int d^2p\,dp^{+} N_{TL}{}^{+}(p)\theta(p^{+})[b_{TL}{}^{+}(p, s)u_{TL}{}^{+}(p, s)e^{-ip\cdot x} + d_{TL}{}^{+\dagger}(p, s)v_{TL}{}^{+}(p, s)e^{+ip\cdot x}] \qquad (17.88)$$

---

[304] There are two quantum fields for each particle in light front quantization. The "-" field is dependent on "+" field and so the "+" field is the quantum field of interest. See Blaha (2007b) for details.

[305] Eqs. 17.88-90 appear in Blaha (2011c).

with

$$N_{TL}^{+}(p) = [2m|\mathbf{p}|/((2\pi)^3(p^+(p^+ - p^-) + p_\perp^{\,2}))]^{1/2} \tag{17.89}$$

where the non-zero anti-commutators of the Fourier coefficient operators are

$$\{b_{TL}^{+}(q,s), b_{TL}^{+\dagger}(p,s')\} = \delta_{ss'}\delta^2(\mathbf{q} - \mathbf{p})\delta(q^+ - p^+)$$
$$\{d_{TL}^{+}(q,s), d_{TL}^{+\dagger}(p,s')\} = \delta_{ss'}\delta^2(\mathbf{q} - \mathbf{p})\delta(q^+ - p^+) \tag{17.90}$$

The spinors are

$$u_{TL}^{+}(p, s) = C^- R^+ S_C(\omega, \mathbf{v_c})w^1(0)$$
$$u_{TL}^{+}(p, -s) = C^- R^+ S_C(\omega, \mathbf{v_c})w^2(0)$$
$$v_{TL}^{+}(p, s) = C^- R^+ S_C(\omega, \mathbf{v_c})w^3(0)$$
$$v_{TL}^{+}(p, -s) = C^- R^+ S_C(\omega, \mathbf{v_c})w^4(0)$$

where the superscript "T" indicates the transpose, $R^\pm = \frac{1}{2}(I \pm \gamma^0\gamma^3)$ are ± field light-front projection operators[306], $S_C(\omega, \mathbf{v_c})$ is an L(C) spinor boost, and the column 4-vector components $[w^i(0)]^k = \delta^{ik}$.

B.2.2 Tachyon CP Eigenstates

From Appendix A we see that

$$\mathscr{C}\mathscr{P}b_T(p, s)\mathscr{P}^{-1}\mathscr{C}^{-1} = d_T(-p, s) \tag{B.1}$$

and

$$\mathscr{C}\mathscr{P}a_T(p, s)\mathscr{P}^{-1}\mathscr{C}^{-1} = -b_T(-p, s) \tag{B.2}$$

up to a common phase factor that we set equal to unity for simplicity. We define for tachyon particles the CP projection operators $(CP)_\pm$ which we express formally as

$$(CP)_\pm = (I \pm CP)/\sqrt{2} \tag{B.3}$$

with the meaning

$$(CP)_\pm b(p, s) \equiv [b(p, s) \pm \mathscr{C}\mathscr{P}b(p, s)\mathscr{P}^{-1}\mathscr{C}^{-1}]/\sqrt{2} \tag{B.4}$$

We further define

$$\begin{aligned} B_{TL\pm}^{+}(p, s) &= (CP)_\pm b_{TL}^{+}(p, s) \\ &= [b_{TL}^{+}(p, s) \pm \mathscr{C}\mathscr{P}b_{TL}^{+}(p, s)\mathscr{P}^{-1}\mathscr{C}^{-1}]/\sqrt{2} \\ &= [b_{TL}^{+}(p, s) \pm d_{TL}^{+}(p, s))]/\sqrt{2} \end{aligned} \tag{B.5}$$

and

306 Blaha (2007b) p. 48.

$$B_{TL\pm}^{+}\dagger(p, s) = (CP)_{\pm} b_{TL}^{+}\dagger(p, s) = [b_{TL}^{+}\dagger(p, s) \pm \mathcal{CP} b_{TL}^{+}\dagger(p, s)\mathcal{P}^{-1}\mathcal{C}^{-1}]/\sqrt{2} = [b_{TL}^{+}\dagger(p, s) \pm d_{TL}^{+}\dagger(-p, s))]/\sqrt{2} \quad (B.6)$$

Eq. B.5 defines CP sharp annihilation operators with CP eigenvalues ±1 and eq. B.6 defines CP sharp creation operators with CP eigenvalues ±1. Having defined creation and annihilation operators that can be used to create or annihilate CP eigenstates, we now define tachyonic CP eigenstates in terms of them.

Each operator in eq. B.6 create a CP eigenstate consisting of a superposition of a positive "energy" particle and antiparticle. The operators in eq. B.5 annihilate a CP eigenstate.

We can define multi-particle CP sharp eigenstates with expressions such as

$$B_{TL+}^{+}\dagger(p_1, s_1) \ldots B_{TL-}^{+}\dagger(p_k, s_k) \ldots B_{TL+}^{+}\dagger(p_n, s_n)|0> \quad (B.7)$$

We can also define CP sharp quantum fields from eq. 17.88 above:

$$\psi_{TL\pm}^{+}(x) = (CP)_{\pm}\psi_{TL}^{+}(x) \quad (B.8)$$

with the corresponding *physical* particle field defined by

$$\psi_{TL}^{+}(x) = \psi_{TL+}^{+}(x) + \psi_{TL-}^{+}(x)$$

$\psi_{TL+}^{+}(x)$ is a CP = +1 quantum field while $\psi_{TL-}^{+}(x)$ is a CP = −1 quantum field. CP eigenstates such as eq. B.7 are not the physical states created and destroyed in interactions.

CP violation originates in parity violation.

B.2.3 Neutrino CP Eigenstates, Neutrino-Anti-Neutrino Oscillations

In this discussion we will assume CP invariance holds in the ElectroWeak and Strong interactions. In a manner reminiscent of the formalism for the $K^0 - \overline{K}^0$ system we define the CP eigenstates:

$$|\nu_1> = B_{TL+}^{+}\dagger(p, s)|0> \qquad |\nu_2> = B_{TL-}^{+}\dagger(p, s)|0> \quad (B.9)$$

If we set the phase to unity then

$$|\nu_1> = (|\nu> + |\bar{\nu}>)/\sqrt{2} \quad (B.10)$$
$$|\nu_2> = (|\nu> - |\bar{\nu}>)/\sqrt{2}$$
$$|\nu> = (|\nu_1> + |\nu_2>)/\sqrt{2}$$
$$<\bar{\nu}| = (<\nu_1| - <\nu_2|)/\sqrt{2}$$

for a single physical left-handed neutrino state ν and physical right-handed anti-neutrino state $\bar{\nu}$. The "handed" particles ν and $\bar{\nu}$ are the physical particles that are created or destroyed in ElectroWeak interactions. $|\nu_1>$ and $|\nu_2>$ are the left-handed CP eigenstates that remain single "particle" states under arbitrary Lorentz transformations.

Because ν_1 and ν_2 have different CP eigenvalues they will have differing interactions so their self-energies will be slightly different and thus their masses will also slightly differ. Due to the CP eigenstates mass difference a neutrino beam will evolve over time.

Consider an initially pure neutrino beam with no admixture of anti-neutrinos. We can specify its initial state as

$$|\nu(0)> = (|\nu_1> + |\nu_2>)/\sqrt{2} \tag{B.11}$$

If the ν_1 and ν_2 masses are m_1 and m_2 respectively then the state at time t is

$$|\nu(t)> = [\exp(-im_1t)|\nu_1> + \exp(-im_2t)|\nu_2>]/\sqrt{2} \tag{B.12}$$

The probability for observing a neutrino at time t is therefore

$$|<\nu|\nu(t)>|^2 = \tfrac{1}{2}\{1 + \cos[(m_1 - m_2)t]\} \tag{B.13}$$

and the probability for observing an anti-neutrino at time t is

$$|<\bar{\nu}|\nu(t)>|^2 = \tfrac{1}{2}\{1 - \cos[(m_1 - m_2)t]\} \tag{B.14}$$

Thus neutrino-anti-neutrino oscillations will occur in neutrino beams due to the tachyonic nature of neutrinos. Since neutrino beams are typically not "pure" there will always be an admixture of anti-neutrinos in the beam. Anti-neutrino beams are also not "pure" either so there will be an admixture of neutrinos. Thus detecting neutrino oscillations is a significant experimental issue.

B.2.4 CP and Tachyonic Quarks

To achieve the form of The Extended Standard Model in the quark sector it was necessary to make one of the quarks in an ElectroWeak quark doublet a tachyon. It seemed reasonable to choose d-type quarks (in all generations) as the tachyons. In discussing the CP situation of the quark sector there are two significant differences: quarks have complex 3-momenta (which leads ultimately to color SU(3)) and d-type quarks are charged ($-\frac{1}{3}$). Neither of these points of difference affect the tachyon discussion of CP. It is similar to that of the prior subsection on neutrinos.

We begin with the free, left-handed, light front +, d-type quark quantum field defined in eqs. 17.99-102 of Blaha (2011c):

$$\psi_{CdL}{}^{+}(x_r, x_i) = \sum_{\pm s} \int d^2p_r dp^+ d^3p_i\, N_{CdL}{}^{+}(p)\theta(p^+)\delta((p_i{}^3(p^+ - p^-)/\sqrt{2} + \mathbf{p}_{r\perp}\cdot\mathbf{p}_{i\perp})/m^2)\cdot$$
$$\cdot[b_{CdL}{}^{+}(p, s)u_{CdL}{}^{+}(p, s)e^{-i(p\cdot x + p^*\cdot x^*)/2} + d_{CdL}{}^{+\dagger}(p, s)v_{CdL}{}^{+}(p, s)e^{+i(p\cdot x + p^*\cdot x^*)/2}] \tag{17.99}$$

where

$$N_{CdL}{}^{+}(p) = (2\pi)^{-3}(2m/p^{+})^{½}$$

and

$$u_{CdL}{}^{+}(p, s) = C^{-} R^{+} S_C(\omega',\mathbf{v_c})w^1(0)$$
$$u_{CdL}{}^{+}(p, -s) = C^{-} R^{+} S_C(\omega',\mathbf{v_c})w^2(0)$$
$$v_{CdL}{}^{+}(p, s) = C^{-} R^{+} S_C(\omega',\mathbf{v_c})w^3(0)$$
$$v_{CdL}{}^{+}(p, -s) = C^{-} R^{+} S_C(\omega',\mathbf{v_c})w^4(0) \quad (17.100)$$

where

$$\omega' = \omega + i\pi/2 \quad (17.101)$$

The momentum 4-vector $p = (p^0, \mathbf{p_r} + i\mathbf{p_i})$ is related to the other quantities by eq. 17.97 with ω replaced by $\omega' = \omega + i\pi/2$ in all the relations of eq. 17.97 of Blaha (2011c):

$$p^0 = m\cosh(\omega) \qquad \mathbf{p} = \mathbf{p_r} + i\mathbf{p_i}$$
$$\mathbf{p_r} = m\omega_r\hat{\mathbf{u}}_r \sinh(\omega)/\omega \qquad \mathbf{p_i} = m\omega_i\hat{\mathbf{u}}_i \sinh(\omega)/\omega$$
$$\omega = (\omega_r{}^2 - \omega_i{}^2)^{½}$$
$$\mathbf{p_r}\cdot\mathbf{p_i} = 0$$
$$\hat{\mathbf{w}} = (\omega_r\hat{\mathbf{u}}_r + i\omega_i\hat{\mathbf{u}}_i)/\omega$$
$$\hat{\mathbf{w}}\cdot\hat{\mathbf{w}} = 1$$
$$\mathbf{v_c} = \hat{\mathbf{w}}\tanh(\omega) \quad (17.97)$$

The non-zero anti-commutators of the fourier coefficient operators are:

$$\{b_{CdL}(p,s), b_{CdL}{}^{\dagger}(p'^*,s')\} = 2^{-½}\delta_{ss'}\delta(p^{+} - p'^{+})\delta^2(\mathbf{p_r} - \mathbf{p'_{r'}})\delta^3(\mathbf{p_i} + \mathbf{p'_{i'}})$$
$$\{d_{CdL}(p,s), d_{CdL}{}^{\dagger}(p'^*,s')\} = 2^{-½}\delta_{ss'}\delta(p^{+} - p'^{+})\delta^2(\mathbf{p_r} - \mathbf{p'_{r'}})\delta^3(\mathbf{p_i} + \mathbf{p'_{i'}})$$
$$(17.102)$$

Following a line of discussion similar to that around eqs. B.1-B.4 we are led to define creation and annihilation operators that are Lorentz covariant:

$$B_{CdL\pm}{}^{+}(p, s) = (b_{CdL}{}^{+}(p, s) \pm d_{CdL}{}^{+}\dagger(-p, s))/\sqrt{2} \quad (B.15)$$
$$B_{CdL\pm}{}^{+}\dagger(p, s) = (b_{CdL}{}^{+}\dagger(p, s) \pm d_{CdL}{}^{+}(-p, s))/\sqrt{2} \quad (B.16)$$

where $-p = (p^0, -\mathbf{p_r} - i\mathbf{p_i})$. S*tates defined using the creation operators* $B_{CdL\pm}{}^{+}\dagger(p, s)$ have a well-defined *particle number and thus charge and baryon number.*

The operators in eq. B.16 create CP eigenstates that are superpositions of a positive "energy" particle and antiparticle. The operators in eq. B.15 are the annihilation operators.

Multi-particle states of sharp charge and baryon number are defined by expressions such as

$$B_{CdL+}{}^{+}(p_1, s_1) \ldots B_{CdL-}{}^{+}\dagger(p_k, s_k) \ldots B_{CdL+}{}^{+}\dagger(p_n, s_n)|0\rangle \quad (B.17)$$

We can define CP sharp, d-type quark, quantum fields from eq. 17.99 above:

$$\psi_{CdL\pm}{}^{+}(x) = (CP)_{\pm}\,\psi_{CdL}{}^{+}(x) \quad (B.18)$$

with a physical quark field defined by

$$\psi_{CdL}^{+}(x) = \psi_{CdL+}^{+}(x) + \psi_{CdL-}^{+}(x)$$

$\psi_{CdL+}^{+}(x)$ is a CP = +1 quantum field while $\psi_{CdL-}^{+}(x)$ is a CP = –1 quantum field. CP eigenstates such as eq. B.17 are not the physical d-type quark states created and destroyed in interactions.

B.3 CPT Symmetry, and Tachyons and Microcausality

B.3.1 Tachyon CPT Non-Invariance

The question of CPT invariance has long been of theoretical and experimental interest. For conventional field theories the CPT Theorem implies CPT invariance under very general conditions. We will examine the case of CPT invariance of a simple free tachyon theory.

For *Dirac* fermions

$$\mathcal{C}\mathcal{P}\mathfrak{T}\psi(\mathbf{x}, t)\, \mathfrak{T}^{-1}\mathcal{P}^{-1}\mathcal{C}^{-1} = i\gamma^5\psi^*(-\mathbf{x}, -t) \quad \text{(B.19)}$$

For spin ½ *tachyons*

$$\mathcal{C}_T\mathcal{P}\mathfrak{T}_T\psi_{Ta}(\mathbf{x}, t)\mathfrak{T}_T^{-1}\mathcal{P}^{-1}\mathcal{C}_T^{-1} = -i[\psi^\dagger(-\mathbf{x}, -t)\gamma^5]_a \quad \text{(B.20)}$$

where a is a spinor index. More succinctly,

$$\mathcal{C}_T\mathcal{P}\mathfrak{T}_T\psi_T(\mathbf{x}, t)\mathfrak{T}_T^{-1}\mathcal{P}^{-1}\mathcal{C}_T^{-1} = i\gamma^5\psi^*(-\mathbf{x}, -t) \quad \text{(B.21)}$$

Eqs. B.19 and B.21 are the same.

The transformation of the free field lagrangian under CPT:

$$\mathcal{L} = \psi_T^\dagger(x)i\gamma^0\gamma^5(\gamma^\mu\partial/\partial x^\mu - m)\psi_T(x) \quad \text{(B.22)}$$

is

$$\mathcal{L}_{CPT}(x) = -\psi_T^\dagger(-x)i\gamma^0\gamma^5(\gamma^\mu\partial/\partial(-x)^\mu - m)\psi_T(-x) = -\mathcal{L}(-x) \quad \text{(B.23)}$$

Thus the free tachyon lagrangian is not CPT invariant due to the non-invariance of the lagrangian under the parity transformation.

B.3.2 Microcausality and Tachyons

Since the CPT Theorem does not hold for spin ½ tachyons it is of interest to consider Jost's Theorem: CPT invariance is equivalent to weak local commutativity, which is a weak form of microcausality. In the case of spin ½ tachyons *weak local commutativity* (or weak microcausality) is defined similarly to Dirac particles:

$$<0|\{\psi_T{}^\dagger(x), \psi_T(y)\}|0> = 0 \quad \text{for} \quad (x - y)^2 < 0 \tag{B.24}$$

i. e. spacelike $(x - y)^2$.

In the past, questions were raised about the locality of tachyon fields due to the incomplete set of states used to define tachyon quantum fields if they were second quantized on the "time front." To avoid the issue of an incomplete set of states we chose to second quantize on the light front. We will now evaluate the strong local commutativity left-handed tachyon field commutator $\{\psi_{TL}{}^{+\dagger}(x), \psi_{TL}{}^{+}(y)\}$ to show that strong microcausality condition holds (and thus weak local commutativity) for left-handed tachyons although it does not hold for "non-handed" tachyon fields:

$$\{\psi_{TL}{}^{+\dagger}(x), \psi_{TL}{}^{+}(y)\} = 0 \quad \text{for} \quad (x - y)^2 < 0 \tag{B.25}$$

We insert the Fourier expansions:

$$\{\psi_{TL}{}^{+}{}_a(x), \psi_{TL}{}^{+}{}^{\dagger}{}_b(y)\} = \sum_{\pm s,s'} \int d^2p dp^+ \int d^2p' dp'^+ N_{TL}{}^{+}(p) N_{TL}{}^{+}(p') \theta(p^+)\theta(p'^+) \cdot$$
$$\cdot[\{b_{TL}{}^{+\dagger}(p',s'), b_{TL}{}^{+}(p,s)\} u_{TL}{}^{+}{}_a(p,s) u_{TL}{}^{+\dagger}{}_b(p',s') e^{+ip'\cdot y - ip\cdot x} +$$
$$+ \{d_{TL}{}^{+}(p',s'), d_{TL}{}^{+\dagger}(p,s)\} v_{TL}{}^{+}{}_a(p,s) v_{TL}{}^{+\dagger}{}_b(p',s') e^{-ip'\cdot y + ip\cdot x}]$$
$$= \sum_{\pm s} \int d^2p dp^+ N_{TL}{}^{+2}(p)\theta(p^+)[u_{TL}{}^{+}{}_a(p,s) u_{TL}{}^{+\dagger}{}_b(p,s) e^{+ip\cdot(y-x)} + v_{TL}{}^{+}{}_a(p,s) v_{TL}{}^{+\dagger}{}_b(p,s) e^{-ip\cdot(y-x)}]$$
$$= i\int d^2p dp^+ \theta(p^+) N_{TL}{}^{+2}(p)(2m|\mathbf{p}|)^{-1}\{[C^-R^+(-i\not{p} + m)\gamma\cdot\mathbf{p}R^+C^-]_{ab} e^{+ip\cdot(y-x)} + [C^-R^+(-i\not{p} - m)\gamma\cdot\mathbf{p}R^+C^-]_{ab} e^{-ip\cdot(y-x)}\}$$
$$= \tfrac{1}{2}[C^-R^+]_{ab} \int d^2p_\perp \int_0^\infty dp^+ (2\pi)^{-3} (e^{+ip\cdot(y-x)} + e^{-ip\cdot(y-x)})$$

where $p\cdot(y-x) = p^-(y^+ - x^+) + p^+(y^- - x^-) - \mathbf{p}_\perp\cdot(\mathbf{y}_\perp - \mathbf{x}_\perp)$. Since $p^2 = -m^2$, the integral can be rewritten, after letting $p^\mu = -p^\mu$, as

$$= \tfrac{1}{2}[C^-R^+]_{ab} \int d^2p_\perp \int_{-\infty}^{\infty} dp^+ (2\pi)^{-3} e^{-ip\cdot(y-x)} \tag{B.26}$$

where $p^- = (p_\perp{}^2 + m^2)/(2p^+)$. For spacelike $(x - y)^2 < 0$ we can always choose a coordinate system where $y^+ - x^+ = 0$ with the result

$$\{\psi_{TL}{}^{+\dagger}(x), \psi_{TL}{}^{+}(y)\} = 2^{-1}C^-R^+\delta(y^- - x^-)\delta^2(\mathbf{y} - \mathbf{x}) \qquad \underline{\text{if}} \;\; y^+ - x^+ = 0 \tag{B.27}$$

Therefore

$$\{\psi_{TL}^{++}(x), \psi_{TL}^{+}(y)\} = 0 \qquad \text{for} \quad (x - y)^2 < 0 \tag{B.28}$$

Consequently, free left-handed (or free right-handed) tachyons with light-front quantization separately satisfy the microcausality condition.

REFERENCES

Akhiezer, N. I., Frink, A. H. (tr), 1962, *The Calculus of Variations* (Blaisdell Publishing, New York, 1962).

Bjorken, J. D., Drell, S. D., 1964, *Relativistic Quantum Mechanics* (McGraw-Hill, New York, 1965).

Bjorken, J. D., Drell, S. D., 1965, *Relativistic Quantum Fields* (McGraw-Hill, New York, 1965).

Blaha, S., 1998, *Cosmos and Consciousness* (Pingree-Hill Publishing, Auburn, NH, 1998).

_______, 2002, *A Finite Unified Quantum Field Theory of the Elementary Particle Standard Model and Quantum Gravity Based on New Quantum Dimensions™ & a New Paradigm in the Calculus of Variations* (Pingree-Hill Publishing, Auburn, NH, 2002).

_______, 2003, *A Finite Unified Quantum Field Theory of the Elementary Particle Standard Model and Quantum Gravity Based on New Quantum Dimensions™ and a New Paradigm in the Calculus of Variations* (Pingree-Hill Publishing, Auburn, NH, 2003).

_______, 2004, *Quantum Big Bang Cosmology: Complex Space-time General Relativity, Quantum Coordinates™ Dodecahedral Universe, Inflation, and New Spin 0, ½, 1 & 2 Tachyons & Imagyons* (Pingree-Hill Publishing, Auburn, NH, 2004).

______, *2005a, Quantum Theory of the Third Kind: A New Type of Divergence-free Quantum Field Theory Supporting a Unified Standard Model of Elementary Particles and Quantum Gravity based on a New Method in the Calculus of Variations* (Pingree-Hill Publishing, Auburn, NH, 2005).

______, 2005b, *The Metatheory of Physics Theories, and the Theory of Everything as a Quantum Computer Language* (Pingree-Hill Publishing, Auburn, NH, 2005).

______, 2005c, *The Equivalence of Elementary Particle Theories and Computer Languages: Quantum Computers, Turing Machines, Standard Model, Superstring Theory, and a Proof that Gödel's Theorem Implies Nature Must Be Quantum* (Pingree-Hill Publishing, Auburn, NH, 2005).

______, 2006a, *The Foundation of the Forces of Nature* (Pingree-Hill Publishing, Auburn, NH, 2006).

______, 2006b, *A Derivation of ElectroWeak Theory based on an Extension of Special Relativity; Black Hole Tachyons; & Tachyons of Any Spin.* (Pingree-Hill Publishing, Auburn, NH, 2006).

_______, 2007a, *Physics Beyond the Light Barrier: The Source of Parity Violation, Tachyons, and A Derivation of Standard Model Features* (Pingree-Hill Publishing, Auburn, NH, 2007).

______, 2007b, *The Origin of the Standard Model: The Genesis of Four Quark and Lepton Species, Parity Violation, the ElectroWeak Sector, Color SU(3), Three Visible Generations of Fermions, and One Generation of Dark Matter with Dark Energy* (Pingree-Hill Publishing, Auburn, NH, 2007).

______, *2008a, A Direct Derivation of the Form of the Standard Model From GL(16) (Pingree-Hill Publishing, Auburn, NH, 2008).*

______, 2008b, *A Complete Derivation of the Form of the Standard Model With a New Method to Generate Particle Masses Second Edition* (Pingree-Hill Publishing, Auburn, NH, 2008)

______, 2009, *The Algebra of Thought & Reality: The Mathematical Basis for Plato's Theory of Ideas, and Reality Extended to Include A Priori Observers and Space-Time Second Edition* (Pingree-Hill Publishing, Auburn, NH, 2009).

______, 2010a, *Operator Metaphysics: A New Metaphysics Based on a New Operator Logic and a New Quantum Operator Logic that Lead to a Mathematical Basis for Plato's Theory of Ideas and Reality* (Pingree-Hill Publishing, Auburn, NH, 2010).

______, 2010b, *The Standard Model's Form Derived from Operator Logic, Superluminal Transformations and GL(16)* (Pingree-Hill Publishing, Auburn, NH, 2010).

______, 2011a, *21st Century Natural Philosophy Of Ultimate Physical Reality* (McMann-Fisher Publishing, Auburn, NH, 2011).

______, 2011b, *All the Universe! Faster Than Light Tachyon Quark Starships & Particle Accelerators with the LHC as a Prototype Starship Drive Scientific Edition* (Pingree-Hill Publishing, Auburn, NH, 2011).

______, 2011c, *From Asynchronous Logic to The Standard Model to Superflight to the Stars* (Blaha Research, Auburn, NH, 2011).

______, 2012a, *From Asynchronous Logic to The Standard Model to Superflight to the Stars volume 2: Superluminal CP and CPT, U(4) Complex General Relativity and The Standard Model, Complex Vierbein General Relativity, Kinetic Theory, Thermodynamics* (Blaha Research, Auburn, NH, 2012).

______, 2012b, *Standard Model Symmetries, And Four And Sixteen Dimension Complex Relativity; The Origin Of Higgs Mass Terms* (Blaha Reasearch, Auburn, NH, 2012).

______, 2013a, *Multi-Stage Space Guns, Micro-Pulse Nuclear Rockets, and Faster-Than-Light Quark-Gluon Ion Drive Starships* (Blaha Research, Auburn, NH, 2013).

______, 2013b, *The Bridge to Dark Matter; A New Sister Universe; Dark Energy; Inflatons; Quantum Big Bang; Superluminal Physics; An Extended Standard Model Based on Geometry* (Blaha Reasearch, Auburn, NH, 2013).

______, 2014a, *Universes and Multiverses: From a New Standard Model to a Physical Multiverse; The Big Bang; Our Sister Universe's Wormhole; Origin of the Cosmological Constant, Spatial Asymmetry of the Universe, and its Web of Galaxies; A Baryonic Field between Universes and Particles; Flatverse Extended Wheeler-DeWitt Equation* (Blaha Reasearch, Auburn, NH, 2014).

______, 2014b, *All the Multiverse! Starships Exploring the Endless Universes of the Cosmos Using the Baryonic Force* (Blaha Research, Auburn, NH, 2014).

______, 2014c, *All the Multiverse! II Between Multiverse Universes: Quantum Entanglement Explained by the Multiverse Coherent Baryonic Radiation Devices – PHASERs Neutron Star Multiverse Slingshot Dynamics Spiritual and UFO Events, and the Multiverse Microscopic Entry into the Multiverse* (Blaha Research, Auburn, NH, 2014).

Eddington, A. S., 1952, *The Mathematical Theory of Relativity* (Cambridge University Press, Cambridge, U.K., 1952).

Fant, Karl M., 2005, *Logically Determined Design: Clockless System Design With NULL Convention Logic* (John Wiley and Sons, Hoboken, NJ, 2005).

Gelfand, I. M., Fomin, S. V., Silverman, R. A. (tr), 2000, *Calculus of Variations* (Dover Publications, Mineola, NY, 2000).

Giaquinta, M., Modica, G., Souchek, J., 1998, *Cartesian Coordinates in the Calculus of Variations* Volumes I and II (Springer-Verlag, New York, 1998).

Giaquinta, M., Hildebrandt, S., 1996, *Calculus of Variations* Volumes I and II (Springer-Verlag, New York, 1996).

Gradshteyn, I. S. and Ryzhik, I. M., 1965, *Table of Integrals, Series, and Products* (Academic Press, New York, 1965).

Heitler, W., 1954, *The Quantum Theory of Radiation* (Claendon Press, Oxford, UK, 1954).

Huang, Kerson, 1992, *Quarks, Leptons & Gauge Fields 2nd Edition* (World Scientific Publishing Company, Singapore, 1992).

Jost, J., Li-Jost, X., 1998, *Calculus of Variations* (Cambridge University Press, New York, 1998).

Rescher, N., 1967, *The Philosophy of Leibniz* (Prentice-Hall, Englewood Cliffs, NJ, 1967).

Sagan, H., 1993, *Introduction to the Calculus of Variations* (Dover Publications, Mineola, NY, 1993).

Sakurai, J. J., 1964, *Invariance Principles and Elementary Particles* (Princeton University Press, Princeton, NJ, 1964).

Streater, R. F. and Wightman, A. S., 2000, *PCT, Spin, Statistics, and All That* (Princeton University Press, Princeton, NJ 2000).

Weinberg, S., 1995, *The Quantum Theory of Fields Volume I* (Cambridge University Press, New York, 1995).

Weyl, H., 1950, *Space, Time, Matter* (Dover, New York, 1950).

Weyl, H., (Tr. S. Pollard et al), 1987, *The Continuum* (Dover Publications, New York, 1987).

INDEX

ϕ^3 theory, 17
ϕ^4, 161, 177, 210, 213, 214, 215, 216, 217, 218, 219, 220, 261, 264, 384, 385
"Periodic" Table of Fermions, 168
16 dimension Reality group, 358
4-valued logic, 5, 6, 367, 368, 371, 391
absolute space, 266, 310
absolute space-time, 266, 310
Adler-Bell-Jackiw anomaly, 267
Age of universe, 318
Aguilar et al, 170
alphabet, 11, 12, 25, 27
alphabets, 11, 28
angular momentum, 282, 283, 288, 289, 290
annihilation of particles, 11, 12, 23, 28
anti-commuting c-number fields, 153, 154, 255, 346
anti-universes, 367, 372
Asynchronicity, 5, 6, 353
Asynchronous Logic, 351, 353, 367, 371, 391, 421, 436
auxiliary field, 224
axial anomaly, 267
baryon "electric" field, 178, 410
baryon "magnetic" field, 178, 410
Baryon density, 318
baryon number conservation, 356, 371
Baryonic "Coulombic" Gauge Field, 179
Baryonic Coupling Constant, 174
baryonic force, 409, 410, 411, 412, 413, 415, 416, 417
baryonic gauge field, 175, 356, 368, 370, 372, 389, 391, 392, 393
baryonic potential, 409
Bergmann, P., 266
Bianchi identities, 249
Big Bang, 210, 296, 297, 299, 304, 305, 306, 307, 308, 311, 316, 318, 319, 321, 326, 328, 329, 336, 337, 338, 339, 340, 420, 435, 446
Big Bang Epoch, 337
Bilenky, S., 144
Bishop Berkeley, 8
Bjorken, J. D., 60, 435
black hole, 250, 259, 263, 264, 267
black hole tachyons, 97
blackbody energy spectrum, 301
blackbody radiation, 335, 337
blackbody temperature, 306
Blaha, 296, 297, 420, 435, 446
Boltzmann, 301
Boltzmann's H theorem, 110, 113
boost transformation, 39, 40, 375
boson, 26
bosonic string, 216, 278
Cabibbo-Kobayashi-Maskawa mass matrix, 428
Calculus of Variations, 271, 272, 273, 277, 420, 435, 437
CERN, 95, 96, 97
Chandrasekhar Limit, 408
Chang, S-J, 60
Charge conjugation, 219
chiral transformation, 267
Chomsky, 11, 13, 15
cloaked propagator, 217
clock, 351
CMB, 311, 326
CODATA, 409
Coherent states, 208, 209, 226, 235, 383
color, 26
color singlets, 168
complex, 202, 203, 205, 208, 213, 220, 278, 382
complex conjugation operator, 56, 68
complex general relativity, 30
complex Lorentz group, 40
complex space-time, 266, 298, 304
Complexon Gauge Groups, 150
complexon quark, 164, 184
Complexon Standard Model, 158
complexon tachyon, 79
complexons, 70, 378, 379

composition of extrema, 250, 272, 273, 274, 276, 277, 278, 280, 285, 286
Compton wavelength, 342
computer language, 11, 12, 14, 27
condensation, 202
confinement, 338
conservation law, 206, 224, 268, 280, 281, 286, 288, 291
conserved axial-vector current, 267
Copenhagen interpretation, 352
Cornish, N., 318, 340
Correspondence Principle, 215, 240, 243
Cosmic Microwave Background, 311, 326
Cosmological Constant, 347, 421, 436
Coulomb gauge, 176, 203, 206, 207, 214, 220, 222, 229, 238, 253, 288, 382
Coulomb interaction, 254
Coulomb potential, 261
coupling constants, 149
CP violation, 428, 430
CPT invariance, 432
creation of particles, 11, 12, 27, 28
Critical density, 318
curvature scalar, 249
curvature tensor, 265
Dark atoms, 168, 170, 171, 172
Dark basic chemistry, 168
Dark chemistry, 170
Dark cluster, 172
Dark dipole effects, 171
Dark electromagnetic force, 170
Dark Energy, 310, 328, 420, 436
Dark gauge bosons, 169
Dark Higgs sector, 170
Dark Jupiter, 172
Dark Lepton Number, 180
Dark Matter, 1, 9, 29, 31, 32, 133, 139, 158, 162, 165, 168, 169, 170, 171, 172, 180, 182, 190, 192, 194, 196, 197, 199, 295, 310, 353, 418, 420, 421, 436
Dark Matter clump, 172
Dark matter density, 318
Dark Matter galaxy, 172
Dark Matter particles, 168
Dark neutrinos, 170
Dark nuclear decays, 172
Dark Periodic Table, 168, 170
Dark quark left-handed covariant derivative, 163, 183
Dark star, 172
data, 12, 28
data packets, 12
decay, 23, 26
degree of divergence, 257, 258
deSitter, 322
Dicke experiments, 173
dilute gas, 110
dimension, 205, 248
Dimensional Interaction, 203
dimensional regularization, 268
dimensions, 202, 204, 205, 209, 210, 250, 264, 267, 306, 337, 383
Dirac equation, 55, 56, 57, 93, 115, 116, 117, 119, 122, 123, 141, 146, 374
Dirac matrices, 374
divergences, 213, 214, 215, 217, 235, 243, 247, 251, 258, 263, 267, 300, 328, 336, 446
Dr. Johnson, 8, 9
Eddington, A., 327
Einstein, 296, 298, 299, 300, 306, 310, 311, 312, 313, 315, 316, 319, 335, 338
electromagnetic interactions, 22
electron, 10, 23, 24, 25, 26, 372, 389, 391, 398
electron number, 140
electrons, 10, 11, 22, 23, 24, 25
ElectroWeak, 125, 130, 131, 134, 164, 165, 184, 185, 420, 435, 436
Electroweak Theory, 213
elementary particles, 11, 12, 25, 27, 28
energy, 10
energy density, 318, 319, 338, 339
energy per particle, 109, 112
energy scale, 263
energy-momentum tensor, 224, 249, 286, 311, 314
energy-time uncertainty relation, 259
entropy, 110, 113
Epictetus, 337
equal time commutation relations, 176, 207, 213, 381

equation of state, 110, 112
error function, 261
Euclid, 1, 4
European Space Agency, 388, 390, 392
Extended Complexon Standard Model, 158
extended Standard Model, 358
extra dimensions, 140
extremum, 271, 273, 276, 277, 278
Eötvös experiments, 173
Faddeev-Popov, 152, 154, 344, 345, 346, 347, 348
Faddeev-Popov ghost terms, 165, 185
Faddeev-Popov mechanism, 163, 164, 183, 184
Faddeev-Popov Mechanism, 165, 185
Fadeev-Popov, 153, 255
Fant, 368
faster than light particles, 95, 97
Fdadeev-Popov, 152
Feinberg, 58, 80
Feinberg, Gerald, 95
fermion species, 373, 375
fermions, 88
fermions, four types of, 88
Feynman propagator, 66, 67, 86, 89
Feynman, Richard, *10*, *12*, *23*, *24*, *25*, *26*, *27*
Feynman-like diagram, 18, 19, 20, 24, 25, 27
fifth force, 173, 389
finite description of a language, 28
first law of thermodynamics, 110, 113
fission, 391, 395, 397
Flatverse, 310, 360
Fock states, 208, 209, 221, 226, 228, 382, 383
forms, 38, 375
four laws of black holes, 356
fuel spheres, 416
gauge boson, 26
gauge field, 152, 245, 246, 282, 290
gauge fields, 25, 26
gauge invariance, 205, 206, 210, 216, 288
Gaussian, 177, 203, 215, 240, 257, 385
General Relativity, 248
generation mixing, 165, 185
generations, 134
ghost, 153, 154, 164, 184, 203, 255, 346
GL(4), 137, 139
globular clusters, 171, 172
gluons, 134, 137
Gödel's Undecidability Theorem, 1, 446
grammar, 11, 12, 13, 14, 25, 27
grammar rules, 11, 12, 13, 15, 17, 21
Gran Sasso Laboratory, 95, 96, 97
gravitational constant, 249, 409
gravitational field fluctuations, 260
gravitational potential, 261
graviton, 246, 250, 254, 256, 257, 260, 271
Gravity, 248, 250, 261, 264
Gupta-Bleuler formulation, QED, 66, 148
Gupta-Bleuler gauge, 206
G_{z16}, 358, 359, 360
Hamilton's equations, 280, 285
Hamilton's Principle, 273, 277, 278
head symbol, 13, 14, 15, 17
heat capacity, 110
Heisenberg Uncertainty Principle, 339
Higgs mechanism, 141, 143, 146, 148, 149
Higgs Mechanism, 29, 97, 181, 182, 186, 187, 189, 190, 191, 193, 194, 195, 196, 197, 199, 200, 367, 390, 446
Higgs particles, 95, 97, 148, 149, 165, 185
higher derivative quantum field theories, 216
homogeneous, 310
homogeneous Lorentz group, 38
Hubble length, 266
Hubble parameter, 318
Hubble rate, 336
Hubble's constant, 316, 319
Hubble's law, 311, 316
hyperbolic motion, 412
hyperplane, 276
imaginary coordinates, 202, 203, 205, 208, 209, 210, 213, 216, 224, 250, 267, 382, 383, 384
inertial reference frames, 266
inflation, 296, 337
input, 11, 12, 13, 14, 17, 18, 19, 20, 25, 27
interactions, 10, 11, 12, 21, 22, 25, 26, 27, 28, 446
Interdimensional Interaction, 203
internal energy, 110, 113
internal symmetry, 283, 290
iota, 2, 3, 8, 54, 55, 71, 91, 117, 123, 140, 141, 144, 195
ISIS, 448

island universes, 350, 355
isotropic, 310
Jacobian, 159, 160, 211, 223, 225, 226, 234, 250, 255, 275, 277, 278, 284, 287
Jost's Theorem, 433
Juan Yin et al, 96
Kaluza-Klein mechanism, 140
Klein-Gordon field, 204, 220, 221, 224, 225, 278
Kogut, J., 60
lagrangian, 10, 11, 12, 13, 17, 25, 27, 153, 154, 205, 206, 220, 223, 224, 226, 228, 235, 249, 250, 251, 252, 255, 256, 258, 271, 273, 282, 290, 346
Landauer mass, 2, 3
Language, *27*
L_C boost transformations, 49
left-handed, 59, 81, 375, 376, 378, 379, 380, 388
left-handed neutrinos, 375
left-handed superluminal boosts, 40
left-handedness, 388
Leibniz, 3, 353, 367, 368, 437
leptoDark, 169
lepton number conservation, 180, 371
letters, 11, 12, 14, 15, 25, 27
LHC, 95, 421, 422, 436
Library of Alexandria, 420, 435
light-front, 60, 61, 62, 63, 64, 66, 67, 75, 82, 83, 84, 86, 87, 89, 428, 434
Linear Hadron Collider, 95
linguistic, 11, 23, 27, 28
lopsidedness, 392
Lorentz force, 410, 411
Lorentz group, 38
Lorentz invariance, 282, 288
Lorentz transformation, 38, 282, 288, 289, 290
Lorentz transformations, 266
LSZ, 224, 229, 230
Luminal symmetry, 140
LVD collaboration, 96
Mach, E., 266
manifold, 273, 275, 277
mass scale, 203, 205, 210, 259, 384
Matter density, 318
matter-dominated, 296, 306, 320, 322, 323, 326
Maxwell-Boltzmann distribution, 106
McDonald, A. B., 144
Megaverse, 350
metric operator, 223, 227, 236, 258
microcausality, 433
microscopic causality, 213
Minimax Principle, 3, 29, 30, 32, 34, 62, 120, 150, 190, 193, 197, 202, 295, 343, 353, 354, 371
Minkowski, 266
momentum, 11, 12
multiverse, 350
Nature, **27**
negative energy states, 65, 119
negative metric states, 216
negative norms, 235
neutrino mass squared, 96
neutrino oscillation experiments, 144
neutrinos, 25, 140, 142, 144, 145
neutron stars, 408
New Complexon Standard Model, 162
Newtonian potential, 254, 261, 264
non-commuting coordinates, 202
Non-deterministic grammars, 15
non-hermitean, 235, 258
nonterminal symbol, 13
nonterminal symbols, 13, 14, 15
normal ordering, 211, 213, 227, 234
nuclear engines, 407
observer, 352
Ockham's Law of Parsimony, 3
Ockham's Razor, 3
orthochronous Lorentz boost, 39
output, 11, 13, 14, 16, 17, 18, 19, 21, 25, 27, 28
parity, 40, 46, 47, 48
Parity, 218, 219
parity transformation, 423
parity violation, 62, 424
Parity Violation, 420, 436
parton model for nucleons, 120
path integral, 153, 154, 254, 255, 256, 257, 346
Pauli matrices, 141
Pauli, W., 268
perfect fluid, 296, 306, 311, 312, 314, 315, 316

Periodic Table of Dark Atoms, 170
peripheral collisions, 390
perturbation theory, 10, 11, 203, 213, 221, 222, 223, 224, 225, 230, 236, 237, 250, 257, 259, 268
photon, 16, 17, 23, 24, 25, 208, 245, 261, 382, 385
photons, 11, 16, 23, 148
physical laws, 12
physical states, 236
piece-wise invertible, 275, 276
Planck, 296, 298, 300, 301, 304, 305, 331, 338, 341, 342
Planck mass, 203, 210, 259, 261, 263, 264, 384
planckton field, 175, 176, 392, 394
Plato's Theory of Ideas, 421, 436
positrons, 22, 23, 119
potential energy, 414
pressure, 109, 112
Principle of Asynchronicity, 18
Principle of Equivalence, 267
Probabilistic Computer, 15
Probabilistic Computer Grammars, 15
probabilistic grammar, 16, 17, 18, 19, 21
probability amplitude, 17, 18, 20, 21
production rules, 11, 12, 13, 14, 15, 22, 23, 24, 25, 26, 27, 28
program, 12
proper distance, 316
pseudo-unitary, 222, 236, 258
QED, 213, 214, 221, 254, 259, 384, 385
q-number coordinate system, 202
quantum computers, 446
quantum coordinate $X^{\mu}(z)$, 158
Quantum Dimensions, 202, 203, 208, 209, 210, 216, 382, 383, 384, 420, 435
Quantum Electrodynamics, 61, 202, 205, 243
quantum entanglement, 96
quantum fluctuations, 203, 251, 259, 337, 338
quantum foam, 250, 259
Quantum Gravity, 203, 214, 216, 235, 248, 249, 251, 252, 254, 257, 258, 259, 260, 261, 263, 264, 296, 420, 435
Quantum Mechanics, 205, 223
quantum numbers, 12, 26
quantum probabilistic grammar, 16, 21
Quantum Turing machine, *11*
quark, 446
quark confinement, 148
quark doublets, 120
quarks, 11, 26
qubits, 8
radiation density, 318
radiation-dominated, 296, 306, 320, 322, 323, 325, 326, 336
radius of the universe, 318, 319, 338, 339, 340, 341, 342
real space-time, 310
Reality group, 29, 31, 32, 33, 34, 41, 52, 126, 132, 133, 139, 140, 141, 143, 146, 150, 178, 181, 199, 201, 295, 306, 307, 308, 327, 329, 330, 331, 335, 338, 341, 344, 346, 350, 353, 354, 357, 358, 359, 360, 361, 367, 372, 404, 406, 410
relative probability, 15, 16, 17, 20, 21, 22
relativistic harmonic oscillator, 216
relativistic Maxwell-Boltzmann distribution, 106
renormalization, 214, 225, 226
Ricci tensor, 249, 311, 312, 313
Riemann-Christoffel curvature tensor, 249
Riemannian manifold, 366
right-handed, 59, 81, 377, 378, 379, 380, 388
right-handed superluminal boosts, 40
right-handed tachyons, 63, 66, 87
Robertson-Walker metric, 296, 298, 304, 307, 308, 310, 311, 312, 314, 315, 316, 330, 331, 334, 335, 337, 338, 339, 446
Root, R., 60
Russia, 448
scalar bosons, 337
scalar field, 203, 208, 213, 214, 221, 228, 238, 274, 277, 278, 279, 282, 284, 288, 382
scalar particles, 264
scale factor, 296, 298, 299, 300, 302, 303, 305, 306, 307, 308, 310, 311, 312, 314, 315, 316, 317, 318, 319, 322, 323, 326, 327, 329, 330, 332, 334, 335, 338
SDSS, 311, 318
second law of thermodynamics, 110, 113
separable hamiltonian, 284

separable Lagrangian, 212, 218, 219, 284, 285, 288, 290
sequence, 24, 25
simplicity, 28
slingshot, 408, 411, 412, 413, 415, 416, 417
Soper, D., 60
Special Relativity, 420, 435
spin, 11, 12, 17, 27, 79, 446
spin 2 particle, 246
spin states, 372, 391
spin, complex, 79
spontaneous symmetry breaking, 140, 142, 143, 149, 200
Standard Cosmological Model, 296, 297, 299, 336, 337, 338
Standard Model, 10, 11, 12, 13, 14, 15, 16, 17, 21, 22, 25, 27, 28, 213, 214, 215, 216, 221, 243, 249, 250, 267, 299, 420, 435, 446
Standard Model Quantum Grammar, 22
Starkman, G. D., 318, 340
start symbol, 13, 17
string, 13, 14, 15, 24, 27
Strong interaction, 25, 26
Strong Interactions, 25, 137
SU(2), 41, 125, 128, 129, 130, 131, 148, 149, 360
SU(3), 134, 135, 136, 137, 138, 139, 146, 148, 165, 185, 420, 436
Sudarshan, E.C.G., 95
superluminal Maxwell-Boltzmann distribution, 107
Superstring, 10, 12, 216, 217, 420, 435
Superstring Theory, 140
Supersymmetric, 216
suspended animation, 406
symbols, 11, 12, 13, 14, 15, 28
Synge, 266
tachyon pole terms, 67
tachyon spinors, 79, 376
tachyon universe, 380, 396, 397
Tegmark, M., 318
temperature, 296, 297, 299, 301, 305, 306, 326, 328, 330, 337, 340
terminal symbols, 13, 14, 15
Theory of Everything, 10, 420, 435
Thorn, C., 60
thrust tubes, 412, 415, 416
time dependent masses, 391
time intervals, 259
time ordering, 18
time reversal, 220, 426
time reversal transformation, 426
time-like photons, 65, 66, 142
time-reversal operator, 56, 68
translational invariance, 206, 280, 286
tree diagrams, 257
triangle diagram, 268
tritium decay, 95, 96, 97
Turing machines, 11
two slit photon experiment, 16
two-tier, 153, 154, 177, 212, 213, 214, 215, 216, 217, 218, 221, 226, 230, 235, 236, 237, 239, 240, 242, 243, 245, 246, 247, 248, 249, 251, 254, 255, 257, 258, 259, 260, 261, 263, 264, 267, 268, 269, 346, 385
two-tier quantum field theory, 298, 337
U matrix, 230, 231, 232, 233
U(4), 32, 34, 134, 173, 181, 182, 183, 185, 186, 187, 188, 189, 190, 191, 192, 193, 196, 199, 306, 347, 393, 421, 436
U(4) Number symmetry, 181
umbrella, 412, 415, 416, 417
umbrella spokes, 416
umbrella uniship, 415
unitarity, 228, 235, 236, 238, 240, 258, 259
unitary, 56, 59, 68, 220, 222, 223, 236, 243, 258, 375, 426, 427
universe particle, 178, 356, 360, 372, 373, 374, 375, 378, 380, 384, 385, 386, 387, 388, 391, 392, 393, 394, 395, 396, 397, 398, 399
unrestricted rewriting system, 13
vacuum fluctuations, 214, 260, 261
vacuum state, 200
variations, 271, 273
vierbein, 251
VLSI, 351, 368
vocabulary, 13, 25, 27, 28
weak field approximation, 251, 252
Weak interaction, 25, 26
weak local commutativity, 433

Web of Galaxies, 421, 436
Weinberg, 310, 319
Weyl, H., 438
Wheeler-DeWitt equation, 344, 345, 347, 348, 349, 350, 356, 364, 365, 366, 367, 370, 371, 373, 390, 394, 447
Wick expansion, 232
Wilkinson Microwave Anisotropy Probe, 390
WIMP Atoms, 169
WIMP Chemistry, 169
WIMP Gauge Bosons, 169
WIMPs, 169, 446
WMAP, 311, 318, 320, 340, 390, 392
Wolfenstein, L., 144
work, 113
Yan, T.-M., 60
Yang-Mills theories, 254
$Y^{\mu}(x)$, 295

About the Author

Stephen Blaha is an internationally known physicist with interests in Science, Society and civilization, the Arts, and Technology. He had an Alfred P. Sloan Foundation scholarship in college. He received his Ph.D. in Physics from Rockefeller University. He has served on the faculties of several major universities. He was also a Member of the Technical Staff at Bell Laboratories, a manager at the Boston Globe Newspaper, a Director at Wang Laboratories, and President of Blaha Software Inc and of Janus Associates Inc. (NH).

Among other achievements he was a co-discoverer of the "r potential" for heavy quark binding developing the first (and still the only demonstrable) non-abelian gauge theory with an "r" potential; first suggested the existence of topological structures in superfluid He-3; first proposed Yang-Mills theories would appear in condensed matter phenomena with non-scalar order parameters; first developed a grammar-based formalism for quantum computers and applied it to elementary particle theories; first developed a new form of quantum field theory without divergences (thus solving a major 60 year old problem that enabled a unified theory of the Standard Model and Quantum Gravity without divergences to be developed); first developed a formulation of complex General Relativity based on analytic continuation from real space-time; first developed a generalized non-homogeneous Robertson-Walker metric that enabled a quantum theory of the Big Bang to be developed without singularities at $t = 0$; first generalized Cauchy's theorem and Gauss' theorem to complex, curved multi-dimensional spaces; received Honorable Mention in the Gravity Research Foundation Essay Competition in 1978; first developed a physically acceptable theory of faster-than-light particles; first showed a universe with three complex spatial dimensions is icosahedral; first derived a composition of extrema method in the Calculus of Variations; first quantitatively suggested that inflationary periods in the history of the universe were not needed; first proved Gödel's Theorem implies Nature must be quantum; provided a new alternative to the Higgs Mechanism, and Higgs particles, to generate masses; first showed how to resolve logical paradoxes including Gödel's Undecidability Theorem by developing Operator Logic and Quantum Operator Logic; first developed a quantitative harmonic oscillator-like model of the life cycle, and interactions, of civilizations; first showed how equations describing superorganisms also apply to civilizations. A recent book shows his theory applies successfully to the past 14 years of history and to *new* archaeological data on Andean and Mayan civilizations as well as Early Anatolian and Egyptian civilizations.

He first developed an axiomatic derivation of the forms of The Standard Model with WIMPs from geometry – space-time properties – The faster than light Standard Model.

He has had a major impact on a succession of elementary particle theories: his Ph.D. thesis (1970), and papers, showed that quantum field theory calculations to all orders in ladder approximations could not give scaling deep inelastic electron-nucleon scattering. He later showed the eigenvalue equation for the fine structure constant α in Johnson-Baker-Willey QED had a zero at $\alpha = 1$ not 1/137 by solving the Schwinger-Dyson equations to all orders in an approximation that agreed with exact results to 8th order in α thus ending interest in this theory. In 1979 at Prof. Ken Johnson's (MIT) suggestion he calculated the proton-neutron mass difference in the MIT bag model and found the result had the wrong sign reducing interest in the bag model. These results all appear in Physical Review papers. In the 2000's he repeatedly pointed out the shortcomings of SuperString theory and showed that The Standard Model's form could be derived from space-time geometry by an extension of Lorentz transformations to faster than light transformations. This deeper space-time basis greatly increases the possibility that it is part of THE fundamental theory.

In graduate school (1966-71) he wrote substantial papers in elementary particles and group theory: The Inelastic E- P Structure Functions in a Gluon Model. Phys.Lett. B40:501-502,1972; Deep-Inelastic E-P Structure Functions In A Ladder Model With Spin 1/2 Nucleons, Phys.Rev. D3:510-523,1971; Continuum Contributions To The Pion Radius, Phys.Rev. 178:2167-2169,1969; Character Analysis of U(N) and SU(N), J. Math. Phys. 10, 2156 (1969); and The Calculation of the Irreducible Characters of the Symmetric Group in Terms of the Compound Characters, (Published as Blaha's Lemma in D. E. Knuth's book: *The Art of Computer Programming Vols. 1 – 4*).

In 1979 he developed a design for a large, ultrastable derigible or helicopter with an on-board computer for the real time measurement and analysis of gravity data to detect anomalies due to oil and metal deposits as it cruises vast regions of interest. This design has recently been implemented (2014). He also developed a method using lasers/masers to position coal boring machines within coal mine seams for maximal results. Seeing is difficult within dusty coal mines.

In the early 1980's Blaha was also a pioneer in the development of UNIX for financial, scientific and Internet applications: benchmarked UNIX versions showing that block size was critical for UNIX performance, developing financial modeling software, starting database benchmarking comparison studies, developing Internet-like UNIX networking (1982) and developing a hybrid shell programming technique (1982) that was a precursor to the PERL programming language. He was also the manager of the AT&T ten-year future products development database. His work helped lead to commercial UNIX on computers such as Sun Micros, IBM AIX minis, and Apple computers.

In the 1980's he pioneered the development of PC Desktop Publishing on laser printers. and was nominated for three "Awards for Technical Excellence" in 1987 by PC Magazine for PC software products that he designed and developed.

Recently he has developed a theory of Megaverses – actual universes of which our universe is one – with quantum particle-like properties based on the Wheeler-DeWitt equation of Quantum Gravity. He has developed a theory of a baryonic force, which had been conjectured many years ago, and estimated the strength of the force based on discrepancies in measurements of the gravitational constant G. This force, operative in 16-dimensinal space, can be used to escape from our universe in "uniships" which are the equivalent of the faster-than-light starships proposed in the author's earlier books. Thus travel to other universes, as well as to other stars is possible.

Blaha also considered the complexified Wheeler-DeWitt equation and showed that its limitation to real-valued coordinates and metrics generated a Cosmological Constant in the Einstein equations.

The author has also recently written a series of books on the serious problems of the United States and their solution as well as a book on the decline of Mankind that will follow from current social and genetic trends in Mankind.

In the past twelve years Dr. Blaha has written over 40 books on a wide range of topics. Some recent major works are: *From Asynchronous Logic to The Standard Model to Superflight to the Stars, All the Universe!, SuperCivilizations: Civilizations as Superorganisms, America's Future: an Islamic Surge, ISIS, al Qaeda, World Epidemics, Ukraine, Russia-China Pact, US Leadership Crisis,The Rises and Falls of Man – Destiny – 3000 AD: New Support for a Superorganism MACRO-THEORY of CIVILIZATIONS From CURRENT WORLD TRENDS and NEW Peruvian, Pre-Mayan, Mayan, Anatolian, and Early Egyptian Data, with a Projection to 3000 AD,* and *Mankind in Decline: Genetic Disasters, Human-Animal Hybrids, Overpopulation, Pollution, Global Warming, Food and Water Shortages, Desertification, Poverty, Rising Violence, Genocide, Epidemics, Wars, Leadership Failure.*

He has taught approximately 4,000 students in undergraduate, graduate, and postgraduate corporate education courses primarily in major universities, and large companies and government agencies.

The above paragraphs summarize much of his work over the past forty seven years. This work is fully documented. He continues to engage in research and writing at Blaha Research.

www.ingramcontent.com/pod-product-compliance
Lightning Source LLC
Chambersburg PA
CBHW062010190326
41458CB00009B/3031
9780989382663